MATHEMATICAL STATISTICS

CHAPMAN & HALL/CRC
Texts in Statistical Science Series

Series Editors
C. Chatfield, *University of Bath, UK*
J. Zidek, *University of British Columbia, Canada*

The Analysis of Time Series — An Introduction, Fifth Edition
C. Chatfield

Applied Bayesian Forecasting and Time Series Analysis
A. Pole, M. West and J. Harrison

Applied Non-parametric Statistical Methods, Second Edition,
P. Sprent

Applied Statistics — Principles and Examples
D.R. Cox and E.J. Snell

Bayesian Data Analysis
A. Gelman, J. Carlin, H. Stern and D. Rubin

Beyond ANOVA — Basics of Applied Statistics
R.G. Miller, Jr.

Computer-Aided Multivariate Analysis Third Edition
A.A. Afifi and V.A. Clark

A Course in Large Sample Theory
T.S. Ferguson

Data Driven Statistical Methods,
P. Sprent

Decision Analysis — A Bayesian Approach
J.Q. Smith

Elementary Applications of Probability Theory, Second Edition
H.C. Tuckwell

Elements of Simulation
B.J.T. Morgan

Epidemiology — Study Design and Data Analysis
M. Woodward

Essential Statistics, Third Edition
D.G. Rees

Interpreting Data — A First Course in Statistics
A.J.B. Anderson

An Introduction to Generalized Linear Models
A.J. Dobson

Introduction to Multivariate Analysis
C. Chatfield and A.J. Collins

Introduction to Optimization Methods and their Applications in Statistics
B.S. Everitt

Large Sample Methods in Statistics
P.K. Sen and J. da Motta Singer

Markov Chain Monte Carlo — Stochastic Simulation for Bayesian Inference
D. Gamerman

Modeling and Analysis of Stochastic Systems
V. Kulkarni

Modelling Binary Data
D. Collett

Modelling Survival Data in Medical Research
D. Collett

Multivariate Analysis of Variance and Repeated Measures — A Practical Approach for Behavioural Scientists
D.J. Hand and C.C. Taylor

Multivariate Statistics — A Practical Approach
B. Flury and H. Riedwyl

Practical Data Analysis for Designed Experiments
B.S. Yandell

Practical Longitudinal Data Analysis
D.J. Hand and M. Crowder

Practical Statistics for Medical Research
D.G. Altman

Probability — Methods and Measurement
A. O'Hagan

Problem Solving — A Statistician's Guide, Second Edition
C. Chatfield

Randomization, Bootstrap and Monte Carlo Methods in Biology, Second Edition
B.F.J. Manly

Readings in Decision Analysis
S. French

Statistical Analysis of Reliability Data
M.J. Crowder, A.C. Kimber,
T.J. Sweeting and R.L. Smith

Statistical Methods for SPC and TQM
D. Bissell

Statistical Methods in Agriculture and Experimental Biology, Second Edition
R. Mead, R.N. Curnow and A.M. Hasted

Statistical Process Control—Theory and Practice, Third Edition
G.B. Wetherill and D.W. Brown

Statistical Theory, Fourth Edition
B.W. Lindgren

Statistics for Accountants, Fourth Edition
S. Letchford

Statistics for Technology—A Course in Applied Statistics, Third Edition
C. Chatfield

Statistics in Engineering—A Practical Approach
A.V. Metcalfe

Statistics in Research and Development, Second Edition
R. Caulcutt

The Theory of Linear Models
B. Jørgensen

Mathematical Statistics
K. Knight

A Course in Categorical Data Analysis
T. Leonard

MATHEMATICAL STATISTICS

Keith Knight

Department of Statistics
University of Toronto
Ontario, Canada

CHAPMAN & HALL/CRC

Boca Raton London New York Washington, D.C.

Library of Congress Cataloging-in-Publication Data

Knight, Keith.
Mathematical statistics / Keith Knight.
p. cm. — (Texts in statistical science series)
Includes bibliographical references and index.
ISBN 1-58488-178-X (alk. paper)
1. Mathematical statistics. I. Title. II. Texts in statistical science.

QA276 .K565 1999
519.5—dc21 99-056997

Direct all inquiries to CRC Press LLC, 2000 N.W. Corporate Blvd., Boca Raton, Florida 33431.

International Standard Book Number 1-58488-178-X
Library of Congress Card Number 99-056997
Printed in the United States of America 2 3 4 5 6 7 8 9 0
Printed on acid-free paper

To my parents

Contents

Preface

This book is intended as a textbook (or reference) for a full year Master's level (or senior level undergraduate) course in mathematical statistics aimed at students in statistics, biostatistics, and related fields.

This book grew from lecture notes and handouts that I developed for a course in mathematical statistics that I first taught in 1992-93 at the University of Toronto. In teaching this course, I realized that many students viewed the course as largely irrelevant to their education. To me this seemed strange since much of mathematical statistics is directly relevant to statistical practice; for example, what statistician has not used a χ^2 approximation at some point in their life? At the same time, I could also sympathize with their point of view. To a student first encountering the subject, the traditional syllabus of a mathematical statistics course does seem heavily weighed down with optimality theory of various flavours that was developed in the 1940s and 1950s; while this is interesting (and certainly important), it does leave the impression that mathematical statistics has little to offer beyond some nice mathematics.

My main objective in writing this book was to provide a set of useful tools that would allow students to understand the theoretical underpinnings of statistical methodology. At the same time, I wanted to be as mathematically rigorous as possible within certain constraints. I have devoted a chapter to convergence for sequences of random variables (and random vectors) since, for better or for worse, these concepts play an important role in the analysis of estimation and other inferential procedures in statistics. I have concentrated on inferential procedures within the framework of parametric models; however, in recognition of the fact that models are typically misspecified, estimation is also viewed from a nonparametric perspective by considering estimation of functional parameters (or statistical functionals, as they are often called). This book also places greater emphasis on "classical" (that is, Frequentist) methodology than it does on Bayesian methodology although this should not be interpreted as a claim of superiority for the Frequentist approach.

The mathematical background necessary for this book is multivariate calculus and linear algebra; some exposure to real analysis (in particular, ϵ-δ proofs) is also useful but not absolutely necessary. I have tried to make the book as self-contained as possible although I have also implicitly assumed that the reader has some familiarity with basic probability theory and, more importantly, has had some exposure to statistical methodology so as to provide some context for this book.

In teaching a course based on drafts of this book, I found it very useful to encourage the use of statistical software packages (such as S-Plus and SAS) as well as other mathematical software packages (such as MATLAB, Maple, and Mathematica). When used appropriately, these packages can greatly enhance the effectiveness of this course by increasing the scope of problems that can be considered by students. To facilitate this to some extent, I have included a few sections on computational issues, in particular, generating random variables and numerical computation of estimates. Moreover, some of the problems given in the book are most easily approached using some sort of mathematical or statistical software.

Unlike many other textbooks in mathematical statistics, I decided not to include tables of the commonly-used distributions in statistics (Normal, χ^2, and so on). My reason for this is simple; most readers will have access to some statistical software that renders obsolete even the most detailed set of tables.

My first exposure to mathematical statistics was as a graduate student at the University of Washington where I was fortunate to have a number of outstanding and inspiring teachers, including Andreas Buja, Peter Guttorp, Doug Martin, Ron Pyke, Paul Sampson, and Jon Wellner. Since then, I have benefited from the collective wisdom of many, including David Andrews, Richard Davis, Andrey Feuerverger, Nancy Heckman, Stephan Morgenthaler, John Petkau, Peter Phillips, Nancy Reid, Sid Resnick, Rob Tibshirani, and Jim Zidek.

In preparing successive drafts of this book, I received many useful comments from anonymous reviewers and from students at the University of Toronto. I would like to acknowledge the assistance of the editors at Chapman and Hall/CRC Press in completing this project. A particular vote of thanks is due Stephanie Harding for her tenacity in getting me to finish this book. I would also like to acknowledge the support of the Natural Sciences and Engineering Research Council of Canada.

Last, but not least, I would like to thank my wife Luisa for her patience and her gentle but constructive criticisms of various drafts of this book.

Keith Knight
Toronto
September, 1999

CHAPTER 1

Introduction to Probability

1.1 Random experiments

In simple terms, a random experiment (or experiment) is a process whose outcome is uncertain. It is often useful to think of this process as being repeatable but, in practice, this is seldom the case. For example, a football game may be regarded as a random experiment in the sense that the outcome of the game is uncertain *a priori*; however, this experiment (the game) is not repeatable as we surely could not guarantee that each game would be played under uniform conditions. Nonetheless, it is often plausible that a given random experiment is conceptually repeatable; for example, we might be willing to assume that team A would win 40% of its games against team B under a certain set of conditions. This "conceptual repeatability" is important as it allows us to interpret probabilities in terms of long-run frequencies.

For a given random experiment, we can define the following terms:

- The sample space is the set of all possible outcomes of a random experiment. We will denote the sample space by Ω.

- A subset of the sample space Ω is called an event. We say that an event A occurs if the true outcome ω lies in A (that is, $\omega \in A$). An event consisting of no outcomes is called the empty set and will be denoted by $\emptyset$.

Operations on events

Let A and B be arbitrary events defined on a sample space Ω.

- The union of A and B (denoted by $A \cup B$) consists of all outcomes that belong to at least one of A and B. That is, $\omega \in A \cup B$ if, and only if, $\omega \in A$ or $\omega \in B$.

- The intersection of A and B (denoted by $A \cap B$) consists of all outcomes that belong to both A and B. That is, $\omega \in A \cap B$ if, and only if, $\omega \in A$ and $\omega \in B$.

- The complement of A (denoted by A^c) consists of all outcomes in Ω that do not belong to A. That is, $\omega \in A^c$ if, and only if, $\omega \notin A$.
- A and B are disjoint (or mutually exclusive) if $A \cap B = \emptyset$.

We can also derive the following properties involving union, intersection and complement.

- $A \cup B \cup C = (A \cup B) \cup C = A \cup (B \cup C)$.
  $A \cap B \cap C = (A \cap B) \cap C = A \cap (B \cap C)$.
- $A \cap (B \cup C) = (A \cap B) \cup (A \cap C)$.
  $A \cup (B \cap C) = (A \cup B) \cap (A \cup C)$.
- $(A \cup B)^c = A^c \cap B^c$.
  $(A \cap B)^c = A^c \cup B^c$.

1.2 Probability measures

Given a random experiment with a sample space Ω, we would like to define a function or measure $P(\cdot)$ on the subsets (events) of Ω that assigns a real number to each event; this number will represent the probability that a given event occurs. Clearly, these probabilities must satisfy certain "consistency" conditions; for example, if $A \subset B$ then we should have $P(A) \leq P(B)$. However, from a mathematical point of view, some care must be taken in defining probability, and there have been a number of axiomatic approaches to defining probability. The approach that we will use in this book is due to Kolmogorov (1933) and effectively defines probability from a measure theoretic point of view; see Billingsley (1995) for more technical details. In fact, Kolmogorov's treatment was facilitated by the ideas of von Mises (1931) who introduced the notion of a sample space.

DEFINITION. $P(\cdot)$ is called a probability measure if the following axioms are satisfied:

1. $P(A) \geq 0$ for any event A.
2. $P(\Omega) = 1$.
3. If $A_1, A_2, \cdots$ are disjoint events then

$$P\left(\bigcup_{i=1}^{\infty} A_i\right) = \sum_{i=1}^{\infty} P(A_i).$$

There are a number of ways of interpreting probabilities. Perhaps the easiest to conceptualize is the interpretation of probabilities as long-run frequencies from a sequence of repeatable experiments. If we assume that a given random experiment is infinitely repeatable then we can interpret $P(A)$ as the relative frequency of occurrences of the event A; that is, if the experiment is repeated N times (where N is large) and A occurs k times then $P(A) \approx k/N$. However, other equally valid interpretations and axiomatic definitions of probability are possible. For example, $P(A)$ could be defined to be a person's degree of belief in the occurrence of the event A; that is, if B is judged more likely to occur than A, we have $P(B) \geq P(A)$. This type of probability is sometimes called subjective probability or personal probability (Savage, 1972).

Consequences of the axioms

The three axioms given above allow us to derive a number of simple but useful properties of probability measures.

PROPOSITION 1.1 *The following are consequence of the axioms of probability:*
(a) $P(A^c) = 1 - P(A)$.
(b) $P(A \cap B) \leq \min(P(A), P(B))$.
(c) $P(A \cup B) = P(A) + P(B) - P(A \cap B)$.
(d) Suppose that $\{A_n\}$ *is a nested, increasing sequence of events (in the sense that* $A_n \subset A_{n+1}$*) and let* $A = \bigcup_{k=1}^{\infty} A_k$. *Then* $P(A_n) \to P(A)$ *as* $n \to \infty$.
(e) Let $A_1, A_2, \cdots$ *be any events. Then*

$$P\left(\bigcup_{k=1}^{\infty} A_k\right) \leq \sum_{k=1}^{\infty} P(A_k).$$

Proof. (a) Since $A^c \cup A = \Omega$ and A and A^c are disjoint, it follows that

$$1 = P(\Omega) = P(A \cup A^c) = P(A) + P(A^c)$$

and so $P(A^c) = 1 - P(A)$.
(b) $B = (B \cap A) \cup (B \cap A^c)$. Since $B \cap A$ and $B \cap A^c$ are disjoint, we have

$$P(B) = P(B \cap A) + P(B \cap A^c) \geq P(B \cap A).$$

A similar argument gives $P(A) \geq P(A \cap B)$.

(c) $A \cup B = A \cup (B \cap A^c)$. Since A and $B \cap A^c$ are disjoint, we have

$$P(A \cup B) = P(A) + P(B \cap A^c)$$

and

$$P(B \cap A^c) = P(B) - P(B \cap A).$$

Thus $P(A \cup B) = P(A) + P(B) - P(A \cap B)$.
(d) Define $B_1 = A_1$ and $B_k = A_k \cap A_{k-1}^c$ for $k \geq 2$. Then $B_1, B_2, \cdots$ are disjoint and

$$A_n = \bigcup_{k=1}^{n} B_k \quad \text{and} \quad A = \bigcup_{k=1}^{\infty} B_k.$$

Hence

$$P(A) = \sum_{k=1}^{\infty} P(B_k) = \lim_{n\to\infty} \sum_{k=1}^{n} P(B_k) = \lim_{n\to\infty} P(A_n).$$

(e) First, it follows from (c) that

$$P(A_1 \cup A_2) \leq P(A_1) + P(A_2)$$

and so for any $n < \infty$, we have

$$P\left(\bigcup_{k=1}^{n} A_k\right) \leq \sum_{k=1}^{n} P(A_k).$$

Now let $B_n = \bigcup_{k=1}^{n} A_k$, $B = \bigcup_{k=1}^{\infty} A_k$ and note that $B_n \subset B_{n+1}$. Thus applying (d),

$$P(B) = \lim_{n\to\infty} P(B_n) \leq \lim_{n\to\infty} \sum_{k=1}^{n} P(A_k) = \sum_{k=1}^{\infty} P(A_k)$$

which completes the proof. □

EXAMPLE 1.1: For any events A_1, A_2, and A_3, we have

$$\begin{aligned} P(A_1 \cup A_2 \cup A_3) = \\ & P(A_1) + P(A_2) + P(A_3) \\ & \quad - P(A_1 \cap A_2) - P(A_1 \cap A_3) - P(A_2 \cap A_3) \\ & \quad + P(A_1 \cap A_2 \cap A_3). \end{aligned}$$

To see this, note that

$$P(A_1 \cup A_2 \cup A_3) = P(A_1 \cup A_2) + P(A_3) - P(A_3 \cap (A_1 \cup A_2))$$

Since $A_3 \cap (A_1 \cup A_2) = (A_1 \cap A_3) \cup (A_2 \cap A_3)$, it follows that

$$P(A_3 \cap (A_1 \cup A_2)) = P(A_1 \cap A_3) + P(A_2 \cap A_3) - P(A_1 \cap A_2 \cap A_3)$$

and the conclusion follows since $P(A_1 \cup A_2) = P(A_1) + P(A_2) - P(A_1 \cap A_2)$. ◇

EXAMPLE 1.2: Suppose that $A_1, A_2, \cdots$ is a collection of events. Then

$$P\left(\bigcap_{i=1}^{\infty} A_i\right) \geq 1 - \sum_{i=1}^{\infty} P(A_i^c).$$

To see this, note that

$$\begin{aligned} P\left(\bigcap_{i=1}^{\infty} A_i\right) &= 1 - P\left(\bigcup_{i=1}^{\infty} A_i^c\right) \\ &\geq 1 - \sum_{i=1}^{\infty} P(A_i^c). \end{aligned}$$

This inequality is known as Bonferroni's inequality. ◇

Finite sample spaces

When an experiment has a finite sample space, it is sometimes possible to assume that each outcome is equally likely. That is, if

$$\Omega = \{\omega_1, \omega_2, \cdots, \omega_N\}$$

then we may be able to assume that $P(\omega_k) = 1/N$ for $k = 1, \cdots, N$. This assumption is particularly appropriate in games of chance (for example, card games and lotteries). However, some care should be taken before assuming equally likely outcomes.

In the event of equally likely outcomes, for any event A we have

$$P(A) = \frac{\text{number of outcomes in } A}{N}.$$

In some cases, it may be possible to enumerate all possible outcomes, but in general such enumeration is physically impossible; for example, enumerating all possible 5 card poker hands dealt from a deck of 52 cards would take several months under the most favourable conditions. Thus we need to develop methods for counting the number of outcomes in a given event A or the sample space itself.

Many experiments can be broken down into a sequence of sub-experiments for which the number of outcomes is easily counted. For example, consider dealing 5 cards from a deck of 52 cards. This particular "experiment" is a sequence of 5 sub-experiments corresponding to the 5 cards being dealt.

EXAMPLE 1.3: Consider a simple experiment that consists of rolling a pair of dice. The outcomes can be denoted by an ordered pair (i, j) with i representing the outcome of the first die and j representing the outcome of the second; these can be thought of as the outcome of two sub-experiments. We can then represent the sample space by

$$\Omega = \{(1,1),(1,2),\cdots,(1,6),(2,1),\cdots,(2,6),\cdots,(6,1),\cdots,(6,6)\}.$$

For each outcome of the first die, there are 6 outcomes of the second die. Since there are 6 outcomes for the first die, it follows that the sample space has $6 \times 6 = 36$ outcomes. ◇

Example 1.3 suggests that if an experiment consists of a sequence of sub-experiments $\mathcal{E}_1, \cdots, \mathcal{E}_k$ having, respectively, $n_1, \cdots, n_k$ possible outcomes then the total number of outcomes is $N = n_1 \times \cdots \times n_k$. However, some care should be taken in defining the numbers $n_1, \cdots, n_k$. In many cases, the outcome of sub-experiment $\mathcal{E}_i$ depends on the outcome of the previous $i-1$ sub-experiments. For example, if we are dealing 5 cards from a deck of 52 cards, once the first card is dealt, there are only 51 possible outcomes for the second card dealt; however, there are in fact 52 possible outcomes for the second card if we ignore the outcome of the first card. In defining n_i for $i \geq 2$, we must take into account the outcomes of the previous $i - 1$ sub-experiments. A general rule for counting the number of outcomes in an experiment can be described as follows: Suppose a random experiment $\mathcal{E}$ consists of sub-experiments $\mathcal{E}_1, \cdots, \mathcal{E}_k$ where the outcome of sub-experiment $\mathcal{E}_i$ may depend on the outcomes of sub-experiments $\mathcal{E}_1, \cdots, \mathcal{E}_{i-1}$ but the number of outcomes of $\mathcal{E}_i$ is n_i independent of the outcomes of $\mathcal{E}_1, \cdots, \mathcal{E}_{i-1}$. Then the number of possible outcomes of $\mathcal{E}$ is

$$N = n_1 \times \cdots \times n_k.$$

This is sometimes called the product rule for determining the number of outcomes in the sample space.

EXAMPLE 1.4: Consider an urn consisting of the integers 0 to

9. We will select three integers from the urn to form a three digit number. The integers can be selected in two possible ways: "with replacement" or "without replacement". If the integers are selected with replacement, an integer is replaced in the urn after it has been selected and so can be selected again in subsequent draws. On the other hand, if the integers are selected without replacement, an integer is removed from the urn after its selection and therefore cannot be selected subsequently. If we draw the integers with replacement, we have $10 \times 10 \times 10 = 1000$ possible three digit numbers since we have 10 choices at each stage. On the other hand, if the selection is done without replacement, we have $10 \times 9 \times 8 = 720$ possible sequences since we have removed one integer from the urn at the second stage and two integers at the third stage. ◇

In many cases, the outcomes of an experiment can be represented as sets of k elements drawn without replacement from a set of n elements; such sets are called either permutations or *combinations* depending on whether we distinguish between distinct orderings of the elements of the sets. Using the product rule stated above, we can determine the number of permutations and combinations. Consider a set consisting of n distinct elements; for convenience, we will represent this set by $S = \{1, 2, \cdots, n\}$. Suppose we look at all sequences of $k \leq n$ elements drawn (without replacement) from S; each of these sequences is called a *permutation* of length k from S. Using the product rule, it follows that the number of permutations of length k from S is

$$n \times (n-1) \times \cdots \times (n-k+1) = \frac{n!}{(n-k)!}.$$

We can also consider all subsets of size k in S; such a subset is called a combination of size k. A combination differs from a permutation in the sense that the ordering of elements is unimportant; that is, the sets $\{1, 2, 3\}$, $\{2, 1, 3\}$ and $\{3, 1, 2\}$ represent the same combination but three distinct permutations. Note that for each subset (or combination) of size k, we have $k!$ permutations of length k and so the number of combinations of size k in S is

$$\frac{n!}{k!(n-k)!} = \binom{n}{k}.$$

We can also think of a combination of size k from S as splitting S into two disjoint subsets, one of size k and one of size $n-k$. Thus

$\binom{n}{k}$ is the number of ways of splitting S into two disjoint subsets of size n and $n-k$ respectively. We can extend the argument above to the problem of finding the number of ways of splitting S into m disjoint subsets of size $k_1, \cdots, k_m$ where $k_1 + \cdots + k_m = n$; this number is simply

$$\frac{n!}{k_1! \cdots k_m!} = \left(\begin{array}{c} n \\ k_1, \cdots, k_m \end{array} \right).$$

The quantities defined above are often called binomial and multinomial coefficients, respectively, due to their presence in the following theorems.

THEOREM 1.2 (Binomial Theorem) *If n is a nonnegative integer and a, b real numbers then*

$$(a+b)^n = \sum_{k=0}^{n} \binom{n}{k} a^k b^{n-k}.$$

THEOREM 1.3 (Multinomial Theorem) *If n is a nonnegative integer and $a_1, \cdots, a_m$ real numbers then*

$$(a_1 + a_2 + \cdots + a_m)^n = \sum_{k_1 + \cdots + k_m = n} \left(\begin{array}{c} n \\ k_1, \cdots, k_m \end{array} \right) a_1^{k_1} \cdots a_m^{k_m};$$

the sum above extends over all nonnegative integers $k_1, \cdots, k_m$ whose sum is n.

EXAMPLE 1.5: Lotto games typically involve "random" selecting k numbers from the integers 1 through n. For example, in the Canadian Lotto 6/49 game, 6 numbers are chosen from the numbers 1 through 49; prizes are awarded if a player selects 3 or more of the 6 numbers. (A 7th "bonus" is also drawn but does not affect the probability calculations given below.) The order of selection of the numbers is not important so the sample space for this "experiment" consists of all combinations of 6 numbers from the set $S = \{1, \cdots, 49\}$; there are $\binom{49}{6} = 13983816$ such combinations. Assuming that each outcome is equally likely, we can determine the probability that a given player who has selected 6 numbers has won a prize. Suppose we want to determine the probability that a player selects k numbers correctly. By choosing 6 numbers, the player has split S into two disjoint subsets one of which consists of the player's 6 numbers and another consisting of the 43 other numbers. Selecting k numbers correctly means that the outcome of the draw consisted of k of the player's 6 numbers while the other $6-k$ numbers were among the 43 he did not select. There are $\binom{6}{k}$ combinations of k numbers from the player's selected 6 and $\binom{43}{6-k}$ combinations of $6-k$ numbers from his unselected 43. Applying the product rule, we find that the number of outcomes with exactly k of the selected numbers is

$$N(k) = \binom{6}{k}\binom{43}{6-k}$$

and so

$$p(k) = P(k \text{ correct numbers}) = \frac{N(k)}{13983816}.$$

Substituting for k, we obtain

$$\begin{aligned} p(3) &= 1.77 \times 10^{-2} \approx 1/57, \\ p(4) &= 9.69 \times 10^{-4} \approx 1/1032, \\ p(5) &= 1.84 \times 10^{-5} \approx 1/54201 \\ \text{and} \quad p(6) &= 1/13983816 \approx 7.15 \times 10^{-8}. \end{aligned}$$

The number of correct selections is an example of a random variable, which will be formally defined in section 1.4. ◇

1.3 Conditional probability and independence

So far in defining probabilities, we have assumed no information over and above that available from specifying the sample space and the probability measure on this space. However, knowledge that a particular event has occurred will change our assessment of the probabilities of other events.

DEFINITION. Suppose that A and B are events defined on some sample space Ω. If $P(B) > 0$ then

$$P(A|B) = \frac{P(A \cap B)}{P(B)}$$

is called the conditional probability of A given B.

The restriction to events B with $P(B) > 0$ in the definition of conditional probability may not seem too restrictive at this point in the discussion. However, as we will see later, it does often make sense to consider $P(A|B)$ where $P(B) = 0$ and $B \neq \emptyset$. One possible approach to defining conditional probability in this case would be to take a sequence of events $\{B_n\}$ decreasing to B (in the sense that $B_{n+1} \subset B_n$ and $B = \bigcap_{n=1}^{\infty} B_n$) such that $P(B_n) > 0$ for all n. We could define $P(A|B)$ by

$$P(A|B) = \lim_{n\to\infty} \frac{P(A \cap B_n)}{P(B_n)}.$$

However, it is not clear that the limit (if indeed it exists) is independent of the sequence of events $\{B_n\}$. A more fundamental problem is the fact that this definition fails the most basic test. Suppose that $A = B$ with $P(B) = 0$; any reasonable definition of conditional probability would seem to give $P(A|B) = 1$. However, for any sequence of events $\{B_n\}$ with $P(B_n) > 0$ for all n, we have $P(A|B_n) = 0$ and so $P(A|B_n) \to 0$.

It is easy to see that, for a fixed B, $P(\cdot|B)$ satisfies the axioms of a probability measure and thus shares the same properties as any other probability measure.

Given $P(A|B)$, we can write $P(A \cap B) = P(B)P(A|B)$. This rearrangement is important because in many situations, it is convenient to specify conditional probabilities and from these derive probabilities of given events. For example, as we mentioned earlier, many experiments can be thought of occurring as a finite (or countable) number of stages with the outcome of the k-th stage dependent (or conditional) on the outcome of the previous $k-1$ stages. If A_k is some event that refers specifically to the k-th stage of the experiment, we can write

$$\begin{aligned} P(A_1 \cap \cdots \cap A_n) &= P(A_1 \cap \cdots \cap A_{n-1}) P(A_n | A_1 \cap \cdots \cap A_{n-1}) \\ &= P(A_1)P(A_2|A_1) \times \cdots \\ &\quad \times P(A_n|A_1 \cap \cdots \cap A_{n-1}). \end{aligned}$$

The following classic example (the "birthday problem") illustrates the application of this formula.

EXAMPLE 1.6: Suppose a room is filled with n people who represent a random sample from the population as a whole. What is the probability that at least two of the people in the room share the same birthday? To answer this question, we must first make a simplifying assumption: we will assume that birthdays are uniformly distributed over the 365 days of a non-leap year.

Define B_n to be the event where no two people (out of n) share the same birthday; the probability we want is simply $P(B_n^c) = 1 - P(B_n)$. In order to evaluate $P(B_n)$, it is convenient to think of the people entering the room one at a time; we can then define the events

$$\begin{aligned} A_2 &= \text{2nd person's birthday is different from 1st person's} \\ A_3 &= \text{3rd person's birthday is different from previous two} \\ &\vdots \\ A_n &= n\text{-th person's birthday is different from previous } n-1. \end{aligned}$$

Clearly now, $B_n = A_2 \cap \cdots \cap A_n$ and so

$$P(A_2)P(A_3|A_2)\cdots P(A_n|A_2 \cap \cdots \cap A_{n-1}).$$

From our assumption of uniformity of birthdays, it follows that $P(A_2) = 364/365$, $P(A_3|A_2) = 363/365$, $P(A_4|A_2 \cap A_3) = 362/365$ and, in general,

$$P(A_k|A_2 \cap \cdots \cap A_{k-1}) = \frac{366-k}{365}$$

since the occurrence of the event $A_2 \cap \cdots \cap A_{k-1}$ implies that the first $k-1$ people in the room have $k-1$ distinct birthdays and hence there are $365-(k-1) = 366-k$ "unclaimed" birthdays. Thus

$$P(B_n) = \frac{364}{365} \times \frac{363}{365} \times \cdots \times \frac{366-n}{365}.$$

Substituting into the expression above, it is straightforward to evaluate the probability of at least one match, namely $P(B_n^c)$. For $n = 5$, $P(B_n^c) = 0.027$ while for $n = 30$, $P(B_n^c) = 0.706$ and $n = 70$, $P(B_n^c) = 0.999$; in fact, it is easy to verify that $P(B_n^c) \geq 0.5$ for $n \geq 23$, a fact that is somewhat counterintuitive to many people. A useful approximation is

$$P(B_n^c) \approx 1 - \exp\left(-\frac{n(n-1)}{730}\right),$$

which is valid if $n/365$ is small. ◇

Bayes' Theorem

As mentioned above, conditional probabilities are often naturally specified by the problem at hand. In many problems, we are given $P(A|B_1), \cdots, P(A|B_k)$ where $B_1, \cdots, B_k$ are disjoint events whose union is the sample space; however, we would like to compute

$P(B_j|A)$ for some B_j. Provided $P(A) > 0$, we have

$$P(B_j|A) = \frac{P(A \cap B_j)}{P(A)}.$$

The following result gives a simple formula for $P(A)$.

PROPOSITION 1.4 (Law of total probability)
If B_1, B_2, $\cdots$ are disjoint events with $P(B_k) > 0$ for all k and $\bigcup_{k=1}^{\infty} B_k = \Omega$ then

$$P(A) = \sum_{k=1}^{\infty} P(B_k)P(A|B_k).$$

Proof. Since $A = A \cap (\bigcup_{k=1}^{\infty} B_k)$, we have

$$\begin{aligned} P(A) &= P\left(A \cap \left(\bigcup_{k=1}^{\infty} B_k\right)\right) \\ &= P\left(\bigcup_{k=1}^{\infty} (A \cap B_k)\right) \\ &= \sum_{k=1}^{\infty} P(A \cap B_k) \\ &= \sum_{k=1}^{\infty} P(B_k)P(A|B_k) \end{aligned}$$

since $A \cap B_1, A \cap B_2, \cdots$ are disjoint. □

A simple corollary of the law of total probability is Bayes' Theorem.

PROPOSITION 1.5 (Bayes' Theorem) *Suppose that B_1, B_2, $B_3, \cdots$ are disjoint sets with $P(B_k) > 0$ for all k and $\bigcup_{k=1}^{\infty} B_k = \Omega$. Then for any event A,*

$$P(B_j|A) = \frac{P(B_j)P(A|B_j)}{\sum_{k=1}^{\infty} P(B_k)P(A|B_k)}.$$

Proof. By definition, $P(B_j|A) = P(A \cap B_j)/P(A)$ and $P(A \cap B_j) = P(B_j)P(A|B_j)$. The conclusion follows by applying the law of total probability to $P(A)$. □

EXAMPLE 1.7: Suppose that the incidence of a certain disease in a population is 0.001. A diagnostic test for this disease exists but has a false positive rate of 0.05 and a false negative rate of

0.01; that is, 5% of tests on non-diseased people will indicate the presence of the disease while 1% of tests on people with the disease will not indicate the presence of the disease. If a person drawn at random from the population tests positive, what is the probability that that person has the disease?

Define the event D to indicate the presence of disease and A to indicate a positive test for the disease. Then $P(D) = 0.001$, $P(A|D^c) = 0.05$ and $P(A^c|D) = 0.01$. By Bayes' Theorem, we have

$$\begin{aligned} P(D|A) &= \frac{P(D)P(A|D)}{P(D)P(A|D) + P(D^c)P(A|D^c)} \\ &= \frac{(0.001)(0.99)}{(0.001)(0.99) + (0.999)(0.05)} \\ &= 0.0194. \end{aligned}$$

This example illustrates the potential danger of mandatory testing in a population where the false positive rate exceeds the incidence of the disease; this danger may be particularly acute in situations where a positive test carries a significant social stigma. ◇

EXAMPLE 1.8: In epidemiology, one is often interested in measuring the relative risk of disease in one group relative to another. (For example, we might be interested in the relative risk of lung cancer in smokers compared to non-smokers.) If we have two distinct groups, A and B, then the relative risk of disease in group A compared to group B is defined to be

$$\mathrm{rr} = \frac{\text{incidence of disease in group } A}{\text{incidence of disease in group } B}.$$

The incidences of disease in groups A and B can be thought of as the conditional probabilities $P(\text{disease}|A)$ and $P(\text{disease}|B)$. If the overall incidence of disease in the population ($P(\text{disease})$) is small then the relative risk can be difficult to estimate if one draws samples from groups A and B. However, using Bayes' Theorem and a little algebra, one obtains

$$\mathrm{rr} = \frac{P(A|\text{disease})[\theta P(B|\text{disease}) + (1-\theta)P(B|\text{disease-free})]}{P(B|\text{disease})[\theta P(A|\text{disease}) + (1-\theta)P(A|\text{disease-free})]}$$

where $\theta = P(\text{disease})$, the overall incidence of disease in the population. If θ is close to 0 then

$$\text{rr} \approx \frac{P(A|\text{disease})P(B|\text{disease-free})}{P(B|\text{disease})P(A|\text{disease-free})}.$$

This suggests that the relative risk can be estimated by drawing samples not from the groups A and B but rather from the "disease" and "disease-free" portions of the population; even though the absolute number of subjects in the "disease" group is small, they can typically be easily sampled (since they are usually undergoing treatment for their illness). Such studies are know as case-control studies and are very common in epidemiology. ◇

Independence

In discussing conditional probability, we noted that our assessment of the probability of an event A may change if we have knowledge about the occurrence of another event B; if our assessment of the probability of A changes then we can say that there is some dependence between A and B. In some cases, this knowledge will not change our assessment of the probability of A, that is, $P(A|B) = P(A)$. In this case, we say that the events A and B are independent.

Since we have defined $P(A|B)$ only for events B with $P(B) > 0$, the "definition" of independence given in the previous paragraph is not completely satisfactory as it does not allow us to deal with events that have probability 0. However, if both $P(A)$ and $P(B)$ are positive then this definition is consistent in the sense that $P(A|B) = P(A)$ is equivalent to $P(B|A) = P(B)$. Also note that if $P(A|B) = P(A)$ then $P(A \cap B) = P(A)P(B)$; since the probabilities in the latter equality are always well-defined, we will use it as a formal definition of independence when at least one of $P(A)$ or $P(B)$ is positive.

DEFINITION. Events A and B are said to be independent if $P(A \cap B) = P(A)P(B)$.

Some care must be taken in interpreting independence of A and B when $P(A)$ or $P(B)$ is equal to 0 or 1. For example, if $B = \Omega$ then $A \cap B = A$ and hence $P(A \cap B) = P(A) = P(A)P(B)$ which implies that A is always independent of the sample space. This is somewhat counterintuitive since in this case, it follows that the event A implies the event $B = \Omega$. This would suggest that A and B are independent but intuitively A should not be independent of the sample space. The following example illustrates the problem.

EXAMPLE 1.9: Suppose that $\Omega = [0,1]$ and that probabilities of events are assigned so that if $A = [a,b]$ (with $0 \le a \le b \le 1$), $P(A) = b-a$. (It turns out that this simple specification is sufficient to define probabilities for virtually all events of interest on Ω.) From this, it follows that for events C consisting of a single outcome (that is, $C = \{c\}$ where c is a real number), we have $P(C) = 0$. Let $A = \{1/4, 3/4\}$ and $B = \{3/4\}$ so $P(A) = P(B) = 0$; clearly, knowing that the outcome of the experiment belongs to A gives us fairly significant information about the occurrence of B. In fact, heuristic considerations suggest that if A is known to have occurred then the two outcomes of A are equally likely, in which case $P(B|A) = 1/2 \neq P(B)$. (In any event, it seems reasonable that $P(B|A) > 0$.) ◇

We can extend our notion of independence to a finite or countably infinite collection of events. It is tempting to define independence of $A_1, \cdots, A_n$ to mean that the probability of the intersection of these events equals the product of their probabilities. However, this definition is defective in the sense that if $P(A_j) = 0$ (for some j) then $P(A_1 \cap \cdots \cap A_n) = 0$ regardless of the other A_j's; this would mean, for example, that independence of $A_1, \cdots, A_n$ does not imply independence of any given pair of events, which seems somewhat illogical. Therefore, in defining independence of a finite or countably infinite collection of events, we look at all possible finite intersections.

DEFINITION. $A_1, \cdots, A_n$ (or $A_1, A_2, \cdots$) are (mutually) independent events if for any finite sub-collection $A_{i_1}, \cdots, A_{i_k}$,

$$P(A_{i_1} \cap A_{i_2} \cap \cdots \cap A_{i_k}) = \prod_{j=1}^{k} P(A_{i_j}).$$

Thus if $A_1, \cdots, A_n$ are independent then A_i is independent of A_j for $i \neq j$; that is, mutual independence implies pairwise independence. However, the converse is not true as the following example indicates.

EXAMPLE 1.10: Consider an experiment with the following

sample space

$$\Omega = \{abc, bac, cab, bca, acb, cba, aaa, bbb, ccc\}$$

where each outcome is equally likely. Define events

$$A_k = a \text{ in } k\text{-th position} \quad (\text{for } k = 1, 2, 3).$$

It is easy to see that $P(A_k) = 1/3$ for $k = 1, 2, 3$ and $P(A_j \cap A_k) = 1/9$ for $j \neq k$; thus A_j and A_k are independent for all $j \neq k$. However, $P(A_1 \cap A_2 \cap A_3) = 1/9 \neq 1/27$ and so A_1, A_2 and A_3 are not independent. This example shows that pairwise independence does not imply mutual independence. ◇

1.4 Random variables

Consider an experiment where a coin is tossed 20 times. We can represent the sample space as the sets of all $2^{20} \approx 10^6$ sequences of heads and tails:

$$\Omega = \{HH \cdots H, THH \cdots H, HTH \cdots H, \cdots, TT \cdots T\}$$

For such experiments, we are usually not interested in every event defined on Ω but rather just those events involving, for example, the number of heads in the 20 tosses. In this case, we could "redefine" to be the set of all possible values of the number of heads:

$$\Omega' = \{0, 1, 2, \cdots, 20\}$$

Note that each outcome $\omega' \in \Omega'$ corresponds to an event defined on Ω; for example, the outcome $\{0\} \in \Omega'$ corresponds to the single outcome $\{HH \cdots H\} \in \Omega$ while the outcome $\{1\} \in \Omega'$ corresponds to the event $\{TH \cdots H, HTH \cdots H, \cdots, H \cdots HT\}$. Thus we can define a function X mapping Ω to Ω' that counts the number of heads in 20 tosses of the coin; such a function is called a random variable.

DEFINITION. A random variable X is a function that maps the sample space to the real line; that is, for each $\omega \in \Omega$, $X(\omega)$ is a real number.

EXAMPLE 1.11: Consider an experiment where a coin is tossed until the first heads comes up. The sample space for this experiment can be represented as

$$\Omega = \{H, TH, TTH, TTTH, \cdots\}$$

where H and T represent heads and tails. We can define a random variable X that counts the number of tails until the first heads:

$$\begin{aligned} X(H) &= 0 \\ X(TH) &= 1 \\ X(TTH) &= 2 \end{aligned}$$

and so on. In this case, each value of the random variable X corresponds to a single outcome in Ω. ◇

Random variables allow us, in some sense, to ignore the sample space; more precisely, we can redefine the sample space to be the range of a random variable or a collection of random variables.

Probability distributions

Suppose that X is a random variable defined on a sample space Ω. If we define the event

$$[a \le X \le b] = \{\omega \in \Omega : a \le X(\omega) \le b\} = A$$

then $P(a \le X \le b) = P(A)$.

DEFINITION. Let X be a random variable (defined on some sample space Ω). The distribution function of X is defined by

$$F(x) = P(X \le x) = P\left(\{\omega \in \Omega : X(\omega) \le x\}\right).$$

(The distribution function is often referred to as the cumulative distribution function.)

Distribution functions satisfy the following basic properties:

- If $x \le y$ then $F(x) \le F(y)$. (F is a non-decreasing function.)
- If $y \downarrow x$ then $F(y) \downarrow F(x)$. (F is a right-continuous function although it is not necessarily a continuous function.)
- $\lim_{x\to-\infty} F(x) = 0$; $\lim_{x\to\infty} F(x) = 1$.

Figure 1.1 shows a generic distribution function on the interval $[0, 1]$. Note that the distribution function has a jump at $x = 0.25$; the height of this jump is the probability that the random variable is equal to 0.25. The distribution function is also flat over $[0.25, 0.5)$ indicating that X can take no values in this interval.

If X has a distribution function F then knowledge of $F(x)$ for all x allows us to compute $P(X \in A)$ for any given set A. It

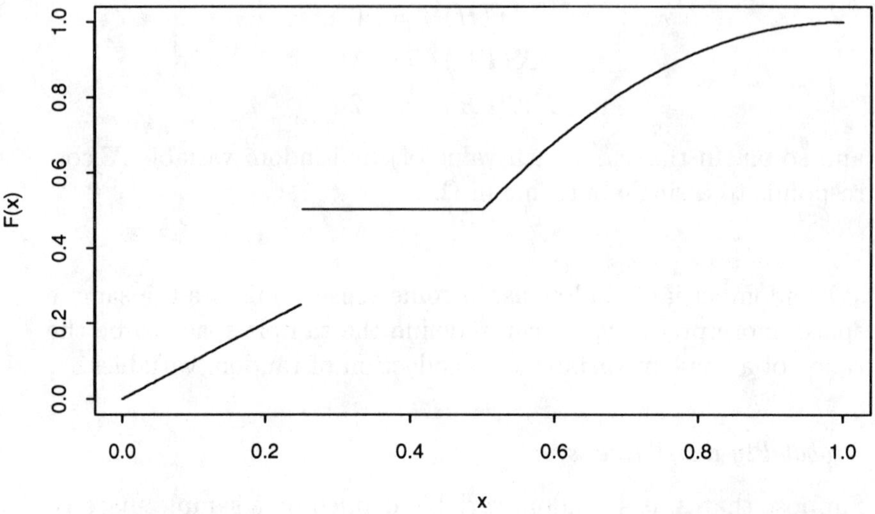

Figure 1.1 *A distribution function on the interval* $[0, 1]$.

is important to note that we could also define the distribution function to be $P(X < x)$; this function would be left-continuous rather than right-continuous.

Suppose that X is a random variable with distribution function F. Then

- $P(a < X \leq b) = F(b) - F(a)$;
- $P(X < a) = F(a-) = \lim_{x \uparrow a} F(x)$; ($F(a-)$ is called the left-hand limit of F at the point a; if F is continuous at a then $F(a-) = F(a)$.)
- $P(X > a) = 1 - F(a)$;
- $P(X = a) = F(a) - F(a-)$. If F is continuous at a then
$$P(X = a) = 0.$$

These simple properties are useful for computing $P(X \in A)$.

Discrete random variables

DEFINITION. A random variable X is discrete if its range is a finite or countably infinite set. That is, there exists a set $S = \{s_1, s_2, \cdots\}$ such that $P(X \in S) = 1$.

From the definition above, we can deduce that the probability distribution of a discrete random variable is completely determined by specifying $P(X = x)$ for all x.

DEFINITION. The frequency function of a discrete random variable X is defined by

$$f(x) = P(X = x).$$

The frequency function of a discrete random variable is known by many other names: some examples are probability mass function, probability function and density function. We will reserve the term "density function" for continuous random variables.

Given the frequency function $f(x)$, we can determine the distribution function:

$$F(x) = P(X \le x) = \sum_{t \le x} f(t).$$

Thus $F(x)$ is a step function with jumps of height $f(x_1), f(x_2), \cdots$ occurring at the points $x_1, x_2, \cdots$. Likewise, we have

$$P(X \in A) = \sum_{x \in A} f(x);$$

in the special case where $A = (-\infty, \infty)$, we obtain

$$1 = P(-\infty < X < \infty) = \sum_{-\infty < x < \infty} f(x).$$

EXAMPLE 1.12: Consider an experiment consisting of independent trials where each trial can result in one of two possible outcomes (for example, success or failure). We will also assume that the probability of success remains constant from trial to trial; we will denote this probability by θ where $0 < \theta < 1$. Such an experiment is sometimes referred to as Bernoulli trials.

We can define several random variables from a sequence of Bernoulli trials. For example, consider an experiment consisting of n

Bernoulli trials. The sample space can be represented as all possible 2^n sequences of successes (S) and failures (F):

$$\Omega = \{F \cdots F, SF \cdots F, FSF \cdots F, \cdots, S \cdots S\}.$$

We can define a random variable X that counts the number of "successes" in the n Bernoulli trials. To find the frequency function of X, we need to

- determine the probability of each outcome in Ω, and
- count the number of outcomes with exactly x successes.

Let ω be any outcome consisting of x successes and $n-x$ failures for some specified x between 0 and n. By independence, it is easy to see that $P(\omega) = \theta^x(1-\theta)^{n-x}$ for each such outcome ω; these outcomes are also disjoint events. To count the number of outcomes with x successes, note that this number is exactly the same as the number of combinations of size x from the set of integers $\{1, 2, \cdots, n\}$ which is

$$\binom{n}{x} = \frac{n!}{x!(n-x)!}.$$

Thus it follows that the frequency function of X is given by

$$\begin{aligned} f_X(x) &= P(X = x) \\ &= \binom{n}{x}\theta^x(1-\theta)^{n-x} \quad \text{for } x = 0, 1, \cdots, n \end{aligned}$$

X is said to have a Binomial distribution with parameters n and θ; we will abbreviate this by $X \sim \text{Bin}(n, \theta)$. When $n = 1$, X has a Bernoulli distribution with parameter θ ($X \sim \text{Bern}(\theta)$).

Next consider an experiment consisting of a sequence of Bernoulli trials, which is stopped as soon as $r \geq 1$ successes are observed. Thus all the outcomes in the sample space consist of exactly r successes and $x \geq 0$ failures. We can define a random variable Y that counts the number of failures before the r successes are observed. To derive the frequency function of Y, we note that if $Y = y$ ($y \geq 0$) then there are $r + y$ Bernoulli trials observed before the termination of the experiment (r successes and y failures) and the outcome of the $(r + y)$-th trial is necessarily a success; thus we observed $r - 1$ successes in the first $r + y - 1$ trials. By the independence of Bernoulli trials, we have

$$f_Y(y) = P\left(r - 1 \text{ } S\text{'s in first } r + y - 1 \text{ trials}\right) P\left(S \text{ on } r\text{-th trial}\right).$$

Using the Binomial frequency function derived earlier, we have

$$P\,(r-1 \text{ } S\text{'s in first } r+y-1 \text{ trials}) = \binom{r+y-1}{r-1}\theta^{r-1}(1-\theta)^y$$

and so

$$f_Y(y) = \binom{r+y-1}{r-1}\theta^r(1-\theta)^y \quad \text{for } y = 0, 1, 2, \cdots.$$

Note that since $\binom{r+y-1}{r-1} = \binom{r+y-1}{y}$, we can also write

$$f_Y(y) = \binom{r+y-1}{y}\theta^r(1-\theta)^y \quad \text{for } y = 0, 1, 2, \cdots.$$

The random variable Y is said to have a Negative Binomial distribution with parameters r and θ; we will sometimes write $Y \sim \text{NegBin}(r, \theta)$. When $r = 1$, we say that Y has a Geometric distribution (with parameter θ) and write $Y \sim \text{Geom}(\theta)$. ◇

EXAMPLE 1.13: Consider a finite population consisting of two distinct groups (group A and group B). Assume that the total population is N with M belonging to group A and the remaining $N - M$ belonging to group B. We draw a random sample of size $n(\leq N)$ without replacement from the population and define the random variable

$$X = \text{number of items from group A in the sample.}$$

Using combinatorial arguments, it follows that the frequency function of X is

$$f(x) = \binom{M}{x}\binom{N}{n} \Big/ \binom{N-M}{n-x}$$

for $x = \max(0, n + M - N), \cdots, \min(M, n)$. The range of X is determined by the fact that we cannot have more than M group A items in the sample nor more than $N - M$ group B items. This distribution is known as the Hypergeometric distribution. ◇

Continuous random variables

DEFINITION. A random variable X is said to be continuous if its distribution function $F(x)$ is continuous at every real number x.

Equivalently, we can say that X is a continuous random variable if $P(X = x) = 0$ for all real numbers x. Thus we have

$$P(a \leq X \leq b) = P(a < X < b)$$

for $a < b$. The fact that $P(X = x) = 0$ for any x means that we cannot usefully define a frequency function as we did for discrete random variables; however, for many continuous distributions, we can define an analogous function that is useful for probability calculations.

DEFINITION. A continuous random variable X has a probability density function $f(x) \geq 0$ if for $-\infty < a < b < \infty$, we have

$$P(a \leq X \leq b) = \int_a^b f(x)\,dx.$$

It is important to note that density functions are not uniquely determined; that is, given a density function f, it is always possible to find another density function f^* such that

$$\int_a^b f(x)\,dx = \int_a^b f^*(x)\,dx$$

for all a and b but for which $f^*(x)$ and $f(x)$ differ at a finite number of points. This non-uniqueness does not pose problems, in general, for probability calculations but can pose subtle problems in certain statistical applications. If the distribution function F is differentiable at x, we can take the density function f to be the derivative of F at x: $f(x) = F'(x)$. For purposes of manipulation, we can assume that $f(x) = F'(x)$ when the derivative is well-defined and define $f(x)$ arbitrarily when it is not.

EXAMPLE 1.14: Suppose that X is a continuous random variable with density function

$$f(x) = \begin{cases} kx^3 & \text{for } 0 \leq x \leq 1 \\ 0 & \text{otherwise} \end{cases}$$

where k is some positive constant. To determine the value of k, we note that

$$1 = P(-\infty < X < \infty) = \int_{-\infty}^{\infty} f(x)\,dx.$$

Since $f(x) = 0$ for $x < 0$ and for $x > 0$, we have

$$1 = \int_{-\infty}^{\infty} f(x)\,dx = \int_0^1 kx^3\,dx = \frac{k}{4}$$

and so $k = 4$. We can also determine the distribution function by integrating the density from $-\infty$ to any point x; we obtain

$$F(x) = \begin{cases} 0 & \text{for } x < 0 \\ x^4 & \text{for } 0 \le x \le 1 \\ 1 & \text{for } x > 1 \end{cases}.$$

As we noted above, we can modify the density function at a countable number of points and still obtain the same distribution function; for example, if we define

$$f^*(x) = \begin{cases} 4x^3 & \text{for } 0 < x < 1 \\ 0 & \text{otherwise} \end{cases}$$

then we obtain the same distribution function although $f^*(x)$ differs from $f(x)$ at $x = 0$ and $x = 1$. ◇

Typically, it turns out to be more convenient to specify the probability distribution of a continuous random variable via its density function rather than via its distribution function. The density function essentially describes the probability that X takes a value in a "neighbourhood" of a point x (and thus it is analogous to the frequency function of a discrete random variable); from the definition above, we have

$$P(x \le X \le x + \Delta) = \int_x^{x+\Delta} f(t)\,dx \approx \Delta f(x)$$

if Δ is small and f is continuous at x. Thus $f(x) > f(y)$ suggests that we are more likely to see values of X close to x than values of X close to y.

EXAMPLE 1.15: (Normal distribution) A random variable X is said to have a Normal distribution with parameters μ and σ^2 ($X \sim N(\mu, \sigma^2)$) if its density is

$$f(x) = \frac{1}{\sigma\sqrt{2\pi}} \exp\left(-\frac{(x-\mu)^2}{2\sigma^2}\right).$$

It is easy to see that $f(x)$ has its maximum value at $x = \mu$ (which is called the mode of the distribution). When $\mu = 0$ and $\sigma^2 = 1$,

X is said to have a standard Normal distribution; we will denote the distribution function of $X \sim N(0,1)$ by

$$\Phi(x) = \frac{1}{\sqrt{2\pi}} \int_{-\infty}^{x} \exp\left(-t^2/2\right)\, dt.$$

If $X \sim N(\mu, \sigma^2)$ then its distribution function is

$$F(x) = \Phi\left(\frac{x-\mu}{\sigma}\right).$$

Tables of $\Phi(x)$ can be found in many books, and $\Phi(x)$ can be evaluated numerically using practically any statistical software package. $\diamond$

EXAMPLE 1.16: (Uniform distribution) Suppose that X is a continuous random variable with density function

$$f(x) = \begin{cases} (b-a)^{-1} & \text{for } a \le x \le b \\ 0 & \text{otherwise} \end{cases}.$$

We say that X has a Uniform distribution on the interval $[a, b]$; we will sometimes write $X \sim \text{Unif}(a, b)$. Simple integration shows that the distribution function is given by $F(x) = (x-a)/(b-a)$ for $a \le x \le b$ ($F(x) = 0$ for $x < a$ and $F(x) = 1$ for $x > b$). $\diamond$

An important special case of the Uniform family of distributions is the Uniform distribution on $[0, 1]$. The following two results are quite useful.

THEOREM 1.6 *Suppose that X is a continuous random variable with distribution function $F(x)$ and let $U = F(X)$. Then $U \sim$ Unif$(0,1)$.*

Proof. Since $0 \le F(x) \le 1$, we have $P(0 \le U \le 1) = 1$. Thus we need to show that $P(U < y) = y$ for $0 < y < 1$. (This, in turn, implies that $P(U \le y) = y$.) For given y, choose x such that $F(x) = y$; such an x exists since F is continuous. Then

$$P(U < y) = P(U < y, X < x) + P(U < y, X \ge x).$$

Since F is a non-decreasing function, $U = F(X)$ and $y = F(x)$, it follows that the event $[X < x]$ is a subset of the event $[U < y]$ and so

$$P(U < y, X < x) = P(X < x) = F(x) = y.$$

Furthermore the event $[U < y, X \ge x]$ is the empty set and so $P(U < y, X \ge x) = 0$. Thus $P(U < y) = y$ for $0 < y < 1$. $\square$

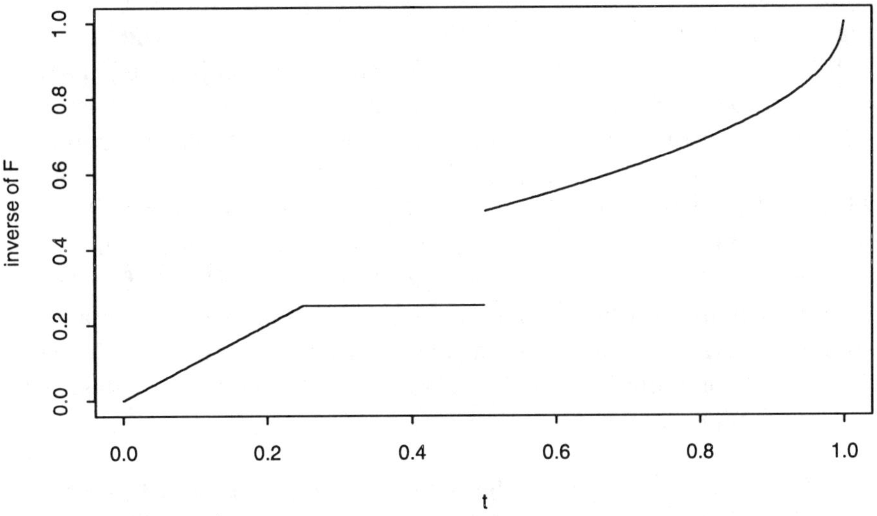

Figure 1.2 *The inverse distribution function* $F^{-1}(t)$ *corresponding to* $F(x)$ *in Figure 1.1.*

An easier proof of Theorem 1.6 can be obtained if we assume that F is strictly increasing. In this case, we can define a continuous inverse F^{-1} of F satisfying $F(F^{-1}(y)) = y$ for $0 < y < 1$ and so

$$P(U \leq y) = P(X \leq F^{-1}(y)) = F(F^{-1}(y)) = y.$$

We can also invert this argument. If $U \sim \text{Unif}(0,1)$ and F is a strictly increasing continuous distribution function then the random variable $X = F^{-1}(U)$ has distribution function F:

$$P(X \leq x) = P(U \leq F(x)) = F(x).$$

In fact, this latter result holds for an arbitrary distribution function F provided that the inverse function is defined appropriately.

THEOREM 1.7 *Let* F *be any distribution function and define* $F^{-1}(t) = \inf\{x : F(x) \geq t\}$ *to be its inverse function for* $0 < t < 1$. *If* $U \sim$ *Unif*$(0,1)$ *and* $X = F^{-1}(U)$ *then the distribution function of* X *is* F.

Before proceeding to the proof of Theorem 1.7, we will briefly discuss the function F^{-1}, which is also called the quantile function of X with $F^{-1}(\alpha)$ being the α quantile (or 100α percentile) of the distribution of X. $F^{-1}(t)$ is simplest to define when $F(x)$ is continuous and strictly increasing; in this case, we can define $F^{-1}(t)$ via the equation $F(F^{-1}(t)) = t$, that is, F^{-1} agrees with the usual definition of inverse function. A somewhat more complicated case occurs when F is the distribution function of a discrete random variable in which case F is piecewise constant with jumps at certain points. If t is such that $F(x) = t$ for $a \le x < b$ then $F^{-1}(t) = a$; if t is such that $F(a) > t$ and $F(a-) < t$ (so that t falls in the middle of a jump in F) then $F^{-1}(t) = a$ since a is the smallest value of x with $F(x) \ge t$. Figure 1.2 illustrates the form of $F^{-1}(t)$ for the generic distribution function F in Figure 1.1. It is also important to note that $F^{-1}(t)$ cannot always be defined for $t = 0$ or $t = 1$; in the context of defining $X = F^{-1}(U)$ (where $U \sim \text{Unif}(0,1)$) this is not important since $P(U = 0) = P(U = 1) = 0$.

Proof. (Theorem 1.7) From the definition of $F^{-1}(t)$ and the fact that $F(x)$ is a right-continuous function, we have that $F^{-1}(t) \le x$ if, and only if, $t \le F(x)$. Thus

$$P(X \le x) = P\left(F^{-1}(U) \le x\right) = P\left(U \le F(x)\right) = F(x)$$

and so the distribution function of X is F. □

A common application of Theorem 1.7 is the simulation of random variables with a particular distribution. If we can simulate a random variable $U \sim \text{Unif}(0,1)$ then given any distribution function F with inverse F^{-1}, $X = F^{-1}(U)$ will have distribution function F. A drawback with this method is the fact that $F^{-1}(t)$ is not necessarily easily computable.

The quantiles are sometimes used to describe various features of a distribution. For example, $F^{-1}(1/2)$ is called the median of the distribution and is sometimes useful as a measure of the centre of the distribution. Another example is the interquartile range, $F^{-1}(3/4) - F^{-1}(1/4)$, which is used as a measure of dispersion.

EXAMPLE 1.17: (Exponential distribution) Suppose that X is a continuous random variable with density

$$f(x) = \begin{cases} \lambda \exp(-\lambda x) & \text{for } x \ge 0 \\ 0 & \text{for } x < 0 \end{cases}.$$

X is said to have an Exponential distribution with parameter λ ($X \sim \text{Exp}(\lambda)$). The distribution function of X is

$$F(x) = \begin{cases} 0 & \text{for } x < 0 \\ 1 - \exp(-\lambda x) & \text{for } x \geq 0 \end{cases}.$$

Since $F(x)$ is strictly increasing over the set where $F(x) > 0$, we can determine the inverse $F^{-1}(t)$ by solving the equation

$$1 - \exp\left(-\lambda F^{-1}(t)\right) = t,$$

which yields

$$F^{-1}(t) = -\frac{1}{\lambda}\ln(1-t).$$

Thus if $U \sim \text{Unif}(0,1)$ then $Y = -\lambda^{-1}\ln(1-U) \sim \text{Exp}(\lambda)$. Note that since $1 - U$ has the same distribution as U, we also have $-\lambda^{-1}\ln(U) \sim \text{Exp}(\lambda)$. ◇

EXAMPLE 1.18: Suppose that X has a Binomial distribution with parameters $n = 3$ and $\theta = 1/2$. We then have $F(x) = 1/8$ for $0 \leq x < 1$, $F(x) = 1/2$ for $1 \leq x < 2$ and $F(x) = 7/8$ for $2 \leq x < 3$ ($F(x) = 0$ and $F(x) = 1$ for $x < 0$ and $x \geq 3$ respectively). The inverse of F is given by

$$F^{-1}(t) = \begin{cases} 0 & \text{for } 0 < t \leq 1/8 \\ 1 & \text{for } 1/8 < t \leq 1/2 \\ 2 & \text{for } 1/2 < t \leq 7/8 \\ 3 & \text{for } 7/8 < t < 1 \end{cases}.$$

If $U \sim \text{Unif}(0,1)$ then it is simple to see that $F^{-1}(U) \sim \text{Bin}(3, 0.5)$. ◇

Hazard functions

Suppose that X is a nonnegative, continuous random variable with density function $f(x)$; it is useful to think of X as a lifetime of some object, for example, a human or an electronic component. Suppose we know that the object is still alive at time x, that is, we know that $X > x$; given this information, we would like to determine the probability of death in the next instant (of length Δ) following time x:

$$P(x < X \leq x + \Delta | X > x) = \frac{P(x < X \leq x + \Delta, X > x)}{P(X > x)}$$

$$= \frac{P(x < X \leq x + \Delta)}{P(X > x)}$$
$$= \frac{F(x + \Delta) - F(x)}{1 - F(x)}$$

Since F is continuous, the conditional probability above tends to 0 as Δ tends to 0. However, by dividing the conditional probability by Δ and letting $\Delta \downarrow 0$, we obtain (if $F'(x) = f(x)$),

$$\lim_{\Delta \downarrow 0} \frac{1}{\Delta} P(x < X \leq x + \Delta | X > x) = \frac{f(x)}{1 - F(x)}.$$

DEFINITION. Suppose that a nonnegative random variable X has distribution function F and density function f. The function

$$\lambda(x) = \frac{f(x)}{1 - F(x)}$$

is called the hazard function of X. (If $F(x) = 1$ then $\lambda(x)$ is defined to be 0.)

The hazard function is useful in applications (such as insurance and reliability) where we are interested in the probability of failure (for example, death) given survival up to a certain point in time. For example, if X represents a lifetime and has hazard function $\lambda(x)$ then

$$P(x \leq X \leq X + \Delta | X \geq x) \approx \Delta \lambda(x)$$

if Δ is sufficiently small.

Given the hazard function of nonnegative continuous random variable X, we can determine both the distribution function and the density of X. Integrating the hazard function from 0 to x, we get

$$\begin{aligned} \int_0^x \lambda(t)\, dt &= \int_0^x \frac{f(t)}{1 - F(t)}\, dt \\ &= \int_0^{F(x)} \frac{du}{1 - u} \quad \text{(where } u = F(x)\text{)} \\ &= -\ln(1 - F(x)) \end{aligned}$$

and so

$$F(x) = 1 - \exp\left(- \int_0^x \lambda(t)\, dt \right).$$

Differentiating $F(x)$, we get

$$f(x) = \lambda(x) \exp\left(-\int_0^x \lambda(t)\, dt\right).$$

Thus the hazard function gives another way of specifying the distribution of a nonnegative continuous random variable.

EXAMPLE 1.19: Suppose that a nonnegative random variable X has a hazard function $\lambda(x) = \lambda\beta x^{\beta-1}$ for some $\lambda > 0$ and $\beta > 0$; when $\beta < 1$, the hazard function is decreasing over time while if $\beta > 1$, the hazard function is increasing over time. The distribution function of X is

$$F(x) = 1 - \exp\left(-\lambda x^\beta\right) \quad (x > 0)$$

while its density function is

$$f(x) = \lambda\beta x^{\beta-1} \exp\left(-\lambda x^\beta\right) \quad (x > 0).$$

The distribution of X is called a Weibull distribution and is commonly used as a model in reliability and survival analysis. The Exponential distribution is a special case of the Weibull with $\beta = 1$; it is easy to see that when $\beta = 1$, the hazard function is constant and X has the "memoryless" property

$$P(X > x + t | X > x) = P(X > t)$$

for any $x > 0$ and $t > 0$. ◇

1.5 Transformations of random variables

Suppose X is a random variable and $Y = g(X)$ for some function g. Given the distribution of X, what is the distribution of Y?

First consider the case where X is discrete. If $Y = g(X)$, it follows that Y is also discrete since Y can assume no more values than X. To determine the frequency function of Y, define

$$A(y) = \{x : g(x) = y\};$$

then if $f_X(x)$ is the frequency function of X, we have

$$f_Y(y) = \sum_{x \in A(y)} f_X(x).$$

The continuous case is somewhat more complicated but the underlying principle is essentially the same as for discrete random variables. First we will assume that g is strictly monotone (that is, either strictly increasing or strictly decreasing) and differentiable (at least over a set A such that $P(X \in A) = 1$). To find the density of Y, we will first find $F_Y(y) = P(Y \leq y)$ and try to express this as

$$F_Y(y) = \int_{-\infty}^{y} f_Y(t)\, dt$$

or, alternatively, differentiate F_Y to determine f_Y. For example, if g is strictly increasing then

$$\begin{aligned} P(Y \leq y) &= P(g(X) \leq y) \\ &= P(X \leq g^{-1}(y)) \\ &= F_X(g^{-1}(y)) \\ &= \int_{-\infty}^{g^{-1}(y)} f_X(x)\, dx \end{aligned}$$

where $g^{-1}(y)$ is the inverse of g ($g(g^{-1}(y)) = y$). Making a change of variables $u = g(x)$, we get

$$F_Y(y) = \int_{-\infty}^{y} f_X(g^{-1}(t)) \frac{d}{dt} g^{-1}(t).$$

The derivative of $g^{-1}(y)$ is $1/g'(g^{-1}(y))$ and so

$$f_Y(y) = \frac{f_X(g^{-1}(y))}{g'(g^{-1}(y))}.$$

The argument in the case where g is strictly decreasing is similar except in this case,

$$P(Y \leq y) = P(X \geq g^{-1}(y)) = 1 - F_X(g^{-1}(y))$$

and so

$$f_Y(y) = -\frac{f_X(g^{-1}(y))}{g'(g^{-1}(y))}.$$

Note that $g'(x) < 0$ when g is strictly decreasing and so the density f_Y will always be nonnegative. If we put the two cases (strictly increasing and strictly decreasing) together, we get

$$f_Y(y) = \frac{f_X(g^{-1}(y))}{|g'(g^{-1}(y))|}.$$

EXAMPLE 1.20: Suppose that $X \sim N(0, 1)$ (see Example 1.15) and define $Y = \mu + \sigma X$ for some $\sigma > 0$. Then

$$f_Y(y) = \frac{1}{\sigma} f_X\left(\frac{y-\mu}{\sigma}\right)$$

and so

$$f_Y(y) = \frac{1}{\sigma\sqrt{2\pi}} \exp\left(-\frac{(x-\mu)^2}{2\sigma^2}\right).$$

Thus $Y \sim N(\mu, \sigma^2)$. ◇

If g is not a monotone function then a somewhat more careful analysis is needed to find the density of $Y = g(X)$. However, the underlying principle is the same as when g is monotone; first, we express $P(Y \le y)$ as $P(X \in A(y))$ where

$$A(y) = \{x : g(x) \le y\}.$$

Then $P(X \in A(y))$ is differentiated with respect to y to obtain $f_Y(y)$. The following examples illustrate the approach.

EXAMPLE 1.21: Suppose that X is a continuous random variable with density

$$f_X(x) = \begin{cases} |x| & \text{for } -1 \le x \le 1 \\ 0 & \text{otherwise} \end{cases}$$

and define $Y = X^2$; note that the function $g(x) = x^2$ is not monotone for $-1 \le x \le 1$. Since $P(-1 \le X \le 1) = 1$, it follows that $P(0 \le Y \le 1) = 1$. Thus for $0 \le y \le 1$, we have

$$\begin{aligned} P(Y \le y) &= P(-\sqrt{y} \le X \le \sqrt{y}) \\ &= F_X(\sqrt{y}) - F_X(-\sqrt{y}). \end{aligned}$$

Now differentiating with respect to y, we get

$$\begin{aligned} f_Y(y) &= f_X(\sqrt{y})\left(\frac{1}{2}y^{-1/2}\right) + f_X(-\sqrt{y})\left(\frac{1}{2}y^{-1/2}\right) \\ &= \frac{1}{2} + \frac{1}{2} = 1 \quad \text{for } 0 \le y \le 1. \end{aligned}$$

Thus $Y \sim \text{Unif}(0, 1)$. ◇

1.6 Expected values

Suppose that X is a random variable with distribution function F; given $F(x)$ for all values of x, we know everything we need to know about X. However, it is sometimes useful to summarize the distribution of X by certain characteristics of the distribution (for example, the median of the distribution). Expected values are one way to characterize a distribution; essentially, the expected value of X is the "centre of mass" of its distribution.

EXAMPLE 1.22: Define X to be a discrete random variable taking values $x_1, \cdots, x_k$ with probabilities $p_1, \cdots, p_k$. Imagine an infinite (weightless) beam with masses $p_1, \cdots, p_k$ suspended at the points $x_1, \cdots, x_k$. Suppose that we try to balance the beam about a fulcrum placed at a point μ; the force exerted downwards by the mass placed at x_i is proportional to $p_i|x_i - \mu|$ and so, in order for the beam to be in balance at the fulcrum μ, we must have

$$\sum_{x_i<\mu} p_i|x_i - \mu| = \sum_{x_i>\mu} p_i|x_i - \mu|$$

or

$$\sum_{x_i<\mu} p_i(\mu - x_i) = \sum_{x_i>\mu} p_i(x_i - \mu).$$

Solving for μ, we get

$$\mu = \sum_{i=1}^{k} x_i p_i = \sum_{i=1}^{k} x_i P(X = x_i)$$

to be the centre of mass; μ is called the expected value of X. ◇

Note if X takes only two values, 0 and 1, then the expected value of X (according to Example 1.22) is simply $P(X = 1)$. Thus given some event A, if we define a random variable $X = I(A)$, where the indicator function $I(A) = 1$ if A occurs and $I(A) = 0$ otherwise, then $E(X) = E[I(A)] = P(A)$. For example, $E[I(Y \le y)] = F(y)$ for any random variable Y where F is the distribution function of Y.

Extending the "centre of mass" definition of expected value to more general distributions is possible although it does involve some mathematical subtleties: it is not clear that the centre of mass will exist if X can take an infinite (whether countable or uncountable) number of values.

EXAMPLE 1.23: Suppose that X is a discrete random variable with frequency function

$$f(x) = \frac{6}{\pi^2 x^2} \quad \text{for } x = 1, 2, 3, \cdots.$$

Then following Example 1.22, we have

$$\mu = \sum_{x=1}^{\infty} x\, f(x) = \frac{6}{\pi^2} \sum_{x=1}^{\infty} x^{-1} = \infty$$

and so this distribution does not have a finite expected value. □

The fact that the expected value in Example 1.23 is infinite does not pose any problems from a mathematical point of view; since X is positive, the infinite summation defining the expected value is unambiguously defined. However, when the random variable takes both positive and negative values, the definition of the expected value becomes somewhat more delicate as an infinite summation with both positive and negative summands need not have a well-defined value. We will split the random variable X into its positive and negative parts X^+ and X^- where

$$X^+ = \max(X, 0) \quad \text{and} \quad X^- = \max(-X, 0);$$

then $X = X^+ - X^-$ and $|X| = X^+ + X^-$. Provided that at least one of $E(X^+)$ and $E(X^-)$ is finite, we can define $E(X) = E(X^+) - E(X^-)$; otherwise (that is, if both $E(X^+)$ and $E(X^-)$ are infinite), $E(X)$ is undefined.

We will now give a (somewhat non-intuitive) definition of the expected value in the general case.

DEFINITION. Suppose that X is a nonnegative random variable with distribution function F. The expected value or mean of X (denoted by $E(X)$) is defined to be

$$E(X) = \int_0^{\infty} (1 - F(x))\, dx,$$

which may be infinite. In general, if $X = X^+ - X^-$, we define $E(X) = E(X^+) - E(X^-)$ provided that at least one of $E(X^+)$ and $E(X^-)$ is finite; if both are infinite then $E(X)$ is undefined. If $E(X)$ is well-defined then

$$E(X) = \int_0^{\infty} (1 - F(x))\, dx - \int_{-\infty}^{0} F(x)\, dx.$$

The definition given above may appear to be inconsistent with our previous discussion of expected values for discrete random variables; in fact, the two definitions are exactly the same in this case.

If X is a discrete random variable with frequency function $f(x)$ then

$$E(X) = \sum_x x\, f(x)$$

provided that at least one of

$$E(X^+) = \sum_{x>0} x\, f(x) \quad \text{and} \quad E(X^-) = -\sum_{x<0} x\, f(x)$$

is finite.

If X is a continuous random variable with density function $f(x)$, we have

$$\begin{aligned} E(X) &= \int_0^\infty (1 - F(x))\, dx - \int_{-\infty}^0 F(x)\, dx \\ &= \int_0^\infty \int_x^\infty f(t)\, dt\, dx - \int_{-\infty}^0 \int_{-\infty}^x f(t)\, dt\, dx \\ &= \int_0^\infty \int_0^t f(t)\, dx\, dt - \int_{-\infty}^0 \int_t^0 f(t)\, dx\, dt \\ &= \int_0^\infty t\, f(t)\, dt + \int_{-\infty}^0 t\, f(t)\, dt \\ &= \int_{-\infty}^\infty t\, f(t)\, dt. \end{aligned}$$

THEOREM 1.8 *Suppose that X has a distribution function F with $F(x) = pF_c(x) + (1-p)F_d(x)$ where F_c is a continuous distribution function with density function f_c and F_d is a discrete distribution function with frequency function f_d. If $E(X)$ is well-defined then*

$$E(X) = p \int_{-\infty}^\infty x\, f_c(x)\, dx + (1-p) \sum_x x\, f_d(x)$$

The expected value (when it is finite) is a measure of the "centre" of a probability distribution. For example, if the distribution of X is symmetric around μ in the sense that

$$P(X \le \mu - x) = P(X \ge \mu + x)$$

for all x then it is easy to show that $E(X) = \mu$ if $E(X)$ is well-defined. We noted earlier that the median is also a measure of the center of a distribution. In the case where the distribution is symmetric around μ then the mean and median will coincide;

however, if the distribution is not symmetric then the mean and median can differ substantially.

We now give a few examples to illustrate the computation of expected values.

EXAMPLE 1.24: Suppose that X has the density function

$$f(x) = \frac{\lambda^\alpha x^{\alpha-1}}{\Gamma(\alpha)} \exp(-\lambda x) \quad \text{for } x > 0$$

where $\lambda > 0$, $\alpha > 0$ and $\Gamma(\alpha)$ is the Gamma function defined by

$$\Gamma(\alpha) = \int_0^\infty t^{\alpha-1} \exp(-t)\, dt.$$

X is said to a Gamma distribution with shape parameter α and scale parameter λ ($X \sim \text{Gamma}(\alpha, \lambda)$). Using properties of the function $\Gamma(\alpha)$, we have

$$\begin{aligned} E(X) &= \int_{-\infty}^\infty x\, f(x)\, dx \\ &= \int_0^\infty \frac{\lambda^\alpha x^\alpha}{\Gamma(\alpha)} \exp(-\lambda x)\, dx \\ &= \frac{\Gamma(\alpha+1)}{\lambda\Gamma(\alpha)} \int_0^\infty \frac{\lambda^{\alpha+1} x^{\alpha+1-1}}{\Gamma(\alpha+1)} \exp(-\lambda x)\, dx \\ &= \frac{\Gamma(\alpha+1)}{\lambda\Gamma(\alpha)} \\ &= \frac{\alpha}{\lambda} \end{aligned}$$

since $\Gamma(\alpha + 1) = \alpha\Gamma(\alpha)$. Note that the Exponential distribution (see Example 1.17) is a special case of the Gamma distribution with $\alpha = 1$ and so if $X \sim \text{Exp}(\lambda)$ then $E(X) = 1/\lambda$. Figure 1.3 shows three Gamma density functions with different shape parameters α with the scale parameters $\lambda = \alpha$ chosen so that each distribution has expected value equal to 1. ◇

EXAMPLE 1.25: Suppose that X has a Binomial distribution with parameters n and θ. Since X is discrete, we have

$$E(X) = \sum_{x=0}^n x \binom{n}{x} \theta^x (1-\theta)^{n-x}$$

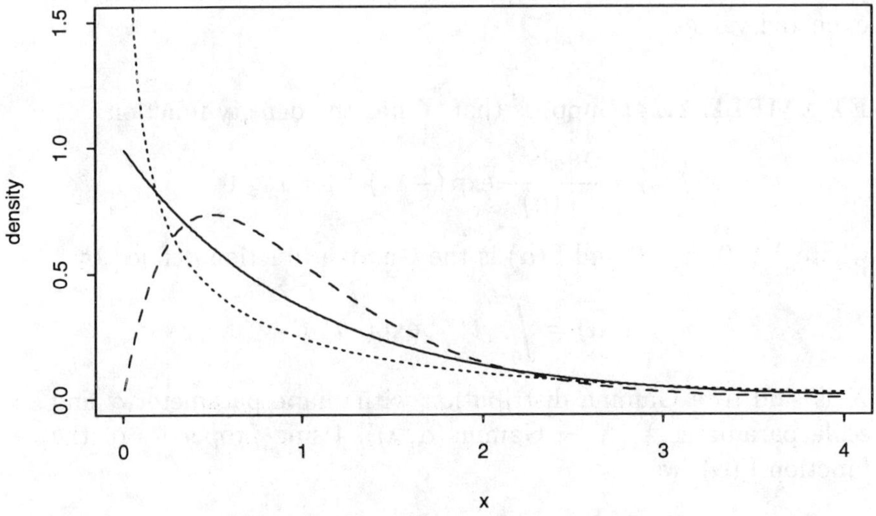

Figure 1.3 *Gamma density functions for* $\alpha = 1$, $\lambda = 1$ *(solid line),* $\alpha = 0.5$, $\lambda = 0.5$ *(dotted line) and* $\alpha = 2$, $\lambda = 2$ *(dashed line).*

$$
\begin{aligned}
&= \sum_{x=1}^{n} \frac{n!}{(n-x)!(x-1)!} \theta^x (1-\theta)^{n-x} \\
&= n\theta \sum_{x=1}^{n} \frac{(n-1)!}{(n-x)!(x-1)!} \theta^{x-1} (1-\theta)^{n-x} \\
&= n\theta \sum_{y=0}^{n-1} \binom{n-1}{y} \theta^y (1-\theta)^{n-1-y} \\
&= n\theta
\end{aligned}
$$

where the last equality follows by applying the Binomial Theorem. ◇

EXAMPLE 1.26: Suppose that X is a continuous random variable with density

$$f(x) = \frac{1}{\pi(1+x^2)} \quad \text{for } -\infty < x < \infty;$$

X is said to have a Cauchy distribution. Note that X is symmetric around 0 so that $E(X)$ should be 0 if $E(X)$ exists. However, note that

$$\begin{aligned} E(X^+) &= \int_0^\infty x f(x)\,dx \\ &= \frac{1}{\pi}\int_0^\infty \frac{x}{1+x^2}\,dx \\ &= \frac{1}{2\pi}\lim_{x\to\infty}\ln(1+x^2) \\ &= \infty. \end{aligned}$$

Likewise $E(X^-) = \infty$ and so it follows that $E(X)$ is not well-defined. ◇

We can also define the expected value of a function of X. If $Y = h(X)$ for some function h then we can define $E[h(X)]$ by applying the definition of expected value for the random variable Y; to do this, of course, requires that we determine the distribution function of Y. However, if X has a distribution function F with $F = pF_c + (1-p)F_d$ as before then

$$E[h(X)] = p\int_{-\infty}^\infty h(x) f_c(x)\,dx + p\sum_x h(x) f_d(x)$$

provided the integral and sum are well-defined.

Suppose that $h(x) = h_1(x) + h_2(x)$. Then if both $E[h_1(X)]$ and $E[h_2(X)]$ are finite, it follows that

$$E[h(X)] = E[h_1(X)] + E[h_2(X)].$$

If $h(x) = a\,g(x)$ where a is constant then $E[h(X)] = aE[g(X)]$; the expected value of $h(X)$ is well-defined if, and only if, the expected value of $g(X)$ is. From these two properties of expected values, we can deduce

$$E[a\,g(X) + b) = aE[g(X)] + b$$

if a and b are constants. In general, $E[g(X)] = g(E(X))$ only if $g(x)$ is a linear function, that is, if $g(x) = ax + b$.

If $g(x)$ is a convex function, that is,

$$g(tx + (1-t)y) \le t\,g(x) + (1-t)\,g(y)$$

for $0 \le t \le 1$, then we have Jensen's inequality, which says that $g(E(X)) \le E[g(X)]$; a simple example of Jensen's inequality is

$|E(X)|^r \le E[|X|^r]$, which holds for $r \ge 1$ since the function $g(x) = |x|^r$ is convex for $r \ge 1$.

At this point, we will introduce a convenient shorthand for writing expected values. Given a random variable X with distribution function F, we will write

$$E[h(X)] = \int_{-\infty}^{\infty} h(x)\, dF(x),$$

that is, $E[h(X)]$ is the integral of $h(x)$ with respect to the distribution function F. For example, if we write $h(x) = I(x \le t)$ then $E[h(X)] = P(X \le t) = F(t)$ and so

$$F(t) = \int_{-\infty}^{\infty} I(x \le t)\, dF(x) = \int_{-\infty}^{t} dF(x).$$

From a mathematical point of view, this integral can be interpreted as a Riemann-Stieltjes integral or (more usefully) as a Lebesgue-Stieltjes integral; precise definitions of these integrals can be found in Rudin (1976). If F is a discrete distribution function then the integral above is interpreted as a summation while if F is a continuous distribution function with a density function then the integral is interpreted as the Riemann integral of calculus.

This integral representation can be manipulated usefully as the following example suggests.

EXAMPLE 1.27: Suppose that X is a positive random variable with distribution function F and set $\gamma = E[w(X)] < \infty$ for some nonnegative function $w(x)$. Given F and γ, we can define a new distribution function G by

$$G(x) = \frac{1}{\gamma} \int_{0}^{x} w(t)\, dF(t) \quad \text{for } x \ge 0.$$

G is sometimes called the selection biased distribution of F. A random variable with distribution G might arise as follows. Suppose that X is a discrete random variable uniformly distributed on the set $S = \{a_1, \cdots, a_n\}$ where the $a_i > 0$ for $i = 1, \cdots, n$. Rather than sampling uniformly from S, we sample so that the probability that a_i is selected is proportional to a_i (this is called length biased sampling); that is, we sample a discrete random variable Y so that $P(Y = a_i) = k a_i$ for some constant k. It is easy to see that for Y to have a valid probability distribution, we must have

$$P(Y = a_i) = \frac{a_i}{\sum_{j=1}^n a_j} = \frac{a_i}{nE(X)}.$$

The distribution function of Y has a jump of $a_i/[nE(X)]$ at a_i (for $i = 1, \cdots, n$) and so can be written as

$$P(Y \le x) = \frac{1}{E(X)} \int_0^x t\, dF(t)$$

where F is the distribution function of X. Returning to the general case, it is fairly easy to see that, given G and the function w, it is possible to determine F and hence evaluate $E[h(X)]$ for any function h. For example, in the case of length biased sampling ($w(x) = x$), we can determine $\mu = E(X)$ by noting that

$$E[g(Y)] = \int_0^\infty g(y)\, dG(y) = \frac{1}{\mu} \int_0^\infty y\, g(y)\, dF(y).$$

Substituting $g(y) = 1/y$ above, we get

$$E(1/Y) = \frac{1}{\mu} \int_0^\infty dF(x) = \frac{1}{\mu}$$

and so $\mu = [E(1/Y)]^{-1}$. ◇

Variance and moment generating function

DEFINITION. Let X be a random variable with $\mu = E(X)$. Then the variance of X, $\text{Var}(X)$, is defined to be

$$\text{Var}(X) = E[(X - \mu)^2].$$

If $E(X^2) < \infty$ then $\text{Var}(X) < \infty$; if $E(X^2) = \infty$ then we will define $\text{Var}(X) = \infty$ (even if $E(X)$ is not finite or is not defined).

Given $\sigma^2 = \text{Var}(X)$, we can define the standard deviation of X to be

$$\text{SD}(X) = \sigma = \sqrt{\text{Var}(X)}.$$

We can also obtain the following properties of the variance and standard deviation of a random variable.

- $\text{Var}(X) \ge 0$ and $\text{Var}(X) = 0$ if, and only if, $P(X = \mu) = 1$ where $\mu = E(X)$.
- $\text{Var}(X) = E(X^2) - \mu^2$.

- For any constants a and b, $\text{Var}(aX + b) = a^2\text{Var}(X)$ and so it follows that $\text{SD}(aX + b) = |a|\text{SD}(X)$.

The last property above is particularly important as it indicates that both the variance and standard deviation somehow measure the amount of "dispersion" in the distribution of a random variable and are unaffected by changes in the "location" of a random variable (since $\text{Var}(X + b) = \text{Var}(X)$). The standard deviation has the same units of measurement as the random variable itself and so has a more natural interpretation than the variance (whose units are the square of the original units). However, as we will see, the variance has very nice algebraic and computational properties. It is important to note that there is nothing particularly special about the variance and standard deviation as measures of dispersion; one problem is the fact that both can be distorted by small amounts of probability occurring in the "tails" of the distribution.

EXAMPLE 1.28: Suppose that X has a Uniform distribution on $[a, b]$. Then

$$E(X) = \frac{1}{b-a}\int_a^b x\,dx = \frac{b^2 - a^2}{2(b-a)} = \frac{1}{2}(a+b)$$

and

$$E(X^2) = \frac{1}{b-a}\int_a^b x^2\,dx = \frac{b^3 - a^3}{3(b-a)} = \frac{1}{3}(a^2 + ab + b^2).$$

Thus $\text{Var}(X) = E(X^2) - [E(X)]^2 = (b-a)^2/12$. ◇

EXAMPLE 1.29: Suppose that X has a Binomial distribution with parameters n and θ. To compute the variance of X, it is convenient to introduce the formula

$$\text{Var}(X) = E[X(X-1)] + E(X) - [E(X)]^2,$$

which follows from the identity $\text{Var}(X) = E(X^2) - [E(X)]^2$. We now have

$$\begin{aligned} E[X(X-1)] &= \sum_{x=0}^{n} x(x-1)\binom{n}{x}\theta^x(1-\theta)^{n-x} \\ &= \sum_{x=2}^{n} \frac{n!}{(x-2)!(n-x)!}\theta^x(1-\theta)^{n-x} \end{aligned}$$

$$= n(n-1)\theta^2 \sum_{y=0}^{n-2} \binom{n-2}{y} \theta^y (1-\theta)^{n-2-y}$$
$$= n(n-1)\theta^2.$$

Using the fact that $E(X) = n\theta$ (as shown in Example 1.25), it follows that $\text{Var}(X) = n\theta(1-\theta)$. ◇

EXAMPLE 1.30: Suppose that X has a standard Normal distribution. Since the density of X is symmetric around 0, it follows that $E(X) = 0$, provided of course that the integral is well-defined. Since

$$\frac{1}{\sqrt{2\pi}} \int_0^\infty x \exp(-x^2/2)\, dx = \frac{1}{\sqrt{2\pi}} \int_0^\infty \exp(-y)\, dy = \frac{1}{\sqrt{2\pi}}$$

is finite, $E(X)$ is well-defined and equals 0. To compute the variance of X, note that $\text{Var}(X) = E(X^2)$ and so

$$\begin{aligned}
\text{Var}(X) &= \frac{1}{\sqrt{2\pi}} \int_{-\infty}^\infty x^2 \exp\left(-x^2/2\right) dx \\
&= \frac{2}{\sqrt{2\pi}} \int_0^\infty x^2 \exp\left(-x^2/2\right) dx \\
&= \frac{2}{\sqrt{2\pi}} \int_0^\infty 2\sqrt{2y} \exp(-y)\, dy \\
&= \frac{1}{\sqrt{\pi}} \int_0^\infty \sqrt{y} \exp(-y)\, dy \\
&= 1
\end{aligned}$$

since

$$\int_0^\infty \sqrt{y} \exp(-y)\, dy = \sqrt{\pi}.$$

If $Y = \mu + \sigma X$ then $Y \sim N(\mu, \sigma^2)$ and it follows that $E(Y) = \mu + \sigma E(X) = \mu$ and $\text{Var}(Y) = \sigma^2 \text{Var}(X) = \sigma^2$. ◇

An useful tool for computing means and variances is the moment generating function, which, when it exists, uniquely characterizes a probability distribution.

DEFINITION. Let X be a random variable and define

$$m(t) = E[\exp(tX)].$$

If $m(t) < \infty$ for $|t| \le b > 0$ then $m(t)$ is called the moment generating function of X.

It is important to note that the moment generating function is not defined for all random variables and, when it is, $m(t)$ is not necessarily finite for all t. However, if the moment generating function $m(t)$ is well-defined then it essentially determines the probability distribution of the random variable.

THEOREM 1.9 *Suppose that X and Y are random variables such that*

$$m_X(t) = E[\exp(tX)] = E[\exp(tY)] = m_Y(t)$$

for all $|t| \leq b > 0$. Then X and Y have the same probability distribution; that is,

$$P(X \leq x) = P(Y \leq x)$$

for all x.

The proof of Theorem 1.9 is quite difficult and will not be pursued here. However, the proof relies on a parallel result to that given here: If we define the characteristic function of X to be the complex-valued function $\phi(t) = E[\exp(itX)]$ where $\exp(is) = \cos(s) + i\sin(s)$ then $\phi(t)$ specifies the probability distribution in the sense that equality of characteristic functions implies equality of distribution functions. It can be shown that under the hypothesis of Theorem 1.9, we obtain equality of the characteristic functions of X and Y and hence equality in distribution. It should be noted that the characteristic function of a random variable is always well-defined.

EXAMPLE 1.31: Suppose that X is an Exponential random variable with parameter λ. The moment generating function (if it exists) is defined by

$$m(t) = E[\exp(tX)] = \int_0^\infty \lambda \exp(-(\lambda - t)x)\, dx.$$

Note that the integral defining $m(t)$ is finite if, and only if, $\lambda - t > 0$ or, equivalently, if $t < \lambda$. Since $\lambda > 0$, we then have $m(t) < \infty$ for t in a neighbourhood around 0 and so the moment generating function exists. Simple integration gives

$$m(t) = \frac{\lambda}{\lambda - t}$$

for $t < \lambda$. ◇

EXAMPLE 1.32: Suppose that X has a Binomial distribution with parameters n and θ. The moment generating function is given by

$$\begin{aligned} m(t) = E[\exp(tX)] &= \sum_{x=0}^{n} \exp(tx) \binom{n}{x} \theta^x (1-\theta)^{n-x} \\ &= \sum_{x=0}^{n} \binom{n}{x} (\theta \exp(t))^x (1-\theta)^{n-x} \\ &= [1 + \theta(\exp(t) - 1)]^n \end{aligned}$$

where the final line follows by the Binomial Theorem. Note that $m(t) < \infty$ for all t. ◇

EXAMPLE 1.33: Suppose that X has a Poisson distribution with parameter $\lambda > 0$ ($X \sim \text{Pois}(\lambda)$); this is a discrete distribution with frequency function

$$f(x) = \frac{\exp(-\lambda)\lambda^x}{x!} \quad \text{for } x = 0, 1, 2, \cdots.$$

The moment generating function of X is

$$\begin{aligned} m(t) &= \sum_{x=0}^{\infty} \exp(tx) \frac{\exp(-\lambda)\lambda^x}{x!} \\ &= \exp(-\lambda) \sum_{x=0}^{\infty} \frac{[\lambda \exp(t)]^x}{x!} \\ &= \exp(-\lambda) \exp[\lambda \exp(t)] \\ &= \exp[\lambda(\exp(t) - 1)] \end{aligned}$$

since $\sum_{x=0}^{\infty} a^x/x! = \exp(a)$ for any a. ◇

EXAMPLE 1.34: Suppose that X has a standard Normal distribution. The moment generating function of X is

$$\begin{aligned} m(t) &= \frac{1}{\sqrt{2\pi}} \int_{-\infty}^{\infty} \exp(tx) \exp\left(-x^2/2\right) dx \\ &= \frac{1}{\sqrt{2\pi}} \int_{-\infty}^{\infty} \exp\left(tx - x^2/2\right) dx \\ &= \frac{1}{\sqrt{2\pi}} \int_{-\infty}^{\infty} \exp\left(-(x-t)^2/2 + t^2/2\right) dx \end{aligned}$$

$$
\begin{aligned}
&= \frac{\exp(t^2/2)}{\sqrt{2\pi}} \int_{-\infty}^{\infty} \exp\left(-(x-t)^2/2\right) dx \\
&= \exp(t^2/2).
\end{aligned}
$$

If $Y = \mu + \sigma X$ so that Y is normally distributed with mean μ and variance σ^2 then the moment generating function of Y is $m_Y(t) = \exp(\mu t + \sigma^2 t^2/2)$. ◇

The moment generating function has no real useful interpretation; it is used almost exclusively as a technical device for computing certain expected values and as a device for proving limit theorems for sequences of distribution functions. Also given the moment generating function of a random variable, it is possible to invert the moment generating function (or the characteristic function) to obtain the distribution or density function; however, using these inversion formulas can involve some very delicate analytical and numerical techniques.

As the name suggests, moment generating functions are useful for computing the moments of a random variable. The moments of X are defined to be the expected values $E(X)$, $E(X^2)$, $E(X^3)$, $\cdots$. $E(X^k)$ is defined to be the k-th moment of X.

THEOREM 1.10 *Suppose that X is a random variable with moment generating function $m(t)$. Then for any $r > 0$, $E(|X|^r) < \infty$ and $E(X) = m'(0)$, $E(X^2) = m''(0)$ and in general $E(X^k) = m^{(k)}(0)$ where $m^{(k)}(t)$ denotes the k-th derivative of m at t.*

Proof. If X has moment generating function then $|X|$ has a moment generating function $m^*(t)$. Then for any $t > 0$, we have $|x|^r \le \exp(t|x|)$ for $|x|$ sufficiently large (say, for $|x| \ge c$). We then have

$$
\begin{aligned}
E[|X|^r] &= E[|X|^r I(|X| \le c)] + E[|X|^r I(|X| > c)] \\
&\le E[|X|^r I(|X| \le c)] + E[\exp(tX) I(|X| > c)] \\
&\le c^r + m^*(t) < \infty
\end{aligned}
$$

and so $E(|X|^r) < \infty$. Now using the expansion

$$\exp(tx) = \sum_{k=0}^{\infty} \frac{t^k x^k}{k!}$$

we obtain

$$m(t) = E[\exp(tX)] = \sum_{k=0}^{\infty} \frac{t^k E(X^k)}{k!}.$$

(It can be shown in this instance that taking expectations inside the infinite summation is valid; however, this is not true in general.) Differentiating with respect to t, we get

$$m'(t) = \sum_{k=1}^{\infty} \frac{kt^{k-1}E(X^k)}{k!} = E(X) + \sum_{k=2}^{\infty} \frac{t^{k-1}E(X^k)}{(k-1)!}$$

and setting $t = 0$, it follows that $E(X) = 0$. Repeatedly differentiating, we get

$$m^{(k)}(t) = E(X^k) + \sum_{j=k+1}^{\infty} \frac{t^{j-k}E(X^j)}{(j-k)!}$$

and so $m^{(k)}(0) = E(X^k)$. (Again, in this case, we are justified in differentiating inside the infinite summation although this is not true in general.) □

EXAMPLE 1.35: Suppose that X is an Exponential random variable with parameter λ. In Example 1.31, we found the moment generating function to be $m(t) = \lambda/(\lambda - t)$ (for $t < \lambda$). Differentiating we obtain $m'(t) = \lambda/(\lambda - t)^2$, $m''(t) = 2\lambda/(\lambda - t)^3$ and, in general,

$$m^{(k)}(t) = \frac{k!\lambda}{(\lambda - t)^{k+1}}.$$

Thus, setting $t = 0$, we get $E(X^k) = k!\lambda^{-k}$; in particular, $E(X) = \lambda^{-1}$ and $\text{Var}(X) = E(X^2) - [E(X)]^2 = \lambda^{-2}$. ◇

EXAMPLE 1.36: Suppose that $X \sim \text{Geom}(\theta)$ (Example 1.12); recall that the frequency function of this random variable is

$$f(x) = \theta(1-\theta)^x \quad \text{for } x = 0, 1, 2, \cdots.$$

Thus the moment generating function is given by

$$\begin{aligned} m(t) = E[\exp(tX)] &= \sum_{x=0}^{\infty} \exp(tx)\theta(1-\theta)^x \\ &= \theta \sum_{x=0}^{\infty} [(1-\theta)\exp(t)]^x. \end{aligned}$$

The infinite series above converges if, and only if, $|(1-\theta)\exp(t)| < 1$ or (since $(1-\theta)\exp(t) > 0$) for $t < -\ln(1-\theta)$. Since $-\ln(1-\theta) > 0$, the moment generating function of X is

$$m(t) = \frac{\theta}{1 - (1-\theta)\exp(t)}$$

for $t < -\ln(1-\theta)$ (which includes a neighbourhood of 0) and differentiating, we get

$$m'(t) = \frac{\theta(1-\theta)\exp(t)}{(1-(1-\theta)\exp(t))^2}$$

$$\text{and} \quad m''(t) = \frac{\theta(1-\theta)\exp(t)(1-\exp(t)\theta(1-\theta))}{(1-(1-\theta)\exp(t))^3}.$$

We obtain $E(X) = (1-\theta)/\theta$ and $\text{Var}(X) = (1-\theta)/\theta^2$ by setting $t = 0$. ◇

The converse to Theorem 1.10 is not true; that is, the finiteness of $E(X^k)$ for all $k > 0$ does not imply the existence of the moment generating function of X. The following example illustrates this.

EXAMPLE 1.37: Define X to be a continuous random variable such that $\ln(X)$ has a standard Normal distribution; the density of X is

$$f(x) = \frac{1}{x\sqrt{2\pi}} \exp\left(-(\ln(x))^2/2\right) \quad \text{for } x > 0$$

(X is said to have a log-Normal distribution). Since $Y = \ln(X)$ is a standard Normal random variable, it follows that

$$E(X^k) = E[\exp(kY)] = \exp(k^2/2)$$

(using the moment generating function of the standard Normal distribution) and so $E(X^k) < \infty$ for any real k. However, X does not have a moment generating function. To see this, note that

$$E[\exp(tX)] = \frac{1}{\sqrt{2\pi}} \int_0^\infty \exp(tx)x^{-1} \exp\left(-(\ln(x))^2/2\right) dx$$

$$= \frac{1}{\sqrt{2\pi}} \int_{-\infty}^\infty \exp\left(t\exp(y) - y^2/2\right) dy;$$

for $t > 0$, the integral is infinite since $t\exp(y) - y^2/2 \to \infty$ as $y \to \infty$ (for $t > 0$). Thus $E[\exp(tX)] = \infty$ for $t > 0$ and so the moment generating function of X does not exist. ◇

Taken together, Theorems 1.9 and 1.10 imply that if X and Y are two random variables such that $E(X^k) = E(Y^k)$ for $k = 1, 2, 3, \cdots$ and X has a moment generating function then X and Y have the same distribution. However, the fact that X and Y have similar moment generating functions does not mean that X and Y will have similar distributions; for example, McCullagh (1994) gives an example of two moment generating functions that are virtually indistinguishable (they differ by at most 3×10^{-9}) but correspond to very different distributions (see also Waller, 1995). Moreover, equality of $E(X^k)$ and $E(Y^k)$ (for $k = 1, 2, \cdots$) does not imply equality of distributions if moment generating functions do not exist as the following example illustrates.

EXAMPLE 1.38: Suppose that X and Y are continuous random variables with density functions

$$f_X(x) = \frac{1}{x\sqrt{2\pi}} \exp\left(-(\ln(x))^2/2\right) \quad \text{for } x > 0$$

and

$$f_Y(x) = \frac{1}{x\sqrt{2\pi}} \exp\left(-(\ln(x))^2/2\right)(1 + \sin(2\pi \ln(x))) \quad \text{for } x > 0.$$

(X has the log-Normal distribution of Example 1.37.) For any $k = 1, 2, 3, \cdots$, we have

$$\begin{aligned} E(Y^k) &= E(X^k) \\ &\quad + \frac{1}{\sqrt{2\pi}} \int_0^\infty x^{k-1} \exp\left(-\frac{(\ln(x))^2}{2}\right) \sin(2\pi \ln(x)))\, dx \end{aligned}$$

and making the substitution $y = \ln(x) - k$, we have

$$\begin{aligned} &\int_0^\infty x^{k-1} \exp\left(-(\ln(x))^2/2\right) \sin(2\pi \ln(x)))\, dx \\ &= \int_{-\infty}^\infty \exp\left(k^2 + ky - (y+k)^2/2\right) \sin(2\pi(y+k))\, dy \\ &= \exp\left(k^2/2\right) \int_{-\infty}^\infty \exp\left(-y^2/2\right) \sin(2\pi y)\, dy \\ &= 0 \end{aligned}$$

since the integrand $\exp\left(-y^2/2\right)\sin(2\pi y)$ is an odd function. Thus $E(X^k) = E(Y^k)$ for $k = 1, 2, 3, \cdots$ even though X and Y have different distributions. ◇

1.7 Problems and complements

1.1: Show that

$$P(A_1 \cup \cdots \cup A_n) = \sum_{i=1}^n P(A_i) - \sum\sum\nolimits_{i<j} P(A_i \cap A_j)$$

$$+\sum\sum\sum_{i<j<k} P(A_i \cap A_j \cap A_k)$$
$$-\cdots-(-1)^n P(A_1 \cap \cdots \cap A_n).$$

1.2: Suppose that A and B are independent events. Determine which of the following pairs of events are always independent and which are always disjoint.

(a) A and B^c.

(b) $A \cap B^c$ and B.

(c) A^c and B^c.

(d) A^c and B.

(e) $A^c \cap B^c$ and $A \cup B$.

1.3: Consider an experiment where a coin is tossed an infinite number of times; the probability of heads on the k-th toss is $(1/2)^k$.

(a) Calculate (as accurately as possible) the probability that at least one head is observed.

(b) Calculate (as accurately as possible) the probability that exactly one head is observed.

1.4: Suppose that $A_1, A_2, \cdots$ are independent events with $P(A_k) = p_k$. Define

$$B = \bigcup_{k=1}^{\infty} A_k.$$

Show that $P(B) = 1$ if, and only if,

$$\sum_{k=1}^{\infty} \ln(1 - p_k) = -\infty.$$

1.5: (a) Suppose that $\{A_n\}$ is a decreasing sequence of events with limit A; that is $A_{n+1} \subset A_n$ for all $n \geq 1$ with

$$A = \bigcap_{n=1}^{\infty} A_n.$$

Using the axioms of probability show that

$$\lim_{n\to\infty} P(A_n) = P(A).$$

(b) Let X be a random variable and suppose that $\{x_n\}$ is a strictly decreasing sequence of numbers (that is, $x_n > x_{n+1}$ for all n) whose limit is x_0. Define $A_n = [X \le x_n]$. Show that

$$\bigcap_{n=1}^{\infty} A_n = [X \le x_0]$$

and hence (using part (a)) that $P(X \le x_n) \to P(X \le x_0)$.

(c) Now let $\{x_n\}$ be a strictly increasing sequence of numbers (that is, $x_n < x_{n+1}$ for all n) whose limit is x_0. Again defining $A_n = [X \le x_n]$ show that

$$\bigcup_{n=1}^{\infty} A_n = [X < x_0]$$

and hence that $P(X \le x_n) \to P(X < x_0)$.

1.6: Let A, B, C be events. We say that A and B are conditionally independent given C if

$$P(A \cap B|C) = P(A|C)P(B|C).$$

Suppose that

- A and B are conditionally independent given C, and
- A and B are conditionally independent given C^c.

Show that A and B are not necessarily independent but that A and B are independent if C is independent of either A or B.

1.7: Suppose that $F_1(x), \cdots, F_k(x)$ are distribution functions.

(a) Show that $G(x) = p_1F_1(x) + \cdots + p_kF_k(x)$ is a distribution function provided that $p_i \ge 0$ $(i = 1, \cdots, k)$ and $p_1 + \cdots + p_k = 1$.

(b) If $F_1(x), \cdots, F_k(x)$ have density (frequency) functions $f_1(x)$, $\cdots$, $f_k(x)$, show that $G(x)$ defined in (a) has density (frequency) function $g(x) = p_1f_1(x) + \cdots + p_kf_k(x)$.

1.8: (a) Let X be a nonnegative discrete random variable taking values $x_1, x_2, \cdots$ with probabilities $f(x_1), f(x_2), \cdots$. Show that

$$E(X) = \int_0^{\infty} P(X > x)\, dx = \sum_{k=1}^{\infty} x_k f(x_k).$$

(Hint: Note that $P(X > x)$ has downward jumps of height $f(x_k)$ at $x = x_k$ and is constant between the jumps.)

(b) Suppose that X only takes nonnegative integer values 0, 1,

2, $\cdots$. Show that

$$E(X) = \sum_{k=0}^{\infty} P(X > k) = \sum_{k=1}^{\infty} P(X \geq k).$$

1.9: Suppose that X is a random variable with distribution function F and inverse (or quantile function) F^{-1}. Show that

$$E(X) = \int_0^1 F^{-1}(t)\, dt$$

if $E(X)$ is well-defined.

1.10: Suppose that X is stochastically greater than Y in the sense that $P(X > x) \geq P(Y > x)$ for all x.

(a) Suppose that at least one of $E(X)$ or $E(Y)$ is finite. Show that $E(X) \geq E(Y)$.

(b) Let $F_X^{-1}(t)$ and $F_Y^{-1}(t)$ be the quantile functions of X and Y, respectively. Show that $F_X^{-1}(t) \geq F_Y^{-1}(t)$ for all t.

1.11: Let X be a random variable with finite expected value $E(X)$ and suppose that $g(x)$ is a convex function:

$$g(tx + (1-t)y) \leq t\, g(x) + (1-t)\, g(y)$$

for $0 \leq t \leq 1$.

(a) Show that for any x_0, there exists a linear function $h(x) = ax + b$ such that $h(x_0) = g(x_0)$ and $h(x) \leq g(x)$ for all x.

(b) Prove Jensen's inequality: $g(E(X)) \leq E[g(X)]$. (Hint: Set $x_0 = E(X)$ in part (a); then $h(X) \leq g(X)$ with probability 1 and so $E[h(X)] \leq E[g(X)]$.)

1.12: The Gamma function $\Gamma(x)$ is defined for $x > 0$ via the integral

$$\Gamma(x) = \int_0^{\infty} t^{x-1} \exp(-t)\, dt$$

Prove the following facts about $\Gamma(x)$.

(a) $\Gamma(x+1) = x\Gamma(x)$ for any $x > 0$. (Hint: Integrate $\Gamma(x+1)$ by parts.)

(b) $\Gamma(k) = (k-1)!$ for $k = 1, 2, 3, \cdots$.

(c) $\Gamma(1/2) = \sqrt{\pi}$. (Hint: Note that

$$\Gamma(1/2) = \int_0^{\infty} t^{-1/2} \exp(-t)\, dt$$

$$= 2\int_0^\infty \exp(-s^2)\, ds$$

and so

$$\Gamma^2(1/2) = 4\int_0^\infty \int_0^\infty \exp\left[-(s^2+t^2)\right]\, ds\, dt.$$

Then make a change of variables to polar coordinates.)

1.13: Suppose that $X \sim \text{Gamma}(\alpha, \lambda)$. Show that
(a) $E(X^r) = \Gamma(r+\alpha)/(\lambda^2\Gamma(\alpha))$ for $r > -\alpha$;
(b) $\text{Var}(X) = \alpha/\lambda^2$.

1.14: Suppose that $X \sim \text{Gamma}(k, \lambda)$ where k is a positive integer. Show that

$$P(X > x) = \sum_{j=0}^{k-1} \frac{\exp(-\lambda x)(\lambda x)^j}{j!}$$

for $x > 0$; thus we can evaluate the distribution function of X in terms of a Poisson distribution. (Hint: Use integration by parts.)

1.15: Suppose that $X \sim N(0,1)$.

(a) Show that $E(X^k) = 0$ if k is odd.

(b) Show that $E(X^k) = 2^{k/2}\Gamma((k+1)/2)/\Gamma(1/2)$ if k is even.

1.16: Suppose that X is a continuous random variable with density function

$$f(x) = k(p)\exp(-|x|^p) \quad \text{for } -\infty < x < \infty$$

where $p > 0$.

(a) Show that $k(p) = p/(2\Gamma(1/p))$.

(b) Show that $E[|X|^r] = \Gamma((r+1)/p)/\Gamma(1/p)$ for $r > -1$.

1.17: Let $m(t) = E[\exp(tX)]$ be the moment generating function of X. $c(t) = \ln m_X(t)$ is often called the cumulant generating function of X.

(a) Show that $c'(0) = E(X)$ and $c''(0) = \text{Var}(X)$.

(b) Suppose that X has a Poisson distribution with parameter λ as in Example 1.33. Use the cumulant generating function of X to show that $E(X) = \text{Var}(X) = \lambda$.

(c) The mean and variance are the first two cumulants of a

distribution; in general, the k-th cumulant is defined to be $c^{(k)}(0)$. Show that the third and fourth cumulants are

$$\begin{aligned} c^{(3)}(0) &= E(X^3) - 3E(X)E(X^2) + 2[E(X)]^3 \\ c^{(4)}(0) &= E(X^4) - 4E(X^3)E(X^3) + 12E(X^2)[E(X)]^2 \\ &\quad -3[E(X^2)]^2 - 6[E(X)]^4. \end{aligned}$$

(d) Suppose that $X \sim N(\mu, \sigma^2)$. Show that all but the first two cumulants of X are exactly 0.

1.18: Suppose that X is an integer-valued random variable with moment generating function $m(t)$. The probability generating function of X is defined to be

$$p(t) = E(t^X) \quad \text{for } t > 0.$$

Note that $p(t) = m(\ln(t))$.

(a) Show that $p'(1) = E(X)$.

(b) Let $p^{(k)}(t)$ be the k-th derivative of p. Show that

$$p^{(k)}(1) = E[X(X-1) \times \cdots \times (X-k+1)].$$

(c) Suppose that X is nonnegative. Show that

$$P(X = k) = p^{(k)}(0)/k!.$$

1.19: The Gompertz distribution is sometimes used as a model for the length of human life; this model is particular good for modeling survival beyond 40 years. Its distribution function is

$$F(x) = 1 - \exp[-\beta(\exp(\alpha x) - 1)] \quad \text{for } x \geq 0$$

where $\alpha, \beta > 0$.

(a) Find the hazard function for this distribution.

(b) Suppose that X has distribution function F. Show that

$$E(X) = \frac{\exp(\beta)}{\alpha} \int_1^\infty \frac{\exp(-\beta t)}{t}\, dt$$

while the median of F is

$$F^{-1}(1/2) = \frac{1}{\alpha} \ln\left(1 + \ln(2)/\beta\right).$$

(c) Show that $F^{-1}(1/2) \geq E(X)$ for all $\alpha > 0$, $\beta > 0$.

1.20: Suppose that X is a nonnegative random variable.

(a) Show that

$$E(X^r) = r\int_0^\infty x^{r-1}(1 - F(x))\,dx.$$

for any $r > 0$.

(b) Since X is nonnegative, it follows that $\phi(t) = E[\exp(-tX)]$ is finite for $t \geq 0$. Show that for any $r > 0$,

$$E(1/X^r) = \frac{1}{\Gamma(r)}\int_0^\infty t^{r-1}\phi(t)\,dt.$$

(Hint: Write $\phi(t)$ as an integral involving the distribution function of X.)

1.21: Suppose that X is a nonnegative random variable where $E(X^r)$ is finite for some $r > 0$. Show that $E(X^s)$ is finite for $0 \leq s \leq r$.

1.22: Let X be a continuous random variable with distribution function $F_X(x)$. Suppose that $Y = g(X)$ where g is a strictly increasing continuous function. If $F_X^{-1}(t)$ is the inverse of F_X show that the inverse of the distribution function of Y is

$$F_Y^{-1}(t) = g(F_X^{-1}(t)).$$

1.23: Suppose that X is a nonnegative random variable with distribution function $F(x) = P(X \leq x)$. Show that

$$E(X^r) = r\int_0^\infty x^{r-1}(1 - F(x))\,dx.$$

for any $r > 0$.

1.24: Let U be a Uniform random variable on $[0, 1]$ and define $X = \tan(\pi(U - 1/2))$. Show that the density of X is

$$f(x) = \frac{1}{\pi(1 + x^2)}.$$

(This is the Cauchy distribution in Example 1.26.)

1.25: Suppose that X has a distribution function $F(x)$ with inverse $F^{-1}(t)$.

(a) Suppose also that $E(|X|) < \infty$ and define $g(t) = E[|X - t|]$. Show that g is minimized at $t = F^{-1}(1/2)$.

(b) The assumption that $E(|X|) < \infty$ in (a) is unnecessary if we redefine $g(t) = E[|X - t| - |X|]$. Show that $g(t)$ is finite for all t and that $t = F^{-1}(1/2)$ minimizes $g(t)$.

(c) Define $\rho_\alpha(x) = \alpha x I(x \geq 0) + (\alpha - 1)x I(x < 0)$ for some $0 < \alpha < 1$. Show that $g(t) = E[\rho_\alpha(X-t) - \rho_\alpha(X)]$ is minimized at $t = F^{-1}(\alpha)$.

1.26: Suppose that X is a random variable with $E[X^2] < \infty$. Show that $g(t) = E[(X-t)^2]$ is minimized at $t = E(X)$.

1.27: Let X be a positive random variable with distribution function F. Show that $E(X) < \infty$ if, and only if,

$$\sum_{k=1}^{\infty} P(X > k\epsilon) < \infty$$

for any $\epsilon > 0$.

CHAPTER 2

Random vectors and joint distributions

2.1 Introduction

To this point, we have considered only a single random variable defined on a sample space. It is, of course, possible to define more than one random variable for a given experiment. Consider, for example, sampling from a finite population consisting of N people. Each person in the population has a number of attributes that can be measured: height, weight, age, and so on. Clearly, we can define random variables that measure each of these attributes for a randomly chosen person. If we take a sample of $n < N$ people from this population, we can define random variables $X_1, \cdots, X_n$ that measure a certain attribute (for example, weight) for each of the n people in the sample.

Suppose we have random variables $X_1, \cdots, X_k$ defined on some sample space. We then call the vector $\boldsymbol{X} = (X_1, \cdots, X_k)$ a random vector.

DEFINITION. The joint distribution function of a random vector $(X_1, \cdots, X_k)$ is

$$F(x_1, \cdots, x_k) = P(X_1 \le x_1, \cdots, X_k \le x_k)$$

where the event $[X_1 \le x_1, \cdots, X_k \le x_k]$ is the intersection of the events $[X_1 \le x_1], \cdots, [X_k \le x_k]$.

Given the joint distribution function of random vector $\boldsymbol{X}$, we can determine $P(\boldsymbol{X} \in A)$ for any set $A \subset R^k$.

Not surprisingly, it is often convenient to think of random vectors as (random) elements of a vector space; this allows us, for example, to manipulate random vectors via the operations of linear algebra. When this is the case, we will assume (by default) that the random vector is, in fact, a column vector unless explicitly stated otherwise. Hopefully, this will be clear from the context.

2.2 Discrete and continuous random vectors

For single random variables, we noted earlier that it is possible to describe the probability distribution by means of a density function or frequency function, depending on whether the random variable is continuous or discrete. It is possible to define analogous functions in the case of random vectors.

DEFINITION. Suppose that $X_1, \cdots, X_k$ are discrete random variables defined on the same sample space. Then the joint frequency function of $X = (X_1, \cdots, X_k)$ is defined to be

$$f(x_1, \cdots, x_k) = P(X_1 = x_1, \cdots, X_k = x_k).$$

If $X_1, \cdots, X_k$ are discrete then the joint frequency function must exist.

DEFINITION. Suppose that $X_1, \cdots, X_n$ are continuous random variables defined on the same sample space and that

$$P[X_1 \le x_1, \cdots, X_k \le x_k] = \int_{-\infty}^{x_k} \cdots \int_{-\infty}^{x_1} f(t_1, \cdots, t_k)\, dt_1 \cdots dt_k$$

for all $x_1, \cdots, x_k$. Then $f(x_1, \cdots, x_k)$ is the joint density function of $(X_1, \cdots, X_k)$ (provided that $f(x_1, \cdots, x_k) \ge 0$).

To avoid confusion, we will sometimes refer to the density or frequency function of a random variable X_i as its marginal density or frequency function. The joint density and frequency functions must satisfy the following conditions:

$$\sum_{(x_1, \cdots, x_k)} f(x_1, \cdots, x_k) = 1$$

if $f(x_1, \cdots, x_k)$ is a frequency function, and

$$\int_{-\infty}^{\infty} \cdots \int_{-\infty}^{\infty} f(x_1, \cdots, x_k)\, dx_1 \cdots dx_k = 1$$

if $f(x_1, \cdots, x_k)$ is a density function. Moreover, we can determine the probability that $(X_1, \cdots, X_n)$ lies in a given set A by summing or integrating $f(x_1, \cdots, x_n)$ over A.

EXAMPLE 2.1: Suppose that X and Y are continuous random variables with joint density function

$$f(x, y) = \frac{1}{\pi} \quad \text{for } x^2 + y^2 \le 1.$$

X and Y thus have a Uniform distribution on a disk of radius 1 centered at the origin. Suppose we wish to determine $P(X \leq u)$ for $-1 \leq u \leq 1$. We can do this by integrating the joint density $f(x, y)$ over the region $\{(x, y) : -1 \leq x \leq u, x^2 + y^2 \leq 1\}$. Thus we have

$$\begin{aligned} P(X \leq u) &= \int_{-1}^{u} \int_{-\sqrt{1-x^2}}^{\sqrt{1-x^2}} \frac{1}{\pi}\, dy\, dx \\ &= \frac{1}{\pi} \int_{-1}^{u} 2\sqrt{1 - x^2}\, dx \\ &= \frac{1}{2} + \frac{1}{\pi} u\sqrt{1 - u^2} + \frac{1}{\pi} \sin^{-1}(u). \end{aligned}$$

Note that to find the marginal density of X, we can differentiate $P(X \leq u)$ with respect to u:

$$f_X(x) = \frac{2}{\pi}\sqrt{1 - x^2} \quad \text{for } |x| \leq 1.$$

It is easy to see (by symmetry) that Y has the same marginal density. ◇

The following result indicates how to obtain the joint density (frequency) function of a subset of $X_1, \cdots, X_k$ or the marginal density function of one of the X_i's.

THEOREM 2.1 *(a) Suppose that* $\boldsymbol{X} = (X_1, \cdots, X_k)$ *has joint frequency function* $f(\boldsymbol{x})$*. For* $\ell < k$*, the joint frequency function of* $(X_1, \cdots, X_\ell)$ *is*

$$g(x_1, \cdots, x_\ell) = \sum_{x_{\ell+1}, \cdots, x_k} f(x_1, \cdots, x_k).$$

(b) Suppose that $\boldsymbol{X} = (X_1, \cdots, X_k)$ *has joint density function* $f(\boldsymbol{x})$*. For* $\ell < k$*, the joint density function of* $(X_1, \cdots, X_\ell)$ *is*

$$g(x_1, \cdots, x_\ell) = \int_{-\infty}^{\infty} \cdots \int_{-\infty}^{\infty} f(x_1, \cdots, x_k)\, dx_{\ell+1} \cdots dx_k.$$

Proof. (a) This result follows trivially since

$$P(X_1 = x_1, \cdots, X_\ell = x_\ell) = \sum_{(x_{\ell+1}, \cdots, x_k)} P(X_1 = x_1, \cdots, X_k = x_k)$$

for any $x_1, \cdots, x_\ell$.

(b) For any $x_1, \cdots, x_\ell$,

$$\begin{aligned} P(X_1 \le x_1, \cdots, X_\ell \le x_\ell) \\ &= \int_{-\infty}^{x_1} \cdots \int_{-\infty}^{x_\ell} \int_{-\infty}^{\infty} \cdots \int_{-\infty}^{\infty} f(t_1, \cdots, t_k)\, dt_k \cdots dt_{\ell+1} dt_\ell \cdots dt_1 \\ &= \int_{-\infty}^{x_1} \cdots \int_{-\infty}^{x_\ell} g(t_1, \cdots, t_\ell)\, dt_\ell \cdots dt_1 \end{aligned}$$

and so $g(x_1, \cdots, x_\ell)$ is the joint density of $(X_1, \cdots, X_\ell)$. □

Theorem 2.1 can be applied to find the marginal density function of X in Example 2.1. Since $f(x, y) = 0$ for $|y| > \sqrt{1 - x^2}$ and $|x| \le 1$, it follows that the marginal density of X is

$$\begin{aligned} f_X(x) &= \frac{1}{\pi} \int_{-\sqrt{1-x^2}}^{\sqrt{1-x^2}} dy \\ &= \frac{2}{\pi} \sqrt{1 - x^2} \end{aligned}$$

for $|x| \le 1$ as given in Example 2.1.

As is the case for continuous random variables, the joint density function of a continuous random vector is not uniquely defined; for example, we could change $f(\boldsymbol{x})$ at a countably infinite number of points and the resulting function would still satisfy the definition. However, just as it is useful to think of the density function as the derivative of the distribution function of a random variable, we can think of the joint density as a partial derivative of the joint distribution function. More precisely, if $F(x_1, \cdots, x_k)$ is differentiable then we can write

$$f(x_1, \cdots, x_k) = \frac{\partial^k}{\partial x_1 \cdots \partial x_k} F(x_1, \cdots, x_k).$$

It is also important to note that the joint density function of $(X_1, \cdots, X_k)$ need not exist even if each of $X_1, \cdots, X_k$ have their own marginal density function. For example, suppose that $X_2 = g(X_1)$ for some continuous function g. Existence of a joint density $f(x_1, x_2)$ of (X_1, X_2) implies that the range of (X_1, X_2) contains an open rectangle of the form

$$\{(x_1, x_2) : a_1 < x_1 < b_1, a_2 < x_2 < b_2\}.$$

However, since the range of (X_1, X_2) is at most

$$\{(x_1, x_2) : -\infty < x_1 < \infty, x_2 = g(x_1)\}$$

which does not contain an open rectangle, we can conclude that (X_1, X_2) does not have a joint density function.

Independent random variables

DEFINITION. Let $X_1, \cdots, X_k$ be random variables defined on the same sample space. $X_1, \cdots, X_k$ are said to be independent if the events $[a_1 < X_1 \le b_1]$, $[a_2 < X_2 \le b_2], \cdots, [a_k < X_k \le b_k]$ are independent for all $a_i < b_i$, $i = 1, \cdots, k$. An infinite collection $X_1, X_2, \cdots$ of random variables are independent if every finite collection of random variables is independent.

If $(X_1, \cdots, X_k)$ have a joint density or joint frequency function then there is a simple equivalent condition for independence.

THEOREM 2.2 *If $X_1, \cdots, X_k$ are independent and have joint density (or frequency) function $f(x_1, \cdots, x_k)$ then*

$$f(x_1, \cdots, x_k) = \prod_{i=1}^{k} f_i(x_i)$$

where $f_i(x_i)$ is the marginal density (frequency) function of X_i. Conversely, if the joint density (frequency) function is the product of marginal density (frequency) functions then $X_1, \cdots, X_k$ are independent.

Proof. We will give the proof only for the case where $f(x_1, \cdots, x_n)$ is a joint density function; the proof for the frequency function case is similar. If $X_1, \cdots, X_k$ are independent,

$$\begin{aligned} & P(a_1 < X_1 \le b_1, \cdots, a_k < X_k \le b_k) \\ = & \prod_{i=1}^{k} P(a_i < X_i \le b_i) \\ = & \prod_{i=1}^{k} \int_{a_i}^{b_i} f_i(x_i)\, dx_i \\ = & \int_{a_k}^{b_k} \cdots \int_{a_1}^{b_1} \prod_{i=1}^{k} f_i(x_i)\, dx_1 \cdots dx_k \end{aligned}$$

and so $f_1(x_1) \times \cdots \times f_k(x_k)$ is the joint density of $(X_1, \cdots, X_k)$. Conversely, if $f(x_1, \cdots, x_k) = f_1(x_1) \cdots f_k(x_k)$ then it is easy to verify that $[a_1 < X_1 \le b_1], \cdots, [a_k < X_k \le b_k]$ are independent for all choices of the a_i's and b_i's. □

Independence is an important assumption for many statistical models. Assuming independence essentially allows us to concentrate on the marginal distributions of random variables; given these marginal distributions together with independence, we can determine the joint distribution of the random variables. If $X_1, \cdots, X_n$ are independent random variables with the same marginal distribution, we say that $X_1, \cdots, X_n$ are independent, identically distributed (i.i.d.) random variables.

EXAMPLE 2.2: Suppose that $X_1, \cdots, X_n$ are i.i.d. continuous random variables with common (marginal) density $f(x)$ and distribution function $F(x)$. Given $X_1, \cdots, X_n$, we can define two new random variables

$$U = \min(X_1, \cdots, X_n) \quad \text{and} \quad V = \max(X_1, \cdots, X_n),$$

which are the minimum and maximum of the i.i.d. sample. It is fairly simple to determine the marginal densities of U and V. We note that $U > x$ if, and only if, $X_i > x$ for all i and also that $V \leq x$ if, and only if, $X_i \leq x$ for all i. Thus

$$\begin{aligned} P(U \leq x) &= 1 - P(U > x) \\ &= 1 - P(X_1 > x, \cdots, X_n > x) \\ &= 1 - [1 - F(x)]^n \end{aligned}$$

and

$$\begin{aligned} P(V \leq x) &= P(X_1 \leq x, \cdots, X_n \leq n) \\ &= [F(x)]^n. \end{aligned}$$

From this it follows that the marginal densities of U and V are

$$f_U(x) = n[1 - F(x)]^{n-1} f(x)$$

and

$$f_V(x) = n[F(x)]^{n-1} f(x).$$

The joint density of (U, V) is somewhat more complicated. Define $f_{U,V}(u, v)$ to be this joint density. Since $U \leq V$, it follows that $f_{U,V}(u, v) = 0$ for $u > v$. For $u \leq v$, we have

$$\begin{aligned} P(U \leq u, V \leq v) &= P(V \leq v) - P(U > u, V \leq v) \\ &= P(V \leq v) \\ &\qquad -P(u < X_1 \leq v, \cdots, u < X_n \leq v) \\ &= [F(v)]^n - [F(v) - F(u)]^n. \end{aligned}$$

Now if $P(U \leq u, V \leq v)$ is twice-differentiable with respect to u and v, we can evaluate the joint density of (U, V) by

$$\begin{aligned} f_{U,V}(u,v) &= \frac{\partial^2}{\partial u \partial v} P(U \leq u, V \leq v) \\ &= n(n-1)[F(v) - F(u)]^{n-2} f(u) f(v). \end{aligned}$$

In general, we have

$$f_{U,V}(u,v) = \begin{cases} n(n-1)[F(v) - F(u)]^{n-2} f(u) f(v) & \text{for } u \leq v \\ 0 & \text{otherwise} \end{cases}$$

as the joint density of (U, V). ◇

Transformations

Suppose that $\boldsymbol{X} = (X_1, \cdots, X_k)$ is a random vector with some joint distribution. Define new random variables $Y_i = h_i(\boldsymbol{X})$ $(i = 1, \cdots, k)$ where $h_1, \cdots, h_k$ are real-valued functions. We would like to determine

- the (marginal) distribution of Y_i, and
- the joint distribution of $\boldsymbol{Y} = (Y_1, \cdots, Y_k)$.

The two questions above are, of course, related; if the joint distribution of $\boldsymbol{Y}$ is known then we can (at least in theory) find the marginal distribution of a single Y_i. On the other hand, it would seem to be easier to determine the marginal distribution of a single Y_i than the joint distribution of $\boldsymbol{Y}$; while this is true in some cases, in many cases (particularly when the Y_i's are continuous), the marginal distribution is most easily determined by first determining the joint distribution.

We will first consider determining the marginal distribution of a single Y_i. Given a random vector $\boldsymbol{X} = (X_1, \cdots, X_k)$, define $Y = h(\boldsymbol{X})$ for some (real-valued) function h. To simplify the discussion somewhat, we will assume that $\boldsymbol{X}$ has a joint density or joint frequency function. The following general algorithm can be used for determining the distribution of Y:

- Find $P(Y \leq y)$ (or $P(Y = y)$ if this is non-zero) by integrating (summing) the joint density (frequency) function over the appropriate region.
- If Y is a continuous random variable, write

$$P(Y \leq y) = \int_{-\infty}^{y} f_Y(t)\, dt$$

to obtain the density function $f_Y(y)$ of Y; typically, this can be done by differentiating $P(Y \leq y)$.

This algorithm is conceptually simple but not always feasible in practice unless h is a fairly simple function. In the case where Y is a sum of two random variables (independent or not), the algorithm is very simple to apply.

EXAMPLE 2.3: Suppose that X_1, X_2 are discrete random variables with joint frequency function $f_X(x_1, x_2)$ and let $Y = X_1 + X_2$. Then

$$\begin{aligned} f_Y(y) = P(Y = y) &= \sum_x P(X_1 = x, X_2 = y - x) \\ &= \sum_x f_X(x, y - x) \end{aligned}$$

is the frequency function of Y. ◇

EXAMPLE 2.4: Suppose that X_1, X_2 be continuous random variables with joint density function $f_X(x_1, x_2)$ and let $Y = X_1 + X_2$. We can obtain $P(Y \leq y)$ by integrating the joint density over the region $\{(x_1, x_2) : x_1 + x_2 \leq y\}$. Thus

$$\begin{aligned} P(Y \leq y) &= \int_{-\infty}^{\infty} \int_{-\infty}^{y - x_2} f_X(x_1, x_2)\, dx_1\, dx_2 \\ &= \int_{-\infty}^{\infty} \int_{-\infty}^{y} f_X(t - x_2, x_2)\, dt\, dx_2 \\ &\qquad \text{(setting } t = x_1 + x_2\text{)} \\ &= \int_{-\infty}^{y} \int_{-\infty}^{\infty} f_X(t - x_2, x_2)\, dx_2\, dt. \end{aligned}$$

Thus it follows that Y has a density function

$$f_Y(y) = \int_{-\infty}^{\infty} f_X(y - u, u)\, du$$

or equivalently

$$f_Y(y) = \int_{-\infty}^{\infty} f_X(u, y - u)\, du.$$

Note that if both X and Y are nonnegative random variables then $f_X(y - u, u) = 0$ if $u < 0$ or if $u > y$. In this case, the density simplifies to

$$f_Y(y) = \int_0^y f_X(y - u, u)\, du.$$

Notice that the expression for the density function of Y is similar to the expression for the frequency function of Y in Example 2.3 with the integral replacing the summation. ◇

We now turn our attention to finding the joint density of $\boldsymbol{Y} = (Y_1, \cdots, Y_k)$ where $Y_i = h_i(X_1, \cdots, X_k)$ $(i = 1, \cdots, k)$ and $\boldsymbol{X} = (X_1, \cdots, X_k)$ has a joint density f_X.

We start by defining a vector-valued function $\boldsymbol{h}$ whose elements are the functions $h_1, \cdots, h_k$:

$$\boldsymbol{h}(\boldsymbol{x}) = \begin{pmatrix} h_1(x_1, \cdots, x_k) \\ h_2(x_1, \cdots, x_k) \\ \vdots \\ h_k(x_1, \cdots, x_k) \end{pmatrix}.$$

We will assume (for now) that $\boldsymbol{h}$ is a one-to-one function with inverse $\boldsymbol{h}^{-1}$ (that is, $\boldsymbol{h}^{-1}(\boldsymbol{h}(\boldsymbol{x})) = \boldsymbol{x}$). Next we will define the Jacobian matrix of $\boldsymbol{h}$ to be a $k \times k$ whose i-th row and j-th column element is

$$\frac{\partial}{\partial x_j} h_i(x_1, \cdots, x_k)$$

with the Jacobian of $\boldsymbol{h}$, $J_h(x_1, \cdots, x_k)$, defined to be the determinant of this matrix.

THEOREM 2.3 *Suppose that $P(\boldsymbol{X} \in S) = 1$ for some open set $S \subset R^k$. If*
(a) $\boldsymbol{h}$ has continuous partial derivatives on S,
(b) $\boldsymbol{h}$ is one-to-one on S,
(c) $J_h(\boldsymbol{x}) \neq 0$ for $\boldsymbol{x} \in S$
then $(Y_1, \cdots, Y_k)$ has joint density function

$$\begin{aligned} f_Y(\boldsymbol{y}) &= \frac{f_X(\boldsymbol{h}^{-1}(\boldsymbol{y}))}{|J_h(\boldsymbol{h}^{-1}(\boldsymbol{y}))|} \\ &= f_X(\boldsymbol{h}^{-1}(\boldsymbol{y}))|J_{h^{-1}}(\boldsymbol{y})| \end{aligned}$$

for $\boldsymbol{y} \in \boldsymbol{h}(S)$. ($J_{h^{-1}}$ is the Jacobian of $\boldsymbol{h}^{-1}$.)

The proof of Theorem 2.3 follows from the standard change-of-variables formula for integrals in multivariate calculus.

EXAMPLE 2.5: Suppose that X_1, X_2 are independent Gamma random variables with common scale parameters:

$$X_1 \sim \text{Gamma}(\alpha, \lambda) \quad \text{and} \quad X_2 \sim \text{Gamma}(\beta, \lambda).$$

Define

$$\begin{aligned} Y_1 &= X_1 + X_2 \\ Y_2 &= \frac{X_1}{X_1 + X_2} \end{aligned}$$

and note that Y_1 takes values in the interval $(0, \infty)$ while Y_2 takes its values in $(0, 1)$. The functions $\boldsymbol{h}$ and $\boldsymbol{h}^{-1}$ in this case are

$$\begin{aligned} \boldsymbol{h}(x_1, x_2) &= \begin{pmatrix} x_1 + x_2 \\ x_1/(x_1 + x_2) \end{pmatrix} \\ \text{and} \quad \boldsymbol{h}^{-1}(y_1, y_2) &= \begin{pmatrix} y_1 y_2 \\ y_1(1 - y_2) \end{pmatrix} \end{aligned}$$

and the Jacobian of $\boldsymbol{h}^{-1}$ is $J_{h^{-1}}(y_1, y_2) = -y_1$. The joint density of X_1 and X_2 can be determined by multiplying the marginal densities together:

$$f_X(x_1, x_2) = \frac{\lambda^{\alpha+\beta} x_1^{\alpha-1} x_2^{\beta-1} \exp[-\lambda(x_1 + x_2)]}{\Gamma(\alpha)\Gamma(\beta)} \quad \text{for } x_1, x_2 > 0$$

Now applying Theorem 2.3 and doing some rearranging, we get

$$f_Y(y_1, y_2) = g_1(y_1) g_2(y_2)$$

where

$$\begin{aligned} g_1(y_1) &= \frac{\lambda^{\alpha+\beta} y_1^{\alpha+\beta-1} \exp(-\lambda y_1)}{\Gamma(\alpha+\beta)} \\ \text{and} \quad g_2(y_2) &= \frac{\Gamma(\alpha+\beta)}{\Gamma(\alpha)\Gamma(\beta)} y_2^{\alpha-1} (1 - y_2)^{\beta-1} \end{aligned}$$

for $0 < y_1 < \infty$ and $0 < y_2 < 1$.

There are several things to note here. First, Y_1 is independent of Y_2; second, Y_1 has a Gamma distribution with shape parameter $\alpha + \beta$ and scale parameter λ; finally, Y_2 has a Beta distribution with parameters α and β ($Y_2 \sim \text{Beta}(\alpha, \beta)$). ◇

Theorem 2.3 can be extended to the case where the transformation $\boldsymbol{h}$ is not one-to-one. Suppose that $P[\boldsymbol{X} \in S] = 1$ for some open set and that S is a disjoint union of open sets $S_1, \cdots, S_m$ where $\boldsymbol{h}$

is one-to-one on each of the S_j's (with inverse $\boldsymbol{h}_j^{-1}$ on S_j). Then defining $(Y_1, \cdots, Y_k)$ as before, the joint density of $(Y_1, \cdots, Y_k)$ is

$$f_Y(\boldsymbol{y}) = \sum_{j=1}^{m} f_X(\boldsymbol{h}_j^{-1}(\boldsymbol{y}))|J_{h_j^{-1}}(\boldsymbol{y})|I(\boldsymbol{h}_j^{-1}(\boldsymbol{y} \in S_j)$$

where $J_{h_j^{-1}}$ is the Jacobian of $\boldsymbol{h}_j^{-1}$.

EXAMPLE 2.6: Suppose that $X_1, \cdots, X_n$ are i.i.d. random variables with density function $f(x)$. Reorder the X_i's so that $X_{(1)} < X_{(2)} < \cdots < X_{(n)}$; these latter random variables are called the order statistics of $X_1, \cdots, X_n$. Define $Y_i = X_{(i)}$ and notice that the transformation taking $X_1, \cdots, X_n$ to the order statistics is not one-to-one; given a particular ordering $y_1 < y_2 < \cdots < y_n$, $(x_1, \cdots, x_n)$ could be any one of the $n!$ permutations of $(y_1, \cdots, y_n)$. To determine the joint density of the order statistics, we divide the range of $(X_1, \cdots, X_n)$ into the $n!$ disjoint subregions

$$\begin{aligned} S_1 &= \{(x_1, \cdots, x_n) : x_1 < x_2 < \cdots < x_n\} \\ S_2 &= \{(x_1, \cdots, x_n) : x_2 < x_1 < x_3 < \cdots < x_n\} \\ \vdots \quad & \vdots \quad \vdots \\ S_{n!} &= \{(x_1, \cdots, x_n) : x_n < x_{n-1} < \cdots < x_1\} \end{aligned}$$

corresponding to the $n!$ orderings of $(x_1, \cdots, x_n)$. Note that the transformation $\boldsymbol{h}$ from the X_i's to the order statistics is one-to-one on each S_j; moreover, the Jacobian $J_{h_j^{-1}} = \pm 1$ for each $j = 1, \cdots, n!$. Since the joint density of $(X_1, \cdots, X_n)$ is

$$f_X(x_1, \cdots, x_n) = f(x_1) \times \cdots \times f(x_n),$$

it follows that the joint density of the order statistics is

$$\begin{aligned} f_Y(y_1, \cdots, y_n) &= \sum_{j=1}^{n!} f_X(\boldsymbol{h}_j^{-1}(y_1, \cdots, y_n)) \\ &= n! \prod_{i=1}^{n} f(y_i) \end{aligned}$$

for $y_1 < y_2 < \cdots < y_n$. ◇

Expected values

Suppose that $X_1, \cdots, X_n$ are random variables defined on some sample space and let $Y = h(X_1, \cdots, X_k)$ for some real-valued function h. In section 1.6, we defined the expected value of Y to be

$$E(Y) = \int_0^\infty P(Y > y)\,dy - \int_{-\infty}^0 P(Y \le y)\,dy.$$

This formula implies that we need to first determine the distribution function of Y (or equivalently its density or frequency function) in order to evaluate $E(Y)$. Fortunately, evaluating $E(Y)$ is typically not so complicated if $\boldsymbol{X} = (X_1, \cdots, X_k)$ has a joint density or frequency function; more precisely, we can define

$$E[h(\boldsymbol{X})] = \sum_{\boldsymbol{x}} h(\boldsymbol{x}) f(\boldsymbol{x})$$

if $\boldsymbol{X}$ has joint frequency function $f(\boldsymbol{x})$ and

$$E[h(\boldsymbol{X})] = \int_{-\infty}^{\infty} \cdots \int_{-\infty}^{\infty} h(\boldsymbol{x}) f(\boldsymbol{x})\,dx_1 \cdots dx_k$$

if $\boldsymbol{X}$ has joint density function $f(\boldsymbol{x})$. (Of course, in order for the expected value to be finite we require that $E[|h(\boldsymbol{X})|] < \infty$.)

PROPOSITION 2.4 *Suppose that $X_1, \cdots, X_k$ are random variables with finite expected values.*
(a) If $X_1, \cdots, X_k$ are defined on the same sample space then

$$E(X_1 + \cdots + X_k) = \sum_{i=1}^k E(X_i).$$

(b) If $X_1, \cdots, X_k$ are independent random variables then

$$E\left(\prod_{i=1}^k X_i\right) = \prod_{i=1}^k E(X_i).$$

The proofs of parts (a) and (b) of Proposition 2.4 are quite simple if we assume that $(X_1, \cdots, X_k)$ has a joint density or frequency function; this is left as an exercise. More generally, the proofs are a bit more difficult but can be found in a more advanced probability text such as Billingsley (1995).

EXAMPLE 2.7: Suppose that $X_1, \cdots, X_n$ are independent random variables with moment generating functions $m_1(t), \cdots, m_n(t)$

respectively. Define $S = X_1 + \cdots + X_n$; the moment generating function of S is

$$\begin{aligned}
m_S(t) &= E[\exp(tS)] \\
&= E[\exp(t(X_1 + \cdots + X_n))] \\
&= E[\exp(tX_1)\exp(tX_2) \times \cdots \times \exp(tX_n)] \\
&= \prod_{i=1}^{n} E[\exp(tX_i)] \\
&= \prod_{i=1}^{n} m_i(t)
\end{aligned}$$

where we use the independence of $X_1, \cdots, X_n$. Thus the moment generating function of a sum of independent random variables is the product of the individual moment generating functions. ◇

EXAMPLE 2.8: Suppose that $X_1, \cdots, X_n$ are independent Normal random variables with $E(X_i) = \mu_i$ and $\text{Var}(X_i) = \sigma_i^2$. Again define $S = X_1 + \cdots + X_n$. The moment generating function of X_i is

$$m_i(t) = \exp\left(\mu_i t + \frac{1}{2}\sigma_i^2 t^2\right)$$

and so the moment generating function of S is

$$m_S(t) = \prod_{i=1}^{n} m_i(t) = \exp\left(t\sum_{i=1}^{n}\mu_i + \frac{t^2}{2}\sum_{i=1}^{n}\sigma_i^2\right).$$

Since moment generating functions characterize distributions, it is easy to see that S has a Normal distribution with mean $\sum_{i=1}^{n}\mu_i$ and variance $\sum_{i=1}^{n}\sigma_i^2$. ◇

EXAMPLE 2.9: Suppose that $X_1, \cdots, X_n$ are i.i.d. random variables with moment generating function $m(t)$. Define

$$\bar{X}_n = \frac{1}{n}\sum_{i=1}^{n} X_i$$

and let

$$m_n(t) = E[\exp(t\bar{X}_n)] = [m(t/n)]^n$$

be its moment generating function. What happens to $m_n(t)$ as $n \to \infty$? Note that we can write

$$m(t) = 1 + tE(X_1) + \frac{t^2}{2}E(X_1^2) + \frac{t^3}{6}E(X_1^3) + \cdots$$

and so

$$\begin{aligned} m(t/n) &= 1 + \frac{t}{n}E(X_1) + \frac{t^2}{2n^2}E(X_1^2) + \frac{t^3}{6n^3}E(X_1^3) + \cdots \\ &= 1 + \frac{t}{n}\left(E(X_1) + \frac{t}{2n}E(X_1^2) + \frac{t^2}{6n^2}E(X_1^3) + \cdots\right) \\ &= 1 + \frac{t}{n}a_n(t) \end{aligned}$$

where $a_n(t) \to E(X_1)$ for all t as $n \to \infty$. Now using the fact that

$$\lim_{n\to\infty}\left(1 + \frac{c_n}{n}\right) = \exp(c)$$

if $c_n \to c$, it follows that

$$\lim_{n\to\infty} m_n(t) = \lim_{n\to\infty}\left(1 + \frac{t a_n(t)}{n}\right)^n = \exp(tE(X_1)).$$

Note that $\exp(tE(X_1))$ is the moment generating function of a random variable that takes the single value $E(X_1)$. This suggests that sequence of random variables $\{\bar{X}_n\}$ converges in some sense to the constant $E(X_1)$. We will elaborate on this in Chapter 3. ◇

Covariance and Correlation

DEFINITION. Suppose X and Y are random variables with $E(X^2)$ and $E(Y^2)$ both finite and let $\mu_X = E(X)$ and $\mu_Y = E(Y)$. The covariance between X and Y is

$$\text{Cov}(X,Y) = E[(X-\mu_X)(Y-\mu_Y)] = E(XY) - \mu_X\mu_Y.$$

The covariance is a measure of linear dependence between two random variables. Using properties of expected values it is quite easy to derive the following properties.

1. For any constants a, b, c, and d,

$$\text{Cov}(aX+b, cY+d) = a\,c\,\text{Cov}(X,Y).$$

2. If X and Y are independent random variables (with $E(X)$ and $E(Y)$ finite) then $\text{Cov}(X,Y) = 0$.

The converse to 2 is not true. In fact, it is simple to find an example where $Y = g(X)$ but $\text{Cov}(X,Y) = 0$.

EXAMPLE 2.10: Suppose that X has a Uniform distribution on the interval $[-1,1]$ and let $Y = -1$ if $|X| < 1/2$ and $Y = 1$ if

$|X| \geq 1/2$. Since $E(X) = E(Y) = 0$, $\text{Cov}(X, Y) = E(XY)$ and

$$XY = \begin{cases} -X & \text{if } |X| < 1/2 \\ X & \text{if } |X| \geq 1/2. \end{cases}$$

Thus

$$E(XY) = \frac{1}{2}\int_{-1}^{-1/2} x\,dx + \frac{1}{2}\int_{-1/2}^{1/2}(-x)\,dx + \frac{1}{2}\int_{1/2}^{1} x\,dx = 0$$

and $\text{Cov}(X, Y) = 0$ even though $Y = g(X)$. ◇

PROPOSITION 2.5 *Suppose that $X_1, \cdots, X_n$ are random variables with $E(X_i^2) < \infty$ for all i. Then*

$$Var\left(\sum_{i=1}^{n} a_i X_i\right) = \sum_{i=1}^{n} a_i^2\, Var(X_i) + 2\sum_{j=2}^{n}\sum_{i=1}^{j-1} a_i a_j\, Cov(X_i, X_j).$$

Proof. Define $\mu_i = E(X_i)$. Then

$$\begin{aligned} \text{Var}\left(\sum_{i=1}^{n} a_i X_i\right) &= E\left[\sum_{i=1}^{n}\sum_{j=1}^{n} a_i a_j (X_i - \mu_i)(X_j - \mu_j)\right] \\ &= \sum_{i=1}^{n}\sum_{j=1}^{n} a_i a_j E[(X_i - \mu_i)(X_j - \mu_j)] \\ &= \sum_{i=1}^{n} a_i^2 \text{Var}(X_i) + 2\sum_{j=2}^{n}\sum_{i=1}^{j-1} a_i a_j \text{Cov}(X_i, X_j) \end{aligned}$$

since $\text{Cov}(X_i, X_j) = \text{Cov}(X_j, X_i)$. □

Note that Proposition 2.5 implies that if $X_1, \cdots, X_n$ are independent random variables then the variance of $X_1 + \cdots + X_n$ is simply the sum of the variances of the X_i's.

EXAMPLE 2.11: Suppose we are sampling without replacement from a finite population consisting of N items $a_1, \cdots, a_N$. Let X_i denote the result of the i-th draw; we then have

$$P(X_i = a_k) = \frac{1}{N} \quad \text{and} \quad P(X_i = a_k, X_j = a_\ell) = \frac{1}{N(N-1)}$$

where $1 \leq i, j, k, \ell \leq N$, $i \neq j$ and $k \neq \ell$. Suppose we define

$$S_n = \sum_{i=1}^{n} X_i$$

where $n \le N$; what are the mean and variance of S_n?

First of all, we define

$$\mu_a = \frac{1}{N}\sum_{k=1}^{N} a_k \quad \text{and} \quad \sigma_a^2 = \frac{1}{N}\sum_{k=1}^{N}(a_k - \mu_a)^2,$$

which are the mean and variance of the population; it is easy to see that $E(X_i) = \mu_a$ and $\text{Var}(X_i) = \sigma_a^2$. Thus it follows that

$$E(S_n) = \sum_{i=1}^{n} E(X_i) = n\mu_a.$$

$\text{Var}(S_n)$ is somewhat trickier; note that

$$\text{Var}(S_n) = n\sigma_a^2 + n(n-1)\text{Cov}(X_1, X_2)$$

since $\text{Cov}(X_i, X_j)$ will be the same for all $i \neq j$. ($\text{Cov}(X_i, X_j)$ depends on the joint distribution of (X_i, X_j) and this is the same for all $i \neq j$.) We can determine $\text{Cov}(X_1, X_2)$ using the joint distribution of X_1 and X_2; however, an easier approach is to consider the case $n = N$. $S_N = \sum_{k=1}^{N} a_k$, a constant, and so $\text{Var}(S_N) = 0$. Thus

$$0 = N\sigma_a^2 + N(N-1)\text{Cov}(X_1, X_2)$$

and so

$$\text{Cov}(X_1, X_2) = -\frac{\sigma_a^2}{N-1}.$$

Substituting into the expression for $\text{Var}(S_n)$ above, we get

$$\text{Var}(S_n) = n\sigma_a^2 \left(\frac{N-n}{N-1}\right).$$

Note that if we sampled with replacement, the random variables $X_1, X_2, \cdots$ would be independent and so $\text{Var}(S_n) = n\sigma_a^2$, which is greater than the variance when the sampling is without replacement (provided $n \ge 2$). The extra factor $(N-n)/(N-1)$ that appears in $\text{Var}(S_n)$ when the sampling is without replacement is called the finite population correction. ◇

Given random variables $X_1, \cdots, X_n$, it is often convenient to represent the variances and covariances of the X_i's via a $n \times n$ matrix. Set $\boldsymbol{X} = (X_1, \cdots, X_n)^T$ (a column vector); then we define the variance-covariance matrix (or covariance matrix) of $\boldsymbol{X}$ to be an $n \times n$ matrix $C = \text{Cov}(\boldsymbol{X})$ whose diagonal elements are $C_{ii} = \text{Var}(X_i)$ $(i = 1, \cdots, n)$ and whose off-diagonal elements are $C_{ij} = \text{Cov}(X_i, X_j)$ $(i \neq j)$. Variance-covariance matrices can be manipulated usefully for linear transformations of $\boldsymbol{X}$: If $\boldsymbol{Y} = B\boldsymbol{X} + \boldsymbol{a}$ for some $m \times n$ matrix B and vector $\boldsymbol{a}$ of length m then

$$\text{Cov}(\boldsymbol{Y}) = B\text{Cov}(\boldsymbol{X})B^T.$$

Likewise, if we define the mean vector of $\boldsymbol{X}$ to be

$$E(\boldsymbol{X}) = \begin{pmatrix} E(X_1) \\ \vdots \\ E(X_n) \end{pmatrix}$$

then $E(\boldsymbol{Y}) = BE(\boldsymbol{X}) + \boldsymbol{a}$.

While the covariance gives some indication of the linear association between two random variables, its value is dependent on the scale of the two random variables.

DEFINITION. Suppose that X and Y are random variables where both $E(X^2)$ and $E(Y^2)$ are finite. Then the correlation between X and Y is

$$\text{Corr}(X, Y) = \frac{\text{Cov}(X, Y)}{[\text{Var}(X)\text{Var}(Y)]^{1/2}}.$$

The advantage of the correlation is the fact that it is essentially invariant to linear transformations (unlike covariance). That is, if $U = aX + b$ and $V = cY + d$ then

$$\text{Corr}(U, V) = \text{Corr}(X, Y)$$

if a and c have the same sign; if a and c have different signs then $\text{Corr}(U, V) = -\text{Corr}(X, Y)$.

PROPOSITION 2.6 *Suppose that X and Y are random variables where both $E(X^2)$ and $E(Y^2)$ are finite. Then*
(a) $-1 \leq Corr(X, Y) \leq 1$;
(b) $Corr(X, Y) = 1$ if, and only if, $Y = aX + b$ for some $a > 0$; $Corr(X, Y) = -1$ if, and only if, $Y = aX + b$ for some $a < 0$.

We will leave the proof of this result as an exercise. If X and Y are independent random variables (with $E(X^2)$ and $E(Y^2)$ finite) then $\text{Corr}(X, Y) = 0$ since $\text{Cov}(X, Y) = 0$. However, as with the covariance, a correlation of 0 does not imply independence. Correlation is merely a measure of linear dependence between random variables; it essentially measures the degree to which we may approximate one random variable by a linear function of another.

PROPOSITION 2.7 *Suppose that X and Y are random variables where both $E(X^2)$ and $E(Y^2)$ are finite and define*

$$g(a, b) = E[(Y - a - bX)^2].$$

Then $g(a, b)$ is minimized at

$$\begin{aligned} b_0 &= \frac{Cov(X,Y)}{Var(X)} = Corr(X,Y)\left(\frac{Var(Y)}{Var(X)}\right)^{1/2} \\ \text{and} \quad a_0 &= E(Y) - b_0 E(X) \end{aligned}$$

with $g(a_0, b_0) = Var(Y)(1 - Corr^2(X, Y))$.

The proof is left as an exercise. Proposition 2.7 can be interpreted by considering predicting Y as a linear function h of X and considering the mean square prediction error $E[(Y - h(X))^2]$. If we take $h(x)$ to be a constant, then as a function of a, $E[(Y - a)^2]$ is minimized at $a = E(Y)$ with $E[(Y - E(Y))^2] = \text{Var}(Y)$. Taking $h(x)$ to be a linear function, the minimum mean square prediction error (according to Proposition 2.7) is $\text{Var}(Y)(1 - \text{Corr}^2(X, Y))$. Thus the reduction in the mean square prediction error when predicting Y by a linear function of X depends explicitly on the correlation.

With some imagination, it is possible to derive a more useful measure of dependence between two random variables. Let X and Y be random variables and consider

$$\text{Corr}(\phi(X), \psi(Y)).$$

If X and Y are independent then this correlation (when well-defined) will always be 0 since $\phi(X)$ and $\psi(Y)$ will always be independent. Other the other hand, if $Y = \phi(X)$ then $\text{Corr}(\phi(X), Y) = 1$ even if $\text{Corr}(X, Y) = 0$. This suggests that we can define the maximal correlation between X and Y to be

$$\text{max-Corr}(X, Y) = \sup_{\phi, \psi} \text{Corr}(\phi(X), \psi(Y))$$

where the supremum is taken over all functions ϕ and ψ with $\text{Var}(\phi(X)) = \text{Var}(\psi(Y)) = 1$. The condition that $\text{Var}(\phi(X)) = \text{Var}(\psi(Y)) = 1$ is needed to rule out constant transformations ϕ and ψ. Clearly, $\text{max-Corr}(X, Y) \geq 0$ with $\text{max-Corr}(X, Y) = 0$ if, and only if, X and Y are independent. (Of course, the functions ϕ and ψ maximizing the correlation are not unique since $\text{Corr}(a\phi(X) + b, c\psi(Y) + d)$ is the same as $\text{Corr}(\phi(X), \psi(Y))$ if a

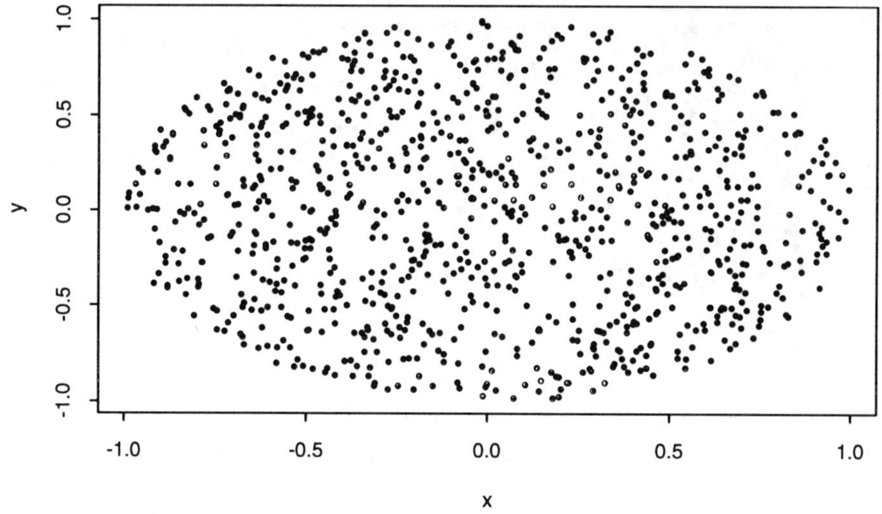

Figure 2.1 *1000 independent pairs of random variables uniformly distributed on the unit circle.*

and c have the same sign.) Unfortunately, the maximal correlation is typically not easy to compute. The following example is an exception.

EXAMPLE 2.12: Suppose that X and Y are random variables with joint density function

$$f(x, y) = \frac{1}{\pi} \quad \text{if } x^2 + y^2 \leq 1.$$

Thus (X, Y) have a Uniform distribution over the region $\{(x, y) : x^2 + y^2 \leq 1\}$; it can be shown that $\text{Cov}(X, Y) = \text{Corr}(X, Y) = 0$. Buja (1990) shows that $\text{Cov}(\phi(X), \psi(Y))$ is maximized when

$$\phi(x) = x^2 \quad \text{and} \quad \psi(y) = -y^2.$$

The maximal correlation between X and Y is $1/3$. Figure 2.1 is a plot of 1000 independent pairs of observations from the density $f(x, y)$; Figure 2.2 shows the observations transformed to give the maximal correlation. ◇

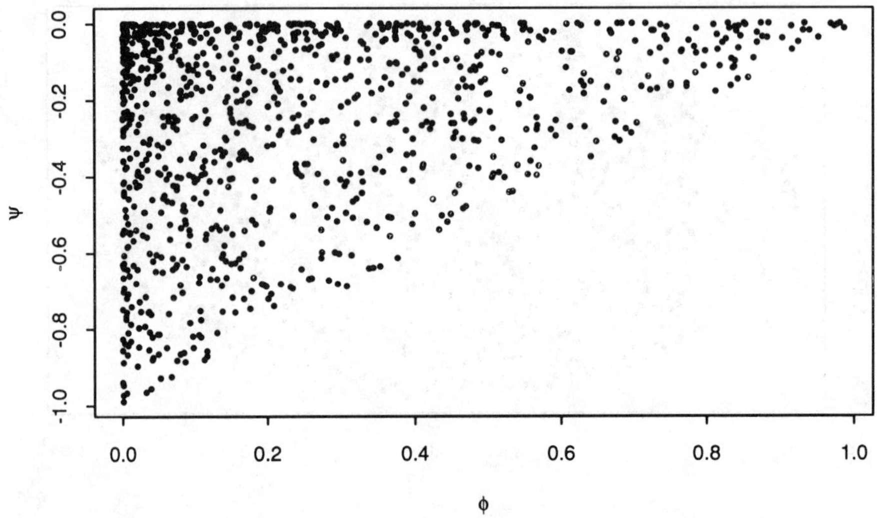

Figure 2.2 *Data in Figure 2.1 transformed to have maximal correlation.*

2.3 Conditional distributions and expected values

We are often interested in the probability distribution of a random variable (or random variables) given knowledge of some event A.

If the conditioning event A has positive probability then we can define conditional distributions, conditional density functions (marginal and joint) and conditional frequency functions using the definition of conditional probability, for example,

$$P(X_1 \leq x_1, \cdots, X_k \leq x_k | A) = \frac{P(X_1 \leq x_1, \cdots, X_k \leq x_k, A)}{P(A)}.$$

In the case of discrete random variables, it is straightforward to define the conditional frequency function of (say) $X_1, \cdots, X_j$ given the event $X_{j+1} = x_{j+1}, \cdots, X_k = x_k$ as

$$\begin{aligned} & f(x_1, \cdots, x_j | x_{j+1}, \cdots, x_k) \\ &= P(X_1 = x_1, \cdots, X_j = x_j | X_{j+1} = x_{j+1}, \cdots, X_k = x_k) \\ &= \frac{P(X_1 = x_1, \cdots, X_j = x_j, X_{j+1} = x_{j+1}, \cdots, X_k = x_k)}{P(X_{j+1} = x_{j+1}, \cdots, X_k = x_k)}, \end{aligned}$$

which is simply the joint frequency function of $X_1, \cdots, X_k$ divided by the joint frequency function of $X_{j+1}, \cdots, X_k$.

In many problems, it is in fact the conditional distributions that are most naturally specified and from these, the joint distribution can be obtained. The following example is a non-trivial illustration of this.

EXAMPLE 2.13: Mark/recapture experiments are used to estimate the size of animal populations. Suppose that the size of the population is N (unknown). Initially, m_0 members of the populations are captured and tagged for future identification before being returned to the population. Subsequently, a similar process is repeated k times: m_i members are captured at stage i and we define a random variable X_i to be the number of captured members who were tagged previously; the $m_i - X_i$ non-tagged members are tagged and all m_i members are returned to the population.

If we assume that the population size does not change over the course of the experiment then it is possible to derive the joint distribution of $(X_1, \cdots, X_k)$. Given our assumptions, it is reasonable to assume that X_1 has a Hypergeometric distribution; that is,

$$P(X_1 = x_1) = \binom{m_0}{x_1}\binom{N - m_0}{m_1 - x_1}\Big/\binom{N}{m_1}.$$

Similarly, given $X_1 = x_1$, X_2 has a Hypergeometric distribution

$$P(X_2 = x_2 | X_1 = x_1) = \binom{n_1}{x_2}\binom{N - n_1}{n_1 - x_2}\Big/\binom{N}{n_1}$$

where $n_1 = m_0 + (m_1 - x_1)$ is the number of tagged members of the population prior to the second sampling stage. Similarly, we can find the conditional distribution of X_{j+1} given $X_1 = x_1, \cdots, X_j = x_j$. Setting $n_j = m_0 + (m_1 - x_1) + \cdots + (m_j - x_j)$, we have

$$P(X_{j+1} = x_{j+1} | X_1 = x_1, \cdots, X_j = x_j)$$
$$= \binom{n_j}{x_{j+1}}\binom{N - n_j}{m_{j+1} - x_{j+1}}\Big/\binom{N}{m_{j+1}}.$$

The joint frequency function of $(X_1, \cdots, X_k)$ is now obtained by multiplying the respective conditional frequency functions:

$$P(X_1 = x_1, X_2 = x_2) = P(X_1 = x_1)P(X_2 = x_2 | X_1 = x_1)$$

and for $k \geq 3$,

$$\begin{aligned} & P(X_1 = x_1, \cdots, X_k = x_k) \\ &= P(X_1 = x_1, \cdots, X_{k-1} = x_{k-1}) \\ &\quad \times P(X_k = x_k | X_1 = x_1, \cdots, X_{k-1} = x_{k-1}). \end{aligned}$$

From a statistical point of view, the goal in this problem is to use the information in $(X_1, \cdots, X_k)$ to estimate the population size N. ◇

Defining conditional distributions when the conditioning event has probability 0 is much more difficult but nonetheless an important problem. For example, if we have two continuous random variables X and Y, we might be interested in the conditional distribution of Y given $X = x$ where (since X is continuous) $P(X = x) = 0$.

DEFINITION. Suppose that $(X_1, \cdots, X_k)$ has the joint density function $g(x_1, \cdots, x_k)$. Then the conditional density function of $X_1, \cdots, X_j$ given $X_{j+1} = x_{j+1}, \cdots, X_k = x_k$ is defined to be

$$f(x_1, \cdots, x_j | x_{j+1}, \cdots, x_k) = \frac{g(x_1, \cdots, x_j, x_{j+1}, \cdots, x_k)}{h(x_{j+1}, \cdots, x_k)}$$

provided that $h(x_{j+1}, \cdots, x_k)$, the joint density of $X_{j+1}, \cdots, X_k$, is strictly positive.

This conditional density function (viewed as a function of x_1, $\cdots, x_j$ for fixed $x_{j+1}, \cdots, x_k$) has the same properties as any other density function; we can use this conditional density function to evaluate conditional probabilities (given $X_{j+1} = x_{j+1}, \cdots, X_k = x_k$) by integration as well as evaluate conditional expected values.

DEFINITION. Given an event A with $P(A) > 0$ and a random variable X with $E[|X|] < \infty$, we define

$$E(X|A) = \int_0^\infty P(X > x|A)\,dx - \int_{-\infty}^0 P(X < x|A)\,dx$$

to be the conditional expected value of X given A.

The assumption that $E[|X|] < \infty$ is not always necessary to ensure that $E(X|A)$ is well-defined; for example, the condition A may imply that X is bounded on A in which case the integral defining $E(X|A)$ would be well-defined.

The following result extends Proposition 1.4 (law of total probability) to conditional expectations.

THEOREM 2.8 *Suppose that $A_1, A_2, \cdots$ are disjoint events with $P(A_k) > 0$ for all k and $\bigcup_{k=1}^{\infty} A_k = \Omega$. Then if $E[|X|] < \infty$,*

$$E(X) = \sum_{k=1}^{\infty} E(X|A_k)P(A_k).$$

Proof. Assume $X \geq 0$ with probability 1. (Otherwise, split X into its positive and negative parts.) Then

$$E(X|A_k) = \int_0^{\infty} P(X > x|A_k)\, dx$$

and so

$$\begin{aligned}\sum_{k=1}^{\infty} E(X|A_k)P(A_k) &= \sum_{k=1}^{\infty} \int_0^{\infty} P(A_k)P(X > x|A_k)\, dx \\ &= \int_0^{\infty} \sum_{k=1}^{\infty} P(A_k)P(X > x|A_k)\, dx \\ &= \int_0^{\infty} P(X > x)\, dx \\ &= E(X).\end{aligned}$$

We can interchange the order of summation and integration because $P(A_k)P(X > x|A_k) \geq 0$. □

Theorem 2.8 also holds if some of the A_k's have $P(A_k) = 0$ provided we take care of the fact that $E(X|A_k)$ is not necessarily well-defined. This can be done by assigning $E(X|A_k)$ an arbitrary (but finite) value, which is then annihilated by multiplying by $P(A_k) = 0$.

There is an interesting interpretation of Theorem 2.8 that will be useful in a more general setting. Given $E(X|A_k)$ $(k = 1, 2, \cdots)$, define a random variable Y such that

$$Y(\omega) = E(X|A_k) \quad \text{if } \omega \in A_k \ (k = 1, 2, \cdots).$$

Y is now a discrete random variable whose expected value is

$$\begin{aligned}E(Y) &= \sum_y y\, P(Y = y) \\ &= \sum_{k=1}^{\infty} E(X|A_k)P(A_k) \\ &= E(X)\end{aligned}$$

by Theorem 2.8 (provided that $E[|X|] < \infty$).

More generally, given a continuous random vector $\boldsymbol{X}$, we would like to define $E(Y|\boldsymbol{X} = \boldsymbol{x})$ for a random variable Y with $E[|Y|] < \infty$. Since the event $[\boldsymbol{X} = \boldsymbol{x}]$ has probability 0, this is somewhat delicate from a technical point of view, although if Y has a conditional density function given $\boldsymbol{X} = \boldsymbol{x}$, $f(y|\boldsymbol{x})$ then we can define

$$E(Y|\boldsymbol{X} = \boldsymbol{x}) = \int_{-\infty}^{\infty} y\, f(y|\boldsymbol{x})\, dy.$$

We can obtain similar expressions for $E[g(\boldsymbol{Y})|\boldsymbol{X} = \boldsymbol{x}]$ provided that we can define the conditional distribution of $\boldsymbol{Y}$ given $\boldsymbol{X} = \boldsymbol{x}$ in a satisfactory way.

In general, $E[g(\boldsymbol{Y})|\boldsymbol{X} = \boldsymbol{x}]$ is a function of $\boldsymbol{x}$. Moreover, if $h(\boldsymbol{x}) = E[g(\boldsymbol{Y})|\boldsymbol{X} = \boldsymbol{x}]$ then

$$E[h(\boldsymbol{X})] = E[E[E(g(\boldsymbol{Y})|\boldsymbol{X}]] = E[g(\boldsymbol{Y})]$$

as we had in Theorem 2.8. The following result records some of the key properties of conditional expected values.

PROPOSITION 2.9 *Suppose that $\boldsymbol{X}$ and $\boldsymbol{Y}$ are random vectors. Then*
(a) if $E[|g_1(\boldsymbol{Y})|]$ and $E[|g_2(\boldsymbol{Y})|]$ are finite,

$$\begin{aligned} &E[a\, g_1(\boldsymbol{Y}) + b\, g_2(\boldsymbol{Y})|\boldsymbol{X} = \boldsymbol{x}] \\ &= \; a\, E[g_1(\boldsymbol{Y})|\boldsymbol{X} = \boldsymbol{x}] + b\, E[g_2(\boldsymbol{Y})|\boldsymbol{X} = \boldsymbol{x}] \end{aligned}$$

(b) $E[g_1(\boldsymbol{X})g_2(\boldsymbol{Y})|\boldsymbol{X} = \boldsymbol{x}] = g_1(\boldsymbol{x})E[g_2(\boldsymbol{Y})|\boldsymbol{X} = \boldsymbol{x}]$ if $E[|g_2(\boldsymbol{Y})|]$ is finite;
(c) If $h(x) = E[g(\boldsymbol{Y})|\boldsymbol{X} = x]$ then $E[h(\boldsymbol{X})] = E[g(\boldsymbol{Y})]$ if $E[|g(\boldsymbol{Y})|]$ is finite.

A rigorous proof of Proposition 2.9 follows from a more technically rigorous definition of conditional expectation; see, for example, Billingsley (1995). In special cases, for example when we assume a conditional density, the proof is straightforward.

The following result provides a useful decomposition for the variance of a random variable.

THEOREM 2.10 *Suppose that Y is a random variable with finite variance. Then*

$$\mathit{Var}(Y) = E[\mathit{Var}(Y|\boldsymbol{X})] + \mathit{Var}[E(Y|\boldsymbol{X})]$$

where $\mathit{Var}(Y|\boldsymbol{X}) = E[(Y - E(Y|\boldsymbol{X}))^2|\boldsymbol{X}]$.

Proof. Define $h(\boldsymbol{X}) = E(Y|\boldsymbol{X})$ and $\mu = E(Y)$. Then

$$\begin{aligned}
\text{Var}(Y) &= E[(Y - h(\boldsymbol{X}) + h(\boldsymbol{X}) - \mu)^2] \\
&= E\left[E[(Y - h(\boldsymbol{X}) + h(\boldsymbol{X}) - \mu)^2|\boldsymbol{X}]\right] \\
&= E\left[E[(Y - h(\boldsymbol{X}))^2|\boldsymbol{X}]\right] + E\left[E[(h(\boldsymbol{X}) - \mu)^2|\boldsymbol{X}]\right] \\
&\quad +2E\left[E[(Y - h(\boldsymbol{X}))(h(\boldsymbol{X}) - \mu)|\boldsymbol{X}]\right].
\end{aligned}$$

Now

$$\begin{aligned}
E\left[E[(Y - h(\boldsymbol{X}))^2|\boldsymbol{X}]\right] &= E[\text{Var}(Y|\boldsymbol{X})] \\
\text{and} \quad E\left[E[(h(\boldsymbol{X}) - \mu)^2|\boldsymbol{X}]\right] &= E[(h(\boldsymbol{X}) - \mu)^2] \\
&= \text{Var}[E(Y|\boldsymbol{X})].
\end{aligned}$$

Finally, for the "cross-product" term, we have

$$\begin{aligned}
&E[(Y - h(\boldsymbol{X}))(h(\boldsymbol{X}) - \mu)|\boldsymbol{X} = \boldsymbol{x}] \\
&\quad = (h(\boldsymbol{x}) - \mu)E[Y - h(\boldsymbol{X})|\boldsymbol{X} = \boldsymbol{x}] \\
&\quad = (h(\boldsymbol{x}) - \mu)[h(\boldsymbol{x}) - h(\boldsymbol{x})] \\
&\quad = 0.
\end{aligned}$$

Thus

$$E\left[E[(Y - h(\boldsymbol{X}))(h(\boldsymbol{X}) - \mu)|\boldsymbol{X}]\right] = 0,$$

which completes the proof. □

EXAMPLE 2.14: Suppose that $X_1, X_2, \cdots$ are i.i.d. random variables with mean μ and variance σ^2; let N be a Poisson random variable (with mean λ) that is independent of the X_i's. Define the random variable

$$S = \sum_{i=1}^{N} X_i$$

where $S = 0$ if $N = 0$. We would like to determine the mean and variance of S.

First of all, note that $E(S) = E[E(S|N)]$ and $E(S|N = n) = n\mu$ since N is independent of the X_i's. Thus

$$E(S) = \mu E(N) = \lambda\mu.$$

Likewise, we have $\text{Var}(S) = \text{Var}[E(S|N)] + E[\text{Var}(S|N)]$. We already know that $E(S|N) = \mu N$ and since the X_i's are i.i.d. and independent of N, we have $\text{Var}(S|N = n) = n\sigma^2$. Thus

$$\text{Var}(S) = \mu^2\text{Var}(N) + \sigma^2 E(N) = \lambda(\mu^2 + \sigma^2).$$

Note that $\mu^2 + \sigma^2 = E(X_i^2)$.

The random variable S is said to have a compound Poisson distribution. Such distributions are often used in actuarial science to model the monetary value of claims against various types of insurance policies in a given period of time; the motivation here is that N represents the number of claims with the X_i's representing the monetary value of each claim. ◇

2.4 Distribution theory for Normal samples

In Example 1.15 we introduced the Normal distribution. Recall that a random variable X has a Normal distribution with mean μ and variance σ^2 if its density is

$$f(x) = \frac{1}{\sigma\sqrt{2\pi}} \exp\left(-\frac{(x-\mu)^2}{2\sigma^2}\right).$$

The Normal distribution is very important in probability and statistics as it is frequently used in statistical models and is often used to approximate the distribution of certain random variables.

A number of distributions arise naturally in connection with samples from a Normal distribution. These distributions include the χ^2 (chi-square) distribution, Student's t distribution and the F distribution.

The Multivariate Normal Distribution

Suppose that $X_1, \cdots, X_p$ are i.i.d. Normal random variables with mean 0 and variance 1. Then the joint density function of $\boldsymbol{X} = (X_1, \cdots, X_p)^T$ is

$$\begin{aligned} f_X(\boldsymbol{x}) &= \frac{1}{(2\pi)^{p/2}} \exp\left(-\frac{1}{2}(x_1^2 + \cdots + x_p^2)\right) \\ &= \frac{1}{(2\pi)^{p/2}} \exp\left(-\frac{1}{2}\boldsymbol{x}^T\boldsymbol{x}\right). \end{aligned}$$

The random vector $\boldsymbol{X}$ has a standard multivariate (or p-variate) Normal distribution.

DEFINITION. Let A be a $p \times p$ matrix and $\boldsymbol{\mu} = (\mu_1, \cdots, \mu_p)^T$ a vector of length p. Given a standard multivariate Normal random vector $\boldsymbol{X}$, define

$$\boldsymbol{Y} = \boldsymbol{\mu} + A\boldsymbol{X}.$$

We say that $\boldsymbol{Y}$ has a multivariate Normal distribution with mean vector μ and variance-covariance matrix $C = AA^T$. (We will abbreviate this $\boldsymbol{Y} \sim N_p(\boldsymbol{\mu}, C)$.)

Note that $C = AA^T$ is a symmetric, nonnegative definite matrix (that is, $C^T = C$ and $\boldsymbol{v}^T C \boldsymbol{v} \geq 0$ for all vectors $\boldsymbol{v}$).

If A is an invertible matrix then the joint density of $\boldsymbol{Y}$ exists and is given by

$$f_Y(\boldsymbol{y}) = \frac{1}{(2\pi)^{p/2} |\det(C)|^{1/2}} \exp\left(-\frac{1}{2}(\boldsymbol{y} - \boldsymbol{\mu})^T C^{-1} (\boldsymbol{y} - \boldsymbol{\mu})\right).$$

Note that C^{-1} exists since A^{-1} exists. On the other hand, if A is not invertible then the joint density of $\boldsymbol{Y}$ does not exist; Y is defined on a hyperplane in R^p of dimension $r = \text{rank}(C) < p$.

We will now state some basic properties of the multivariate Normal distribution. Assume that $\boldsymbol{Y} \sim N_p(\boldsymbol{\mu}, C)$ and let C_{ij} be the element of C in the i-th row and j-th column of C. Then

1. if B is an $r \times p$ matrix then
$$\boldsymbol{BY} \sim N_r(B\boldsymbol{\mu}, BCB^T);$$
2. any subcollection $(Y_{i_1}, \cdots, Y_{i_k})^T$ of $\boldsymbol{Y}$ has a multivariate Normal distribution;
3. $Y_i \sim N(\mu_i, \sigma_i^2)$ where $\sigma_i^2 = C_{ii}$;
4. $\text{Cov}(Y_i, Y_j) = C_{ij}$.

Properties 2 and 3 follow from property 1 by choosing the matrix B appropriately. Property 4 follows by writing $\boldsymbol{Y} = \boldsymbol{\mu} + A\boldsymbol{X}$ where $\boldsymbol{X} \sim N_p(\boldsymbol{0}, I)$ and $C = AA^T$; if the i-th and j-th rows of A are $\boldsymbol{a}_i^T$ and $\boldsymbol{a}_j^T$ then $Y_i = \mu_i + \boldsymbol{a}_i^T \boldsymbol{X}$ and $Y_j = \mu_j + \boldsymbol{a}_j^T \boldsymbol{X}$, and so it is easy to verify that

$$\text{Cov}(Y_i, Y_j) = \boldsymbol{a}_i^T \boldsymbol{a}_j = C_{ij}.$$

The χ^2 distribution and orthogonal transformations

Suppose that $X_1, \cdots, X_p$ are i.i.d. $N(0,1)$ random variables so that $\boldsymbol{X} = (X_1, \cdots, X_p)^T \sim N_p(0, I)$ where I is the identity matrix. In some statistical applications, we are interested in the squared "length" of $\boldsymbol{X}$, $\|\boldsymbol{X}\|^2 = \boldsymbol{X}^T \boldsymbol{X}$.

DEFINITION. Let $\boldsymbol{X} \sim N_p(0, I)$ and define $V = \|\boldsymbol{X}\|^2$. The random variable V is said to have a χ^2 (chi-square) distribution with p degrees of freedom. ($V \sim \chi^2(p)$.)

The density of a χ^2 distribution is quite easy to determine. We start with the case where $p = 1$; in this case, $V = X_1^2$ and so

$$\begin{aligned} P(V \leq x) &= P(-\sqrt{x} \leq X_1 \leq \sqrt{x}) \\ &= \Phi(\sqrt{x}) - \Phi(-\sqrt{x}) \end{aligned}$$

where Φ is the $N(0,1)$ distribution function. Differentiating, we get the density

$$f_V(x) = \frac{x^{-1/2}}{\sqrt{2\pi}} \exp(-x/2) \quad (x > 0),$$

which is simply the density of a Gamma(1/2, 1/2) distribution; thus a $\chi^2(1)$ distribution is simply a Gamma(1/2, 1/2) distribution. The general case now is simple since $V = X_1^2 + \cdots + X_p^2$, a sum of p independent $\chi^2(1)$ random variables. It follows (from Example 2.5) that the density of $V \sim \chi^2(p)$ is

$$f_V(x) = \frac{x^{p/2-1}}{2^{p/2}\Gamma(p/2)} \exp(-x/2) \quad (x > 0).$$

Orthogonal transformations are effectively transformations that preserve the length of vectors; for this reason, they turn out to be a useful tool in connection with the multivariate Normal distribution.

To define an orthogonal matrix, we start by defining vectors (of length p) $\boldsymbol{a}_1, \boldsymbol{a}_2, \cdots, \boldsymbol{a}_p$ such that

$$\boldsymbol{a}_k^T \boldsymbol{a}_k = 1 \quad \text{for } k = 1, \cdots, p \quad \text{and} \quad \boldsymbol{a}_j^T \boldsymbol{a}_k = 0 \quad \text{for } j \neq k$$

The vectors $\boldsymbol{a}_1, \cdots, \boldsymbol{a}_p$ are said to be orthogonal vectors and, in fact, form an orthonormal basis for R^p; that is, any vector $\boldsymbol{v} \in R^p$, we have

$$\boldsymbol{v} = \sum_{i=1}^{p} c_i \boldsymbol{a}_i.$$

Now define a $p \times p$ matrix O such that the k-th row of O is $\boldsymbol{a}_k^T$:

$$O = \begin{pmatrix} \boldsymbol{a}_1^T \\ \vdots \\ \boldsymbol{a}_p^T \end{pmatrix}$$

Note that

$$OO^T = \begin{pmatrix} \boldsymbol{a}_1^T \\ \vdots \\ \boldsymbol{a}_p^T \end{pmatrix} (\boldsymbol{a}_1 \cdots \boldsymbol{a}_p) = I$$

and so $O^{-1} = O^T$; thus we also have $O^T O = I$. The matrix O is called an orthogonal (or orthonormal) matrix.

EXAMPLE 2.15: If $p = 2$, orthogonal matrices rotate vectors by some angle θ. To see this, let

$$O = \begin{pmatrix} \cos(\theta) & -\sin(\theta) \\ \sin(\theta) & \cos(\theta) \end{pmatrix}$$

for some θ between 0 and 2π; it is easy to verify that O is an orthogonal matrix and, in fact, an 2×2 orthogonal matrix can be written in this form. Now let $\boldsymbol{v}$ be any vector and write

$$\boldsymbol{v} = \begin{pmatrix} v_1 \\ v_2 \end{pmatrix} = \begin{pmatrix} r\cos(\phi) \\ r\sin(\phi) \end{pmatrix}$$

where $r^2 = \boldsymbol{v}^T\boldsymbol{v} = v_1^2 + v_2^2$ and ϕ is the angle between $\boldsymbol{v}$ and the vector $(1,0)^T$. Then

$$O\boldsymbol{v} = \begin{pmatrix} r\cos(\theta)\cos(\phi) - r\sin(\theta)\sin(\phi) \\ r\sin(\theta)\cos(\phi) + \cos(\theta)\sin(\phi) \end{pmatrix} = \begin{pmatrix} r\cos(\theta+\phi) \\ r\sin(\theta+\phi) \end{pmatrix}.$$

Thus the orthogonal matrix O rotates the vector $\boldsymbol{v}$ by an angle θ. ◇

Now take $\boldsymbol{X} \sim N_p(\boldsymbol{0}, I)$ and define define $\boldsymbol{Y} = O\boldsymbol{X}$ where O is an orthogonal matrix. It follows from the properties of the multivariate Normal distribution that $\boldsymbol{Y} \sim N_p(0, I)$ since $OO^T = I$. Moreover,

$$\begin{aligned} \sum_{i=1}^p Y_i^2 &= \boldsymbol{Y}^T\boldsymbol{Y} \\ &= \boldsymbol{X}^T O^T O \boldsymbol{X} \\ &= \boldsymbol{X}^T\boldsymbol{X} \\ &= \sum_{i=1}^p X_i^2 \end{aligned}$$

and both $\sum_{i=1}^p X_i^2$ and $\sum_{i=1}^p Y_i^2$ will have χ^2 distributions with p degrees of freedom.

Consider the following application of this theory. Suppose that $X_1, \cdots, X_n$ are i.i.d. Normal random variables with mean μ and variance σ^2 and define

$$\bar{X} = \frac{1}{n}\sum_{i=1}^n X_i$$

and

$$S^2 = \frac{1}{n-1}\sum_{i=1}^{n}(X_i - \bar{X})^2;$$

$\bar{X}$ and S^2 are called the sample mean and sample variance respectively. We know already that $\bar{X} \sim N(\mu, \sigma^2/n)$. The following results indicates that $\bar{X}$ is independent of S^2 and that the distribution of S^2 is related to a χ^2 with $n-1$ degrees of freedom.

PROPOSITION 2.11 $(n-1)S^2/\sigma^2 \sim \chi^2(n-1)$ *and is independent of* $\bar{X} \sim N(\mu, \sigma^2)$.

Proof. First note that

$$\begin{aligned}\frac{(n-1)}{\sigma^2}S^2 &= \frac{1}{\sigma^2}\sum_{i=1}^{n}(X_i - \bar{X})^2 \\ &= \sum_{i=1}^{n}\left[(X_i - \mu)/\sigma - (\bar{X} - \mu)/\sigma\right]^2\end{aligned}$$

and so we can assume (without loss of generality) that $\mu = 0$ and $\sigma^2 = 1$. Define an orthogonal matrix O whose first row consists of $n^{-1/2}$ repeated n times; the remaining rows can be determined by some orthogonalization procedure (such as the Gram-Schmidt procedure) but do not need to be specified here. Now let $\boldsymbol{Y} = O\boldsymbol{X}$ where $\boldsymbol{X} = (X_1, \cdots, X_n)^T$ and note that

$$Y_1 = \frac{1}{\sqrt{n}}\sum_{i=1}^{n} X_i$$

is independent of the remaining elements of $\boldsymbol{Y}$, $Y_2, \cdots, Y_n$. We now have

$$\begin{aligned}\sum_{i=1}^{n}(X_i - \bar{X})^2 &= \sum_{i=1}^{n} X_i^2 - \left(\frac{1}{\sqrt{n}}\sum_{i=1}^{n} X_i\right)^2 \\ &= \sum_{i=1}^{n} Y_i^2 - Y_1^2 \\ &= \sum_{i=2}^{n} Y_i^2 \\ &\sim \chi^2(n-1).\end{aligned}$$

Moreover, $\sum_{i=1}^{n}(X_i - \bar{X})^2$ depends only on $Y_2, \cdots, Y_n$ and hence is independent of $Y_1 = \sqrt{n}\bar{X}$. Thus in general we have $(n-1)S^2/\sigma^2$ independent of $\bar{X}$ (and S^2 is independent of $\bar{X}$). □

The t and F distributions

DEFINITION. Let $Z \sim N(0,1)$ and $V \sim \chi^2(n)$ be independent random variables. Define $T = Z/\sqrt{V/n}$; the random variable T is said to have Student's t distribution with n degrees of freedom. ($T \sim \mathcal{T}(n)$.)

DEFINITION. Let $V \sim \chi^2(n)$ and $W \sim \chi^2(m)$ be independent random variables. Define $F = (V/n)/(W/m)$; the random variable F is said to have an F distribution with n and m degrees of freedom. ($F \sim \mathcal{F}(n,m)$.)

Alternatively, we could define the t and F distributions via their density functions. However, the representation of t and F random variables as functions of Normal and χ^2 random variables turns out to be convenient in many situations (see Example 2.17 below). Nonetheless, it is fairly easy to find the densities of the t and F distributions using the representations as well as some the techniques developed earlier for finding densities of functions of random variables.

EXAMPLE 2.16: Suppose that $Z \sim N(0,1)$ and $V \sim \chi^2(n)$ are independent random variables, and define $T = Z/\sqrt{V/n}$. To determine the density of T, we will introduce another random variable S, determine the joint density of (S,T) and then integrate out this joint density to determine the marginal density of T. The choice of S is somewhat arbitrary; we will take $S = V$ so that $(S,T) = \boldsymbol{h}(V,Z)$ for some function $\boldsymbol{h}$ that is one-to-one over the range of the random variables V and Z. The inverse of $\boldsymbol{h}$ is $\boldsymbol{h}^{-1}(s,t) = (s, t\sqrt{s/n})$ and the Jacobian of $\boldsymbol{h}^{-1}$ is $\sqrt{s/n}$. Thus the joint density of (S,T) is (after substituting the appropriate terms into the joint density of (V,Z))

$$g(s,t) = \frac{s^{(n-1)/2}}{2^{(n+1)/2}\sqrt{n\pi}\,\Gamma(n/2)} \exp\left[-\frac{s}{2}(1+t^2/n)\right]$$

for $s > 0$ and $-\infty < t < \infty$. Integrating out over s, we get

$$f_T(t) = \int_0^\infty g(s,t)\,ds$$

$$= \frac{\Gamma((n+1)/2)}{\sqrt{n\pi}\Gamma(n/2)} \left(1 + \frac{t^2}{n}\right)^{-(n+1)/2}$$

Note that when $n = 1$, the t distribution is simply the Cauchy distribution given in Example 1.26. ◇

We can go through a similar procedure to determine the density of the F distribution but that will be left as an exercise. The density of an F distribution with n and m degrees of freedom is

$$f(x) = \frac{n^{n/2}\Gamma((n+m)/2)}{m^{n/2}\Gamma(m/2)\Gamma(n/2)} x^{(n-2)/2} (1 + ns/m)^{-(n+m)/2}$$

for $x > 0$.

EXAMPLE 2.17: Suppose that $X_1, \cdots, X_n$ are i.i.d. Normal random variables with mean μ and variance σ^2. Define the sample mean and variance of the X_i's:

$$\bar{X} = \frac{1}{n}\sum_{i=1}^{n} X_i$$

$$S^2 = \frac{1}{n-1}\sum_{i=1}^{n}(X_i - \bar{X})^2.$$

Now define $T = \sqrt{n}(\bar{X} - \mu)/S$; it is well-known that $T \sim \mathcal{T}(n-1)$. To see this, note that we can rewrite T as

$$T = \frac{\sqrt{n}(\bar{X} - \mu)/\sigma}{\sqrt{S^2/\sigma^2}}$$

where $\sqrt{n}(\bar{X} - \mu)/\sigma \sim N(0, 1)$ and independent of $(n-1)S^2/\sigma^2 \sim \chi^2(n-1)$ by Proposition 2.11. Thus using the definition of Student's t distribution, $T \sim \mathcal{T}(n - 1)$. ◇

Projection matrices

Orthogonal matrices effectively "rotate" vectors without changing their length. Another class of matrices that has important applications to probability and statistics is projection matrices.

DEFINITION. Let H be a symmetric $p \times p$ matrix with $H^2 = H$. Then H is called a projection matrix.

As its name suggests, a projection matrix "projects" onto a subspace S of R^p. More precisely, if $\boldsymbol{v} \in R^p$ is a vector then $H\boldsymbol{v}$ lies in the subspace S. Moreover, $H\boldsymbol{v}$ is the "closest" vector in S to $\boldsymbol{v}$ in the sense that $H\boldsymbol{v}$ minimizes $\|\boldsymbol{v} - \boldsymbol{u}\|^2$ over all $\boldsymbol{u} \in S$.

EXAMPLE 2.18: Suppose that $\boldsymbol{x}_1, \cdots, \boldsymbol{x}_r$ be $r \le p$ linearly independent vectors in R^p. (Linear independence means that $c_1\boldsymbol{x}_1 + \cdots + c_r\boldsymbol{x}_r = 0$ implies $c_1 = \cdots = c_r = 0$.) Define the matrices

$$B = (\boldsymbol{x}_1 \cdots \boldsymbol{x}_r)$$

and

$$H = B(B^T B)^{-1} B^T.$$

Then H is a projection matrix onto the space spanned by the vectors $\boldsymbol{x}_1, \cdots, \boldsymbol{x}_r$. To see this, note that

$$H^T = \left(B(B^T B)^{-1} B^T\right)^T = B(B^T B)^{-1} B^T = H$$

and

$$H^2 = B(B^T B)^{-1} B^T B (B^T B)^{-1} B^T = B(B^T B)^{-1} B^T = H.$$

Moreover, for any vector $\boldsymbol{v}$, $H\boldsymbol{v}$ clearly lies in the space spanned by $\boldsymbol{x}_1, \cdots, \boldsymbol{x}_r$ since we can write

$$H\boldsymbol{v} = B(B^T B)^{-1} B^T \boldsymbol{v} = B\boldsymbol{v}^\star$$

and $B\boldsymbol{v}^\star$ lies in the space spanned by $\boldsymbol{x}_1, \cdots, \boldsymbol{x}_r$. ◇

The projection matrix in Example 2.18 turns out to have special significance in statistics, for example in linear regression analysis (see Chapter 8).

If H is a projection matrix onto the subspace S and $\boldsymbol{v} \in S$ then $H\boldsymbol{v} = \boldsymbol{v}$ which implies that 1 is an eigenvalue of H. The following result shows that 0 and 1 are the only possible eigenvalues of H.

PROPOSITION 2.12 *Suppose that H is a projection matrix. Then 0 and 1 are the only possible eigenvalues of H.*

Proof. Let λ be an eigenvalue of H and $\boldsymbol{v}$ be a corresponding eigenvector; thus $H\boldsymbol{v} = \lambda\boldsymbol{v}$. Multiplying both sides by H, we get

$$H^2\boldsymbol{v} = H\lambda\boldsymbol{v} = \lambda H\boldsymbol{v} = \lambda^2\boldsymbol{v}.$$

However, $H^2 = H$ and so λ^2 is an eigenvalue of H with eigenvector $\boldsymbol{v}$; thus $\lambda^2 = \lambda$ and so $\lambda = 0$ or 1. □

Since H is symmetric , we can find eigenvectors $\boldsymbol{a}_1, \cdots, \boldsymbol{a}_p$ of H, which form an orthonormal basis for R^p. Since 0 and 1 are the only possible eigenvalues of H, we have

$$H\boldsymbol{a}_k = \begin{cases} \boldsymbol{a}_k & k = 1, \cdots, r \\ 0 & k = r+1, \cdots, p \end{cases}$$

where $\boldsymbol{a}_1, \cdots, \boldsymbol{a}_r$ are the eigenvectors of H with eigenvalues equal to 1. (r is the rank of H.)

Now take any vector $\boldsymbol{v} \in R^p$. Since the eigenvectors of H, $\boldsymbol{a}_1, \cdots, \boldsymbol{a}_p$ form an orthonormal basis for R^p, we have

$$\boldsymbol{v} = \sum_{k=1}^{p} c_k \boldsymbol{a}_k \quad \text{where } c_k = \boldsymbol{v}^T \boldsymbol{a}_k.$$

From this it follows that

$$H\boldsymbol{v} = \sum_{k=1}^{p} c_k H\boldsymbol{a}_k = \sum_{k=1}^{r} c_k \boldsymbol{a}_k.$$

The space S onto which the matrix H projects is spanned by the eigenvectors $\boldsymbol{a}_1, \cdots, \boldsymbol{a}_r$. The space spanned by the vectors $\boldsymbol{a}_{r+1}, \cdots, \boldsymbol{a}_p$ is called the orthogonal complement of S. If $S^\perp$ is the orthogonal complement of S, $\boldsymbol{v} \in S$ and $\boldsymbol{v}^\perp \in S^\perp$ then $\boldsymbol{v}^T\boldsymbol{v}^\perp = 0$; that is, $\boldsymbol{v}$ and $\boldsymbol{v}^\perp$ are orthogonal vectors.

PROPOSITION 2.13 *Suppose that H is a projection matrix that projects onto the subspace S. Then $I - H$ is a projection matrix projecting onto the orthogonal complement of S.*

Proof. $I - H$ is a projection matrix since $(I - H)^T = I - H^T = H^T$ and $(I - H)^2 = I - 2H + H^2 = I - H$. To see that $I - H$ projects onto the orthogonal complement of S, write

$$\boldsymbol{v} = \sum_{k=1}^{p} c_k \boldsymbol{a}_k$$

where $\boldsymbol{a}_1, \cdots, \boldsymbol{a}_p$ are the eigenvectors of H. Then

$$(I - H)\boldsymbol{v} = \sum_{k=1}^{p} c_k \boldsymbol{a}_k - \sum_{k=1}^{r} c_k \boldsymbol{a}_k = \sum_{k=r+1}^{p} c_k \boldsymbol{a}_k,$$

which lies in the orthogonal complement of S. □

From this result, we obtain the almost trivial decomposition of a vector $\boldsymbol{v}$ into a sum of orthogonal vectors:

$$\begin{aligned} \boldsymbol{v} &= H\boldsymbol{v} + (I - H)\boldsymbol{v} \\ &= \boldsymbol{u} + \boldsymbol{u}^\perp \end{aligned}$$

where $\boldsymbol{u} \in S$ and $\boldsymbol{u}^{\perp} \in S^{\perp}$.

We will now consider an application of projection matrices using the multivariate Normal distribution. Suppose that $\boldsymbol{X} \sim N_p(0, I)$ and define the random variable

$$V = \boldsymbol{X}^T H \boldsymbol{X}$$

for some projection matrix H. The following result shows that V has a χ^2 distribution.

PROPOSITION 2.14 *V has a χ^2 distribution with $r = rank(H)$ degrees of freedom.*

Proof. First of all, note that

$$\boldsymbol{X}^T H \boldsymbol{X} = \boldsymbol{X}^T H^2 \boldsymbol{X} = (H\boldsymbol{X})^T (H\boldsymbol{X}) = \|H\boldsymbol{X}\|^2$$

and

$$H\boldsymbol{X} = \sum_{k=1}^{r} (\boldsymbol{a}_k^T \boldsymbol{X}) \boldsymbol{a}_k$$

where $\boldsymbol{a}_1, \cdots, \boldsymbol{a}_p$ are the (orthonormal) eigenvectors of H. Now we can define the orthogonal matrix

$$O = \begin{pmatrix} \boldsymbol{a}_1^T \\ \vdots \\ \boldsymbol{a}_p^T \end{pmatrix}$$

and define $\boldsymbol{Y} = O\boldsymbol{X}$; $\boldsymbol{Y} \sim N_p(0, I)$. If $\boldsymbol{Y} = (Y_1, \cdots, Y_p)^T$ then $Y_k = \boldsymbol{a}_k^T \boldsymbol{X}$ where $Y_1, \cdots, Y_p$ are i.i.d. $N(0,1)$ random variables. Thus we have

$$H\boldsymbol{X} = \sum_{k=1}^{r} Y_k \boldsymbol{a}_k$$

and so

$$\begin{aligned} (H\boldsymbol{X})^T (H\boldsymbol{X}) &= \boldsymbol{X}^T H \boldsymbol{X} \\ &= \sum_{k=1}^{r} \sum_{j=1}^{r} Y_j Y_k \boldsymbol{a}_j^T \boldsymbol{a}_k \\ &= \sum_{k=1}^{r} Y_k^2 \\ &\sim \chi^2(r) \end{aligned}$$

since $\boldsymbol{a}_k^T \boldsymbol{a}_k = 1$ and $\boldsymbol{a}_k^T \boldsymbol{a}_j = 0$ for $j \neq k$. $\square$

EXAMPLE 2.19: Suppose that $\boldsymbol{X} \sim N_n(\boldsymbol{0}, I)$. If H is a projection matrix with rank$(H) = r < n$ then (by Proposition 2.14) $\|H\boldsymbol{X}\|^2 \sim \chi^2(r)$ and likewise $\|(I-H)\boldsymbol{X}\|^2 \sim \chi^2(n-r)$; moreover, these two random variables are independent. Thus if we define

$$W = \frac{\|H\boldsymbol{X}\|^2/r}{\|(I-H)\boldsymbol{X}\|^2/(n-r)}$$

then $W \sim \mathcal{F}(r, n-r)$. ◇

2.5 Poisson processes

Suppose we are interested in the arrival patterns of customers entering a grocery store at different times of the day. Let $T_1, T_2, \cdots$ be random variables representing the arrival times of the customers and, for a given time interval A, define

$$N(A) = \sum_{i=1}^{\infty} I(T_i \in A)$$

to be the number of arrivals in the interval A. Clearly, $N(A)$ is a nonnegative integer-valued random variable that gives the number of arrivals in the interval A. If B is another interval that is disjoint of A (that is, $A \cap B = \emptyset$) then clearly

$$N(A \cup B) = N(A) + N(B).$$

We can now specify a model for the "point process" $N(\cdot)$ by specifying the joint distribution of the random vector $(N(A_1), N(A_2), \cdots, N(A_k))$ for any sets $A_1, \cdots, A_k$.

DEFINITION. Let $S \subset R^k$ and suppose that for any $A \subset S$, $N(A)$ is a nonnegative integer-valued random variable. Then $N(\cdot)$ is called a point process (on S) if
(a) $N(\emptyset) = 0$;
(b) $N(A \cup B) = N(A) + N(B)$ for any disjoint sets A and B.

DEFINITION. A point process $N(\cdot)$ defined on $S \subset R^k$ is called a (homogeneous) Poisson process if
(a) For any $A = [a_1, b_1] \times \cdots \times [a_k, b_k] \subset S$, $N(A)$ has a Poisson distribution with mean

$$\lambda \prod_{i=1}^{k} (b_i - a_i)$$

for some $\lambda > 0$;

(b) For any two disjoint sets A and B, $N(A)$ and $N(B)$ are independent random variables.

The parameter λ is called the intensity of the Poisson process.

Although the definition gives the distribution of $N(A)$ only for "rectangular" sets A, it is quite easy to extend this to a general set A. Any set A can be expressed as a countable union of disjoint rectangles $B_1, B_2, \cdots$. Since $N(B_1), N(B_2), \cdots$ are independent Poisson random variables, we have that

$$N(A) = \sum_{i=1}^{\infty} N(B_i)$$

is Poisson. The mean of $N(A)$ is $\lambda \operatorname{vol}(A)$ where $\operatorname{vol}(A)$ is the "volume" of the set A.

EXAMPLE 2.20: A Poisson process on the positive real line can be constructed as follows. Let $X_1, X_2, \cdots$ be independent Exponential random variables with parameter λ and define

$$T_k = \sum_{i=1}^{k} X_i;$$

note that T_k has a Gamma distribution with shape parameter k and scale parameter λ. Then we can define a point process $N(\cdot)$ such that

$$N(A) = \sum_{k=1}^{\infty} I(T_k \in A).$$

To see that $N(\cdot)$ is a Poisson process, note that $N(A)$ and $N(B)$ will be independent for disjoint A and B because of the independence of the X_i's and the "memorylessness" property of the Exponential distribution. Moreover, it follows from the memorylessness property that if $A = [s, s+t]$, the distribution of $N(A)$ is the same for all $s \geq 0$. Thus it suffices to show that $N(A)$ has a Poisson distribution for $A = [0, t]$.

First of all, note that $N(A) \leq k$ if, and only if, $Y_{k+1} > t$. Since Y_{k+1} has a Gamma distribution with parameters $k+1$ and λ, we have

$$P(N(A) \leq k) = \frac{\lambda^{k+1}}{k!} \int_t^{\infty} x^k \exp(-\lambda x)\, dx$$

$$= \frac{(\lambda t)^k \exp(-\lambda t)}{k!} + \frac{\lambda^k}{(k-1)!} \int_t^\infty x^{k-1} \exp(-\lambda x)\, dx$$

after integrating by parts. Repeating this procedure (reducing the power of x by one at each stage), we get

$$P(N(A) \leq k) = \sum_{j=0}^{k} \frac{(\lambda t)^j \exp(-\lambda t)}{j!}$$

and so $N(A)$ has a Poisson distribution with mean λt for $A = [0, t]$. Thus the point process $N(\cdot)$ is a Poisson process with intensity λ. ◇

The homogeneous Poisson process assumes a constant intensity λ. One can also define a non-homogeneous Poisson process $N(\cdot)$ on $S \subset R^k$ whose intensity varies over S. Given a function $\lambda(\boldsymbol{t})$ defined for $\boldsymbol{t} \in S$, $N(\cdot)$ is a non-homogeneous Poisson process with intensity $\lambda(\boldsymbol{t})$ if $N(A)$ and $N(B)$ are independent for disjoint A and B with

$$N(A) \sim \text{Pois}\left(\int_A \lambda(\boldsymbol{t})\, d\boldsymbol{t}\right).$$

An interesting type of non-homogeneous Poisson process is the marked Poisson process. Let $T_1, T_2, \cdots$ be the points of a homogeneous Poisson process on S with intensity λ and define i.i.d. random variables $X_1, X_2, \cdots$ that are also independent of the T_i's. Then for $A \subset S$ and $B \subset R$, define

$$N^*(A \times B) = \sum_{i=1}^{\infty} I(T_i \in A, X_i \in B).$$

Then $N^*(A \times B) \sim \text{Pois}(\lambda\, \text{vol}(A) P(X_1 \in B))$. Moreover, if A_1 and A_2 are disjoint sets then $N^*(A_1 \times B_1)$ and $N^*(A_2 \times B_2)$ are independent for any B_1 and B_2; the same is true if B_1 and B_2 are disjoint. N^* is typically on non-homogeneous Poisson process on $S \times R$ whose intensity depends on the distribution of the X_i's.

EXAMPLE 2.21: In forestry, one measure of the density of a "stand" of trees is the basal area proportion, which is roughly defined to be the proportion of the forest actually covered by the bases of trees. (In fact, the basal area of a given tree is typically not measured at a ground level but rather at a certain height above ground level.) A commonly-used method of estimating the basal area proportion is the angle-count method, which is due to the Austrian forester Walter Bitterlich (see Bitterlich, 1956; Holgate, 1967).

Suppose that the forest under consideration is a "Poisson" forest; that is, we assume that the tree centers form a Poisson process with intensity λ trees/meter^2. Furthermore, we will assume that each tree is circular in cross section. The angle-count method counts the number of trees whose diameter subtends an angle greater than θ from some randomly chosen point O. (θ is specified and typically quite small.) Thus a tree whose diameter is x will be sampled if its distance d from the point O satisfies

$$d < \frac{x}{2\sin(\theta/2)};$$

that is, a tree with diameter x must lie within a circle of radius $r(x) = x/(2\sin(\theta/2))$ around O. The number of trees in a circle of radius $r(x)$ is a Poisson random variable with mean $\lambda\pi r^2(x)$.

Suppose that the tree diameters are represented by i.i.d. random variables $X_1, X_2, \cdots$ with distribution function F. Thus we have a marked Poisson process. Assume first that F is a discrete distribution putting probability p_k at x_k for $k = 1, \cdots, m$. Define N_k to be the number of sampled trees with diameter x_k and $N = N_1 + \cdots + N_m$ to be the total number of sampled trees. Now $N_1, \cdots, N_m$ are independent Poisson random variables with

$$E(N_k) = p_k \lambda\pi r^2(x_k) = \lambda\pi p_k \frac{x_k^2}{4\sin^2(\theta/2)}$$

and so N is a Poisson random variable with mean

$$E(N) = \frac{\lambda\pi}{4\sin^2(\theta/2)} \sum_{k=1}^{m} p_k x_k^2 = \frac{\lambda\pi}{4\sin^2(\theta/2)} E(X_i^2).$$

More generally, it can be shown that N (the total number of sampled trees) has a Poisson distribution regardless of the distribution of the X_i's. What is interesting here is the proportionality between $E(X_i^2)$ and $E(N)$, which can be exploited to estimate $E(X_i^2)$.

For our Poisson forest, we can define the "average" basal area proportion to be $B = \lambda\pi E(X_i^2)/4$. (Note that $\pi E(X_i^2)/4$ is the mean basal area for a tree while λ is the density of trees.) Using the relationship between $E(X_i^2)$ and $E(N)$, we have that $B =$

$E(N)\sin^2(\theta/2)$. Using this fact, it is possible to estimate B based on N. The angle count method is, in fact, a special case of biased sampling (Example 1.27) as we are more likely to sample larger trees than smaller trees. ◇

EXAMPLE 2.22: Poisson process models are often used to model the amount of traffic on a network such as a telephone system. Suppose that telephone calls arrive as a homogeneous Poisson process (on the entire real line) with intensity λ calls/minute. In addition, assume that the lengths of each call are i.i.d. continuous random variables with density function $f(x)$. If $S_1, S_2, \cdots$ represent the starting times of the calls and $X_1, X_2, \cdots$ the lengths of the calls then for $a < b$, we have

$$\sum_{i=1}^{\infty} I(a < S_i \le b, X_i \in B) \sim \text{Pois}\left(\lambda(b-a)\int_B f(x)\,dx\right).$$

What is of most interest in this example is the number of calls being made at a given point in time t; we will call this random variable $N(t)$. A given call represented by (S_i, X_i) (its starting time and length) will be active at time t if both $S_i \le t$ and $S_i + X_i \ge t$. Thus

$$N(t) = \sum_{i=1}^{\infty} I(S_i \le t, S_i + X_i \ge t) = \sum_{i=1}^{\infty} I((S_i, X_i) \in B(t))$$

where $B(t) = \{(s,x) : s \le t, s + x \ge t\}$, and so $N(t) \sim \text{Pois}(\mu(t))$ where

$$\begin{aligned}\mu(t) &= \int_0^{\infty}\int_{t-x}^{t} \lambda f(x)\,ds\,dx \\ &= \lambda\int_0^{\infty} x\,f(x)\,dx \\ &= \lambda E(X_i).\end{aligned}$$

Thus the distribution of $N(t)$ is independent of t. Figure 2.3 shows a plot of $N(t)$ versus t for simulated data; the calls arrive as a Poisson process with rate 10 calls/minute while the call lengths are Exponential random variables with mean 5. Based on the calculations above the mean of $N(t)$ is 50 while the standard deviation is $\sqrt{50} \approx 7.1$; note that the simulated number of calls $N(t)$ is, for the most part, within 2 standard deviations of the mean. ◇

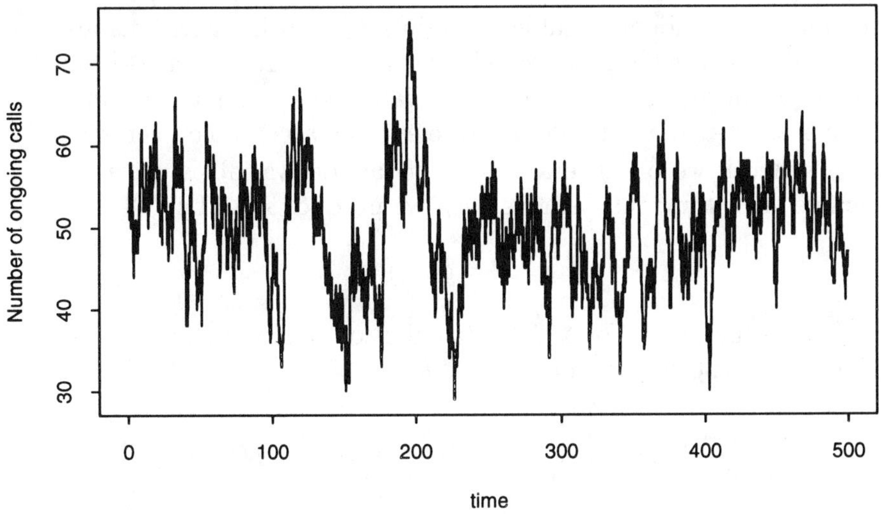

Figure 2.3 *Plot of simulated data over a 500 minute period; the call arrival rate is 10 calls/minute while the call lengths are Exponential with mean 5.*

2.6 Generating random variables

In complex problems in probability and statistics (as well as in other disciplines), computer simulation is necessary because exact analytic properties of certain random processes may be difficult or even impossible to derive exactly. Some examples of problems where computer simulation is used are

- evaluation or approximation of the probability distributions;
- examination of sample path properties of random processes (such as a Poisson process);
- evaluation of integrals.

In order to simulate random processes on a computer, we must be able to generate (by computer) random variables having a specified joint distribution. However, for simplicity, we will consider generating independent random variables with a common distribution F.

It is important to note that computer-generated random variables are not really random as they are typically produced by deterministic algorithms (and so cannot be truly random). For this reason, the "random variables" produced by such algorithms are referred to as pseudo-random. The key lies in finding an algorithm which produces (pseudo-) random variables possessing all the analytic properties of the random variables to an acceptable tolerance.

In general, we produce a stream of random variables via a recursive algorithm; that is, given x_n, we produce x_{n+1} by

$$x_{n+1} = g(x_n)$$

for some function g. For a given distribution function F, we would like to choose g so that for any $a < b$

$$\frac{1}{N}\sum_{k=1}^{n} I(a < x_k \leq b) \to F(b) - F(a) \quad \text{as } N \to \infty$$

independently of the starting value x_0. In fact, this condition is quite easy to satisfy. A much more stringent requirement is that the pseudo-random variables $\{x_1, x_2, x_3, \cdots\}$ behave like outcomes of i.i.d. random variables $X_1, X_2, X_3, \cdots$ (again independently of x_0). Clearly this is impossible since x_{n+1} depends explicitly on x_n; nonetheless, if g is chosen appropriately, it may be possible to achieve "pseudo-independence".

We will first consider generating independent Uniform random variables on the interval $[0, 1]$. A Uniform random variable is a continuous random variable and so any real number between 0 and 1 is a possible outcome of the random variable. However, real numbers cannot be represented exactly on a computer but instead are represented as floating point numbers. Since floating point numbers are countable, it is impossible to generate random variables having exactly a Uniform distribution. What is typically done is to generate a random variable R that is uniformly distributed on the set $\{0, 1, 2, \cdots, N-1\}$ and define $U = R/N$. If N is large then U will be approximately Uniform (since $P(R = k) = 1/N$ for $k = 0, \cdots, N-1$).

There are a number of methods for generating the integer random variable $r_1, \cdots, r_n$ so that the resulting pseudo-random variables are approximately independent. Perhaps the most commonly used generator is the linear congruential generator. We define our sequence $r_1, \cdots, r_n$ via the recursive relationship

$$r_i = \text{mod}(\lambda r_{i-1} + \alpha, N)$$

where $\text{mod}(a, b)$ is the remainder when a is divided by b (for example, $\text{mod}(5, 3) = 2$) and λ, α are integers. (The number r_0 used to start the generator is sometimes called the seed.) It is easy to see that the sequence of r_i's will repeat after $P \leq N$ steps (that is, $r_i = r_{i+P}$) with P called the period of the generator. Clearly, we would like the period of the generator to be as long as possible.

However, if λ and α are chosen appropriately then the period $P = N$ and the integers $r_1, r_2, \cdots, r_n$ will behave more orless like outcomes of independent random variables $R_1, \cdots, R_n$ when n is much smaller than N. (Hence $u_1, \cdots, u_n$ (where $u_i = r_i/N$) will behave like outcomes of independent Uniform random variables.) Finding good linear congruential generators is much more complicated than finding a generator with period $P = N$ (which is quite easy to achieve); in fact, linear congruential and other Uniform generators should be subjected to a battery of tests to ensure that the dependence in $U_1, \cdots, U_n$ is minimal.

EXAMPLE 2.23: One reliable linear congruential generator uses $\lambda = 25211$, $\alpha = 0$ and $N = 2^{15} = 32768$. The period of this generator is N. Using the seed $R_0 = 29559$, we obtain

$$\begin{aligned} U_1 &= 0.06387329 \\ U_2 &= 0.30953979 \\ U_3 &= 0.80776978 \\ U_4 &= 0.68380737 \\ U_5 &= 0.46768188 \end{aligned}$$

and so on.

An example of a bad linear congruential generator is the so-called "randu" generator that was first introduced by IBM in 1963 and was widely used in the 1960s and 1970s. This generator uses $\lambda = 65539$, $\alpha = 0$ and $N = 2^{31} = 2147483648$. This generator has period N but produces $U_1, U_2, \cdots$ that are fairly strongly dependent even though the dependence is not obvious. Figure 2.4 shows a plot of U_{3i-1} versus $U_{3i-2} + 0.1\,U_{3i}$ (for $i = 1, \cdots, 1000$); note that the points fall along diagonal lines, which would not happen if the U_i's were truly independent. ◇

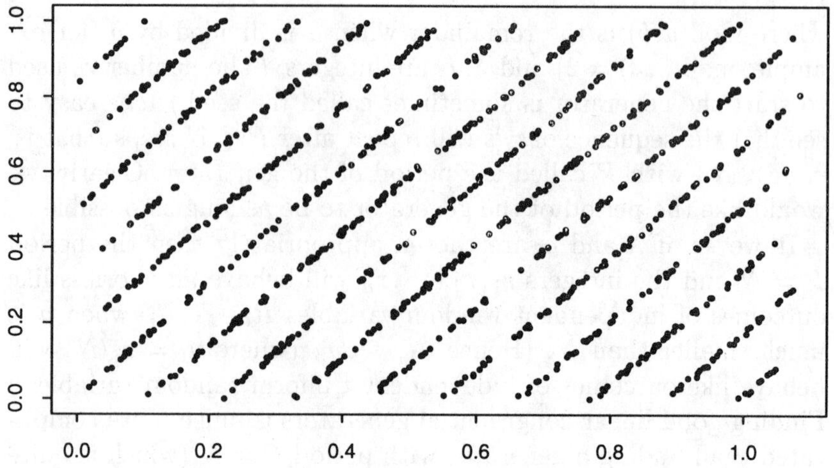

Figure 2.4 *Plot of* U_{3i-1} *versus* $U_{3i-2}+0.1U_{3i}$ *for the "randu" generator; the initial seed* $r_0 = 12345$.

We now move to the more general problem. Suppose we want to generate outcomes of independent random variables $X_1, \cdots, X_n$ with common distribution function F. Several techniques exist for generating random variables. These techniques generally exploit known properties of the distributions. However, generation of Uniform random variables is inevitably involved (to some degree) in virtually all methods as these random variables are typically manipulated to produce $X_1, \cdots, X_n$. Thus given a stream of independent Uniform random variables $U_1, U_2, \cdots$, X_i and X_j $(i \neq j)$ will be independent provided they do not depend on the same Uniform random variables.

Perhaps the most obvious method for generating random variables with a distribution function F is the inverse method. Given a random variable U that has a Uniform distribution on $[0, 1]$, we can transform U to obtain a random variable X with an arbitrary distribution function F. This can be done by using the inverse distribution function

$$F^{-1}(t) = \inf\{x : F(x) \geq t\}.$$

We showed in Chapter 1 that if $X = F^{-1}(U)$ then the distribution function of X is F. (Since U and $1-U$ have the same distribution, $F^{-1}(1-U)$ will also have distribution function F.) The inverse method is easy to implement provided the $F^{-1}(t)$ is easily computable as in the following example.

EXAMPLE 2.24: Suppose we want to generate Exponential random variables with parameter λ; for this distribution, we have $F(x) = \lambda \exp(-\lambda x)$ for $x \geq 0$. Solving the equation $F(F^{-1}(t)) = t$, we get

$$F^{-1}(t) = -\frac{1}{\lambda}\ln(1-t).$$

Thus $X = -\lambda^{-1}\ln(1-U)$ has an Exponential distribution. Alternatively, we could also define $X = -\lambda^{-1}\ln(U)$. ◇

The inverse method becomes less feasible when F^{-1} is difficult to compute. In such cases, there are a variety of other methods that may be used; some of these will be investigated in the exercises.

2.7 Problems and complements

2.1: Suppose that X and Y are independent Geometric random variables with frequency function

$$f(x) = \theta(1-\theta)^x \quad \text{for } x = 0, 1, 2, \cdots$$

(a) Show that $Z = X+Y$ has a Negative Binomial distribution and identify the parameters of Z.

(b) Extend the result of part (a): If $X_1, \cdots, X_n$ are i.i.d. Geometric random variables, show that $S = X_1 + X_2 + \cdots + X_n$ has a Negative Binomial distribution and identify the parameters of S. (Hint: Use mathematical induction or, alternatively, the result of Problem 2.5 below.)

2.2: (a) Suppose that Y is a nonnegative random variable. Show that $X+Y$ is always stochastically greater than X (for any random variable X). (Hint: See Problem 1.10; you need to show that $P(X+Y > x) \geq P(X > x)$ for all x.)

(b) Suppose that X is stochastically greater than Y and suppose that X and Y are defined on the same sample space. Show that $X - Y$ is not necessarily stochastically greater than 0.

2.3: If $f_1(x), \cdots, f_k(x)$ are density (frequency) functions then

$$g(x) = p_1 f_1(x) + \cdots + p_k f_k(x)$$

is also a density (frequency) function provided that $p_i \geq 0$ $(i = 1, \cdots, k)$ and $p_1 + \cdots + p_k = 1$. We can think of sampling from $g(x)$ as first sampling a discrete random variable Y taking values 1 through k with probabilities $p_1, \cdots, p_k$ and then, conditional on $Y = i$, sampling from $f_i(x)$. The distribution whose density or frequency function is $g(x)$ is called a mixture distribution.

(a) Suppose that X has frequency function $g(x)$. Show that

$$P(Y = i|X = x) = \frac{p_i f_i(x)}{g(x)}$$

provided that $g(x) > 0$.

(b) Suppose that X has density function $g(x)$. Show that we can reasonably define

$$P(Y = i|X = x) = \frac{p_i f_i(x)}{g(x)}$$

in the sense that $P(Y_i = i) = E[P(Y = i|X)]$.

2.4: Suppose that $X_1, \cdots, X_5$ are i.i.d. random variables such that

$$\begin{aligned} P(X_i = 0) &= 0.5 \\ \text{and} \quad P(X_i > x) &= 0.5 \exp(-x) \quad \text{for } x \geq 0. \end{aligned}$$

Define $S = X_1 + \cdots + X_5$.

(a) Find the distribution function of S. (Hint: Note that the distribution of S can be written as Binomial mixture of Gamma distributions.)

(b) Evaluate $P(Y \leq 5)$.

2.5: Mixture distributions can be extended in the following way. Suppose that $f(x;\theta)$ is a density or frequency function where θ lies in some set $\Theta \subset R$. Let $p(\theta)$ be a density function on Θ and define

$$g(x) = \int_{\Theta} f(x;\theta)p(\theta)\, d\theta.$$

Then $g(x)$ is itself a density or frequency function. As before, we can view sampling from $g(x)$ as first sampling from $p(\theta)$ and then given θ, sampling from $f(x;\theta)$.

(a) Suppose that X has the mixture density or frequency function $g(x)$. Show that

$$E(X) = E[E(X|\theta)]$$

and

$$\text{Var}(X) = \text{Var}[E(X|\theta)] + E[\text{Var}(X|\theta)]$$

where $E(X|\theta)$ and $\text{Var}(X|\theta)$ are the mean and variance of a random variable with density or frequency function $f(x;\theta)$.

(b) The Negative Binomial distribution introduced in Example 1.12 can be obtained as a Gamma mixture of Poisson distributions. Let $f(x;\lambda)$ be a Poisson frequency function with mean λ and $p(\lambda)$ be a Gamma distribution with mean μ and variance μ^2/α. Show that the mixture distribution has frequency function

$$g(x) = \frac{\Gamma(x+\alpha)}{x!\Gamma(\alpha)}\left(\frac{\alpha}{\alpha+\mu}\right)^{\alpha}\left(\frac{\mu}{\alpha+\mu}\right)^{x}$$

for $x = 0, 1, 2, \cdots$. Note that this form of the Negative Binomial is richer than the form given in Example 1.12.

(c) Suppose that X has a Negative Binomial distribution as given in part (b). Find the mean and variance of X. (Hint: Use the approach outlined in part (a).)

(d) Show that the moment generating function of the Negative Binomial distribution in (b) is

$$m(t) = \left(\frac{\alpha}{\alpha+\mu(1-\exp(t))}\right)^{\alpha}$$

for $t < \ln(1 + \alpha/\mu)$.

2.6: A distribution F is said to be infinitely divisible if for each $n = 1, 2, \cdots$ there exist i.i.d. random variables $X_1, \cdots, X_n$ such that $X_1 + \cdots + X_n \sim F$. For example, a Normal distribution with mean μ and variance σ^2 is infinitely divisible; to see this, we take $X_1, \cdots, X_n$ to be i.i.d. Normal random variables with mean μ/n and variance σ^2/n.

(a) Suppose that the distribution F has moment generating function $m(t)$. Show that F is infinitely divisible if, and only if, $[m(t)]^{1/n}$ is a moment generating function for each n.

(b) Show that all Poisson distributions are infinitely divisible.

(c) Show that all Negative Binomial distributions are infinitely divisible.

2.7: Suppose that $X_1, X_2, \cdots$ are i.i.d. random variables with moment generating function $m(t) = E[\exp(tX_i)]$. Let N be a Poisson random variable (independent of the X_i's) with parameter λ and define the compound Poisson random variable

$$S = \sum_{i=1}^{N} X_i$$

where $S = 0$ if $N = 0$.

(a) Show that the moment generating function of S is

$$E[\exp(tS)] = \exp[\lambda(m(t) - 1)].$$

(b) Suppose that the X_i's are Exponential with $E(X_i) = 1$ and $\lambda = 5$. Evaluate $P(S > 5)$. (Hint: First evaluate $P(S > 5|N)$ using the result of Problem 1.14.)

2.8: Suppose that $X_1, \cdots, X_n$ are independent nonnegative integer-valued random variables with probability generating functions $p_1(t), \cdots, p_n(t)$. Show that the probability generating function of $S = X_1 + \cdots + X_n$ is $p(t) = p_1(t) \times \cdots \times p_n(t)$.

2.9: Consider the experiment in Problem 1.3 where a coin is tossed an infinite number of times where the probability of heads on the k-th toss is $(1/2)^k$. Define X to be the number of heads observed in the experiment.

(a) Show that the probability generating function of X is

$$p(t) = \prod_{k=1}^{\infty} \left(1 - \frac{1-t}{2^k}\right).$$

(Hint: Think of X as a sum of independent Bernoulli random variables.)

(b) Use the result of part (a) to evaluate $P(X = x)$ for $x = 0, \cdots, 5$.

2.10: Consider the following method (known as the rejection method) for generating random variables with a density $f(x)$. Suppose that $\gamma(x)$ be a function such that $\gamma(x) \geq f(x)$ for all x, and

$$\int_{-\infty}^{\infty} \gamma(x)\, dx = \alpha < \infty.$$

Then $g(x) = \gamma(x)/\alpha$ is a probability density function. Suppose we generate a random variable X by the following algorithm:

- Generate a random variable T with density function $g(x)$.
- Generate a random variable $U \sim \text{Unif}(0,1)$, independent of T. If $U \leq f(T)/\gamma(T)$ then set $X = T$; if $U > f(T)/\gamma(T)$ then repeat steps I and II.

(a) Show that the generated random variable X has density $f(x)$. (Hint: you need to evaluate $P(T \leq x | U \leq f(T)/\gamma(T))$.)

(b) Show that the number of "rejections" before X is generated has a Geometric distribution. Give an expression for the parameter of this distribution.

(c) Show that the rejection method also works if we want to generate from a joint density $f(\boldsymbol{x})$. (In this case, $U \sim \text{Unif}(0,1)$ as before but now $\boldsymbol{T}$ is a random vector with density $g(\boldsymbol{x})$.)

2.11: Suppose we want to generate random variables with a Cauchy distribution. As an alternative to the method described in Problem 1.24, we can generate independent random variables V and W where $P(V = 1) = P(V = -1) = 1/2$ and W has density

$$g(x) = \frac{2}{\pi(1+x^2)} \quad \text{for } |x| \leq 1.$$

(W can be generated by using the rejection method in Problem 2.10) Then we define $X = W^V$; show that X has a Cauchy distribution.

2.12: Suppose that X and Y are independent Uniform random variables on $[0,1]$.

(a) Find the density function of $X + Y$.

(b) Find the density function of XY.

2.13: Suppose that $X_1, \cdots, X_n$ are i.i.d. Uniform random variables on $[0,1]$. Define $S_n = (X_1 + \cdots + X_n) \bmod 1$; S_n is simply the "decimal" part of $X_1 + \cdots + X_n$.

(a) Show that $S_n = (S_{n-1} + X_n) \bmod 1$ for all $n \geq 2$.

(b) Show that $S_n \sim \text{Unif}(0,1)$ for all $n \geq 1$. (Hint: Prove the result for $n = 2$ and apply the result of part (a).)

2.14: Suppose that X and Y are independent Exponential random variables with parameter λ. Define $Z = X - Y$.

(a) Show that the density function of Z is

$$f_Z(x) = \frac{\lambda}{2}\exp(-\lambda|x|).$$

(Hint: Evaluate $P(Z \leq x)$ for $x < 0$ and $x > 0$.)

(b) Find the moment generating function of Z. (Hint: use the fact that $Z = X - Y$ for independent Exponential random variables.

2.15: Suppose that $X_1, \cdots, X_n$ are independent, nonnegative continuous random variables where X_i has hazard function $\lambda_i(x)$ $(i = 1, \cdots, n)$.

(a) If $U = \min(X_1, \cdots, X_n)$, show that the hazard function of U is $\lambda_U(x) = \lambda_1(x) + \cdots + \lambda_n(x)$.

(b) If $V = \max(X_1, \cdots, X_n)$, show that the hazard function of V satisfies $\lambda_V(x) \leq \min(\lambda_1(x), \cdots, \lambda_n(x))$.

(c) Show that the result of (b) holds even if the X_i's are not independent.

2.16: Jensen's inequality (see Problem 1.11) can be extended to convex functions in higher dimensions. $g(\boldsymbol{x})$ is a convex function if

$$g(t\boldsymbol{x} + (1 - t)\boldsymbol{y}) \leq t\, g(\boldsymbol{x}) + (1 - t)g(\boldsymbol{y})$$

for $0 \leq t \leq 1$.

(a) Let $\boldsymbol{X}$ be a random vector with well-defined expected value $E(\boldsymbol{X})$. Show that $E[g(\boldsymbol{X})] \geq g(E(\boldsymbol{X}))$ for any convex function g. (Hint: Repeat the approach used in Problem 1.11 making appropriate changes.)

(b) Let $g(\boldsymbol{x}) = \max(x_1, \cdots, x_k)$. Show that g is a convex function and so

$$E[\max(X_1, \cdots, X_k)] \geq \max(E(X_1), \cdots, E(X_k))$$

for any random variables $X_1, \cdots, X_k$.

2.17: Suppose that X and Y are random variables such that both $E(X^2)$ and $E(Y^2)$ are finite. Define $g(t) = E[(Y + tX)^2]$.

(a) Show that $g(t)$ is minimized at $t = -E(XY)/E(X^2)$.

(b) Show that $[E(XY)]^2 \leq E(X^2)E(Y^2)$; this is called the Cauchy-Schwarz inequality. (Hint: Note that $g(t) \geq 0$ for all t.)

(c) Use part (b) to show that $|\text{Corr}(X, Y)| \leq 1$.

2.18: Suppose that R and U are independent continuous random variables where U has a Uniform distribution on $[0, 1]$ and R has the density function

$$f_R(x) = x \exp(-x^2/2) \quad \text{for } x \geq 0.$$

(a) Show that R^2 has an Exponential distribution.

(b) Define $X = R\cos(2\pi U)$ and $Y = R\sin(2\pi U)$. Show that X and Y are independent standard Normal random variables.

(c) Suggest a method for generating Normal random variables based on the results in part (a) and (b).

2.19: Suppose that X and Y are independent random variables with X discrete and Y continuous. Define $Z = X + Y$.

(a) Show that Z is a continuous random variable with

$$P(Z \le z) = \sum_x P(Y \le z - x)P(X = x).$$

(b) If Y has a density function $f_Y(y)$, show that the density of Z is

$$f_Z(z) = \sum_x f_Y(z - x) f_X(x)$$

where $f_X(x)$ is the frequency function of X.

2.20: Suppose that X and Y are independent Normal random variables each with mean μ and variance σ^2 and let $U = Y$ and $V = X(1 + Y)$. Evaluate the following:

(a) $E(V|U = u)$.

(b) $\text{Var}(V|U = u)$.

(c) $\text{Var}(V)$.

(d) $\text{Cov}(U, V)$ and $\text{Corr}(U, V)$.

(Hint: You do not need to evaluate any integrals.)

2.21: (a) Show that

$$\text{Cov}(X, Y) = E[\text{Cov}(X, Y|\boldsymbol{Z})] + \text{Cov}(E(X|\boldsymbol{Z}), E(Y|\boldsymbol{Z})).$$

(Hint: Follow the proof for the similar result involving variances.)

(b) Suppose that $X_1, X_2, \cdots$ be i.i.d. Exponential random variables with parameter 1 and take N_1, N_2 to be independent Poisson random variables with parameters λ_1, λ_2 that are independent of the X_i's. Define compound Poisson random variables

$$S_1 = \sum_{i=1}^{N_1} X_i$$

$$S_2 = \sum_{i=1}^{N_2} X_i$$

and evaluate $\text{Cov}(S_1, S_2)$ and $\text{Corr}(S_1, S_2)$. When is this correlation maximized?

2.22: Suppose that $\boldsymbol{X} = (X_1, \cdots, X_k)$ is a random vector and define the (joint) moment generating function of $\boldsymbol{X}$;

$$m(\boldsymbol{t}) = E[\exp(t_1 X_1 + \cdots + t_k X_k)];$$

we say that this exists if $m(\boldsymbol{t}) < \infty$ for $\|\boldsymbol{t}\| < b$ where $b > 0$.

(a) Show that

$$E(X_i^k X_j^\ell) = \left.\frac{\partial^{k+\ell}}{\partial t_i^k \partial t_j^\ell} m(\boldsymbol{t})\right|_{t=0}$$

for $k, \ell = 0, 1, 2, \cdots$

(b) Show that

$$\text{Cov}(X_i, X_j) = \left.\frac{\partial^2}{\partial t_i \partial t_j} \ln m(\boldsymbol{t})\right|_{t=0}.$$

(c) Suppose that $\boldsymbol{X}_1, \cdots, \boldsymbol{X}_n$ are independent random vectors with moment generating functions $m_1(\boldsymbol{t}), \cdots, m_n(\boldsymbol{t})$, respectively. Show that the moment generating function of $\boldsymbol{S} = \boldsymbol{X}_1 + \cdots + \boldsymbol{X}_n$ is

$$m_S(\boldsymbol{t}) = m_1(\boldsymbol{t}) \times \cdots \times m_n(\boldsymbol{t}).$$

2.23: The mean residual life function $r(t)$ of a nonnegative random variable X is defined to be

$$r(t) = E(X - t | X \geq t).$$

($r(t)$ would be of interest, for example, to a life insurance company.)

(a) Suppose that F is the distribution function of X. Show that

$$r(t) = \frac{1}{1 - F(t)} \int_t^\infty (1 - F(x))\, dx.$$

(b) Show that $r(t)$ is constant if, and only if, X has an Exponential distribution.

(c) Show that

$$E(X^2) = 2 \int_0^\infty r(t)(1 - F(t))\, dt.$$

(Hint: Show that $E(X^2)$ can be written as

$$2\int_0^\infty \int_0^t (1 - F(t))\, ds\, dt$$

and change the order of integration.)

(d) Suppose that X has a density function $f(x)$ that is differentiable and $f(x) > 0$ for $x > 0$. Show that

$$\lim_{t\to\infty} r(t) = \lim_{t\to\infty}\left(-\frac{f(t)}{f'(t)}\right).$$

(e) Suppose that X has a Gamma distribution:

$$f(x) = \frac{1}{\Gamma(\alpha)}\lambda^\alpha x^{\alpha-1}\exp(-\lambda x) \quad \text{for } x > 0.$$

Evaluate the limit in part (c) for this distribution. Give an interpretation of this result.

2.24: Suppose that X is a nonnegative random variable with mean $\mu > 0$ and variance $\sigma^2 < \infty$. The coefficient of variation of X is defined to be $\mathrm{CV}(X) = \sigma/\mu$.

(a) Suppose that X and Y are independent nonnegative random variables with $\mathrm{CV}(X)$ and $\mathrm{CV}(Y)$ finite. Show that

$$\mathrm{CV}(X+Y) \le \mathrm{CV}(X) + \mathrm{CV}(Y).$$

(b) Define $r(t) = E(X - t | X \ge t)$ to be the mean residual life function of X. Show that $\mathrm{CV}(X) \le 1$ if $r(t) \le r(0) = E(X)$ and $\mathrm{CV}(X) \ge 1$ if $r(t) \ge r(0)$. (Hint: Note that $\mathrm{CV}(X) \le 1$ if, and only if, $E(X^2)/[E(X)]^2 \le 2$ and use the result of Problem 2.23(c).)

2.25: Suppose that $X_1, \cdots, X_n$ are i.i.d. continuous random variables with distribution function $F(x)$ and density function $f(x)$; let $X_{(1)} < X_{(2)} < \cdots < X_{(n)}$ be the order statistics.

(a) Show that the distribution function of $X_{(k)}$ is

$$G_k(x) = \sum_{j=k}^{n} \binom{n}{j} F(x)^j (1 - F(x))^{n-j}.$$

(Hint: Let $Y_i = I(X_i \le x)$ and define $S = Y_1 + \cdots + Y_n$; S has a Binomial distribution and $P(X_{(k)} \le x) = P(S \ge k)$.)

(b) Show that the density function of $X_{(k)}$ is

$$g_k(x) = \frac{n!}{(n-k)!(k-1)!} F(x)^{k-1}(1 - F(x))^{n-k} f(x).$$

(Hint: Assume that $F'(x) = f(x)$ and differentiate $G_k(x)$.)

2.26: Suppose that $X_1, \cdots, X_n$ are i.i.d. Exponential random variables with parameter λ. Let $X_{(1)} < \cdots < X_{(n)}$ be the order statistics and define

$$\begin{aligned} Y_1 &= nX_{(1)} \\ Y_2 &= (n-1)(X_{(2)} - X_{(1)}) \\ Y_3 &= (n-2)(X_{(3)} - X_{(2)}) \\ \vdots &\ \vdots\ \vdots \\ Y_n &= X_{(n)} - X_{(n-1)}. \end{aligned}$$

Show that $Y_1, \cdots, Y_n$ are i.i.d. Exponential random variables with parameter λ. (Note that the "Jacobian" matrix here is triangular and so the Jacobian itself can be computed as the product of the diagonal elements.)

2.27: Suppose that $X_1, \cdots, X_{n+1}$ be i.i.d. Exponential random variables with parameter λ and define

$$U_k = \frac{1}{T} \sum_{i=1}^{k} X_i \quad \text{for } k = 1, \cdots, n$$

where $T = X_1 + \cdots + X_{n+1}$.

(a) Find the joint density of $(U_1, \cdots, U_n, T)$. (Note that $0 < U_1 < U_2 < \cdots < U_n < 1$.)

(b) Show that the joint distribution of $(U_1, \cdots, U_n)$ is exactly the same as the joint distribution of the order statistics of an i.i.d. sample of n observations from a Uniform distribution on $[0, 1]$.

2.28: A discrete random vector $\boldsymbol{X}$ is said to have a Multinomial distribution if its joint frequency function is

$$f(\boldsymbol{x}) = \frac{n!}{x_1! \times \cdots \times x_k!} \theta_1^{x_1} \times \cdots \times \theta_k^{x_k}$$

for nonnegative integers $x_1, \cdots, x_k$ with $x_1 + \cdots + x_k = n$ and nonnegative parameters $\theta_1, \cdots, \theta_k$ with $\theta_1 + \cdots + \theta_k = 1$. (We will write $\boldsymbol{X} \sim \text{Mult}(n, \boldsymbol{\theta})$ where $\boldsymbol{\theta} = (\theta_1, \cdots, \theta_k)$.)

(a) Show that the marginal distribution of X_i is Binomial with parameters n and θ_i.

(b) Show that $E(\boldsymbol{X}) = n\boldsymbol{\theta}$ and

$$\text{Cov}(\boldsymbol{X}) = n \begin{pmatrix} \theta_1(1-\theta_1) & -\theta_1\theta_2 & \cdots & -\theta_1\theta_k \\ -\theta_1\theta_2 & \theta_2(1-\theta_2) & \cdots & -\theta_2\theta_k \\ \vdots & \vdots & \ddots & \vdots \\ -\theta_1\theta_k & -\theta_2\theta_k & \cdots & \theta_k(1-\theta_k) \end{pmatrix}.$$

(c) Suppose that $\boldsymbol{X}_1, \cdots, \boldsymbol{X}_n$ are independent Multinomial random vectors with $\boldsymbol{X}_i \sim \text{Mult}(n_i, \boldsymbol{\theta})$ $(i = 1, \cdots, n)$. Show that

$$\sum_{i=1}^{n} \boldsymbol{X}_i \sim \text{Mult}\left(\sum_{i=1}^{n} n_i, \boldsymbol{\theta}\right).$$

(Hint: For parts (b) and (c), evaluate the moment generating function of a Multinomial random vector using the Multinomial Theorem.)

2.29: Suppose that X and Y are independent Exponential random variables with parameters λ and μ respectively. Define random variables

$$T = \min(X, Y) \quad \text{and} \quad \Delta = \begin{cases} 1 & \text{if } X < Y \\ 0 & \text{otherwise.} \end{cases}$$

Note that T has a continuous distribution while Δ is discrete. (This is an example of type I censoring in reliability or survival analysis.)

(a) Find the density of T and the frequency function of Δ.

(b) Find the joint distribution function of (T, Δ).

2.30: Suppose that X has a Beta distribution with parameters α and β (see Example 2.5).

(a) Show that for $r > 0$,

$$E(X^r) = \frac{\Gamma(\alpha + r)\Gamma(\alpha + \beta)}{\Gamma(\alpha)\Gamma(\alpha + \beta + r)}.$$

(b) Use part (a) and properties of $\Gamma(x)$ to evaluate $E(X)$ and $\text{Var}(X)$.

2.31: Suppose that X has a Beta distribution with parameters α and β.

(a) Find the density function of $Y = (1 - X)^{-1}$.

(b) Suppose that $\alpha = m/2$ and $\beta = n/2$ and define Y as in part (a). Using the definition of the F distribution, show that $nY/m \sim \mathcal{F}(m, n)$. (Hint: Take $U \sim \chi^2(m)$ and $V \sim \chi^2(n)$ to be independent random variables. Show that $U/(U + V)$ has a Beta distribution and then apply part (a).)

2.32: Suppose that $\boldsymbol{X} \sim N_p(\boldsymbol{0}, C)$ where C^{-1} exists. Show that

$$\boldsymbol{X}^T C^{-1} \boldsymbol{X} \sim \chi^2(p).$$

(Hint: Write $C = O^T \Lambda O$ where O is an orthogonal matrix and Λ is a diagonal matrix whose entries are the eigenvalues of C; then define $C^{1/2} = O^T \Lambda^{1/2} O$ to be a symmetric root of C.)

2.33: Suppose that $X \sim \chi^2(n)$.

(a) Show that $E(X^r) = 2^{-r}\Gamma(r + n/2)/\Gamma(n/2)$ if $r > -n/2$.

(b) Using part (a), show that $E(X) = n$ and $\text{Var}(X) = 2n$.

2.34: Suppose that $T \sim \mathcal{T}(n)$. Show that

(a) $E(|T|^r) < \infty$ if, and only if, $n > r \geq 0$.

(b) $E(T) = 0$ if $n > 1$; $\text{Var}(T) = n/(n - 2)$ if $n > 2$.

(Hint: You don't need to do any integration here. Write $T = Z/\sqrt{V/n}$ where $Z \sim N(0, 1)$ and $V \sim \chi^2(n)$ are independent.)

2.35: Suppose that $W \sim \mathcal{F}(m, n)$. Show that

$$E(W^r) = \left(\frac{n}{m}\right)^r \frac{\Gamma(r + m/2)\Gamma(-r + n/2)}{\Gamma(m/2)\Gamma(n/2)}$$

if $-m/2 < r < n/2$.

2.36: Suppose that X is a Normal random variable with mean θ and variance 1 and define $Y = X^2$.

(a) Show that the density of Y is

$$f_Y(y) = \frac{1}{2\sqrt{2\pi y}} \exp\left(\frac{1}{2}(y + \theta^2)\right)(\exp(\theta\sqrt{y}) + \exp(-\theta\sqrt{y}))$$

for $y > 0$. (Y is said to have a non-central χ^2 distribution with 1 degree of freedom and non-centrality parameter θ^2.)

(b) Show that the density of Y can be written as

$$f_Y(y) = \sum_{k=0}^{\infty} \frac{\exp(-\theta^2/2)(\theta^2/2)^k}{k!} f_{2k+1}(y)$$

where $f_{2k+1}(y)$ is the density function of a χ^2 random variable with $2k+1$ degrees of freedom. (Hint: Expand $\exp(\theta\sqrt{y})$ and $\exp(-\theta\sqrt{y})$ as power series; note that the odd terms in the two expansions will cancel each other out.)

2.37: Suppose that $\boldsymbol{X} \sim N_n(\boldsymbol{\mu}, I)$; the elements of $\boldsymbol{X}$ are independent Normal random variables with variances equal to 1.

(a) Suppose that O is an orthogonal matrix whose first row is $\boldsymbol{\mu}^T/\|\boldsymbol{\mu}\|$ and let $\boldsymbol{Y} = O\boldsymbol{X}$. Show that $E(Y_1) = \|\boldsymbol{\mu}\|$ and $E(Y_k) = 0$ for $k \geq 2$.

(b) Using part (a), show that the distribution of $\|\boldsymbol{X}\|^2$ is the same as that of $\|\boldsymbol{Y}\|^2$ and hence depends on $\boldsymbol{\mu}$ only through its norm $\|\boldsymbol{\mu}\|$.

(c) Let $\theta^2 = \|\boldsymbol{\mu}\|^2$. Show that the density of $V = \|\boldsymbol{X}\|^2$ is

$$f_V(x) = \sum_{k=0}^{\infty} \frac{\exp(-\theta^2/2)(\theta^2/2)^k}{k!} f_{2k+n}(x)$$

where $f_{2k+n}(x)$ is the density function of a χ^2 random variable with $2k+n$ degrees of freedom. (V has a non-central χ^2 distribution with n degrees of freedom and non-centrality parameter θ^2.)

2.38: Consider a marked Poisson process similar to that given in Example 2.22 such that the call starting times arrive as a homogeneous Poisson process (with rate λ calls/minute) on the positive real line. Assume that the call lengths are continuous random variables with density function $f(x)$ and define $N(t)$ to be the number of calls active at time t for $t \geq 0$.
Show that $N(t) \sim \text{Pois}(\mu(t))$ where

$$\mu(t) = \lambda \int_0^t (1 - F(s))\, ds$$

and F is the distribution function of the call lengths.

2.39: Consider the marked Poisson process in Example 2.22 where the call starting times arrive as a homogeneous Poisson process (with rate λ calls/minute) on the entire real line and the call lengths are continuous random variables with density function $f(x)$. In Example 2.22, we showed that the distribution of $N(t)$ is independent of t.

(a) Show that for any r,

$$\text{Cov}[N(t), N(t+r)]$$

$$
\begin{aligned}
&= \lambda \int_{|r|}^{\infty} x\, f(x)\, dx \\
&= \lambda \left[|r|(1 - F(|r|) + \int_{|r|}^{\infty} (1 - F(x))\, dx \right]
\end{aligned}
$$

and hence is independent of t and depends only on $|r|$. (Hint: Assume that $r > 0$. Then

$$
\begin{aligned}
N(t) &= \sum_{i=1}^{\infty} I(S_i \le t, t \le S_i + X_i < t + r) \\
&\quad + \sum_{i=1}^{\infty} I(S_i \le t, S_i + X_i \ge t + r) \\
N(t + r) &= \sum_{i=1}^{\infty} I(S_i \le t, S_i + X_i \ge t + r) \\
&\quad + \sum_{i=1}^{\infty} I(t < S_i \le t + r, S_i + X_i \ge t + r)
\end{aligned}
$$

and use the independence of Poisson processes on disjoint sets.)

(b) Suppose that the call lengths are Exponential random variables with mean μ. Evaluate $\text{Cov}[N(t), N(t+r)]$. (This is called the autocovariance function of $N(t)$.)

(c) Suppose that the call lengths have a density function

$$f(x) = \alpha x^{-\alpha-1} \quad \text{for } x \ge 1.$$

Show that $E(X_i) < \infty$ if, and only if, $\alpha > 1$ and evaluate $\text{Cov}[N(t), N(t + r)]$ in this case.

(d) Compare the autocovariance functions obtained in parts (b) and (c). For which distribution does $\text{Cov}[N(t), N(t + r)]$ decay to 0 more slowly as $|r| \to \infty$?

CHAPTER 3

Convergence of Random Variables

3.1 Introduction

In probability and statistics, it is often necessary to consider the distribution of a random variable that is itself a function of several random variables, for example, $Y = g(X_1, \cdots, X_n)$; a simple example is the sample mean of random variables $X_1, \cdots, X_n$. Unfortunately, finding the distribution exactly is often very difficult or very time-consuming even if the joint distribution of the random variables is known exactly. In other cases, we may have only partial information about the joint distribution of $X_1, \cdots, X_n$ in which case it is impossible to determine the distribution of Y. However, when n is large, it may be possible to obtain approximations to the distribution of Y even when only partial information about $X_1, \cdots, X_n$ is available; in many cases, these approximations can be remarkably accurate.

The standard approach to approximating a distribution function is to consider the distribution function as part of an infinite sequence of distribution functions; we then try to find a "limiting" distribution for the sequence and use that limiting distribution to approximate the distribution of the random variable in question. This approach, of course, is very common in mathematics. For example, if n is large compared to x, one might approximate $(1 + x/n)^n$ by $\exp(x)$ since

$$\lim_{n\to\infty} \left(1 + \frac{x}{n}\right)^n = \exp(x).$$

(However, this approximation may be very poor if x/n is not close to 0.) A more interesting example is Stirling's approximation, which is used to approximate $n!$ for large values of n:

$$n! \approx \sqrt{2\pi} \exp(-n) n^{n+1/2} = s(n)$$

where the approximation holds in the sense that $n!/s(n) \to 1$ as

Table 3.1 *Comparison of $n!$ and its Stirling approximation $s(n)$.*

n	$n!$	$s(n)$
1	1	0.92
2	2	1.92
3	6	5.84
4	24	23.51
5	120	118.02
6	720	710.08

$n \to \infty$. In fact, Stirling's approximation is not too bad even for small n as Table 3.1 indicates.

In a sense, Stirling's approximation shows that asymptotic approximations can be useful in a more general context. In statistical practice, asymptotic approximations (typically justified for large sample sizes) are very commonly used even in situations where the sample size is small. Of course, it is not always clear that the use of such approximations is warranted but nonetheless there is a sufficiently rich set of examples where it is warranted to make the study of convergence of random variables worthwhile.

To motivate the notion of convergence of random variables, consider the following example. Suppose that $X_1, \cdots, X_n$ are i.i.d. random variables with mean μ and variance σ^2 and define

$$\bar{X}_n = \frac{1}{n}\sum_{i=1}^{n} X_i$$

to be their sample mean; we would like to look at the behaviour of the distribution of $\bar{X}_n$ when n is large. First of all, it seems reasonable that $\bar{X}_n$ will be close to μ if n is sufficiently large; that is, the random variable $\bar{X}_n - \mu$ should have a distribution that, for large n, is concentrated around 0 or, more precisely,

$$P[|\bar{X}_n - \mu| \le \epsilon] \approx 1$$

when ϵ is small. (Note that $\text{Var}(\bar{X}_n) = \sigma^2/n \to 0$ as $n \to \infty$.) This latter observation is, however, not terribly informative about the distribution of $\bar{X}_n$. However, it is also possible to look at the difference between $\bar{X}_n$ and μ on a "magnified" scale; we do this by multiplying the difference $\bar{X}_n - \mu$ by $\sqrt{n}$ so that the mean and variance are constant. Thus define

$$Z_n = \sqrt{n}(\bar{X}_n - \mu)$$

and note that $E(Z_n) = 0$ and $\text{Var}(Z_n) = \sigma^2$. We can now consider the behaviour of the distribution function of Z_n as n increases. If this sequence of distribution functions has a limit (in some sense) then we can use the limiting distribution function to approximate the distribution function of Z_n (and hence of $\bar{X}_n$). For example, if we have

$$P(Z_n \leq x) = P\left(\sqrt{n}(\bar{X}_n - \mu) \leq x\right) \approx F_0(x)$$

then

$$\begin{aligned} P(\bar{X}_n \leq y) &= P\left(\sqrt{n}(\bar{X}_n - \mu) \leq \sqrt{n}(y - \mu)\right) \\ &\approx F_0\left(\sqrt{n}(y - \mu)\right) \end{aligned}$$

provided that n is sufficiently large to make the approximation valid.

3.2 Convergence in probability and distribution

In this section, we will consider two different types of convergence for sequences of random variables, convergence in probability and convergence in distribution.

DEFINITION. Let $\{X_n\}$, X be random variables. Then $\{X_n\}$ converges in probability to X as $n \to \infty$ ($X_n \to_p X$) if for each $\epsilon > 0$,

$$\lim_{n\to\infty} P(|X_n - X| > \epsilon) = 0.$$

If $X_n \to_p X$ then for large n we have that $X_n \approx X$ with probability close to 1. Frequently, the limiting random variable X is a constant; $X_n \to_p \theta$ (a constant) means that for large n there is almost no variation in the random variable X_n. (A stronger form of convergence, convergence with probability 1, is discussed in section 3.7.)

DEFINITION. Let $\{X_n\}$, X be random variables. Then $\{X_n\}$ converges in distribution to X as $n \to \infty$ ($X_n \to_d X$) if

$$\lim_{n\to\infty} P(X_n \leq x) = P(X \leq x) = F(x)$$

for each continuity point of the distribution function $F(x)$.

It is important to remember that $X_n \to_d X$ implies convergence of distribution functions and not of the random variables themselves. For this reason, it is often convenient to replace $X_n \to_d X$ by $X_n \to_d F$ where F is the distribution function of X, that is, the limiting distribution; for example, $X_n \to_d N(0, \sigma^2)$ means that $\{X_n\}$ converges in distribution to a random variable that has a Normal distribution (with mean 0 and variance σ^2).

If $X_n \to_d X$ then for sufficiently large n we can approximate the distribution function of X_n by that of X; thus, convergence in distribution is potentially useful for approximating the distribution function of a random variable. However, the statement $X_n \to_d X$ does not say how large n must be in order for the approximation to be practically useful. To answer this question, we typically need a further result dealing explicitly with the approximation error as a function of n.

EXAMPLE 3.1: Suppose that $X_1, \cdots, X_n$ are i.i.d. Uniform random variables on the interval $[0, 1]$ and define

$$M_n = \max(X_1, \cdots, X_n).$$

Intuitively, M_n should be approximately 1 for large n. We will first show that $M_n \to_p 1$ and then find the limiting distribution of $n(1 - M_n)$. The distribution function of M_n is

$$F_n(x) = x^n \quad \text{for} \quad 0 \leq x \leq 1.$$

Thus for $0 < \epsilon < 1$,

$$\begin{aligned} P(|M_n - 1| > \epsilon) &= P(M_n < 1 - \epsilon) \\ &= (1 - \epsilon)^n \to 0 \end{aligned}$$

as $n \to \infty$ since $|1 - \epsilon| < 1$. To find the limiting distribution of $n(1 - M_n)$, note that

$$\begin{aligned} P(n(1 - M_n) \leq x) &= P(M_n \geq 1 - x/n) \\ &= 1 - \left(1 - \frac{x}{n}\right)^n \\ &\to 1 - \exp(-x) \end{aligned}$$

as $n \to \infty$ for $x \geq 0$. Thus $n(1 - M_n)$ has a limiting Exponential distribution with parameter 1. In this example, of course, there is no real advantage in knowing the limiting distribution of $n(1 - M_n)$ as its exact distribution is known. ◇

EXAMPLE 3.2: Suppose that $X_1, \cdots, X_n$ are i.i.d. random variables with

$$P(X_i = j) = \frac{1}{10} \quad \text{for } j = 0, 1, 2, \cdots, 9$$

and define

$$U_n = \sum_{k=1}^{n} \frac{X_k}{10^k}.$$

U_n can be thought of as the first n digits of a decimal representation of a number between 0 and 1 ($U_n = 0.X_1X_2X_3X_4 \cdots X_n$). It turns out that U_n tends in distribution to a Uniform on the interval $[0, 1]$. To see this, note that each outcome of $(X_1, \cdots, X_n)$ produces a unique value of U_n; these possible values are $j/10^n$ for $j = 0, 1, 2, \cdots, 10^n - 1$, and so it follows that

$$P(U_n = j/10^n) = \frac{1}{10^n} \quad \text{for } j = 0, 1, 2, \cdots, 10^n - 1.$$

If $j/10^n \le x < (j+1)/10^n$ then

$$P(U_n \le x) = \frac{j+1}{10^n}$$

and so

$$|P(U_n \le x) - x| \le 10^{-n} \to 0 \quad \text{as } n \to \infty$$

and so $P(U_n \le x) \to x$ for each x between 0 and 1. ◇

Some important results

We noted above that convergence in probability deals with convergence of the random variables themselves while convergence in distribution deals with convergence of the distribution functions. The following result shows that convergence in probability is stronger than convergence in distribution unless the limiting random variable is a constant in which case the two are equivalent.

THEOREM 3.1 *Let $\{X_n\}$, X be random variables.*
(a) If $X_n \to_p X$ then $X_n \to_d X$.
(b) If $X_n \to_d \theta$ (a constant) then $X_n \to_p \theta$.

Proof. (a) Let x be a continuity point of the distribution function of X. Then for any $\epsilon > 0$,

$$\begin{aligned} P(X_n \le x) &= P(X_n \le x, |X_n - X| \le \epsilon) \\ &\quad + P(X_n \le x, |X_n - X| > \epsilon) \\ &\le P(X \le x + \epsilon) + P(|X_n - X| > \epsilon) \end{aligned}$$

where the latter inequality follows since $[X_n \leq x, |X_n - X| \leq \epsilon]$ implies $[X \leq x + \epsilon]$. Similarly,

$$P(X \leq x - \epsilon) \leq P(X_n \leq x) + P(|X_n - X| > \epsilon)$$

and so

$$P(X_n \leq x) \geq P(X \leq x - \epsilon) - P(|X_n - X| > \epsilon).$$

Thus putting the two inequalities for $P(X_n \leq x)$ together, we have

$$\begin{aligned} P(X \leq x - \epsilon) &- P(|X_n - X| > \epsilon) \\ &\leq P(X_n \leq x) \\ &\leq P(X \leq x + \epsilon) + P(|X_n - X| > \epsilon). \end{aligned}$$

By hypothesis, $P(|X_n - X| > \epsilon) \to 0$ as $n \to \infty$ for any $\epsilon > 0$. Moreover, since x is a continuity point of the distribution function of X, $P(X \leq x \pm \epsilon)$ can be made arbitrarily close to $P(X \leq x)$ by taking ϵ close to 0. Hence,

$$\lim_{n \to \infty} P(X_n \leq x) = P(X \leq x).$$

(b) Define $F(x)$ to be the distribution function of the degenerate random variable taking the single value θ; thus, $F(x) = 0$ for $x < \theta$ and $F(x) = 1$ for $x \geq \theta$. Note that F is continuous at all but one point. Then

$$\begin{aligned} P(|X_n - \theta| > \epsilon) &= P(X_n > \theta + \epsilon) + P(X_n < \theta - \epsilon) \\ &\leq 1 - P(X_n \leq \theta + \epsilon) + P(X_n \leq \theta - \epsilon). \end{aligned}$$

However, since $X_n \to_d \theta$, it follows that $P(X_n \leq \theta + \epsilon) \to 1$ and $P(X_n \leq \theta - \epsilon) \to 0$ as $n \to \infty$ and so $P(|X_n - \theta| > \epsilon) \to 0$. □

It is often difficult (if not impossible) to verify the convergence of a sequence of random variables using simply its definition. Theorems 3.2, 3.3 and 3.4 are sometimes useful for showing convergence in such cases.

THEOREM 3.2 (Continuous Mapping Theorem) *Suppose that $g(x)$ is a continuous real-valued function.*
(a) If $X_n \to_p X$ then $g(X_n) \to_p g(X)$.
(b) If $X_n \to_d X$ then $g(X_n) \to_d g(X)$.

The proofs will not be given here. The proof of (a) is sketched as an exercise. The proof of (b) is somewhat more technical; however, if we further assume g to be strictly increasing or decreasing (so that g has an inverse function), a simple proof of (b) can be given.

(Also see Example 3.16 for a simple proof assuming more technical machinery.) The assumption of continuity can also be relaxed somewhat. For example, Theorem 3.2 will hold if g has a finite or countable number of discontinuities provided that these discontinuity points are continuity points of the distribution function of X. For example, if $X_n \to_d \theta$ (a constant) and $g(x)$ is continuous at $x = \theta$ then $g(X_n) \to_d g(\theta)$.

THEOREM 3.3 (Slutsky's Theorem) *Suppose that $X_n \to_d X$ and $Y_n \to_p \theta$ (a constant). Then*
(a) $X_n + Y_n \to_d X + \theta$.
(b) $X_n Y_n \to_d \theta X$.

Proof. (a) Without loss of generality, let $\theta = 0$. (If $\theta \neq 0$ then $X_n + Y_n = (X_n + \theta) + (Y_n - \theta)$ and $Y_n - \theta \to_p 0$.) Let x be a continuity point of the distribution function of X. Then

$$\begin{aligned} P(X_n + Y_n \le x) &= P(X_n + Y_n \le x, |Y_n| \le \epsilon) \\ &\quad + P(X_n + Y_n \le x, |Y_n| > \epsilon) \\ &\le P(X_n \le x + \epsilon) + P(|Y_n| > \epsilon). \end{aligned}$$

Also,

$$\begin{aligned} P(X_n \le x - \epsilon) &= P(X_n \le x - \epsilon, |Y_n| \le \epsilon) \\ &\quad + P(X_n \le x - \epsilon, |Y_n| > \epsilon) \\ &\le P(X_n + Y_n \le x) + P(|Y_n| > \epsilon) \end{aligned}$$

(since $[X_n \le x - \epsilon, |Y_n| \le \epsilon]$ implies $[X_n + Y_n \le x]$). Hence,

$$\begin{aligned} P(X_n \le x - \epsilon) - P(|Y_n| > \epsilon) &\le P(X_n + Y_n \le x) \\ &\le P(X_n \le x + \epsilon) + P(|Y_n| > \epsilon). \end{aligned}$$

Now take $x \pm \epsilon$ to be continuity points of the distribution function of X. Then

$$\lim_{n \to \infty} P(X_n \le x \pm \epsilon) = P(X \le x \pm \epsilon)$$

and the limit can be made arbitrarily close to $P(X \le x)$ by taking ϵ to 0. Since $P(|Y_n| > \epsilon) \to 0$ as $n \to \infty$ the conclusion follows.

(b) Again we will assume that $\theta = 0$. (To see that it suffices to consider this single case, note that $X_n Y_n = X_n(Y_n - \theta) + \theta X_n$. Since $\theta X_n \to_d \theta X$ the conclusion will follow from part (a) if we show that $X_n(Y_n - \theta) \to_p 0$.) We need to show that $X_n Y_n \to_p 0$. Taking $\epsilon > 0$ and $M > 0$, we have

$$P(|X_n Y_n| > \epsilon) = P(|X_n Y_n| > \epsilon, |Y_n| \le 1/M)$$

$$\begin{aligned} & +P(|X_nY_n| > \epsilon, |Y_n| > 1/M) \\ \leq \; & P(|X_nY_n| > \epsilon, |Y_n| \leq 1/M) + P(|Y_n| > 1/M) \\ \leq \; & P(|X_n| > \epsilon M) + P(|Y_n| > 1/M). \end{aligned}$$

Since $Y_n \to_p 0$, $P(|Y_n| > 1/M) \to 0$ as $n \to \infty$ for any fixed $M > 0$. Now take ϵ and M such that $\pm\epsilon M$ are continuity points of the distribution function of X; then $P(|X_n| > \epsilon M) \to P(|X| > \epsilon M)$ and the limit can be made arbitrarily close to 0 by making M sufficiently large. □

Since $Y_n \to_p \theta$ is equivalent to $Y_n \to_d \theta$ when θ is a constant, we could replace "$Y_n \to_p \theta$" by "$Y_n \to_d \theta$" in the statement of Slutsky's Theorem. We can also generalize this result as follows. Suppose that $g(x, y)$ is a continuous function and that $X_n \to_d X$ and $Y_n \to_p \theta$ for some constant θ. Then it can be shown that $g(X_n, Y_n) \to_d g(X, \theta)$. In fact, this result is sometimes referred to as Slutsky's Theorem with Theorem 3.3 a special case for $g(x, y) = x + y$ and $g(x, y) = xy$.

THEOREM 3.4 (The Delta Method) *Suppose that*

$$a_n(X_n - \theta) \to_d Z$$

where θ is a constant and $\{a_n\}$ is a sequence of constants with $a_n \uparrow \infty$. If $g(x)$ is a function with derivative $g'(\theta)$ at $x = \theta$ then

$$a_n(g(X_n) - g(\theta)) \to_d g'(\theta)Z.$$

Proof. We'll start by assuming that g is continuously differentiable at θ. First, note that $X_n \to_p \theta$. (This follows from Slutsky's Theorem.) By a Taylor series expansion of $g(x)$ around $x = \theta$, we have

$$g(X_n) = g(\theta) + g'(\theta_n^*)(X_n - \theta)$$

where θ_n^* lies between X_n and θ; thus $|\theta_n^* - \theta| \leq |X_n - \theta|$ and so $\theta_n^* \to_p \theta$. Since $g'(x)$ is continuous at $x = \theta$, it follows that $g'(\theta_n^*) \to_p g'(\theta)$. Now,

$$\begin{aligned} a_n(g(X_n) - g(\theta)) &= g'(\theta_n^*)a_n(X_n - \theta) \\ &\to_d g'(\theta)Z \end{aligned}$$

by Slutsky's Theorem. For the more general case (where g is not necessarily continuously differentiable at θ), note that

$$g(X_n) - g(\theta) = g'(\theta)(X_n - \theta) + R_n$$

where $R_n/(X_n - \theta) \to_p 0$. Thus

$$a_n R_n = a_n(X_n - \theta)\frac{R_n}{a_n(X_n - \theta)} \to_p 0$$

and so the conclusion follows by Slutsky's Theorem. □

A neater proof of the Delta Method is given in Example 3.17. Also note that if $g'(\theta) = 0$, we would have that $a_n(g(X_n) - g(\theta)) \to_p 0$. In this case, we may have

$$a_n^k(g(X_n) - g(\theta)) \to_d \text{some } V$$

for some $k \geq 2$; see Problem 3.10 for details.

If $X_n \to_d X$ (or $X_n \to_p X$), it is tempting to say that $E(X_n) \to E(X)$; however, this statement is not true in general. For example, suppose that $P(X_n = 0) = 1 - n^{-1}$ and $P(X_n = n) = n^{-1}$. Then $X_n \to_p 0$ but $E(X_n) = 1$ for all n (and so converges to 1). To ensure convergence of moments, additional conditions are needed; these conditions effectively bound the amount of probability mass in the distribution of X_n concentrated near $\pm\infty$ for large n. The following result deals with the simple case where the random variables $\{X_n\}$ are uniformly bounded; that is, there exists a constant M such that $P(|X_n| \leq M) = 1$ for all n.

THEOREM 3.5 *If $X_n \to_d X$ and $|X_n| \leq M$ (finite) then $E(X)$ exists and $E(X_n) \to E(X)$.*

Proof. For simplicity, assume that X_n is nonnegative for all n; the general result will follow by considering the positive and negative parts of X_n. From Chapter 1, we have that

$$\begin{aligned}
|E(X_n) - E(X)| &= \left|\int_0^\infty (P(X_n > x) - P(X > x))\,dx\right| \\
&= \left|\int_0^M (P(X_n > x) - P(X > x))\,dx\right| \\
&\qquad \text{(since } P(X_n > M) = P(X > M) = 0\text{)} \\
&\leq \int_0^M |P(X_n > x) - P(X > x)|\,dx \\
&\to 0
\end{aligned}$$

since $P(X_n > x) \to P(X > x)$ for all but a countable number of x's and the interval of integration is bounded. □

3.3 Weak Law of Large Numbers

An important result in probability theory is the Weak Law of Large Numbers (WLLN), which deals with the convergence of the sample mean to the population mean as the sample size increases. We start by considering the simple case where $X_1, \cdots, X_n$ are i.i.d. Bernoulli random variables with $P(X_i = 1) = \theta$ and $P(X_i = 0) = 1 - \theta$ so that $E(X_i) = \theta$. Define $S_n = X_1 + \cdots + X_n$, which has a Binomial distribution with parameters n and θ. We now consider the behaviour of S_n/n as $n \to \infty$; S_n/n represents the proportion of 1's in the n Bernoulli trials. Our intuition tells us that for large n, this proportion should be approximately equal to θ, the probability that any $X_i = 1$. Indeed, since the distribution of S_n/n is known, it is possible to show the following law of large numbers:

$$\frac{S_n}{n} \to_p \theta$$

as $n \to \infty$.

In general, the WLLN applies to any sequence of independent, identical distributed random variables whose mean exists. The result can be stated as follows:

THEOREM 3.6 (Weak Law of Large Numbers) *Suppose that $X_1, X_2, \cdots$ are i.i.d. random variables with $E(X_i) = \mu$ (where $E(|X_i|) < \infty$). Then*

$$\bar{X}_n = \frac{1}{n} \sum_{i=1}^{n} X_i \to_p \mu$$

as $n \to \infty$.

While this result certainly agrees with intuition, a rigorous proof of the result is certainly not obvious. However, before proving the WLLN, we will give a non-trivial application of it by proving that the sample median of i.i.d. random variables $X_1, \cdots, X_n$ converges in probability to the population median.

EXAMPLE 3.3: Suppose that $X_1, \cdots, X_n$ are i.i.d. random variables with a distribution function $F(x)$. Assume that the X_i's have a unique median μ ($F(\mu) = 1/2$); in particular, this implies that for any $\epsilon > 0$, $F(\mu + \epsilon) > 1/2$ and $F(\mu - \epsilon) < 1/2$.

Let $X_{(1)}, \cdots, X_{(n)}$ be the order statistics of the X_i's and define $Z_n = X_{(m_n)}$ where $\{m_n\}$ is a sequence of positive integers with $m_n/n \to 1/2$ as $n \to \infty$. For example, we could take $m_n = n/2$

if n is even and $m_n = (n+1)/2$ if n is odd; in this case, Z_n is essentially the sample median of the X_i's. We will show that $Z_n \to_p \mu$ as $n \to \infty$.

Take $\epsilon > 0$. Then we have

$$P(Z_n > \mu + \epsilon) = P\left(\frac{1}{n}\sum_{i=1}^{n} I(X_i > \mu + \epsilon) \geq \frac{m_n}{n}\right)$$

and

$$P(Z_n < \mu - \epsilon) = P\left(\frac{1}{n}\sum_{i=1}^{n} I(X_i \geq \mu - \epsilon) \leq \frac{n - m_n}{n}\right).$$

By the WLLN, we have

$$\frac{1}{n}\sum_{i=1}^{n} I(X_i > \mu + \epsilon) \to_p 1 - F(\mu + \epsilon) < 1/2$$

and

$$\frac{1}{n}\sum_{i=1}^{n} I(X_i > \mu - \epsilon) \to_p 1 - F(\mu - \epsilon) > 1/2.$$

Since $m_n/n \to_p 1/2$, it follows that $P(Z_n > \mu + \epsilon) \to 0$ and $P(Z_n < \mu - \epsilon) \to 0$ as $n \to \infty$ and so $Z_n \to_p \mu$. ◇

Proving the WLLN

The key to proving the WLLN lies in finding a good bound for $P[|\bar{X}_n - \mu| > \epsilon]$; one such bound is Chebyshev's inequality.

THEOREM 3.7 (Chebyshev's inequality) *Suppose that X is a random variable with $E(X^2) < \infty$. Then for any $\epsilon > 0$,*

$$P[|X| > \epsilon] \leq \frac{E(X^2)}{\epsilon^2}.$$

Proof. The key is to write $X^2 = X^2 I(|X| \leq \epsilon) + X^2 I(|X| > \epsilon)$. Then

$$\begin{aligned} E(X^2) &= E[X^2 I(|X| \leq \epsilon)] + E[X^2 I(|X| > \epsilon)] \\ &\geq E[X^2 I(|X| > \epsilon)] \\ &\geq \epsilon^2 P(|X| > \epsilon) \end{aligned}$$

where the last inequality holds since $X^2 \geq \epsilon^2$ when $|X| > \epsilon$ and $E[I(|X| > \epsilon)] = P(|X| > \epsilon)$. □

From the proof, it is quite easy to see that Chebyshev's inequality remains valid if $P[|X| > \epsilon]$ is replaced by $P[|X| \geq \epsilon]$.

Chebyshev's inequality is primarily used as a tool for proving various convergence results for sequences of random variables; for example, if $\{X_n\}$ is a sequence of random variables with $E(X_n^2) \to 0$ then Chebyshev's inequality implies that $X_n \to_p 0$. However, Chebyshev's inequality can also be used to give probability bounds for random variables. For example, let X be a random variable with mean μ and variance σ^2. Then by Chebyshev's inequality, we have

$$P[|X - \mu| \leq k\sigma] \geq 1 - \frac{E[(X-\mu)^2]}{k^2\sigma^2} = 1 - \frac{1}{k^2}.$$

However, the bounds given by Chebyshev's inequality are typically very crude and are seldom of any practical use. Chebyshev's inequality can be also generalized in a number of ways; these generalizations are examined in Problem 3.8.

We will now sketch the proof of the WLLN. First of all, we will assume that $E(X_i^2) < \infty$. In this case, the WLLN follows trivially since (by Chebyshev's inequality)

$$P[|\bar{X}_n - \mu| > \epsilon] \leq \frac{\text{Var}(\bar{X}_n)}{\epsilon^2} = \frac{\text{Var}(X_1)}{n\epsilon^2}$$

and the latter quantity tends to 0 as $n \to \infty$ for each $\epsilon > 0$.

How can the weak law of large numbers be proved if we assume only that $E(|X_i|) < \infty$? The answer is to write

$$X_k = U_{nk} + V_{nk}$$

where $U_{nk} = X_k$ if $|X_k| \leq \delta n$ (for some $0 < \delta < 1$) and $U_{nk} = 0$ otherwise; it follows that $V_{nk} = X_k$ if $|X_k| > \delta n$ and 0 otherwise. Then

$$\bar{X}_n = \frac{1}{n}\sum_{i=1}^n U_{ni} + \frac{1}{n}\sum_{i=1}^n V_{ni} = \bar{U}_n + \bar{V}_n$$

and so it suffices to show that $\bar{U}_n \to_p \mu$ and $\bar{V}_n \to_p 0$. First, we have

$$\begin{aligned}
E[(\bar{U}_n - \mu)^2] &= \frac{\text{Var}(U_{n1})}{n} + (E(U_{n1}) - \mu)^2 \\
&\leq \frac{E(U_{n1}^2)}{n} + (E(U_{n1}) - \mu)^2 \\
&\leq E[|U_{n1}|]\delta + E^2(U_{n1} - X_1) \\
&\leq E[|X_1|]\delta + E^2[|X_1| I(|X_1| > \delta n)]
\end{aligned}$$

and so by Chebyshev's inequality

$$P[|\bar{U}_n - \mu| > \epsilon] \leq \frac{E[|X_1|]\delta + E^2[|X_1|I(|X_1| > \delta n)]}{\epsilon^2},$$

which can be made close to 0 by taking $n \to \infty$ and then δ to 0. Second,

$$\begin{aligned} P[|\bar{V}_n| > \epsilon] &\leq P\left(\bigcup_{i=1}^{n}[V_{ni} \neq 0]\right) \\ &\leq \sum_{i=1}^{n} P[V_{ni} \neq 0] \\ &= nP[X_1 > \delta n] \end{aligned}$$

and the latter can be shown to tend to 0 as $n \to \infty$ (for any $\delta > 0$). The details of the proof are left as exercises.

The WLLN can be strengthened to a strong law of large numbers (SLLN) by introducing another type of convergence known as convergence with probability 1 (or "almost sure" convergence). This is discussed in section 3.7.

The WLLN for Bernoulli random variables was proved by Jacob Bernoulli in 1713 and strengthened to random variables with finite variance by Chebyshev in 1867 using the inequality that bears his name. Chebyshev's result was extended by Khinchin in 1928 to sums of i.i.d. random variables with finite first moment.

3.4 Proving convergence in distribution

Recall that a sequence of random variables $\{X_n\}$ converges in distribution to a random variable X if the corresponding sequence of distribution functions $\{F_n(x)\}$ converges to $F(x)$, the distribution function of X, at each continuity point of F. It is often difficult to verify this condition directly for a number of reasons. For example, it is often difficult to work with the distribution functions $\{F_n\}$. Also, in many cases, the distribution function F_n may not be specified exactly but may belong to a wider class; we may know, for example, the mean and variance corresponding to F_n but little else about F_n. (From a practical point of view, the cases where F_n is not known exactly are most interesting; if F_n is known exactly, there is really no reason to worry about a limiting distribution F unless F_n is difficult to work with computationally.)

For these reasons, we would like to have alternative methods for establishing convergence in distribution. Fortunately, there are several other sufficient conditions for convergence in distribution that are useful in practice for verifying that a sequence of random variables converges in distribution and determining the distribution of the limiting random variable (the limiting distribution).

- Suppose that X_n has density function f_n (for $n \geq 1$) and X has density function f. Then $f_n(x) \to f(x)$ (for all but a countable number of x) implies that $X_n \to_d X$. Similarly, if X_n has frequency function f_n and X has frequency function f then $f_n(x) \to f(x)$ (for all x) implies that $X_n \to_d X$. (This result is known as Scheffé's Theorem.) The converse of this result is not true; in fact, a sequence of discrete random variables can converge in distribution to a continuous variable (see Example 3.2) and a sequence of continuous random variables can converge in distribution to a discrete random variable.
- If X_n has moment generating function $m_n(t)$ and X has moment generating function $m(t)$ then $m_n(t) \to m(t)$ (for all $|t| \leq$ some $b > 0$) implies $X_n \to_d X$. Convergence of moment generating functions is actually quite strong (in fact, it implies that $E(X_n^k) \to E(X^k)$ for integers $k \geq 1$); convergence in distribution does not require convergence of moment generating functions. It is also possible to substitute other generating functions for the moment generating function to prove convergence in distribution. For example, if X_n has characteristic function $\varphi_n(t) = E[\exp(i\,tX)]$ and X has characteristic function $\varphi(t)$ then $\varphi_n(t) \to \varphi(t)$ (for all t) implies $X_n \to_d X$; in fact, $X_n \to_d X$ if, and only if, $\varphi_n(t) \to \varphi(t)$ for all t.

In addition to the methods described above, we can also use some of the results given earlier (for example, Slutsky's Theorem and the Delta Method) to help establish convergence in distribution.

EXAMPLE 3.4: Suppose that $\{X_n\}$ is a sequence of random variables where X_n has Student's t distribution with n degrees of freedom. The density function of X_n is

$$f_n(x) = \frac{\Gamma((n+1)/2)}{\sqrt{\pi n}\,\Gamma(n/2)} \left(1 + \frac{x^2}{n}\right)^{-(n+1)/2}.$$

Stirling's approximation, which may be stated as

$$\lim_{y\to\infty} \frac{\sqrt{y}\,\Gamma(y)}{\sqrt{2\pi}\exp(-y)y^y} = 1$$

allows us to approximate $\Gamma((n+1)/2)$ and $\Gamma(n/2)$ for large n. We then get

$$\lim_{n\to\infty} \frac{\Gamma((n+1)/2)}{\sqrt{\pi n}\Gamma(n/2)} = \frac{1}{\sqrt{2\pi}}.$$

Also

$$\lim_{n\to\infty} \left(1+\frac{x^2}{n}\right)^{-(n+1)/2} = \exp\left(-\frac{x^2}{2}\right)$$

and so

$$\lim_{n\to\infty} f_n(x) = \frac{1}{\sqrt{2\pi}} \exp\left(-\frac{x^2}{2}\right)$$

where the limit is a standard Normal density function. Thus $X_n \to_d Z$ where Z has a standard Normal distribution.

An alternative (and much simpler) approach is to note that the t distributed random variable X_n has the same distribution as $Z/\sqrt{V_n/n}$ where Z and V_n are independent random variables with $Z \sim N(0,1)$ distribution and $V_n \sim \chi^2(n)$. Since V_n can be thought of as a sum of n i.i.d. χ^2 random variables with 1 degree of freedom, it follows from the WLLN and the Continuous Mapping Theorem that $\sqrt{V_n/n} \to_p 1$ and hence from Slutsky's Theorem that $Z/\sqrt{V_n/n} \to_d N(0,1)$. The conclusion follows since X_n has the same distribution as $Z/\sqrt{V_n/n}$. ◇

EXAMPLE 3.5: Suppose that $U_1, \cdots, U_n$ are i.i.d. Uniform random variables on the interval $[0,1]$ and let $U_{(1)}, \cdots, U_{(n)}$ be their order statistics. Define $Z_n = U_{(m_n)}$ where $m_n \approx n/2$ in the sense that $\sqrt{n}(m_n/n - 1/2) \to 0$ as $n \to \infty$; note that we are requiring that m_n/n converge to $1/2$ at a faster rate than in Example 3.3. (Note that taking $m_n = n/2$ for n even and $m_n = (n+1)/2$ for n odd will satisfy this condition.) We will consider the asymptotic distribution of the sequence of random variables $\{\sqrt{n}(Z_n - 1/2)\}$ by computing its limiting density. The density of $\sqrt{n}(Z_n - 1/2)$ is

$$f_n(x) = \frac{n!}{(m_n-1)!(n-m_n)!\sqrt{n}} \left(\frac{1}{2}+\frac{x}{\sqrt{n}}\right)^{m_n-1} \left(\frac{1}{2}-\frac{x}{\sqrt{n}}\right)^{n-m_n}$$

for $-\sqrt{n}/2 \le x \le \sqrt{n}/2$. We will show that $f_n(x)$ converges to a Normal density with mean 0 and variance $1/4$. First using Stirling's approximation (as in Example 3.4, noting that $n! = n\Gamma(n)$), we obtain

$$\frac{n!}{\sqrt{n}(m_n-1)!(n-m_n)!} \approx \frac{2^n}{\sqrt{2\pi}}$$

in the sense that the ratio of the right-hand to left-hand side tends to 1 as $n \to \infty$. We also have

$$\begin{aligned}&\left(\frac{1}{2}+\frac{x}{\sqrt{n}}\right)^{m_n-1}\left(\frac{1}{2}-\frac{x}{\sqrt{n}}\right)^{n-m_n}\\&= \frac{1}{2^{n-1}}\left(1-\frac{4x^2}{n}\right)^{m_n-1}\left(1-\frac{2x}{\sqrt{n}}\right)^{n-2m_n+1}.\end{aligned}$$

We now obtain

$$\left(1-\frac{4x^2}{n}\right)^{m_n-1} \to \exp(-2x^2)$$

and

$$\left(1-\frac{2x}{\sqrt{n}}\right)^{n-2m_n+1} \to 1$$

where, in both cases, we use the fact that $(1+t/a_n)^{c_n} \to \exp(kt)$ if $a_n \to \infty$ and $c_n/a_n \to k$. Putting the pieces from above together, we get

$$f_n(x) \to \frac{2}{\sqrt{2\pi}}\exp(-2x^2)$$

for any x. Thus $\sqrt{n}(Z_n - 1/2) \to_d N(0, 1/4)$. ◇

EXAMPLE 3.6: We can easily extend Example 3.5 to the case where $X_1, \cdots, X_n$ are i.i.d. random variables with distribution function F and unique median μ where $F(x)$ is differentiable at $x = \mu$ with $F'(\mu) > 0$; if F has a density f then $F'(\mu) = f(\mu)$ typically. Defining $F^{-1}(t) = \inf\{x : F(x) \geq t\}$ to be the inverse of F, we note that the order statistic $X_{(k)}$ has the same distribution as $F^{-1}(U_{(k)})$ where $U_{(k)}$ is an order statistic from an i.i.d. sample of Uniform random variables on $[0,1]$ and also that $F^{-1}(1/2) = \mu$. Thus

$$\sqrt{n}(X_{(m_n)} - \mu) =_d \sqrt{n}(F^{-1}(U_{(m_n)}) - F^{-1}(1/2))$$

and so by the Delta Method, we have

$$\sqrt{n}(X_{(m_n)} - \mu) \to_d N(0, (F'(\mu))^{-2}/4)$$

if $\sqrt{n}(m_n/n - 1/2) \to 0$. The limiting variance follows from the fact that $F^{-1}(t)$ is differentiable at $t = 1/2$ with derivative $1/F'(\mu)$. Note that existence of a density is not sufficient to imply existence of the derivative of $F(x)$ at $x = \mu$; however, if F is continuous but not differentiable at $x = \mu$ then $\sqrt{n}(X_{(m_n)} - \mu)$ may still converge in distribution but the limiting distribution will be different.

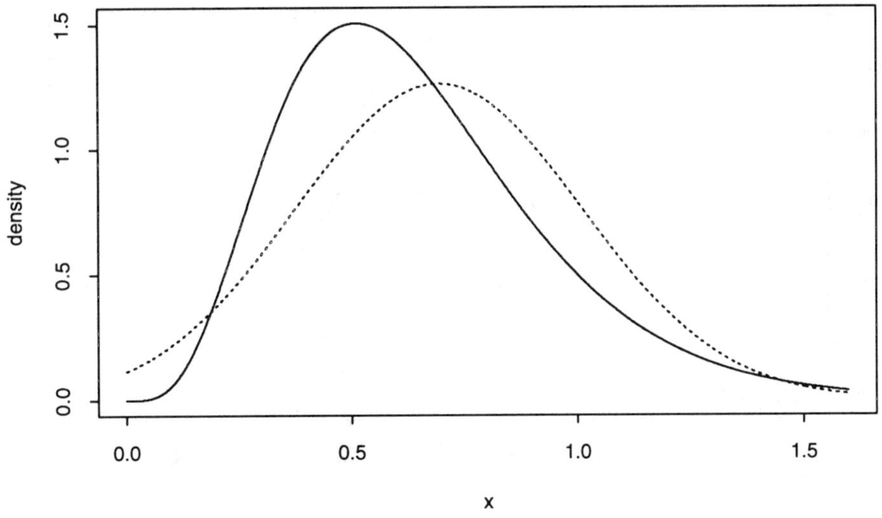

Figure 3.1 *Density of $X_{(n/2)}$ for $n = 10$ Exponential random variables; the dotted line is the approximating Normal density.*

As an illustration of the convergence of the distribution of the sample median to a Normal distribution, we will consider the density of the order statistic $X_{(n/2)}$ for i.i.d. Exponential random variables $X_1, \cdots, X_n$ with density

$$f(x) = \exp(-x) \quad \text{for } x \geq 0.$$

Figures 3.1, 3.2, and 3.3 give the densities of $X_{(n/2)}$ for $n = 10, 50$, and 100 respectively; the corresponding approximating Normal density is indicated with dotted lines. ◇

EXAMPLE 3.7: Suppose that $\{X_n\}$ is a sequence of Binomial random variables with X_n having parameters n and θ_n where $n\theta_n \to \lambda > 0$ as $n \to \infty$. The moment generating function of X_n is

$$\begin{aligned} m_n(t) &= (1 + \theta_n(\exp(t) - 1))^n \\ &= \left(1 + \frac{n\theta_n(\exp(t) - 1)}{n}\right)^n. \end{aligned}$$

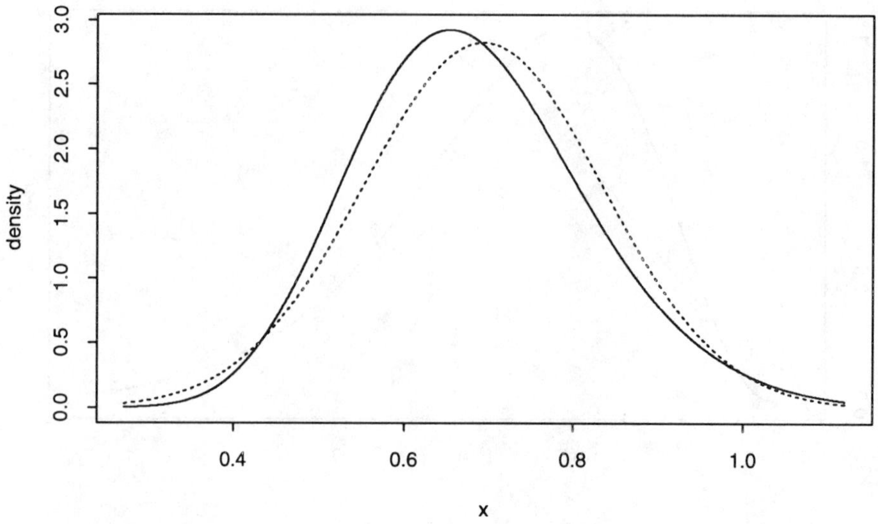

Figure 3.2 *Density of* $X_{(n/2)}$ *for* $n = 50$ *Exponential random variables; the dotted line is the approximating Normal density.*

Since $n\theta_n \to \lambda$, it follows that

$$\lim_{n\to\infty} m_n(t) = \exp[\lambda(\exp(t) - 1)]$$

where the limiting moment generating function is that of a Poisson distribution with parameter λ. Thus $X_n \to_d X$ where X has a Poisson distribution with parameter λ. This result can be used to compute Binomial probabilities when n is large and θ is small so that $n\theta \approx n\theta(1-\theta)$. For example, suppose that X has a Binomial distribution with $n = 100$ and $\theta = 0.05$. Then using the Poisson approximation

$$P[a \le X \le b] \approx \sum_{a \le x \le b} \frac{\exp(-n\theta)(n\theta)^x}{x!}$$

we get, for example, $P[4 \le X \le 6] \approx 0.497$ compared to the exact probability $P[4 \le X \le 6] = 0.508$. $\diamond$

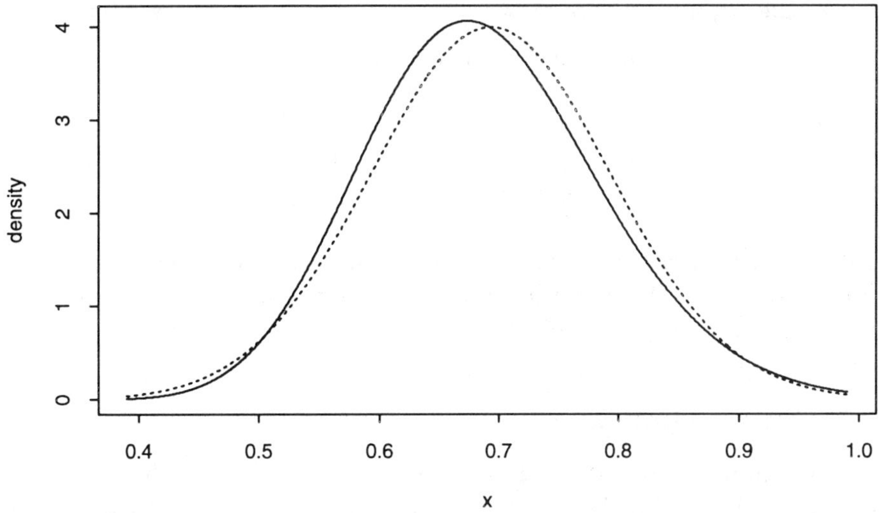

Figure 3.3 *Density of* $X_{(n/2)}$ *for* $n = 100$ *Exponential random variables; the dotted line is the approximating Normal density.*

EXAMPLE 3.8: As in Example 3.2, define U_n by

$$U_n = \sum_{k=1}^{n} \frac{X_k}{10^k}$$

where $X_1, X_n, \cdots$ are i.i.d. discrete random variables uniformly distributed on the integers $0, 1, \cdots, 9$. We showed in Example 3.2 that $U_n \to_d \text{Unif}(0, 1)$ by showing convergence of the distribution functions. In this example, we will do the same using moment generating functions. The moment generating function of each X_k is

$$m(t) = \frac{1}{10}(1 + \exp(t) + \cdots + \exp(9t)) = \frac{\exp(10t) - 1}{10(\exp(t) - 1)}$$

and so the moment generating function of U_n is

$$\begin{aligned} \phi_n(t) &= \prod_{k=1}^{n} m(t/10^k) \\ &= \prod_{k=1}^{n} \left(\frac{\exp(t/10^{k-1}) - 1}{10(\exp(t/10^k) - 1} \right) \end{aligned}$$

$$= \frac{\exp(t) - 1}{10^n(\exp(t/10^n) - 1)}.$$

Using the expansion $\exp(x) = 1 + x + x^2/2 + \cdots$, it follows that

$$\lim_{n\to\infty} 10^n(\exp(t/10^n) - 1) = t$$

and so

$$\lim_{n\to\infty} \phi_n(t) = \frac{1}{t}(\exp(t) - 1) = \int_0^1 \exp(tx)\,dx,$$

which is the moment generating function of the Uniform distribution on $[0, 1]$. Thus we have shown (using moment generating functions) that $U_n \to_d \text{Unif}(0, 1)$. ◇

3.5 Central Limit Theorems

In probability theory, central limit theorems (CLTs) establish conditions under which the distribution of a sum of random variables may be approximated by a Normal distribution. (We have seen already in Examples 3.4, 3.5, and 3.6 cases where the limiting distribution is Normal.) A wide variety of CLTs have been proved; however, we will consider CLTs only for sums and weighted sums of i.i.d. random variables.

THEOREM 3.8 (CLT for i.i.d. random variables) *Suppose that $X_1, X_2, \cdots$ are i.i.d. random variables with mean μ and variance $\sigma^2 < \infty$ and define*

$$S_n = \frac{1}{\sigma\sqrt{n}} \sum_{i=1}^{n} (X_i - \mu) = \frac{\sqrt{n}(\bar{X}_n - \mu)}{\sigma}.$$

Then $S_n \to_d Z \sim N(0, 1)$ as $n \to \infty$.

(In practical terms, the CLT implies that for "large" n, the distribution of $\bar{X}_n$ is approximately Normal with mean μ and variance σ^2/n.)

Before discussing the proof of this CLT, we will give a little of the history behind the result. The French-born mathematician de Moivre is usually credited with proving the first CLT (in the 18th century); this CLT dealt with the special case that the X_i's were Bernoulli random variables (that is, $P[X_i = 1] = \theta$ and $P[X_i = 0] = 1 - \theta$). His work (de Moivre, 1738) was not significantly improved until Laplace (1810) extended de Moivre's work to sums of independent bounded random variables. The Russian mathematician Chebyshev extended Laplace's work to sums of random variables with finite moments $E(|X_i|^k)$ for all $k \geq 1$. However, it was not until the early twentieth century that Markov and Lyapunov (who were students of Chebyshev) removed nearly all unnecessary moment restrictions. Finally, Lindeberg (1922) proved the CLT assuming only finite variances. It should be noted that most of the work subsequent to Laplace dealt with sums of independent (but not necessarily identically distributed) random variables; it turns out that this added generality does not pose any great technical complications.

Proving the CLT

We will consider two proofs of the CLT for i.i.d. random variables. In the first proof, we will assume the existence of moment generating function of the X_i's and show that the moment generating function of S_n converges to the moment generating function of Z. Of course, assuming the existence of moment generating functions implies that $E(X_i^k)$ exists for all integers $k \geq 1$. The second proof will require only that $E(X_i^2)$ is finite and will show directly that $P[S_n \leq x] \to P[Z \leq x]$.

We can assume (without loss of generality) that $E(X_i) = 0$ and $\text{Var}(X_i) = 1$. Let $m(t) = E[\exp(tX_i)]$ be the moment generating function of the X_i's. Then

$$m(t) = 1 + \frac{t^2}{2} + \frac{t^3 E(X_i^3)}{6} + \cdots$$

(since $E(X_i) = 0$ and $E(X_i^2) = 1$) and the moment generating function of S_n is

$$\begin{aligned}
m_n(t) &= [m_n(t/\sqrt{n})]^n \\
&= \left[1 + \frac{t^2}{2n} + \frac{t^3 E(X_i^3)}{6n^{3/2}} + \cdots\right]^n \\
&= \left[1 + \frac{t^2}{2n}\left(1 + \frac{tE(X_i^3)}{3n^{1/2}} + \frac{t^2 E(X_i^4)}{12n} + \cdots\right)\right]^n \\
&= \left[1 + \frac{t^2}{2n} r_n(t)\right]^n
\end{aligned}$$

where $|r_n(t)| < \infty$ and $r_n(t) \to 1$ as $n \to \infty$ for each $|t| \leq$ some

$b > 0$. Thus

$$\lim_{n\to\infty} m_n(t) = \exp\left(\frac{t^2}{2}\right)$$

and the limit is the moment generating function of a standard Normal random variable. It should be noted that a completely rigorous proof of Theorem 3.8 can be obtained by replacing moment generating functions by characteristic functions in the proof above.

The second proof we give shows directly that

$$P[S_n \leq x] \to \int_{-\infty}^{x} \frac{1}{\sqrt{2\pi}} \exp\left(-\frac{t^2}{2}\right) dt;$$

we will first assume that $E(|X_i|^3)$ is finite and then indicate what modifications are necessary if we assume only that $E(X_i^2)$ is finite. The method used in this proof may seem at first somewhat complicated but, in fact, is extremely elegant and is actually the method used by Lindeberg to prove his 1922 CLT.

The key to the proof directly lies in approximating $P[S_n \leq x]$ by $E[f_\delta^+(S_n)]$ and $E[f_\delta^-(S_n)]$ where f_δ^+ and f_δ^- are two bounded, continuous functions. In particular, we define $f_\delta^+(y) = 1$ for $y \leq x$, $f_\delta^+(y) = 0$ for $y \geq x + \delta$ and $0 \leq f_\delta^+(y) \leq 1$ for $x < y < x + \delta$; we define $f_\delta^-(y) = f_\delta^+(y + \delta)$. If

$$g(y) = I(y \leq x),$$

it is easy to see that

$$f_\delta^-(y) \leq g(y) \leq f_\delta^+(y)$$

and

$$P[S_n \leq x] = E[g(S_n)].$$

Then if Z is a standard Normal random variable, we have

$$\begin{aligned} P[S_n \leq x] &\leq E[f_\delta^+(S_n)] \\ &\leq E[f_\delta^+(S_n)] - E[f_\delta^+(Z)] + E[f_\delta^+(Z)] \\ &\leq |E[f_\delta^+(S_n)] - E[f_\delta^+(Z)]| + P[Z \leq x + \delta] \end{aligned}$$

and similarly,

$$P[S_n \leq x] \geq P[Z \leq x - \delta] - |E[f_\delta^-(S_n)] - E[f_\delta^-(Z)]|.$$

Thus we have

$$\begin{aligned} &|E[f_\delta^+(S_n)] - E[f_\delta^+(Z)]| + P[Z \leq x + \delta] \\ &\geq P[S_n \leq x] \\ &\geq P[Z \leq x - \delta] - |E[f_\delta^-(S_n)] - E[f_\delta^-(Z)]|; \end{aligned}$$

since $P[Z \le x \pm \delta]$ can be made arbitrarily close to $P(Z \le x)$ (because Z has a continuous distribution function), it suffices to show that $E[f_\delta^+(S_n)] \to E[f_\delta^+(Z)]$ and $E[f_\delta^-(S_n)] \to E[f_\delta^-(Z)]$ for suitable choices of f_δ^+ and f_δ^-. In particular, we will assume that f_δ^+ (and hence f_δ^-) has three bounded continuous derivatives.

Let f be a bounded function (such as f_δ^+ or f_δ^-) with three bounded continuous derivatives and let $Z_1, Z_2, Z_3, \cdots$ be a sequence of i.i.d. standard Normal random variables that are also independent of $X_1, X_2, \cdots$; note that $n^{-1/2}(Z_1 + \cdots + Z_n)$ is also standard Normal. Now define random variables $T_{n1}, \cdots, T_{nn}$ where

$$T_{nk} = \frac{1}{\sqrt{n}} \sum_{j=1}^{k-1} Z_j + \frac{1}{\sqrt{n}} \sum_{j=k+1}^{n} X_j$$

(where the sum is taken to be 0 if the upper limit is less than the lower limit). Then

$$\begin{aligned} & E[f(S_n)] - E[f(Z)] \\ = & \sum_{k=1}^{n} E[f(T_{nk} + n^{-1/2} X_k) - f(T_{nk} + n^{-1/2} Z_k)]. \end{aligned}$$

Expanding $f(T_{nk} + n^{-1/2} X_k)$ in a Taylor series around T_{nk}, we get

$$f(T_{nk} + n^{-1/2} X_k) = f(T_{nk}) + \frac{X_k}{\sqrt{n}} f'(T_{nk}) + \frac{X_k^2}{2n} f''(T_{nk}) + R_k^X$$

where R_k^X is a remainder term (whose value will depend on the third derivative of f); similarly,

$$f(T_{nk} + n^{-1/2} Z_k) = f(T_{nk}) + \frac{Z_k}{\sqrt{n}} f'(T_{nk}) + \frac{Z_k^2}{2n} f''(T_{nk}) + R_k^Z.$$

Taking expected values (and noting that T_{nk} is independent of both X_k and Z_k), we get

$$E[f(S_n)] - E[f(Z)] = \sum_{k=1}^{n} [E(R_k^X) - E(R_k^Z)].$$

We now try to find bounds for R_k^X and R_k^Z; it follows that

$$|R_k^X| \le \frac{K}{6} \frac{|X_k|^3}{n^{3/2}} \quad \text{and} \quad |R_k^Z| \le \frac{K}{6} \frac{|Z_k|^3}{n^{3/2}}$$

where K is an upper bound on $|f'''(y)|$, which (by assumption) is finite. Thus

$$\begin{aligned}|E[f(S_n)] - E[f(Z)]| &\leq \sum_{k=1}^{n} |E(R_k^X) - E(R_k^Z)| \\ &\leq \sum_{k=1}^{n} [E(|R_k^X|) + E(|R_k^Z|)] \\ &\leq \frac{K}{6\sqrt{n}} (E[|X_1|^3] + E[|Z_1|^3]) \\ &\to 0\end{aligned}$$

as $n \to \infty$ since both $E[|X_1|^3]$ and $E[|Z_1|^3]$ are finite.

Applying the previous result to f_δ^+ and f_δ^- having three bounded continuous derivatives, it follows that $E[f_\delta^+(S_n)] \to E[f_\delta^+(Z)]$ and $E[f_\delta^-(S_n)] \to E[f_\delta^-(Z)]$. Now since

$$P[Z \leq x \pm \delta] \to P[Z \leq x]$$

as $\delta \to 0$, it follows from above that for each x,

$$P[S_n \leq x] \to P[Z \leq x].$$

We have, of course, assumed that $E[|X_i|^3]$ is finite; to extend the result to the case where we assume only that $E(X_i^2)$ is finite, it is necessary to find a more accurate bound on $|R_k^X|$. Such a bound is given by

$$|R_k^X| \leq K' \left(\frac{|X_k|^3}{6n^{3/2}} I(|X_k| \leq \epsilon\sqrt{n}) + \frac{X_k^2}{n} I(|X_k| > \epsilon\sqrt{n}) \right)$$

where K' is an upper bound on both $|f''(y)|$ and $|f'''(y)|$. It then can be shown that

$$\sum_{k=1}^{n} E[|R_k^X|] \to 0$$

and so $E[f(S_n)] \to E[f(Z)]$.

Using the CLT as an approximation theorem

In mathematics, a distinction is often made between limit theorems and approximation theorems; the former simply specifies the limit of a sequence while the latter provides an estimate or bound on the difference between an element of the sequence and its limit. For example, it is well-known that $\ln(1 + x)$ can be approximated by x when x is small; a crude bound on the absolute difference

$|\ln(1+x) - x|$ is $x^2/2$ when $x \geq 0$ and $x^2/[2(1+x)^2]$ when $x < 0$. For practical purposes, approximation theorems are more useful as they allow some estimate of the error made in approximating by the limit.

The CLT as stated here is not an approximation theorem. That is, it does not tell us how large n should be in order for a Normal distribution to approximate the distribution of S_n. Nonetheless, with additional assumptions, the CLT can be restated as an approximation theorem. Let F_n be the distribution function of S_n and Φ be the standard Normal distribution function. To gain some insight into the factors affecting the speed of convergence of F_n to Φ, we will use Edgeworth expansions. Assume that F_n is a continuous distribution function and that $E(X_i^4) < \infty$ and define

$$\gamma = \frac{E[(X_i - \mu)^3]}{\sigma^3} \quad \text{and} \quad \kappa = \frac{E[(X_i - \mu)^4]}{\sigma^4} - 3;$$

γ and κ are, respectively, the skewness and kurtosis of the distribution of X_i both of which are 0 when X_i is normally distributed. It is now possible to show that

$$F_n(x) = \Phi(x) - \phi(x)\left(\frac{\gamma}{6\sqrt{n}}p_1(x) + \frac{\kappa}{24n}p_2(x) + \frac{\gamma^2}{72n}p_3(x)\right) + r_n(x)$$

where $\phi(x)$ is the standard Normal density function and $p_1(x)$, $p_2(x)$, $p_3(x)$ are the polynomials

$$\begin{aligned} p_1(x) &= x^2 - 1 \\ p_2(x) &= x^3 - 3x \\ p_3(x) &= x^5 - 10x^3 + 15x; \end{aligned}$$

the remainder term $r_n(x)$ satisfies $nr_n(x) \to 0$. From this expansion, it seems clear that the approximation error $|F_n(x) - \Phi(x)|$ depends on the skewness and kurtosis (that is, γ and κ) of the X_i's. The skewness and kurtosis are simple measures of how a particular distribution differs from normality; skewness is a measure of the asymmetry of a distribution ($\gamma = 0$ if the distribution is symmetric around its mean) while kurtosis is a measure of the thickness of the tails of a distribution ($\kappa > 0$ indicates heavier tails than a Normal distribution while $\kappa < 0$ indicates lighter tails). For example, a Uniform distribution has $\gamma = 0$ and $\kappa = -1.2$ while an Exponential distribution has $\gamma = 2$ and $\kappa = 6$. Thus we should expect convergence to occur more quickly for sums of Uniform random variables than for sums of Exponential random variables.

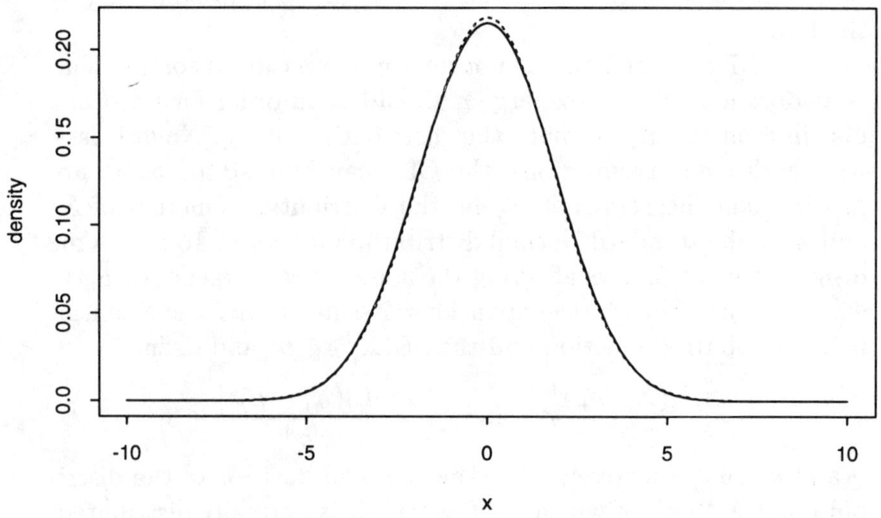

Figure 3.4 *Density of the sum of 10 Uniform random variables; the dotted curve is the approximating Normal density.*

Indeed, this is true; in fact, the distribution of a sum of as few as ten Uniform random variables is sufficiently close to a Normal distribution to allow generation of Normal random variables on a computer by summing Uniform random variables. To illustrate the difference in the accuracy of the Normal approximation, we consider the distribution of $X_1 + \cdots + X_{10}$ when the X_i's are Uniform and Exponential; Figures 3.4 and 3.5 give the exact densities and their Normal approximations in these two cases.

The speed of convergence of the CLT (and hence the goodness of approximation) can often be improved by applying transformations to reduce the skewness and kurtosis of $\bar{X}_n$. Recall that if $\sqrt{n}(\bar{X}_n - \mu) \to_d Z$ then

$$\sqrt{n}(g(\bar{X}_n) - g(\mu)) \to_d g'(\mu)Z.$$

If g is chosen so that the distribution of $g(\bar{X}_n)$ is more symmetric and has lighter tails than that of $\bar{X}_n$ then the CLT should provide a more accurate approximation for the distribution of $\sqrt{n}(g(\bar{X}_n) - g(\mu))$ than it does for the distribution of $\sqrt{n}(\bar{X}_n - \mu)$.

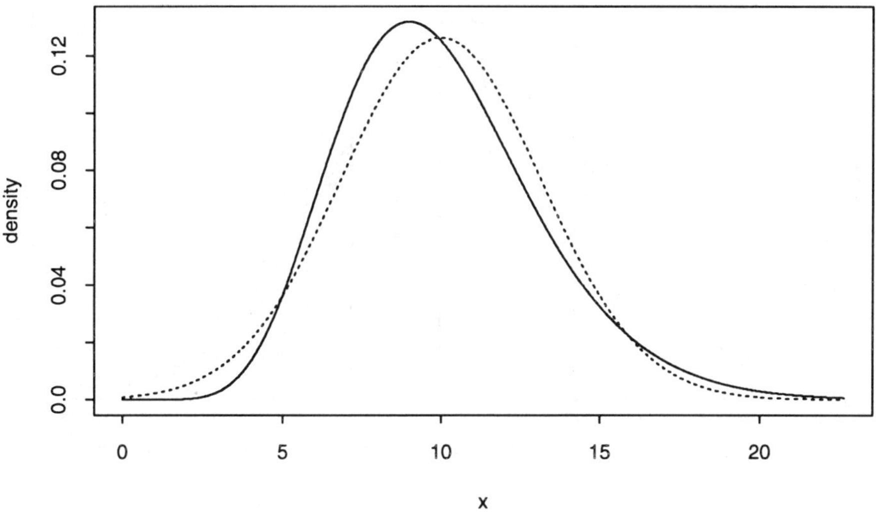

Figure 3.5 *Density of the sum of 10 Exponential random variables; the dotted curve is the approximating Normal density.*

Although the Edgeworth expansion above does not always hold when the X_i's are discrete, the preceding comments regarding speed of convergence and accuracy of the Normal approximation are still generally true. However, when the X_i's are discrete, there is a simple technique that can improve the accuracy of the Normal approximation. We will illustrate this technique for the Binomial distribution. Suppose that X is a Binomial random variable with parameters n and θ; X can be thought of as a sum of n i.i.d. Bernoulli random variables so the distribution of X can be approximated by a Normal distribution if n is sufficiently large. More specifically, the distribution of

$$\frac{X - n\theta}{\sqrt{n\theta(1-\theta)}}$$

is approximately standard Normal for large n. Suppose we want to evaluate $P[a \leq X \leq b]$ for some integers a and b. Ignoring the fact that X is a discrete random variable and the Normal distribution

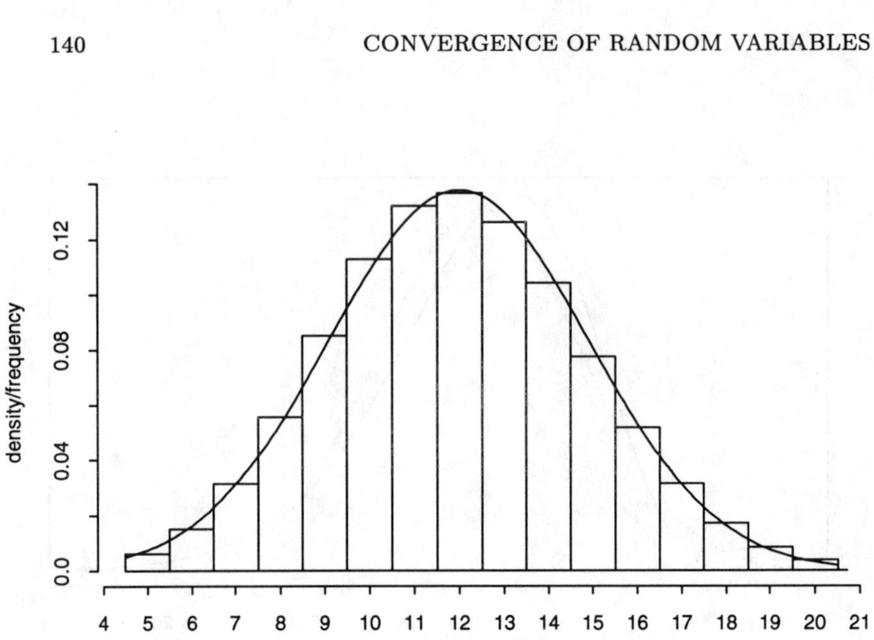

Figure 3.6 *Binomial distribution* ($n = 40$, $\theta = 0.3$) *and approximating Normal density*

is a continuous distribution, a naive application of the CLT gives

$$\begin{aligned} & P[a \le X \le b] \\ &= P\left[\frac{a - n\theta}{\sqrt{n\theta(1-\theta)}} \le \frac{X - n\theta}{\sqrt{n\theta(1-\theta)}} \le \frac{b - n\theta}{\sqrt{n\theta(1-\theta)}}\right] \\ &\approx \Phi\left(\frac{b - n\theta}{\sqrt{n\theta(1-\theta)}}\right) - \Phi\left(\frac{a - n\theta}{\sqrt{n\theta(1-\theta)}}\right). \end{aligned}$$

How can this approximation be improved? The answer is clear if we compare the exact distribution of X to its Normal approximation. The distribution of X can be conveniently represented as a probability histogram as in Figure 3.6 with the area of each bar representing the probability that X takes a certain value. The naive Normal approximation given above merely integrates the approximating Normal density from $a = 8$ to $b = 17$; this probability is represented by the shaded area in Figure 3.7. It seems that the naive Normal approximation will underestimate the true probability and Figures 3.7 and 3.8 suggest that a better approximation

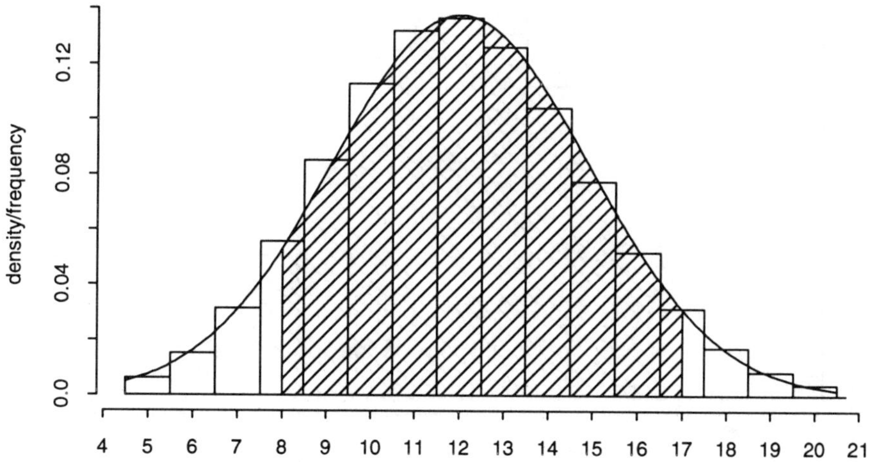

Figure 3.7 *Naive Normal approximation of* $P(8 \leq X \leq 17)$

may be obtained by integrating from $a - 0.5 = 7.5$ to $b + 0.5 = 17.5$. This corrected Normal approximation is

$$\begin{aligned} P[a \leq X \leq b] &= P[a - 0.5 \leq X \leq b + 0.5] \\ &\approx \Phi\left(\frac{b + 0.5 - n\theta}{\sqrt{n\theta(1-\theta)}}\right) - \Phi\left(\frac{a - 0.5 - n\theta}{\sqrt{n\theta(1-\theta)}}\right). \end{aligned}$$

The correction used here is known as a continuity correction and can be applied generally to improve the accuracy of the Normal approximation for sums of discrete random variables. (In Figures 3.6, 3.7, and 3.8, X has a Binomial distribution with parameters $n = 40$ and $\theta = 0.3$.)

Some other Central Limit Theorems

CLTs can be proved under a variety of conditions; neither the assumption of independence nor that of identical distributions are necessary. In this section, we will consider two simple modifications of the CLT for sums of i.i.d. random variables. The first modification deals with weighted sums of i.i.d. random variables while

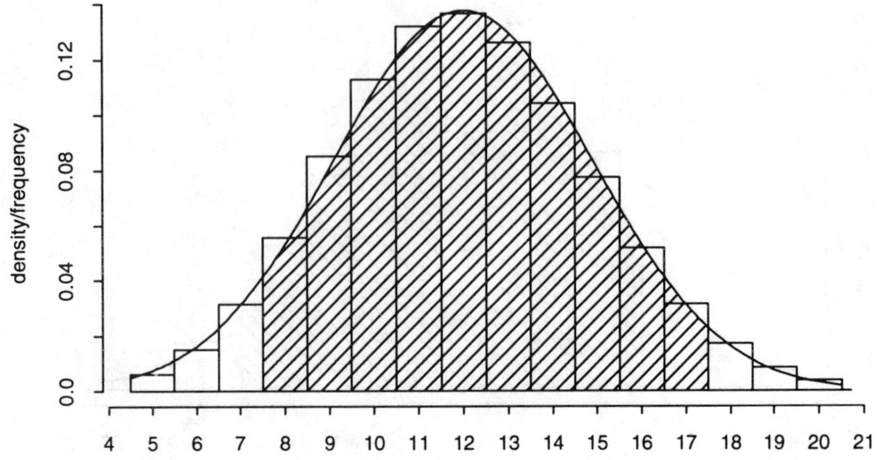

Figure 3.8 *Normal approximation of* $P(8 \leq X \leq 17)$ *with continuity correction*

the second deals with sums of independent but not identically distributed random variables with finite third moment.

THEOREM 3.9 (CLT for weighted sums) *Suppose that* X_1, $X_2, \cdots$ *are i.i.d. random variables with* $E(X_i) = 0$ *and* $Var(X_i) = 1$ *and let* $\{c_i\}$ *be a sequence of constants. Define*

$$S_n = \frac{1}{s_n} \sum_{i=1}^{n} c_i X_i \quad \textit{where} \quad s_n^2 = \sum_{i=1}^{n} c_i^2.$$

Then $S_n \to_d Z$*, a standard Normal random variable, provided that*

$$\max_{1 \leq i \leq n} \frac{c_i^2}{s_n^2} \to 0$$

as $n \to \infty$.

What is the practical meaning of the "negligibility" condition on the constants $\{c_i\}$ given above? For each n, it is easy to see that

$\text{Var}(S_n) = 1$. Now writing

$$S_n = \sum_{i=1}^{n} \frac{c_i}{s_n} X_i = \sum_{i=1}^{n} Y_{ni}$$

and noting that $\text{Var}(Y_{ni}) = c_i^2/s_n^2$, it follows that this condition implies that no single component of the sum S_n contributes an excessive proportion of the variance of S_n. For example, the condition rules out situations where S_n depends only on a negligible proportion of the Y_{ni}'s. An extreme example of this occurs when $c_1 = c_2 = \cdots = c_k = 1$ (for some fixed k) and all other c_i's are 0; in this case,

$$\max_{1 \le i \le n} \frac{c_i^2}{s_n^2} = \frac{1}{k},$$

which does not tend to 0 as $n \to \infty$ since k is fixed. On the other hand, if $c_i = i$ then

$$s_n^2 = \sum_{i=1}^{n} i^2 = \frac{1}{6} n(2n^2 + 3n + 1)$$

and

$$\max_{1 \le i \le n} \frac{c_i^2}{s_n^2} = \frac{6n^2}{n(2n^2 + 3n + 1)} \to 0$$

as $n \to \infty$ and so the negligibility condition holds. Thus if the X_i's are i.i.d. random variables with $E(X_i) = 0$ and $\text{Var}(X_i) = 1$, it follows that

$$\frac{1}{s_n} \sum_{i=1}^{n} i X_i \to_d Z$$

where Z has a standard Normal distribution.

When the negligibility condition of Theorem 3.9 fails, it may still be possible to show that the weighted sum S_n converges in distribution although the limiting distribution will typically be non-Normal.

EXAMPLE 3.9: Suppose that $X_1, X_2, \cdots$ are i.i.d. random variables with common density function

$$f(x) = \frac{1}{2} \exp(-|x|)$$

(called a Laplace distribution) and define

$$S_n = \frac{1}{s_n} \sum_{k=1}^{n} \frac{X_k}{k}$$

where $s_n^2 = \sum_{i=1}^n k^{-2}$. Note that $s_n^2 \to \sum_{k=1}^\infty k^{-2} = \pi^2/6$ as $n \to \infty$ and the negligibility condition does not hold. However, it can be shown that $S_n \to_d S$. Since $s_n \to \pi/\sqrt{6}$, we will consider the limiting distribution of $V_n = s_n S_n$; if $V_n \to_d V$ then $S_n \to_d \sqrt{6}/\pi V = S$. The moment generating function of X_i is

$$m(t) = \frac{1}{1-t^2} \quad \text{(for } |t| < 1\text{)}$$

and so the moment generating function of V_n is

$$m_n(t) = \prod_{k=1}^n m(t/k) = \prod_{k=1}^n \left(\frac{k^2}{k^2 - t^2}\right).$$

As $n \to \infty$, $m_n(t) \to m_V(t)$ where

$$m_V(t) = \prod_{k=1}^\infty \left(\frac{k^2}{k^2 - t^2}\right) = \Gamma(1+t)\Gamma(1-t)$$

for $|t| < 1$. The limiting moment generating function $m_V(t)$ is not immediately recognizable. However, note that

$$\begin{aligned}\int_{-\infty}^\infty \exp(tx)\frac{\exp(x)}{(1+\exp(x))^2}\,dx &= \int_0^1 u^t(1-u)^{-t}\,du \\ &= \frac{\Gamma(1+t)\Gamma(1-t)}{\Gamma(2)} \\ &= \Gamma(1+t)\Gamma(1-t)\end{aligned}$$

and so the density of V is

$$f_V(x) = \frac{\exp(x)}{(1+\exp(x))^2}$$

(this distribution is often called the Logistic distribution). The density of S is thus

$$f_S(x) = \frac{\pi}{\sqrt{6}} \frac{\exp(\pi x/\sqrt{6})}{(1+\exp(\pi x/\sqrt{6}))^2}.$$

The limiting density $f_S(x)$ is shown in Figure 3.9. ◇

Another useful CLT for sums of independent (but not identically distributed) random variables is the Lyapunov CLT. Like the CLT for weighted sums of i.i.d. random variables, this CLT depends on a condition that can be easily verified.

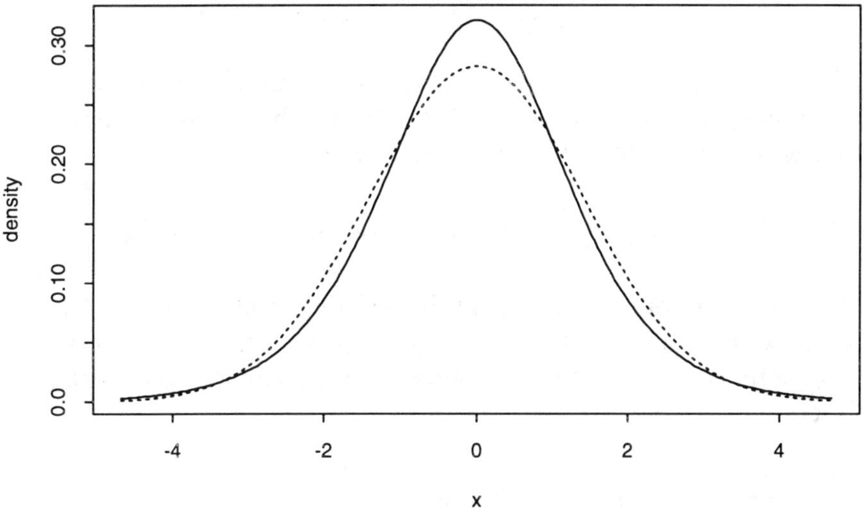

Figure 3.9 *Density of S; the dotted curve is a Normal density with the same mean and variance as S.*

THEOREM 3.10 (Lyapunov CLT) *Suppose that $X_1, X_2, \cdots$ are independent random variables with $E(X_i) = 0$, $E(X_i^2) = \sigma_i^2$ and $E(|X_i|^3) = \gamma_i$ and define*

$$S_n = \frac{1}{s_n} \sum_{i=1}^{n} X_i$$

where $s_n^2 = \sum_{i=1}^{n} \sigma_i^2$. If

$$\lim_{n \to \infty} \frac{1}{s_n^{3/2}} \sum_{i=1}^{n} \gamma_i = 0$$

then $S_n \to_d Z$, a standard Normal random variable.

It is possible to adapt the proof of the CLT for sums of i.i.d. random variables to the two CLTs given in this section. In the case of the CLT for weighted sums, the key modification lies in redefining T_{nk} to be

$$T_{nk} = \frac{1}{s_n} \sum_{j=1}^{k-1} c_j Z_j + \frac{1}{s_n} \sum_{j=k+1}^{n} c_j X_j$$

where $Z_1, Z_2, \cdots$ are independent standard Normal random variables independent of the X_i's. Then letting f be a bounded function with three bounded derivatives, we have

$$E[f(S_n)] - E[f(Z)] = \sum_{k=1}^{n} E[f(T_{nk} + c_k X_k / s_n) - f(T_{nk} + c_k Z_k / s_n)].$$

The remainder of the proof is much the same as before and is left as an exercise. It is also possible to give a proof using moment generating functions assuming, of course, that the moment generating function of X_i exists.

Multivariate Central Limit Theorem

CLTs for sums of random variables can be generalized to deal with sums of random vectors. For example, suppose that $\boldsymbol{X}_1, \boldsymbol{X}_2, \cdots$ are i.i.d. random vectors with mean vector $\boldsymbol{\mu}$ and variance-covariance matrix C; define

$$\bar{\boldsymbol{X}}_n = \frac{1}{n} \sum_{i=1}^{n} \boldsymbol{X}_i$$

to be the (coordinate-wise) sample mean of $\boldsymbol{X}_1, \cdots, \boldsymbol{X}_n$. The logical extension of the CLT for i.i.d. random variables is to consider the limiting behaviour of the distributions of $\sqrt{n}(\bar{\boldsymbol{X}}_n - \boldsymbol{\mu})$.

Before considering any multivariate CLT, we need to extend the notion of convergence in distribution to sequences of random vectors. This extension is fairly straightforward and involves the joint distribution functions of the random vectors; given random vectors $\{\boldsymbol{X}_n\}$ and $\boldsymbol{X}$, we say that $\boldsymbol{X}_n \to_d \boldsymbol{X}$ if

$$\lim_{n\to\infty} P[\boldsymbol{X}_n \le \boldsymbol{x}] = P[\boldsymbol{X} \le \boldsymbol{x}] = F(\boldsymbol{x})$$

at each continuity point $\boldsymbol{x}$ of the joint distribution function F of $\boldsymbol{X}$. ($\boldsymbol{X} \le \boldsymbol{x}$ means that each coordinate of $\boldsymbol{X}$ is less than or equal to the corresponding coordinate of $\boldsymbol{x}$.) This definition, while simple enough, is difficult to prove analytically. Fortunately, convergence in distribution of random vectors can be cast in terms of their one-dimensional projections.

THEOREM 3.11 *Suppose that $\{\boldsymbol{X}_n\}$ and $\boldsymbol{X}$ are random vectors. Then $\boldsymbol{X}_n \to_d \boldsymbol{X}$ if, and only if,*

$$\boldsymbol{t}^T \boldsymbol{X}_n \to_d \boldsymbol{t}^T \boldsymbol{X}$$

for all vectors $\boldsymbol{t}$.

Theorem 3.11 is called the Cramér-Wold device. The proof of this result will not be given here. The result is extremely useful for proving multivariate CLTs since it essentially reduces multivariate CLTs to special cases of univariate CLTs. We will only consider a multivariate CLT for sums of i.i.d. random vectors but more general multivariate CLTs can also be deduced from appropriate univariate CLTs.

THEOREM 3.12 (Multivariate CLT) *Suppose that $\boldsymbol{X}_1$, $\boldsymbol{X}_2$, $\boldsymbol{X}_3, \cdots$ are i.i.d. random vectors with mean vector $\boldsymbol{\mu}$ and variance-covariance matrix C and define*

$$\boldsymbol{S}_n = \frac{1}{\sqrt{n}} \sum_{i=1}^{n} (\boldsymbol{X}_i - \boldsymbol{\mu}) = \sqrt{n}(\bar{\boldsymbol{X}}_n - \boldsymbol{\mu}).$$

Then $\boldsymbol{S}_n \to_d \boldsymbol{Z}$ where $\boldsymbol{Z}$ has a multivariate Normal distribution with mean $\boldsymbol{0}$ and variance-covariance matrix C.

Proof. It suffices to show that $\boldsymbol{t}^T \boldsymbol{S}_n \to_d \boldsymbol{t}^T \boldsymbol{Z}$; note that $\boldsymbol{t}^T \boldsymbol{Z}$ is Normally distributed with mean 0 and variance $\boldsymbol{t}^T C \boldsymbol{t}$. Now

$$\begin{aligned} \boldsymbol{t}^T \boldsymbol{S}_n &= \frac{1}{\sqrt{n}} \sum_{i=1}^{n} \boldsymbol{t}^T (\boldsymbol{X}_i - \boldsymbol{\mu}) \\ &= \frac{1}{\sqrt{n}} \sum_{i=1}^{n} Y_i \end{aligned}$$

where the Y_i's are i.i.d. with $E(Y_i) = 0$ and $\text{Var}(Y_i) = \boldsymbol{t}^T C \boldsymbol{t}$. Thus by the CLT for i.i.d. random variables,

$$\boldsymbol{t}^T \boldsymbol{S}_n \to_d N(0, \boldsymbol{t}^T C \boldsymbol{t})$$

and the theorem follows. □

The definition of convergence in probability can be extended quite easily to random vectors. We will say that $\boldsymbol{X}_n \to_p \boldsymbol{X}$ if each coordinate of $\boldsymbol{X}_n$ converges in probability to the corresponding coordinate of $\boldsymbol{X}$. Equivalently, we can say that $\boldsymbol{X}_n \to_p \boldsymbol{X}$ if

$$\lim_{n \to \infty} P[\|\boldsymbol{X}_n - \boldsymbol{X}\| > \epsilon] = 0$$

where $\|\cdot\|$ is the Euclidean norm of a vector.

It is possible to generalize many of the results proved above. For example, suppose that $\boldsymbol{X}_n \to_d \boldsymbol{X}$; then if g is a continuous real-valued function, $g(\boldsymbol{X}_n) \to_d g(\boldsymbol{X})$. (The same is true if $\to_d$ is replaced by $\to_p$.) This multivariate version of the Continuous Mapping Theorem can be used to obtain a generalization of Slutsky's Theorem. Suppose that $X_n \to_d X$ and $Y_n \to_p \theta$. By using the Cramér-Wold device (Theorem 3.11) and Slutsky's Theorem, it follows that $(X_n, Y_n) \to_d (X, \theta)$. Thus if $g(x, y)$ is a continuous function, we have

$$g(X_n, Y_n) \to_d g(X, \theta).$$

EXAMPLE 3.10: Suppose that $\{\boldsymbol{X}_n\}$ is a sequence of random vectors with $\boldsymbol{X}_n \to_d \boldsymbol{Z}$ where $\boldsymbol{Z} \sim N_p(\boldsymbol{0}, C)$ and C is non-singular. Define the function

$$g(\boldsymbol{x}) = \boldsymbol{x}^T C^{-1} \boldsymbol{x},$$

which is a continuous function of $\boldsymbol{x}$. Then we have

$$g(\boldsymbol{X}_n) = \boldsymbol{X}_n^T C^{-1} \boldsymbol{X}_n \to_d \boldsymbol{Z}^T C^{-1} \boldsymbol{Z} = g(\boldsymbol{Z}).$$

It follows from Chapter 2 that the random variable $\boldsymbol{Z}^T C^{-1} \boldsymbol{Z}$ has a χ^2 distribution with p degrees of freedom. Thus for large n, $\boldsymbol{X}_n^T C^{-1} \boldsymbol{X}_n$ is approximately χ^2 with p degrees of freedom. ◇

It is also possible to extend the Delta Method to the multivariate case. Let $\{a_n\}$ be a sequence of constants tending to infinity and suppose that

$$a_n(\boldsymbol{X}_n - \boldsymbol{\theta}) \to_d \boldsymbol{Z}.$$

If $\boldsymbol{g}(\boldsymbol{x}) = (g_1(\boldsymbol{x}), \cdots, g_k(\boldsymbol{x}))$ is a vector-valued function that is continuously differentiable at $\boldsymbol{x} = \boldsymbol{\theta}$, we have

$$a_n(\boldsymbol{g}(\boldsymbol{X}_n) - \boldsymbol{g}(\boldsymbol{\theta})) \to_d D(\boldsymbol{\theta})\boldsymbol{Z}$$

where $D(\boldsymbol{\theta})$ is a matrix of partial derivatives of $\boldsymbol{g}$ with respect to $\boldsymbol{x}$ evaluated at $\boldsymbol{x} = \boldsymbol{\theta}$; more precisely, if $\boldsymbol{x} = (x_1, \cdots, x_p)$,

$$D(\boldsymbol{\theta}) = \begin{pmatrix} \frac{\partial}{\partial x_1} g_1(\boldsymbol{\theta}) & \cdots & \frac{\partial}{\partial x_p} g_1(\boldsymbol{\theta}) \\ \frac{\partial}{\partial x_1} g_2(\boldsymbol{\theta}) & \cdots & \frac{\partial}{\partial x_p} g_2(\boldsymbol{\theta}) \\ \cdots & \cdots & \cdots \\ \frac{\partial}{\partial x_1} g_k(\boldsymbol{\theta}) & \cdots & \frac{\partial}{\partial x_p} g_k(\boldsymbol{\theta}) \end{pmatrix}.$$

The proof of this result parallels that of the Delta Method given earlier and is left as an exercise.

EXAMPLE 3.11: Suppose that $(X_1, Y_1), \cdots, (X_n, Y_n)$ are i.i.d. pairs of random variables with $E(X_i) = \mu_X > 0$, $E(Y_i) = \mu_Y > 0$, and $E(X_i^2)$ and $E(Y_i^2)$ both finite. By the multivariate CLT, we have

$$\sqrt{n}\left(\begin{pmatrix}\bar{X}_n \\ \bar{Y}_n\end{pmatrix} - \begin{pmatrix}\mu_X \\ \mu_Y\end{pmatrix}\right) \to_d \boldsymbol{Z} \sim N_2(\boldsymbol{0}, C)$$

where C is the variance-covariance matrix of (X_i, Y_i). We want to consider the asymptotic distribution of $\bar{X}_n/\bar{Y}_n$. Applying the Delta Method (with $g(x, y) = x/y$), we have

$$\begin{aligned}\sqrt{n}\left(\frac{\bar{X}_n}{\bar{Y}_n} - \frac{\mu_X}{\mu_Y}\right) &\to_d D(\mu_X, \mu_Y)\boldsymbol{Z} \\ &\sim N(0, D(\mu_X, \mu_Y)CD(\mu_X, \mu_Y)^T)\end{aligned}$$

where

$$D(\mu_X, \mu_Y) = \left(\frac{1}{\mu_Y}, -\frac{\mu_X}{\mu_Y^2}\right).$$

Letting $\text{Var}(X_i) = \sigma_X^2$, $\text{Var}(Y_i) = \sigma_Y^2$, and $\text{Cov}(X_i, Y_i) = \sigma_{XY}$, it follows that the variance of the limiting Normal distribution is

$$D(\mu_X, \mu_Y)CD(\mu_X, \mu_Y)^T = \frac{\mu_Y^2\sigma_X^2 - 2\mu_X\mu_Y\sigma_{XY} + \mu_X^2\sigma_Y^2}{\mu_Y^4}. \qquad \diamond$$

3.6 Some applications

In subsequent chapters, we will use many of the concepts and results developed in this chapter to characterize the large sample properties of statistical estimators. In this section, we will give some applications of the concepts and results given so far in this chapter.

Variance stabilizing transformations

The CLT states that if $X_1, X_2, \cdots$ are i.i.d. random variables with mean μ and variance σ^2 then

$$\sqrt{n}(\bar{X}_n - \mu) \to_d Z$$

where Z has a Normal distribution with mean 0 and variance σ^2. For many distributions, the variance σ^2 depends only on the mean

μ (that is, $\sigma^2 = V(\mu)$). In statistics, it is often desirable to find a function g such that the limit distribution of

$$\sqrt{n}(g(\bar{X}_n) - g(\mu))$$

does not depend on μ. (We could then use this result to find an approximate confidence interval for μ; see Chapter 7.) If g is differentiable, we have

$$\sqrt{n}(g(\bar{X}_n) - g(\mu)) \to_d g'(\mu)Z$$

and $g'(\mu)Z$ is Normal with mean 0 and variance $[g'(\mu)]^2 V(\mu)$; in order to make the limiting distribution independent of μ, we need to find g so that this variance is 1 (or some other constant). Thus, given $V(\mu)$, we would like to find g such that

$$[g'(\mu)]^2 V(\mu) = 1$$

or

$$g'(\mu) = \pm \frac{1}{V(\mu)^{1/2}}.$$

The function g can be the solution of either of the two differential equations depending on whether one wants g to be an increasing or a decreasing function of μ; g is called a variance stabilizing transformation.

EXAMPLE 3.12: Suppose that $X_1, \cdots, X_n$ are i.i.d. Bernoulli random variables with parameter θ. Then

$$\sqrt{n}(\bar{X}_n - \theta) \to_d Z \sim N(0, \theta(1-\theta)).$$

To find g such that

$$\sqrt{n}(g(\bar{X}_n) - g(\theta)) \to_d N(0, 1)$$

we solve the differential equation

$$g'(\theta) = \frac{1}{\sqrt{\theta(1-\theta)}}.$$

The general form of the solutions to this differential equation is

$$g(\theta) = \sin^{-1}(2\theta - 1) + c$$

where c is an arbitrary constant that could be taken to be 0. (The solutions to the differential equation can also be written $g(\theta) = 2\sin^{-1}(\sqrt{\theta}) + c$.) ◇

Variance stabilizing transformations often improve the speed of convergence to normality; that is, the distribution of $g(\bar{X}_n)$ can be better approximated by a Normal distribution than that of $\bar{X}_n$ if g is a variance stabilizing transformation. However, there may be other transformations that result in a better approximation by a Normal distribution.

A CLT for dependent random variables

Suppose that $\{U_i\}$ is an infinite sequence of i.i.d. random variables with mean 0 and variance σ^2 and define

$$X_i = \sum_{j=0}^{p} c_j U_{i-j}$$

where $c_0, \cdots, c_p$ are constants. Note that $X_1, X_2, \cdots$ are not necessarily independent since they can depend on common U_i's. (In time series analysis, $\{X_i\}$ is called a moving average process.) A natural question to ask is whether a CLT holds for sample means $\bar{X}_n$ based on $X_1, X_2, \cdots$

We begin by noting that

$$\begin{aligned}
\frac{1}{\sqrt{n}}\sum_{i=1}^{n} X_i &= \frac{1}{\sqrt{n}}\sum_{i=1}^{n}\sum_{j=0}^{p} c_j U_{i-j} \\
&= \frac{1}{\sqrt{n}}\sum_{j=0}^{p} c_j \sum_{i=1}^{n} U_{i-j} \\
&= \left(\sum_{j=0}^{p} c_j\right) \frac{1}{\sqrt{n}}\sum_{i=1}^{n} U_i + \sum_{j=1}^{p} c_j R_{nj}
\end{aligned}$$

where

$$\begin{aligned}
R_{nj} &= \frac{1}{\sqrt{n}}\left(\sum_{i=1}^{n} U_{i-j} - \sum_{i=1}^{n} U_i\right) \\
&= \frac{1}{\sqrt{n}}\left(U_{1-j} + \cdots + U_0 - U_{n-j+1} - \cdots - U_n\right).
\end{aligned}$$

Now $E(R_{nj}) = 0$, $\text{Var}(R_{nj}) = E(R_{nj}^2) = 2j\sigma^2/n$ and so by Chebyshev's inequality $R_{nj} \to_p 0$ as $n \to \infty$ for $j = 1, \cdots, p$; thus,

$$\sum_{j=1}^{p} c_j R_{nj} \to_p 0.$$

Finally,

$$\frac{1}{\sqrt{n}}\sum_{i=1}^{n} U_i \to_d N(0, \sigma^2)$$

and so applying Slutsky's Theorem, we get

$$\frac{1}{\sqrt{n}}\sum_{i=1}^{n} X_i \to_d Z$$

where Z has a Normal distribution with mean 0 and variance

$$\sigma^2 \left(\sum_{j=0}^{p} c_j\right)^2.$$

When $\sum_{j=0}^{p} c_j = 0$, the variance of the limiting Normal distribution is 0; this suggests that

$$\frac{1}{\sqrt{n}}\sum_{i=1}^{n} X_i \to_p 0$$

(if $\sum_{j=0}^{p} c_j = 0$). This is, in fact, the case. It follows from above that

$$\frac{1}{\sqrt{n}}\sum_{i=1}^{n} X_i = \sum_{j=1}^{p} c_j R_{nj},$$

which tends to 0 in probability. An extension to infinite moving averages is considered in Problem 3.22.

In general, what conditions are necessary to obtain a CLT for sums of dependent random variables $\{X_i\}$? Loosely speaking, it may be possible to approximate the distribution of $X_1 + \cdots + X_n$ by a Normal distribution (for sufficiently large n) if both

- the dependence between X_i and X_{i+k} becomes negligible as $k \to \infty$ (for each i), and
- each X_i accounts for a negligible proportion of the variance of $X_1 + \cdots + X_n$.

However, the conditions above provide only a very rough guideline for the possible existence of a CLT; much more specific technical conditions are typically needed to establish CLTs for sums of dependent random variables.

Monte Carlo integration.

Suppose we want to evaluate the multi-dimensional integral

$$\int \cdots \int g(\boldsymbol{x})\,d\boldsymbol{x}$$

where the function g is sufficiently complicated that the integral cannot be evaluated analytically. A variety of methods exist for evaluating integrals numerically. The most well-known of these involve deterministic approximations of the form

$$\int \cdots \int g(\boldsymbol{x})\,d\boldsymbol{x} \approx \sum_{i=1}^{m} a_i g(\boldsymbol{x}_i)$$

where $a_1, \cdots, a_n$, $\boldsymbol{x}_1, \cdots, \boldsymbol{x}_n$ are fixed points that depend on the method used; for a given function g, it is usually possible to give an explicit upper bound on the approximation error

$$\left| \int \cdots \int g(\boldsymbol{x})\,d\boldsymbol{x} - \sum_{i=1}^{m} a_i g(\boldsymbol{x}_i) \right|$$

and so the points $\{\boldsymbol{x}_i\}$ can be chosen to make this error acceptably small. However, as the dimension of domain of integration B increases, the number of points m needed to obtain a given accuracy increases exponentially. An alternative is to use so-called "Monte Carlo" sampling; that is, we evaluate g at random (as opposed to fixed) points. The resulting approximation is of the form

$$\sum_{i=1}^{m} A_i g(\boldsymbol{X}_i)$$

where the $\boldsymbol{X}_i$'s and possibly the A_i's are random. One advantage of using Monte Carlo integration is the fact that the order of the approximation error depends only on m and not the dimension of B. Unfortunately, Monte Carlo integration does not give a guaranteed error bound; hence, for a given value of m, we can never be absolutely certain that the approximation error is sufficiently small.

Why does Monte Carlo integration work? Monte Carlo integration exploits the fact that any integral can be expressed as the expected value of some real-valued function of a random variable or random vector. Since the WLLN says that sample means approximate population means (with high probability) if the sample size is sufficiently large, we can use the appropriate sample mean to approximate any given integral. To illustrate, we will consider evaluating the integral

$$\mathcal{I} = \int_0^1 g(x)\,dx.$$

Suppose that a random variable U has a Uniform distribution on $[0, 1]$. Then

$$E[g(U)] = \int_0^1 g(x)\,dx.$$

If $U_1, U_2, \cdots$ are i.i.d. Uniform random variables on $[0, 1]$, the WLLN says that

$$\frac{1}{n}\sum_{i=1}^{n} g(U_i) \to_p E[g(U)] = \int_0^1 g(x)\,dx \quad \text{as} \quad n \to \infty,$$

which suggests that $\int_0^1 g(x)\,dx$ may be approximated by the Monte Carlo estimate

$$\widehat{\mathcal{I}} = \frac{1}{n}\sum_{i=1}^{n} g(U_i)$$

if n is sufficiently large. (In practice, $U_1, \cdots, U_n$ are pseudo-random variables and so are not truly independent.)

Generally, it is not possible to obtain a useful absolute bound on the approximation error

$$\left|\frac{1}{n}\sum_{i=1}^{n} g(U_i) - \int_0^1 g(x)\,dx\right|$$

since this error is random. However, if $\int_0^1 g^2(x)\,dx < \infty$, it is possible to use the CLT to make a probability statement about the approximation error. Defining

$$\sigma_g^2 = \text{Var}[g(U_i)] = \int_0^1 g^2(x)\,dx - \left(\int_0^1 g(x)\,dx\right)^2,$$

it follows that

$$P\left(\left|\frac{1}{n}\sum_{i=1}^{n} g(U_i) - \int_0^1 g(x)\,dx\right| < \frac{a\sigma_g}{\sqrt{n}}\right) \approx \Phi(a) - \Phi(-a)$$

where $\Phi(x)$ is the standard Normal distribution function.

The simple Monte Carlo estimate of $\int_0^1 g(x)\,dx$ can be improved in a number of ways. We will mention two such methods: importance sampling and antithetic sampling.

Importance sampling exploits the fact that

$$\int_0^1 g(x)\,dx = \int_0^1 \frac{g(x)}{f(x)} f(x)\,dx = E\left(\frac{g(X)}{f(X)}\right)$$

where the random variable X has density f on $[0,1]$. Thus if $X_1, \cdots, X_n$ are i.i.d. random variables with density f, we can approximate $\int_0^1 g(x)\,dx$ by

$$\widetilde{\mathcal{I}} = \frac{1}{n}\sum_{i=1}^{n} \frac{g(X_i)}{f(X_i)}.$$

How do $\widehat{\mathcal{I}}$ and $\widetilde{\mathcal{I}}$ compare as estimates of $\mathcal{I} = \int_0^1 g(x)\,dx$? In simple terms, the estimates can be assessed by comparing their variances or, equivalently, $\text{Var}[g(U_i)]$ and $\text{Var}[g(X_i)/f(X_i)]$. It can be shown that $\text{Var}[g(X_i)/f(X_i)]$ is minimized by sampling the X_i's from the density

$$f(x) = \frac{|g(x)|}{\int_0^1 |g(x)|\,dx};$$

in practice, however, it may be difficult to generate random variables with this density. However, a significant reduction in variance can be obtained if the X_i's are sampled from a density f that is approximately proportional to $|g|$. (More generally, we could approximate the integral $\int_{-\infty}^{\infty} g(x)\,dx$ by

$$\frac{1}{n}\sum_{i=1}^{n} \frac{g(X_i)}{f(X_i)}$$

where $X_1, \cdots, X_n$ are i.i.d. random variables with density f.)

Antithetic sampling exploits the fact that

$$\int_0^1 g(x)\,dx = \frac{1}{2}\int_0^1 (g(x) + g(1-x))\,dx.$$

Hence if $U_1, U_2, \cdots$ are i.i.d. Uniform random variables on $[0,1]$, we can approximate $\int_0^1 g(x)\,dx$ by

$$\frac{1}{2n}\sum_{i=1}^{n}(g(U_i) + g(1-U_i)).$$

Antithetic sampling is effective when g is a monotone function (either increasing or decreasing); in this case, it can be shown that $\text{Cov}[g(U_i), g(1 - U_i)] \le 0$. Now comparing the variances of

$$\widehat{\mathcal{I}}_s = \frac{1}{2n} \sum_{i=1}^{2n} g(U_i) \quad \text{and} \quad \widehat{\mathcal{I}}_a = \frac{1}{2n} \sum_{i=1}^{n} (g(U_i) + g(1 - U_i)),$$

we obtain

$$\text{Var}(\widehat{\mathcal{I}}_s) = \frac{\text{Var}[g(U_1)]}{2n}$$

and

$$\begin{aligned} \text{Var}(\widehat{\mathcal{I}}_a) &= \frac{\text{Var}[g(U_1) + g(1 - U_1)]}{4n} \\ &= \frac{\text{Var}[g(U_1)] + \text{Cov}[g(U_1), g(1 - U_1)]}{2n}. \end{aligned}$$

Since $\text{Cov}[g(U_1), g(1 - U_1)] \le 0$ when g is monotone, it follows that $\text{Var}(\widehat{\mathcal{I}}_a) \le \text{Var}(\widehat{\mathcal{I}}_s)$.

EXAMPLE 3.13: Consider evaluating the integral

$$\int_0^1 \exp(-x) \cos(\pi x/2)\, dx;$$

the integrand is shown in Figure 3.10. This integral can be evaluated in closed-form as

$$\frac{2}{4 + \pi^2} (2 + \pi \exp(-1)) \approx 0.4551.$$

We will evaluate this integral using four Monte Carlo approaches. Let $U_1, \cdots, U_{1000}$ be i.i.d. Uniform random variables on the interval $[0, 1]$ and define the following four Monte Carlo estimates of the integral:

$$\begin{aligned} \widehat{\mathcal{I}}_1 &= \frac{1}{1000} \sum_{i=1}^{1000} \exp(-U_i) \cos(\pi U_i/2) \\ \widehat{\mathcal{I}}_2 &= \frac{1}{1000} \sum_{i=1}^{500} \exp(-U_i) \cos(\pi U_i/2) \\ &\quad + \frac{1}{1000} \sum_{i=1}^{500} \exp(U_i - 1) \cos(\pi (1 - U_i)/2) \\ \widehat{\mathcal{I}}_3 &= \frac{1}{1000} \sum_{i=1}^{1000} \frac{\exp(-V_i) \cos(\pi V_i/2)}{2(1 - V_i)} \end{aligned}$$

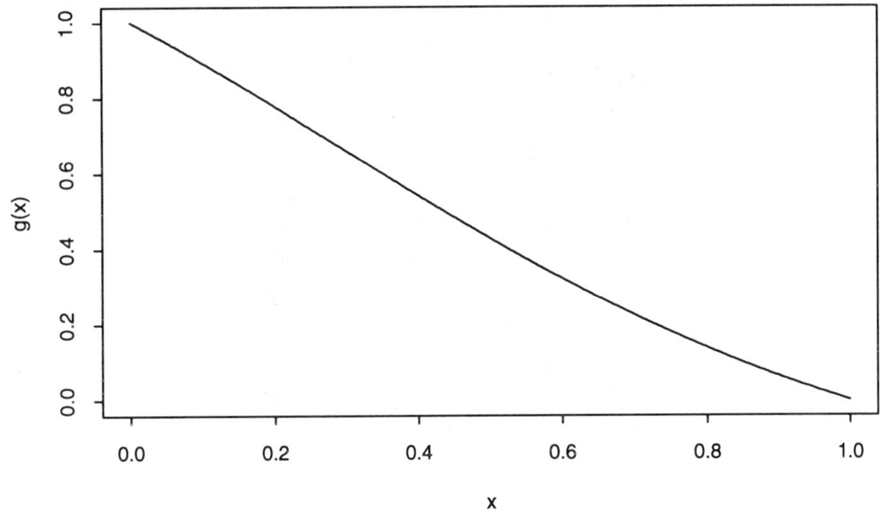

Figure 3.10 *Graph of the function* $g(x) = \exp(-x)\cos(\pi x/2)$.

$$\widehat{\mathcal{I}}_4 = \frac{1}{1000}\sum_{i=1}^{1000}\frac{\exp(-W_i)\cos(\pi W_i/2)}{2W_i}$$

where $V_i = 1-(1-U_i)^{1/2}$ and $W_i = U_i^{1/2}$. $\widehat{\mathcal{I}}_2$ is an antithetic sampling estimate of the integral while $\widehat{\mathcal{I}}_3$ and $\widehat{\mathcal{I}}_4$ are both importance sampling estimates; note that the density of V_i is $f_V(x) = 2(1-x)$ while the density of W_i is $f_W(x) = 2x$ (for $0 \le x \le 1$ in both cases). Each of the four estimates was evaluated for 10 samples of $U_1, \cdots, U_{1000}$ and the results presented in Table 3.2.

A glance at Table 3.2 reveals that the estimates $\widehat{\mathcal{I}}_2$ and $\widehat{\mathcal{I}}_3$ are the best while $\widehat{\mathcal{I}}_4$ is the clear loser; $\widehat{\mathcal{I}}_2$ comes the closet to the true value 4 times while $\widehat{\mathcal{I}}_3$ comes the closest the other 6 times. It is not surprising that $\widehat{\mathcal{I}}_3$ does so well; from Figure 3.10, we can see that the integrand $g(x) = \exp(-x)\cos(\pi x/2)$ is approximately $1 - x$ so that $g(x)/f_V(x) \approx 2$ and so $\widehat{\mathcal{I}}_3$ should be close to the optimal importance sampling estimate. The fact that $g(x) \approx 1 - x$ also explains the success of the antithetic sampling estimate $\widehat{\mathcal{I}}_2$: $g(U_i)+g(1-U_i) \approx 1$, which suggests that $\text{Var}(g(U_i)+g(1-U_i)) \approx 0$.

Table 3.2 *Monte Carlo estimates of* $\int_0^1 \exp(-x)\cos(\pi x/2)\,dx$.

$\widehat{\mathcal{I}}_1$	$\widehat{\mathcal{I}}_2$	$\widehat{\mathcal{I}}_3$	$\widehat{\mathcal{I}}_4$
0.4596	0.4522	0.4555	0.4433
0.4706	0.4561	0.4563	0.5882
0.4653	0.4549	0.4560	0.6569
0.4600	0.4559	0.4559	0.4259
0.4496	0.4551	0.4546	0.4907
0.4412	0.4570	0.4527	0.4206
0.4601	0.4546	0.4552	0.4289
0.4563	0.4555	0.4549	0.4282
0.4541	0.4538	0.4555	0.4344
0.4565	0.4534	0.4556	0.4849

(In fact, $\text{Var}(\widehat{\mathcal{I}}_2) = 0.97 \times 10^{-6}$ while $\text{Var}(\widehat{\mathcal{I}}_3) = 2.25 \times 10^{-6}$ and $\text{Var}(\widehat{\mathcal{I}}_1) = 90.95 \times 10^{-6}$; the variance of $\widehat{\mathcal{I}}_4$ is infinite.) ◇

3.7 Convergence with probability 1

Earlier in this chapter, we mentioned the existence of another type of convergence for sequences of random variables, namely convergence with probability 1. In the interest of completeness, we will discuss this type of convergence although we will not make use of it subsequently in the text; therefore, this section can be skipped without loss of continuity.

DEFINITION. A sequence of random variables $\{X_n\}$ converges with probability 1 (or almost surely) to a random variable X ($X_n \to_{wp1} X$) if

$$P\left(\left[\omega : \lim_{n\to\infty} X_n(\omega) = X(\omega)\right]\right) = 1.$$

What exactly does convergence with probability 1 mean? By the definition above, if $X_n \to_{wp1} X$ then $X_n(\omega) \to X(\omega)$ for all outcomes $\omega \in A$ with $P(A) = 1$. For a given $\omega \in A$ and $\epsilon > 0$, there exists a number $n_\epsilon(\omega)$ such that $|X_n(\omega) - X(\omega)| \leq \epsilon$ for all

$n \geq n_\epsilon(\omega)$. Now consider the sequence of sets $\{B_n(\epsilon)\}$ with

$$B_n(\epsilon) = \bigcup_{k=n}^{\infty} [|X_k - X| > \epsilon];$$

$\{B_n(\epsilon)\}$ is a decreasing sequence of sets (that is, $B_{n+1}(\epsilon) \subset B_n(\epsilon)$) and its limit will contain all ω's lying in infinitely many of the $B_n(\epsilon)$'s. If $\omega \in A$ then for n sufficiently large $|X_k(\omega) - X(\omega)| \leq \epsilon$ for $k \geq n$ and so $\omega \notin B_n(\epsilon)$ (for n sufficiently large). Thus $B_n(\epsilon) \cap A \downarrow \emptyset$ as $n \to \infty$. Likewise, if $\omega \notin A$ then ω will lie in infinitely many of the B_n's and so $B_n \cap A^c \downarrow A^c$. Thus

$$\begin{aligned} P(B_n(\epsilon)) &= P(B_n(\epsilon) \cap A) + P(B_n(\epsilon) \cap A^c) \\ &\to \lim_{n\to\infty} P(B_n(\epsilon) \cap A) + \lim_{n\to\infty} P(B_n(\epsilon) \cap A^c) \\ &= 0 \end{aligned}$$

Thus $X_n \to_{wp1} X$ implies that $P(B_n(\epsilon)) \to 0$ as $n \to \infty$ for all $\epsilon > 0$. Conversely, if $P(B_n(\epsilon)) \to 0$ then using the argument given above, it follows that $X_n \to_{wp1} X$. Thus $X_n \to_{wp1} X$ is equivalent to

$$\lim_{n\to\infty} P\left(\bigcup_{k=n}^{\infty} [|X_k - X| > \epsilon]\right) = 0$$

for all $\epsilon > 0$.

Using the condition given above, it is easy to see that if $X_n \to_{wp1} X$ then $X_n \to_p X$; this follows since

$$[|X_n - X| > \epsilon] \subset \bigcup_{k=n}^{\infty} [|X_k - X| > \epsilon]$$

and so

$$P(|X_n - X| > \epsilon) \leq P\left(\bigcup_{k=n}^{\infty} [|X_k - X| > \epsilon]\right) \to 0.$$

Note that if $[|X_{n+1} - X| > \epsilon] \subset [|X_n - X| > \epsilon]$ for all n then

$$P\left(\bigcup_{k=n}^{\infty} [|X_k - X| > \epsilon]\right) = P(|X_n - X| > \epsilon)$$

in which case $X_n \to_p X$ implies that $X_n \to_{wp1} X$.

EXAMPLE 3.14: Suppose that $X_1, X_2, \cdots$ are i.i.d. Uniform random variables on the interval $[0,1]$ and define

$$M_n = \max(X_1, \cdots, X_n).$$

In Example 3.1, we showed that $M_n \to_p 1$. However, note that $1 \geq M_{n+1}(\omega) \geq M_n(\omega)$ for all ω and so

$$\begin{aligned} [|M_{n+1} - 1| > \epsilon] &= [M_{n+1} < 1 - \epsilon] \\ &\subset [M_n < 1 - \epsilon] \\ &= [|M_n - 1| > \epsilon]. \end{aligned}$$

Thus

$$P\left(\bigcup_{k=n}^{\infty} [|M_k - 1| > \epsilon]\right) = P(|M_n - 1| > \epsilon) \to 0 \quad \text{as } n \to \infty$$

as shown in Example 3.1. Thus $M_n \to_{wp1} 1$. ◇

Example 3.14 notwithstanding, it is, in general, much more difficult to prove convergence with probability 1 than it is to prove convergence in probability. However, assuming convergence with probability 1 rather than convergence in probability in theorems (for example, in Theorems 3.2 and 3.3) can sometimes greatly facilitate the proofs of these results.

EXAMPLE 3.15: Suppose that $X_n \to_{wp1} X$ and $g(x)$ is a continuous function. Then $g(X_n) \to_{wp1} g(X)$. To see this, let A be the set of ω's for which $X_n(\omega) \to X(\omega)$. Since g is a continuous function $X_n(\omega) \to X(\omega)$ implies that $g(X_n(\omega)) \to g(X(\omega))$; this occurs for all ω's in the set A with $P(A) = 1$ and so $g(X_n) \to_{wp1} g(X)$. ◇

We can also extend the WLLN to the so-called Strong Law of Large Numbers (SLLN).

THEOREM 3.13 (Strong Law of Large Numbers) *Suppose that $X_1, X_2, \cdots$ are i.i.d. random variables with $E(|X_i|) < \infty$ and $E(X_i) = \mu$. Then*

$$\bar{X}_n = \frac{1}{n}\sum_{i=1}^{n} X_i \to_{wp1} \mu$$

as $n \to \infty$.

The SLLN was proved by Kolmogorov (1930). Its proof is more difficult than that of the WLLN but similar in spirit; see, for example, Billingsley (1995) for details.

There is a very interesting connection between convergence with probability 1 and convergence in distribution. It follows that convergence with probability 1 implies convergence in distribution; the converse is not true (since $X_n \to_d X$ means that the distribution functions converge and so all the X_n's can be defined on different sample spaces). However, there is a partial converse that is quite useful technically.

Suppose that $X_n \to_d X$; thus $F_n(x) \to F(x)$ for all continuity points of F. Then it is possible to define random variables $\{X_n^*\}$ and X^* with $X_n^* \sim F_n$ and $X^* \sim F$ such that $X_n^* \to_{wp1} X^*$. Constructing these random variables is remarkably simple. Let U be a Uniform random variable on the interval $[0,1]$ and define

$$X_n^* = F_n^{-1}(U) \quad \text{and} \quad X^* = F^{-1}(U)$$

where F_n^{-1} and F^{-1} are the inverses of the distribution functions of X_n and X respectively. It follows now that $X_n^* \sim F_n$ and $X^* \sim F$. Moreover, $X_n^* \to_{wp1} X^*$; the proof of this fact is left as an exercise but seems reasonable given that $F_n(x) \to F(x)$ for all but (at most) a countable number of x's. This representation is due to Skorokhod (1956).

As mentioned above, the ability to construct these random variables $\{X_n^*\}$ and X^* is extremely useful from a technical point of view. In the following examples, we give elementary proofs of the Continuous Mapping Theorem (Theorem 3.2) and the Delta Method (Theorem 3.4).

EXAMPLE 3.16: Suppose that $X_n \to_d X$ and $g(x)$ is a continuous function. We can then construct random variables $\{X_n^*\}$ and X^* such that $X_n^* =_d X_n$ and $X^* =_d X$ and $X_n^* \to_{wp1} X^*$. Since g is continuous, we have

$$g(X_n^*) \to_{wp1} g(X^*).$$

However, $g(X_n^*) =_d g(X_n)$ and $g(X^*) =_d g(X)$ and so it follows that

$$g(X_n) \to_d g(X)$$

since $g(X_n^*) \to_d g(X^*)$. ◇

EXAMPLE 3.17: Suppose that $Z_n = a_n(X_n - \theta) \to_d Z$ where $a_n \uparrow \infty$. Let $g(x)$ be a function that is differentiable at $x = \theta$. We construct the random variables $\{Z_n^*\}$ and Z^* having the same distributions as $\{Z_n\}$ and Z with $Z_n^* \to_{wp1} Z^*$. We can also define $X_n^* = a_n^{-1} Z_n^* + \theta$ which will have the same distribution as X_n; clearly, $X_n^* \to_{wp1} \theta$. Thus

$$\begin{aligned} a_n(g(X_n^*) - g(\theta)) &= \left(\frac{g(X_n^*) - g(\theta)}{X_n^* - \theta}\right) a_n(X_n^* - \theta) \\ &\to_{wp1} g'(\theta)Z^* \end{aligned}$$

since

$$\frac{g(X_n^*) - g(\theta)}{X_n^* - \theta} \to_{wp1} g'(\theta).$$

Now since $a_n(g(X_n^*) - g(\theta))$ and $a_n(g(X_n) - g(\theta))$ have the same distribution as do $g'(\theta)Z^*$ and $g'(\theta)Z$, it follows that

$$a_n(g(X_n) - g(\theta)) \to_d g'(\theta)Z.$$

Note that we have required only existence (and not continuity) of the derivative of $g(x)$ at $x = \theta$. ◇

3.8 Problems and complements

3.1: (a) Suppose that $\{X_n^{(1)}\}, \cdots, \{X_n^{(k)}\}$ are sequences of random variables with $X_n^{(i)} \to_p 0$ as $n \to \infty$ for each $i = 1, \cdots, k$. Show that

$$\max_{1 \le i \le k} |X_n^{(i)}| \to_p 0$$

as $n \to \infty$.

(b) Find an example to show that the conclusion of (a) is not necessarily true if the number of sequences $k = k_n \to \infty$.

3.2: Suppose that $X_1, \cdots, X_n$ are i.i.d. random variables with a distribution function $F(x)$ satisfying

$$\lim_{x \to \infty} x^\alpha (1 - F(x)) = \lambda > 0$$

for some $\alpha > 0$. Let $M_n = \max(X_1, \cdots, X_n)$. We want to show that $n^{-1/\alpha} M_n$ has a non-degenerate limiting distribution.

(a) Show that $n[1 - F(n^{1/\alpha} x)] \to \lambda x^{-\alpha}$ as $n \to \infty$ for any $x > 0$.

(b) Show that

$$\begin{aligned} P\left(n^{-1/\alpha}M_n \le x\right) &= [F(n^{1/\alpha})]^n \\ &= [1-(1-F(n^{1/\alpha}))]^n \\ &\to 1-\exp\left(-\lambda x^{-\alpha}\right) \end{aligned}$$

as $n \to \infty$ for any $x > 0$.

(c) Show that $P(n^{-1/\alpha}M_n \le 0) \to 0$ as $n \to \infty$.

(d) Suppose that the X_i's have a Cauchy distribution with density function

$$f(x) = \frac{1}{\pi(1+x^2)}.$$

Find the value of α such that $n^{-1/\alpha}M_n$ has a non-degenerate limiting distribution and give the limiting distribution function.

3.3: Suppose that $X_1, \cdots, X_n$ are i.i.d. Exponential random variables with parameter λ and let $M_n = \max(X_1, \cdots, X_n)$. Show that $M_n - \ln(n)/\lambda \to_d V$ where

$$P(V \le x) = \exp[-\exp(-\lambda x)]$$

for all x.

3.4: Suppose that $X_1, \cdots, X_n$ are i.i.d. nonnegative random variables with distribution function F. Define

$$U_n = \min(X_1, \cdots, X_n).$$

(a) Suppose that $F(x)/x \to \lambda$ as $x \to 0$. Show that $nU_n \to_d \text{Exp}(\lambda)$.

(b) Suppose that $F(x)/x^\alpha \to \lambda$ as $x \to 0$ for some $\alpha > 0$. Find the limiting distribution of $n^{1/\alpha}U_n$.

3.5: Suppose that X_N has a Hypergeometric distribution (see Example 1.13) with the following frequency function

$$f_N(x) = \binom{M_N}{x}\binom{N-M_N}{r_N - x}\Big/\binom{N}{r_N}$$

for $x = \max(0, r_N + M_N - N), \cdots, \min(M_N, r_N)$. When the population size N is large, it becomes somewhat difficult to compute probabilities using $f_N(x)$ so that it is desirable to find approximations to the distribution of X_N as $N \to \infty$.

(a) Suppose that $r_N \to r$ (finite) and $M_N/N \to \theta$ for $0 < \theta < 1$. Show that $X_N \to_d \text{Bin}(r, \theta)$ as $N \to \infty$

(b) Suppose that $r_N \to \infty$ with $r_N M_N/N \to \lambda > 0$. Show that $X_N \to_d \text{Pois}(\lambda)$ as $N \to \infty$.

3.6: Suppose that $\{X_n\}$ and $\{Y_n\}$ are two sequences of random variables such that

$$a_n(X_n - Y_n) \to_d Z$$

for a sequence of numbers $\{a_n\}$ with $a_n \to \infty$ (as $n \to \infty$).

(a) Suppose that $X_n \to_p \theta$. Show that $Y_n \to_p \theta$.

(b) Suppose that $X_n \to_p \theta$ and $g(x)$ is a function continuously differentiable at $x = \theta$. Show that

$$a_n(g(X_n) - g(Y_n)) \to_d g'(\theta)Z.$$

3.7: (a) Let $\{X_n\}$ be a sequence of random variables. Suppose that $E(X_n) \to \theta$ (where θ is finite) and $\text{Var}(X_n) \to 0$. Show that $X_n \to_p \theta$.

(b) A sequence of random variables $\{X_n\}$ converges in probability to infinity ($X_n \to_p \infty$) if for each $M > 0$,

$$\lim_{n\to\infty} P(X_n \leq M) = 0.$$

Suppose that $E(X_n) \to \infty$ and $\text{Var}(X_n) \leq kE(X_n)$ for some $k < \infty$. Show that $X_n \to_p \infty$. (Hint: Use Chebyshev's inequality to show that

$$P[|X_n - E(X_n)| > \epsilon E(X_n)] \to 0$$

for each $\epsilon > 0$.)

3.8: (a) Let g be a nonnegative even function ($g(x) = g(-x)$) that is increasing on $[0, \infty)$ and suppose that $E[g(X)] < \infty$. Show that

$$P[|X| > \epsilon] \leq \frac{E[g(X)]}{g(\epsilon)}$$

for any $\epsilon > 0$. (Hint: Follow the proof of Chebyshev's inequality making the appropriate changes.)

(b) Suppose that $E[|X_n|^r] \to 0$ as $n \to \infty$. Show that $X_n \to_p 0$.

3.9: Suppose that $X_1, \cdots, X_n$ are i.i.d. Poisson random variables with mean λ. By the CLT,

$$\sqrt{n}(\bar{X}_n - \lambda) \to_d N(0, \lambda).$$

(a) Find the limiting distribution of $\sqrt{n}(\ln(\bar{X}_n) - \ln(\lambda))$.

(b) Find a function g such that

$$\sqrt{n}(g(\bar{X}_n) - g(\lambda)) \to_d N(0,1).$$

3.10: Let $\{a_n\}$ be a sequence of constants with $a_n \to \infty$ and suppose that

$$a_n(X_n - \theta) \to_d Z$$

where θ is a constant.

(a) Let g be a function that is twice differentiable at θ and suppose that $g'(\theta) = 0$. Show that

$$a_n^2(g(X_n) - g(\theta)) \to_d \frac{1}{2}Z^2.$$

(b) Now suppose that g is k times differentiable at θ with $g'(\theta) = \cdots = g^{(k-1)}(\theta) = 0$. Find the limiting distribution of $a_n^k(g(X_n) - g(\theta))$. (Hint: Expand $g(X_n)$ in a Taylor series around θ.)

3.11: The sample median of i.i.d. random variables is asymptotically Normal provided that the distribution function F has a positive derivative at the median; when this condition fails, an asymptotic distribution may still exist but will be non-Normal. To illustrate this, let $X_1, \cdots, X_n$ be i.i.d. random variables with density

$$f(x) = \frac{1}{6}|x|^{-2/3} \quad \text{for } |x| \leq 1.$$

(Notice that this density has a singularity at 0.)

(a) Evaluate the distribution function of X_i and its inverse (the quantile function).

(b) Let M_n be the sample median of $X_1, \cdots, X_n$. Find the limiting distribution of $n^{3/2}M_n$. (Hint: use the extension of the Delta Method in Problem 3.10 by applying the inverse transformation from part (a) to the median of n i.i.d. Uniform random variables on $[0, 1]$.)

3.12: Suppose that $X_1, \cdots, X_n$ are i.i.d. random variables with common density

$$f(x) = \alpha x^{-\alpha-1} \quad \text{for } x \geq 1$$

where $\alpha > 0$. Define

$$S_n = \left(\prod_{i=1}^{n} X_i\right)^{1/n}.$$

(a) Show that $\ln(X_i)$ has an Exponential distribution.

(b) Show that $S_n \to_p \exp(1/\alpha)$. (Hint: Consider $\ln(S_n)$.)

(c) Suppose $\alpha = 10$ and $n = 100$. Evaluate $P(S_n > 1.12)$ using an appropriate approximation.

3.13: Suppose that $X_1, \cdots, X_n$ be i.i.d. discrete random variables with frequency function

$$f(x) = \frac{x}{21} \quad \text{for } x = 1, 2, \cdots, 6.$$

(a) Let $S_n = \sum_{k=1}^{n} kX_k$. Show that

$$\frac{(S_n - E(S_n))}{\sqrt{\text{Var}(S_n)}} \to_d N(0, 1).$$

(b) Suppose $n = 20$. Use a Normal approximation to evaluate $P(S_{20} \geq 1000)$.

(c) Suppose $n = 5$. Compute the exact distribution of S_n using the probability generating function of S_n (see Problems 1.18 and 2.8).

3.14: Suppose that $X_{n1}, X_{n2}, \cdots, X_{nn}$ are independent random variables with

$$P(X_{ni} = 0) = 1 - p_n \quad \text{and} \quad P(X_{ni} \leq x | X_{ni} \neq 0) = F(x).$$

Suppose that $\gamma(t) = \int_{-\infty}^{\infty} \exp(tx)\, dF(x) < \infty$ for t in a neighbourhood of 0.

(a) Show that the moment generating function of X_{ni} is

$$m_n(t) = p_n \gamma(t) + (1 - p_n)$$

(b) Let $S_n = \sum_{i=1}^{n} X_{ni}$ and suppose that $np_n \to \lambda > 0$ as $n \to \infty$. Show that S_n converges in distribution to a random variable S that has a compound Poisson distribution. (Hint: See Problem 2.7 for the moment generating function of a compound Poisson distribution.)

3.15: Suppose that $X_{n1}, X_{n2}, \cdots, X_{nn}$ are independent Bernoulli random variables with parameters $\theta_{n1}, \cdots, \theta_{nn}$ respectively. Define $S_n = X_{n1} + X_{n2} + \cdots + X_{nn}$.

(a) Show that the moment generating function of S_n is

$$m_n(t) = \prod_{i=1}^{n} (1 - \theta_{ni} + \theta_{ni} \exp(t)) .$$

(b) Suppose that

$$\sum_{i=1}^{n} \theta_{ni} \quad \to \quad \lambda > 0 \quad \text{and}$$
$$\max_{1 \le i \le n} \theta_{ni} \quad \to \quad 0$$

as $n \to \infty$. Show that

$$\ln m_n(t) = \lambda[\exp(t) - 1] + r_n(t)$$

where for each t, $r_n(t) \to 0$ as $n \to \infty$. (Hint: Use the fact that $\ln(1 + x) = x - x^2/2 + x^3/3 + \cdots$ for $|x| < 1$.)

(c) Deduce from part (b) that $S_n \to_d \text{Pois}(\lambda)$.

3.16: Suppose that $\{X_n\}$ is a sequence of nonnegative continuous random variables and suppose that X_n has hazard function $\lambda_n(x)$. Suppose that for each x, $\lambda_n(x) \to \lambda_0(x)$ as $n \to \infty$ where $\int_0^\infty \lambda_0(x)\, dx = \infty$. Show that $X_n \to_d X$ where

$$P(X > x) = \exp\left(- \int_0^x \lambda_0(t)\, dt \right)$$

3.17: Suppose that $X_1, \cdots, X_n$ are independent nonnegative random variables with hazard functions $\lambda_1(x), \cdots, \lambda_n(x)$ respectively. Define $U_n = \min(X_1, \cdots, X_n)$.

(a) Suppose that for some $\alpha > 0$,

$$\lim_{n\to\infty} \frac{1}{n^\alpha} \sum_{i=1}^{n} \lambda_i(t/n^\alpha) = \lambda_0(t)$$

for all $t > 0$ where $\int_0^\infty \lambda_0(t)\, dt = \infty$. Show that $n^\alpha U_n \to_d V$ where

$$P(V > x) = \exp\left(- \int_0^x \lambda_0(t)\, dt \right) .$$

(b) Suppose that $X_1, \cdots, X_n$ are i.i.d. Weibull random variables (see Example 1.19) with density function

$$f(x) = \lambda \beta x^{\beta - 1} \exp\left(-\lambda x^\beta\right) \quad (x > 0)$$

where $\lambda, \alpha > 0$. Let $U_n = \min(X_1, \cdots, X_n)$ and find α such that $n^\alpha U_n \to_d V$.

3.18: Suppose that $X_n \sim \chi^2(n)$.

(a) Show that $\sqrt{X_n} - \sqrt{n} \to_d N(0, 1/2)$ as $n \to \infty$. (Hint: Recall that X_n can be represented as a sum of n i.i.d. random variables.)

(b) Suppose that $n = 100$. Use the result in part (a) to approximate $P(X_n > 110)$.

3.19: Suppose that $\{X_n\}$ is a sequence of random variables such that $X_n \to_d X$ where $E(X)$ is finite. We would like to investigate sufficient conditions under which $E(X_n) \to E(X)$ (assuming that $E(X_n)$ is well-defined). Note that in Theorem 3.5, we indicated that this convergence holds if the X_n's are uniformly bounded.

(a) Let $\delta > 0$. Show that

$$E(|X_n|^{1+\delta}) = (1+\delta)\int_0^\infty x^\delta P(|X_n| > x)\,dx.$$

(b) Show that for any $M > 0$ and $\delta > 0$,

$$\begin{aligned}\int_0^M P(|X_n| > x)\,dx &\leq E(|X_n|) \\ &\leq \int_0^M P(|X_n| > x)\,dx \\ &\quad + \frac{1}{M^\delta}\int_M^\infty x^\delta P(|X_n| > x)\,dx.\end{aligned}$$

(c) Again let $\delta > 0$ and suppose that $E(|X_n|^{1+\delta}) \leq K < \infty$ for all n. Assuming that $X_n \to_d X$, use the results of parts (a) and (b) to show that $E(|X_n|) \to E(|X|)$ and $E(X_n) \to E(X)$ as $n \to \infty$. (Hint: Use the fact that

$$\int_0^M |P(|X_n| > x) - P(|X| > x)|\,dx \to 0$$

as $n \to \infty$ for each finite M.)

3.20: A sequence of random variables $\{X_n\}$ is said to be bounded in probability if for every $\epsilon > 0$, there exists $M_\epsilon < \infty$ such that $P(|X_n| > M_\epsilon) < \epsilon$ for all n.

(a) If $X_n \to_d X$, show that $\{X_n\}$ is bounded in probability.

(b) If $E(|X_n|^r) \le M < \infty$ for some $r > 0$, show that $\{X_n\}$ is bounded in probability.

(c) Suppose that $Y_n \to_p 0$ and $\{X_n\}$ is bounded in probability. Show that $X_n Y_n \to_p 0$.

3.21: If $\{X_n\}$ is bounded in probability, we often write $X_n = O_p(1)$. Likewise, if $X_n \to_p 0$ then $X_n = o_p(1)$. This useful shorthand notation generalizes the big-oh and little-oh notation that is commonly used for sequences of numbers to sequences of random variables. If $X_n = O_p(Y_n)$ ($X_n = o_p(Y_n)$) then $X_n/Y_n = O_p(1)$ ($X_n/Y_n = o_p(1)$).

(a) Suppose that $X_n = O_p(1)$ and $Y_n = o_p(1)$. Show that $X_n + Y_n = O_p(1)$.

(b) Let $\{a_n\}$ and $\{b_n\}$ be sequences of constants where $a_n/b_n \to 0$ as $n \to \infty$ (that is, $a_n = o(b_n)$) and suppose that $X_n = O_p(a_n)$. Show that $X_n = o_p(b_n)$.

3.22: Suppose that $\{U_i\}$ is an infinite sequence of i.i.d. random variables with mean 0 and variance 1, and define $\{X_i\}$ by

$$X_i = \sum_{j=0}^{\infty} c_j U_{i-j}$$

where we assume that $\sum_{j=0}^{\infty} |c_j| < \infty$ to guarantee that the infinite summation is well-defined.

(a) Define $\widetilde{c}_j = \sum_{k=j+1}^{\infty} c_k$ and define

$$Z_i = \sum_{j=0}^{\infty} \widetilde{c}_j U_{i-j}$$

and assume that $\sum_{j=0}^{\infty} \widetilde{c}_j^2 < \infty$ (so that Z_i is well-defined). Show that

$$X_i = \left(\sum_{j=0}^{\infty} c_j \right) U_i + Z_i - Z_{i-1}.$$

This decomposition is due to Beveridge and Nelson (1981).

(b) Using the decomposition in part (a), show that

$$\frac{1}{\sqrt{n}} \sum_{i=1}^{n} X_i \to_d N(0, \sigma^2)$$

where

$$\sigma^2 = \left(\sum_{j=0}^{\infty} c_j \right)^2 .$$

3.23: Suppose that $A_1, A_2, \cdots$ is a sequence of events. We are sometimes interested in determining the probability that infinitely many of the A_k's occur. Define the event

$$B = \bigcap_{n=1}^{\infty} \bigcup_{k=n}^{\infty} A_k.$$

It is possible to show that an outcome lies in B if, and only if, it belongs to infinitely many of the A_k's. (To see this, first suppose that an outcome ω lies in infinitely many of the A_k's. Then it belongs to $B_n = \bigcup_{k=n}^{\infty} A_k$ for each $n \geq 1$ and hence in $B = \bigcap_{n=1}^{\infty} B_n$. On the other hand, suppose that ω lies in B; then it belongs to B_n for all $n \geq 1$. If ω were in only a finite number of A_k's, there would exist a number m such that A_k did not contain ω for $k \geq m$. Hence, ω would not lie in B_n for $n \geq m$ and so ω would not lie in B. This is a contradiction, so ω must lie in infinitely many of the A_k's.)

(a) Prove the first Borel-Cantelli Lemma: If $\sum_{k=1}^{\infty} P(A_k) < \infty$ then

$$P(A_k \text{ infinitely often}) = P(B) = 0.$$

(Hint: note that $B \subset B_n$ for any n and so $P(B) \leq P(B_n)$.)

(b) When the A_k's are mutually independent, we can strengthen the first Borel-Cantelli Lemma. Suppose that

$$\sum_{k=1}^{\infty} P(A_k) = \infty$$

for mutually independent events $\{A_k\}$. Show that

$$P(A_k \text{ infinitely often}) = P(B) = 1;$$

this result is called the second Borel-Cantelli Lemma. (Hint: Note that

$$B^c = \bigcup_{n=1}^{\infty} \bigcap_{k=n}^{\infty} A_k^c$$

and so

$$P(B^c) \le P\left(\bigcap_{k=n}^{\infty} A_k^c\right) = \prod_{k=n}^{\infty}(1 - P(A_k)).$$

Now use the fact that $\ln(1 - P(A_k)) \le -P(A_k)$.)

3.24: Suppose that $\{X_k\}$ is an infinite sequence of identically distributed random variables with $E(|X_k|) < \infty$.

(a) Show that for $\epsilon > 0$,

$$P\left(\left|\frac{X_k}{k}\right| > \epsilon \text{ infinitely often}\right) = 0.$$

(From this, it follows that $X_n/n \to 0$ with probability 1 as $n \to \infty$.)

(b) Suppose that the X_k's are i.i.d. Show that $X_n/n \to 0$ with probability 1 if, and only if, $E(|X_k|) < \infty$.

3.25: Suppose that $X_1, X_2, \cdots$ are i.i.d. random variables with $E(X_i) = 0$ and $E(X_i^4) < \infty$. Define

$$\bar{X}_n = \frac{1}{n}\sum_{i=1}^{n} X_i.$$

(a) Show that $E[|\bar{X}_n|^4] \le k/n^2$ for some constant k. (Hint: To evaluate

$$\sum_{i=1}^{n}\sum_{j=1}^{n}\sum_{k=1}^{n}\sum_{\ell=1}^{n} E\left[X_i X_j X_k X_\ell\right]$$

note that most of the n^4 terms in the fourfold summation are exactly 0.)

(b) Using the first Borel-Cantelli Lemma, show that

$$\bar{X}_n \to_{wp1} 0.$$

(This gives a reasonably straightforward proof of the SLLN albeit under much stronger than necessary conditions.)

CHAPTER 4

Principles of Point Estimation

4.1 Introduction

To this point, we have assumed (implicitly or explicitly) that all the parameters necessary to make probability calculations for a particular probability model are available to us. Thus, for example, we are able to calculate the probability that a given event occurs either exactly or approximately (with the help of limit theorems). In statistics, however, the roles of parameters (of the probability model) and outcomes (of the experiment) are somewhat reversed; the outcome of the experiment is observed by the experimenter while the true value of the parameter (or more generally, the true probability distribution) is unknown to the experimenter. In very broad terms, the goal of statistics is to use the outcome of the experiment (that is, the data from the experiment) to make inference about the values of the unknown parameters of the assumed underlying probability distribution.

The previous paragraph suggests that no ambiguity exists regarding the probability model for a given experiment. However, in "real life" statistical problems, there may be considerable uncertainty as to the choice of the appropriate probability model and the model is only chosen after the data have been observed. Moreover, in many (perhaps almost all) problems, it must be recognized that any model is, at best, an approximation to reality; it is important for a statistician to verify that any assumed model is more or less close to reality and to be aware of the consequences of misspecifying a model.

A widely recognized philosophy in statistics (and in science more generally) is that a model should be as simple as possible. This philosophy is often expressed by the maxim known as Occam's razor (due to the philosopher William of Occam): "explanations should not be multiplied beyond necessity". In terms of statistical modeling, Occam's razor typically means that we should prefer a model with few parameters to one with many parameters if the data are explained equally well by both.

There are several philosophies of statistical inference; we will crudely classify these into two schools, the Frequentist school and the Bayesian school. The Frequentist approach to inference is perhaps the most commonly used in practice but is, by no means, superior (or inferior) to the Bayesian approach. Frequentist methods assume (implicitly) that any experiment is infinitely repeatable and that we must consider all possible (but unobserved) outcomes of the experiment in order to carry out statistical inference. In other words, the uncertainty in the outcome of the experiment is used to describe the uncertainty about the parameters of the model. In contrast, Bayesian methods depend only on the observed data; uncertainty about the parameters is described via probability distributions that depend on these data. However, there are Frequentist methods that have a Bayesian flavour and vice versa. In this book, we will concentrate on Frequentist methods although some exposure will be given to Bayesian methods.

4.2 Statistical models

Let $X_1, \cdots, X_n$ be random variables (or random vectors) and suppose that we observe $x_1, \cdots, x_n$, which can be thought of as outcomes of the random variables $X_1, \cdots, X_n$. Suppose that the joint distribution of $\boldsymbol{X} = (X_1, \cdots, X_n)$ is unknown but belongs to some particular family of distributions. Such a family of distributions is called a statistical model. Although we usually assume that $\boldsymbol{X}$ is observed, it is also possible to talk about a model for $\boldsymbol{X}$ even if some or all of the X_i's are not observable.

It is convenient to index the distributions belonging to a statistical model by a parameter θ; θ typically represents the unknown or unspecified part of the model. We can then write

$$\boldsymbol{X} = (X_1, \cdots, X_n) \sim F_\theta \quad \text{for} \quad \theta \in \Theta$$

where F_θ is the joint distribution function of $\boldsymbol{X}$ and Θ is the set of possible values for the parameter θ; we will call the set Θ the parameter space. In general, θ can be either a single real-valued parameter or a vector of parameters; in this latter case, we will often write $\boldsymbol{\theta}$ to denote a vector of parameters $(\theta_1, \cdots, \theta_p)$ to emphasize that we have a vector-valued parameter.

Whenever it is not notationally cumbersome to do so, we will write (for example) $P_\theta(A)$, $E_\theta(X)$, and $\text{Var}_\theta(X)$ to denote (respectively) probability, expected value, and variance with respect to a distribution with unknown parameter θ. The reasons for doing this are purely stylistic and mainly serve to emphasize the dependence of these quantities on the parameter θ.

We usually assume that Θ is a subset of some Euclidean space so that the parameter θ is either real- or vector-valued (in the vector case, we will write $\boldsymbol{\theta} = (\theta_1, \cdots, \theta_k)$); such a model is often called a parametric model in the sense that the distributions belonging to the model can be indexed by a finite dimensional parameter. Models whose distributions cannot be indexed by a finite dimensional parameter are often (somewhat misleadingly) called non-parametric models; the parameter space for such models is typically infinite dimensional. However, for some non-parametric models, we can express the parameter space $\Theta = \Theta_1 \times \Theta_2$ where Θ_1 is a subset of a Euclidean space. (Such models are sometimes called semi-parametric models.)

For a given statistical model, a given parameter θ corresponds to a single distribution F_θ. However, this does not rule out the possibility that there may exist distinct parameter values θ_1 and θ_2 such that $F_{\theta_1} = F_{\theta_2}$. To rule out this possibility, we often require that a given model, or more precisely, its parametrization be identifiable; a model is said to have an identifiable parametrization (or to be an identifiable model) if $F_{\theta_1} = F_{\theta_2}$ implies that $\theta_1 = \theta_2$. A non-identifiable parametrization can lead to problems in estimation of the parameters in the model; for this reason, the parameters of an identifiable model are often called estimable. Henceforth unless stated otherwise, we will assume implicitly that any statistical model with which we deal is identifiable.

EXAMPLE 4.1: Suppose that $X_1, \cdots, X_n$ are i.i.d. Poisson random variables with mean λ. The joint frequency function of $\boldsymbol{X} = (X_1, \cdots, X_n)$ is

$$f(\boldsymbol{x}; \lambda) = \prod_{i=1}^{n} \frac{\exp(-\lambda)\lambda^{x_i}}{x_i!}$$

for $x_1, \cdots, x_n = 0, 1, 2, \cdots$. The parameter space for this parametric model is $\{\lambda : \lambda > 0\}$. ◇

EXAMPLE 4.2: Suppose that $X_1, \cdots, X_n$ are i.i.d. random variables with a continuous distribution function F that is unknown.

The parameter space for this model consists of all possible continuous distributions. These distributions cannot be indexed by a finite dimensional parameter and so this model is non-parametric. We may also assume that $F(x)$ has a density $f(x - \theta)$ where θ is an unknown parameter and f is an unknown density function satisfying $f(x) = f(-x)$. This model is also non-parametric but depends on the real-valued parameter θ. (This might be considered a semi-parametric model because of the presence of θ.) ◇

EXAMPLE 4.3: Suppose that $X_1, \cdots, X_n$ are independent Normal random variables with $E(X_i) = \beta_0 + \beta_1 t_i + \beta_2 s_i$ (where $t_1, \cdots, t_n$ and $s_1, \cdots, s_n$ are known constants) and $\text{Var}(X_i) = \sigma^2$; the parameter space is

$$\{(\beta_0, \beta_1, \beta_2, \sigma) : -\infty < \beta_0, \beta_1, \beta_2 < \infty, \sigma > 0\}.$$

We will see that the parametrization for this model is identifiable if, and only if, the vectors

$$\boldsymbol{z}_0 = \begin{pmatrix} 1 \\ \vdots \\ 1 \end{pmatrix}, \boldsymbol{z}_1 = \begin{pmatrix} t_1 \\ \vdots \\ t_n \end{pmatrix}, \quad \text{and} \quad \boldsymbol{z}_2 = \begin{pmatrix} s_1 \\ \vdots \\ s_n \end{pmatrix}$$

are linearly independent, that is, $a_0\boldsymbol{z}_0 + a_1\boldsymbol{z}_1 + a_2\boldsymbol{z}_2 = \boldsymbol{0}$ implies that $a_0 = a_1 = a_2 = 0$. To see why this is true, let

$$\boldsymbol{\mu} = \begin{pmatrix} E(X_1) \\ \vdots \\ E(X_n) \end{pmatrix}$$

and note that the parametrization is identifiable if there is a one-to-one correspondence between the possible values of $\boldsymbol{\mu}$ and the parameters $\beta_0, \beta_1, \beta_2$. Suppose that $\boldsymbol{z}_0$, $\boldsymbol{z}_1$, and $\boldsymbol{z}_2$ are linearly dependent; then $a_0\boldsymbol{z}_0 + a_1\boldsymbol{z}_1 + a_2\boldsymbol{z}_2 = \boldsymbol{0}$ where at least one of a_0, a_1, or a_2 is non-zero. In this case, we would have

$$\begin{aligned} \boldsymbol{\mu} &= \beta_0\boldsymbol{z}_0 + \beta_1\boldsymbol{z}_1 + \beta_2\boldsymbol{z}_2 \\ &= (\beta_0 + a_0)\boldsymbol{z}_0 + (\beta_1 + a_1)\boldsymbol{z}_1 + (\beta_2 + a_2)\boldsymbol{z}_2 \end{aligned}$$

and thus there is not a one-to-one correspondence between $\boldsymbol{\mu}$ and $(\beta_0, \beta_1, \beta_2)$. However, when $\boldsymbol{z}_0$, $\boldsymbol{z}_1$, and $\boldsymbol{z}_2$ are linearly dependent, it is possible to obtain an identifiable parametrization by restricting the parameter space; this is usually achieved by putting constraints on the parameters β_0, β_1, and β_2. ◇

Exponential families

One important class of statistical models is exponential family models. Suppose that $X_1, \cdots, X_n$ have a joint distribution F_θ where $\boldsymbol{\theta} = (\theta_1, \cdots, \theta_p)$ is an unknown parameter. We say that the family of distributions $\{F_\theta\}$ is a k-parameter exponential family if the joint density or joint frequency function of $(X_1, \cdots, X_n)$ is of the form

$$f(\boldsymbol{x}; \boldsymbol{\theta}) = \exp\left[\sum_{i=1}^{k} c_i(\boldsymbol{\theta}) T_i(\boldsymbol{x}) - d(\boldsymbol{\theta}) + S(\boldsymbol{x})\right]$$

for $\boldsymbol{x} = (x_1, \cdots, x_n) \in A$ where A does not depend on the parameter $\boldsymbol{\theta}$. It is important to note that k need not equal p, the dimension of $\boldsymbol{\theta}$, although, in many cases, they are equal.

EXAMPLE 4.4: Suppose that X has a Binomial distribution with parameters n and θ where θ is unknown. Then the frequency function of X is

$$\begin{aligned} f(x; \theta) &= \binom{n}{x} \theta^x (1-\theta)^{n-x} \\ &= \exp\left[\ln\left(\frac{\theta}{1-\theta}\right) x + n \ln(1-\theta) + \ln\binom{n}{x}\right] \end{aligned}$$

for $xinA = \{0, 1, \cdots, n\}$ and so the distribution of X has a one-parameter exponential family. ◇

EXAMPLE 4.5: Suppose that $X_1, \cdots, X_n$ are i.i.d. Gamma random variables with unknown shape parameter α and unknown scale parameter λ. Then the joint density function of $\boldsymbol{X} = (X_1, \cdots, X_n)$ is

$$\begin{aligned} & f(\boldsymbol{x}; \alpha, \lambda) \\ &= \prod_{i=1}^{n} \left[\frac{\lambda^\alpha x_i^{\alpha-1} \exp(-\lambda x_i)}{\Gamma(\alpha)}\right] \\ &= \exp\left[(\alpha - 1)\sum_{i=1}^{n} \ln(x_i) - \lambda \sum_{i=1}^{n} x_i + n\alpha \ln(\lambda) - n \ln(\Gamma(\alpha))\right] \end{aligned}$$

(for $x_1, \cdots, x_n > 0$) and so the distribution of $\boldsymbol{X}$ is a two-parameter exponential family. ◇

EXAMPLE 4.6: Suppose that $X_1, \cdots, X_n$ are i.i.d. Normal random variables with mean θ and variance θ^2 where $\theta > 0$. The joint density function of $(X_1, \cdots, X_n)$ is

$$\begin{aligned} f(\boldsymbol{x}; \theta) & \\ &= \prod_{i=1}^{n} \left[\frac{1}{\theta\sqrt{2\pi}} \exp\left(-\frac{1}{2\theta^2}(x_i - \theta)^2 \right) \right] \\ &= \exp\left[-\frac{1}{2\theta^2} \sum_{i=1}^{n} x_i^2 + \frac{1}{\theta} \sum_{i=1}^{n} x_i - \frac{n}{2}(1 + \ln(\theta^2) + \ln(2\pi)) \right] \end{aligned}$$

and so $A = R^n$. Note that this is a two-parameter exponential family despite the fact that the parameter space is one-dimensional. ◇

EXAMPLE 4.7: Suppose that $X_1, \cdots, X_n$ are independent Poisson random variables with $E(X_i) = \exp(\alpha + \beta t_i)$ where $t_1, \cdots, t_n$ are known constants. Setting $\boldsymbol{X} = (X_1, \cdots, X_n)$, the joint frequency function of $\boldsymbol{X}$ is

$$\begin{aligned} f(\boldsymbol{x}; \alpha, \beta) & \\ &= \prod_{i=1}^{n} \left[\frac{\exp(-\exp(\alpha + \beta t_i)) \exp(\alpha x_i + \beta x_i t_i)}{x!} \right] \\ &= \exp\left[\alpha \sum_{i=1}^{n} x_i + \beta \sum_{i=1}^{n} x_i t_i + \sum_{i=1}^{n} \exp(\alpha + \beta t_i) - \sum_{i=1}^{n} \ln(x_i!) \right]. \end{aligned}$$

This is a two-parameter exponential family model; the set A is simply $\{0, 1, 2, 3, \cdots\}^n$. ◇

EXAMPLE 4.8: Suppose that $X_1, \cdots, X_n$ are i.i.d. Uniform random variables on the interval $[0, \theta]$. The joint density function of $\boldsymbol{X} = (X_1, \cdots, X_n)$ is

$$f(\boldsymbol{x}; \theta) = \frac{1}{\theta^n} \quad \text{for } 0 \le x_1, \cdots, x_n \le \theta.$$

The region on which $f(\boldsymbol{x}; \theta)$ is positive clearly depends on θ and so this model is *not* an exponential family model. ◇

The following result will prove to be useful in the sequel.

PROPOSITION 4.1 *Suppose that $\boldsymbol{X} = (X_1, \cdots, X_n)$ has a one-parameter exponential family distribution with density or frequency function*

$$f(\boldsymbol{x};\theta) = \exp\left[c(\theta)T(\boldsymbol{x}) - d(\theta) + S(\boldsymbol{x})\right]$$

for $\boldsymbol{x} \in A$ where
(a) the parameter space Θ is open,
(b) $c(\theta)$ is a one-to-one function on Θ,
(c) $c(\theta)$, $d(\theta)$ are twice differentiable functions on Θ.
Then

$$\begin{aligned} E_\theta[T(\boldsymbol{X})] &= \frac{d'(\theta)}{c'(\theta)} \\ \text{and} \quad Var_\theta[T(\boldsymbol{X})] &= \frac{d''(\theta)c'(\theta) - d'(\theta)c''(\theta)}{[c'(\theta)]^3}. \end{aligned}$$

Proof. Define $\phi = c(\theta)$; ϕ is called the natural parameter of the exponential family. Let $d_0(\phi) = d(c^{-1}(\phi))$ where c^{-1} is well-defined since c is a one-to-one continuous function on Θ. Then for s sufficiently small (so that $\phi + s$ lies in the natural parameter space), we have (Problem 4.1)

$$E_\phi[\exp(sT(\boldsymbol{X}))] = \exp[d_0(\phi + s) - d_0(\phi)],$$

which is the moment generating function of $T(\boldsymbol{X})$. Differentiating and setting $s = 0$, we get

$$E_\phi[T(\boldsymbol{X})] = d_0'(\phi) \quad \text{and} \quad \text{Var}_\phi[T(\boldsymbol{X})] = d_0''(\phi).$$

Now note that

$$\begin{aligned} d_0'(\phi) &= \frac{d'(\theta)}{c'(\theta)} \\ \text{and} \quad d_0''(\phi) &= \frac{d''(\theta)c'(\theta) - d'(\theta)c''(\theta)}{[c'(\theta)]^3} \end{aligned}$$

and so the conclusion follows. □

Proposition 4.1 can be extended to find the means, variances and covariances of the random variables $T_1(\boldsymbol{X}), \cdots, T_k(\boldsymbol{X})$ in k-parameter exponential family models; see Problem 4.2.

Statistics

Suppose that the model for $\boldsymbol{X} = (X_1, \cdots, X_n)$ has a parameter space Θ. Since the true value of the parameter θ (or, equivalently, the true distribution of $\boldsymbol{X}$) is unknown, we would like to summarize the available information in $\boldsymbol{X}$ without losing too much information about the unknown parameter θ. At this point, we are not interested in estimating θ *per se* but rather in determining how to best use the information in $\boldsymbol{X}$.

We will start by attempting to summarize the information in $\boldsymbol{X}$. Define a statistic $T = T(\boldsymbol{X})$ to be a function of $\boldsymbol{X}$ that does not depend on any unknown parameter; that is, the statistic T depends only on observable random variables and known constants. A statistic can be real- or vector-valued.

EXAMPLE 4.9: $T(\boldsymbol{X}) = \bar{X} = n^{-1}\sum_{i=1}^{n} X_i$. Since n (the sample size) is known, T is a statistic. ◇

EXAMPLE 4.10: $T(\boldsymbol{X}) = (X_{(1)}, \cdots, X_{(n)})$ where $X_{(1)} \le X_{(2)} \le \cdots \le X_{(n)}$ are the order statistics of $\boldsymbol{X}$. Since T depends only on the values of $\boldsymbol{X}$, T is a statistic. ◇

It is important to note that any statistic is itself a random variable and so has its own probability distribution; this distribution may or may not depend on the parameter θ. Ideally, a statistic $T = T(\boldsymbol{X})$ should contain as much information about θ as $\boldsymbol{X}$ does. However, this raises several questions. For example, how does one determine if T and $\boldsymbol{X}$ contain the same information about θ? How do we find such statistics? Before attempting to answer these questions, we will define the concept of ancillarity.

DEFINITION. A statistic T is an ancillary statistic (for θ) if its distribution is independent of θ; that is, for all $\theta \in \Theta$, T has the same distribution.

EXAMPLE 4.11: Suppose that X_1 and X_2 are independent Normal random variables each with mean μ and variance σ^2 (where σ^2 is known). Let $T = X_1 - X_2$; then T has a Normal distribution with mean 0 and variance $2\sigma^2$. Thus T is ancillary for the unknown parameter μ. However, if both μ and σ^2 were unknown, T would not be ancillary for $\theta = (\mu, \sigma^2)$. (The distribution of T depends on σ^2 so T contains some information about σ^2.) ◇

EXAMPLE 4.12: Suppose that $X_1, \cdots, X_n$ are i.i.d. random variables with density function

$$f(x; \mu, \theta) = \frac{1}{2\theta} \quad \text{for } \mu - \theta \le x \le \mu + \theta.$$

Define a statistic $R = X_{(n)} - X_{(1)}$, which is the sample range of $X_1, \cdots, X_n$. The density function of R is

$$f_R(x) = \frac{n(n-1)x^{n-2}}{(2\theta)^{n-1}}\left(1 - \frac{x}{2\theta}\right) \quad \text{for } 0 \leq x \leq 2\theta,$$

which depends on θ but not μ. Thus R is ancillary for μ. ◇

Clearly, if T is ancillary for θ then T contains no information about θ. In other words, if T is to contain any useful information about θ, its distribution must depend explicitly on θ. Moreover, intuition also tells us that the amount of information contained will increase as the dependence of the distribution on θ increases.

EXAMPLE 4.13: Suppose that $X_1, \cdots, X_n$ are i.i.d. Uniform random variables on the interval $[0, \theta]$ where $\theta > 0$ is an unknown parameter. Define two statistics, $S = \min(X_1, \cdots, X_n)$ and $T = \max(X_1, \cdots, X_n)$. The density of S is

$$f_S(x; \theta) = \frac{n}{\theta}\left(1 - \frac{x}{\theta}\right)^{n-1} \quad \text{for} \quad 0 \leq x \leq \theta$$

while the density of T is

$$f_T(x; \theta) = \frac{n}{\theta}\left(\frac{x}{\theta}\right)^{n-1} \quad \text{for} \quad 0 \leq x \leq \theta.$$

Note that the densities of both S and T depend on θ and so neither is ancillary for θ. However, as n increases, it becomes clear that the density of S is concentrated around 0 for all possible values of θ while the density of T is concentrated around θ. This seems to indicate that T provides more information about θ than does S. ◇

Example 4.13 suggests that not all non-ancillary statistics are created equal. In the next section, we will elaborate on this observation.

4.3 Sufficiency

The notion of sufficiency was developed by R.A. Fisher in the early 1920s. The first mention of sufficiency was made by Fisher (1920) in which he considered the estimation of the variance σ^2 of a Normal distribution based on i.i.d. observations $X_1, \cdots, X_n$. (This is formalized in Fisher (1922).) In particular, he considered estimating

σ^2 based on the statistics

$$T_1 = \sum_{i=1}^{n} |X_i - \bar{X}| \quad \text{and} \quad T_2 = \sum_{i=1}^{n} (X_i - \bar{X})^2$$

(where $\bar{X}$ is the average of $X_1, \cdots, X_n$). Fisher showed that the distribution of T_1 conditional on $T_2 = t$ does not depend on the parameter σ while the distribution of T_2 conditional on $T_1 = t$ does depend on σ. He concluded that all the information about σ^2 in the sample was contained in the statistic T_2 and that any estimate of σ^2 should be based on T_2; that is, any estimate of σ^2 based on T_1 could be improved by using the information in T_2 while T_2 could not be improved by using T_1.

We will now try to elaborate on Fisher's argument in a more general context. Suppose that $\boldsymbol{X} = (X_1, \cdots, X_n) \sim F_\theta$ for some $\theta \in \Theta$ and let $T = T(\boldsymbol{X})$ be a statistic. For each t in the range of T, define the level sets of T

$$A_t = \{\boldsymbol{x} : T(\boldsymbol{x}) = t\}.$$

Now look at the distribution of $\boldsymbol{X}$ on the set A_t, that is, the conditional distribution of $\boldsymbol{X}$ given $T = t$. If this conditional distribution is independent of θ then $\boldsymbol{X}$ contains no information about θ on the set A_t; that is, $\boldsymbol{X}$ is an ancillary statistic on A_t. If this is true for each t in the range of the statistic T, it follows that T contains the same information about θ as $\boldsymbol{X}$ does; in this case, T is called a sufficient statistic for θ. The precise definition of sufficiency follows.

DEFINITION. A statistic $T = T(\boldsymbol{X})$ is a sufficient statistic for a parameter θ if for all sets A, $P[\boldsymbol{X} \in A | T = t]$ is independent of θ for all t in the range of T.

Sufficient statistics are not unique; from the definition of sufficiency, it follows that if g is a one-to-one function over the range of the statistic T then $g(T)$ is also sufficient. This emphasizes the point that it is not the sufficient statistic itself that is important but rather the partition of the sample space induced by the statistic (that is, the level sets of the statistic). It also follows that if T is sufficient for θ then the distribution of any other statistic $S = S(\boldsymbol{X})$ conditional on $T = t$ is independent of θ.

How can we check if a given statistic is sufficient? In some cases, sufficiency can be verified directly from the definition.

EXAMPLE 4.14: Suppose that $X_1, \cdots, X_k$ are independent random variables where X_i has a Binomial distribution with parameters n_i (known) and θ (unknown). Let $T = X_1 + \cdots + X_k$; T will also have a Binomial distribution with parameters $m = n_1 + \cdots + n_k$ and θ. To show that T is sufficient, we need to show that

$$P_\theta[\boldsymbol{X} = \boldsymbol{x} | T = t]$$

is independent of θ (for all $x_1, \cdots, x_k$ and t). First note that if $t \neq x_1 + \cdots + x_k$ then this conditional probability is 0 (and hence independent of θ). If $t = x_1 + \cdots + x_k$ then

$$\begin{aligned} P_\theta[\boldsymbol{X} = \boldsymbol{x} | T = t] &= \frac{P_\theta[\boldsymbol{X} = \boldsymbol{x}]}{P_\theta[T = t]} \\ &= \frac{\prod_{i=1}^k \binom{n_i}{x_i} \theta^{x_i} (1-\theta)^{n_i - x_i}}{\binom{m}{t} \theta^t (1-\theta)^{m-t}} \\ &= \frac{\prod_{i=1}^k \binom{n_i}{x_i}}{\binom{m}{t}}, \end{aligned}$$

which is independent of θ. Thus T is a sufficient statistic for θ. ◇

Unfortunately, there are two major problems with using the definition to verify that a given statistic is sufficient. First, the condition given in the definition of sufficiency is sometimes very difficult to verify; this is especially true when $\boldsymbol{X}$ has a continuous distribution. Second, the definition of sufficiency does not allow us to identify sufficient statistics easily. Fortunately, there is a simple criterion due to Jerzy Neyman that gives a necessary and sufficient condition for T to be a sufficient statistic when $\boldsymbol{X}$ has a joint density or frequency function.

THEOREM 4.2 (Neyman Factorization Criterion)
Suppose that $\boldsymbol{X} = (X_1, \cdots, X_n)$ has a joint density or frequency function $f(\boldsymbol{x}; \theta)$ ($\theta \in \Theta$). Then $T = T(\boldsymbol{X})$ is sufficient for θ if, and only if,

$$f(\boldsymbol{x}; \theta) = g(T(\boldsymbol{x}); \theta) h(\boldsymbol{x}).$$

(Both T and θ can be vector-valued.)

A rigorous proof of the Factorization Criterion in its full generality is quite technical and will not be pursued here; see Billingsley (1995) or Lehmann (1991) for complete details. However, the proof when $\boldsymbol{X}$ is discrete is quite simple and will be sketched here.

Suppose first that T is sufficient. Then

$$\begin{aligned} f(\boldsymbol{x};\theta) &= P_\theta[\boldsymbol{X}=\boldsymbol{x}] \\ &= \sum_t P_\theta[\boldsymbol{X}=\boldsymbol{x}, T=t] \\ &= P_\theta[\boldsymbol{X}=\boldsymbol{x}, T=T(\boldsymbol{x})] \\ &= P_\theta[T=T(\boldsymbol{x})]P[\boldsymbol{X}=\boldsymbol{x}|T=T(\boldsymbol{x})]. \end{aligned}$$

Since T is sufficient, $P[\boldsymbol{X}=\boldsymbol{x}|T=T(\boldsymbol{x})]$ is independent of θ and so $f(\boldsymbol{x};\theta)=g(T(\boldsymbol{x});\theta)h(\boldsymbol{x})$.

Now suppose that $f(\boldsymbol{x};\theta)=g(T(\boldsymbol{x});\theta)h(\boldsymbol{x})$. Then if $T(\boldsymbol{x})=t$,

$$\begin{aligned} P_\theta[\boldsymbol{X}=\boldsymbol{x}|T=t] &= \frac{P_\theta[\boldsymbol{X}=\boldsymbol{x}]}{P_\theta[T=t]} \\ &= \frac{g(T(\boldsymbol{x});\theta)h(\boldsymbol{x})}{\sum_{T(\boldsymbol{y})=t} g(T(\boldsymbol{y});\theta)h(\boldsymbol{y})} \\ &= \frac{h(\boldsymbol{x})}{\sum_{T(\boldsymbol{y})=t} h(\boldsymbol{y})}, \end{aligned}$$

which does not depend on θ. If $T(\boldsymbol{x})\neq t$ then $P_\theta[\boldsymbol{X}=\boldsymbol{x}|T=t]=0$. In both cases, $P_\theta[\boldsymbol{X}=\boldsymbol{x}|T=t]$ is independent of θ and so T is sufficient.

EXAMPLE 4.15: Suppose that $X_1,\cdots,X_n$ are i.i.d. random variables with density function

$$f(x;\theta)=\frac{1}{\theta} \quad \text{for } 0\le x\le\theta$$

where $\theta>0$. The joint density function of $\boldsymbol{X}=(X_1,\cdots,X_n)$ is

$$\begin{aligned} f(\boldsymbol{x};\theta) &= \frac{1}{\theta^n} \quad \text{for } 0\le x_1,\cdots,x_n\le\theta \\ &= \frac{1}{\theta^n} I(0\le x_1,\cdots,x_n\le\theta) \\ &= \frac{1}{\theta^n} I\left(\max_{1\le i\le n} x_i\le\theta\right) I\left(\min_{1\le i\le n} x_i\ge 0\right) \\ &= g(\max(x_1,\cdots,x_n);\theta)h(\boldsymbol{x}) \end{aligned}$$

and so $X_{(n)}=\max(X_1,\cdots,X_n)$ is sufficient for θ. ◇

EXAMPLE 4.16: Suppose that $\boldsymbol{X}=(X_1,\cdots,X_n)$ have a distribution belonging to a k-parameter exponential family with joint density or frequency function satisfying

$$f(\boldsymbol{x};\theta) = \exp\left[\sum_{i=1}^{k} c_i(\theta)T_i(\boldsymbol{x}) - d(\theta) + S(\boldsymbol{x})\right] I(\boldsymbol{x} \in A).$$

Then (taking $h(\boldsymbol{x}) = \exp[S(\boldsymbol{x})]I(\boldsymbol{x} \in A)$), it follows from the Factorization Criterion that the statistic

$$T = (T_1(\boldsymbol{X}), \cdots, T_k(\boldsymbol{X}))$$

is sufficient for θ. ◇

From the definition of sufficiency, it is easy to see that the data $\boldsymbol{X}$ is itself always sufficient. Thus sufficiency would not be a particularly useful concept unless we could find sufficient statistics that truly represent a reduction of the data; however, from the examples given above, we can see that this is indeed possible. Thus, the real problem lies in determining whether a sufficient statistic represents the best possible reduction of the data.

There are two notions of what is meant by the "best possible" reduction of the data. The first of these is minimal sufficiency; a sufficient statistic T is minimal sufficient if for any other sufficient statistic S, there exists a function g such that $T = g(S)$. Thus a minimal sufficient statistic is the sufficient statistic that represents the maximal reduction of the data that contains as much information about the unknown parameter as the data itself. A second (and stronger) notion is completeness which will be discussed in more depth in Chapter 6. If $\boldsymbol{X} \sim F_\theta$ then a statistic $T = T(\boldsymbol{X})$ is complete if $E_\theta(g(T)) = 0$ for all $\theta \in \Theta$ implies that $P_\theta(g(T) = 0) = 1$ for all $\theta \in \Theta$. In particular, if T is complete then $g(T)$ is ancillary for θ only if $g(T)$ is constant; thus a complete statistic T contains no ancillary information.

It can be shown that if a statistic T is sufficient and complete then T is also minimal sufficient; however, the converse is not true. For example, suppose that $X_1, \cdots, X_n$ are i.i.d. random variables whose density function is

$$f(x;\theta) = \frac{\exp(x-\theta)}{[1+\exp(x-\theta)]^2}.$$

For this model, a one-dimensional sufficient statistic for θ does not exist and, in fact, the order statistics $(X_{(1)}, \cdots, X_{(n)})$ can be shown to be minimal sufficient. However, the statistic $T = X_{(n)} - X_{(1)}$ is ancillary and so the order statistics are not complete. Thus despite the fact that $(X_{(1)}, \cdots, X_{(n)})$ is a minimal sufficient statistic, it still contains "redundant" information about θ.

How important is sufficiency in practice? The preceding discussion suggests that any statistical procedure should depend only on the minimal sufficient statistic. In fact, we will see in succeeding chapters that optimal statistical procedures (point estimators, hypothesis tests and so on discussed in these chapters) almost invariably depend on minimal sufficient statistics. Nonetheless, statistical models really serve only as approximations to reality and so procedures that are nominally optimal can fail miserably in practice. For example, suppose $X_1, \cdots, X_n$ are i.i.d. random variables with mean μ and variance σ^2. It is common to assume that the X_i's have a Normal distribution in which case $(\sum_{i=1}^n X_i, \sum_{i=1}^n X_i^2)$ is a minimal sufficient statistic for (μ, σ^2). However, optimal procedures for the Normal distribution can fail miserably if the X_i's are not Normal. For this reason, it is important to be flexible in developing statistical methods.

4.4 Point estimation

A point estimator or estimator is a statistic whose primary purpose is to estimate the value of a parameter. That is, if $\boldsymbol{X} \sim F_\theta$ for $\theta \in \Theta$, then an estimator $\widehat{\theta}$ is equal to some statistic $T(\boldsymbol{X})$.

Assume that θ is a real-valued parameter and that $\widehat{\theta}$ is an estimator of θ. The probability distribution of an estimator $\widehat{\theta}$ is often referred to as the sampling distribution of $\widehat{\theta}$. Ideally, we would like the sampling distribution of $\widehat{\theta}$ to be concentrated closely around the true value of the parameter, θ. There are several simple measures of the quality of an estimator based on its sampling distribution.

DEFINITION. The bias of an estimator $\widehat{\theta}$ is defined to be

$$b_\theta(\widehat{\theta}) = E_\theta(\widehat{\theta}) - \theta.$$

An estimator is said to be unbiased if $b_\theta(\widehat{\theta}) = 0$, that is, $E_\theta(\widehat{\theta}) = \theta$.

DEFINITION. The mean absolute error (MAE) of $\widehat{\theta}$ is defined to be

$$\text{MAE}_\theta(\widehat{\theta}) = E_\theta[|\widehat{\theta} - \theta|].$$

DEFINITION. The mean square error (MSE) of $\widehat{\theta}$ is defined to be

$$\text{MSE}_\theta(\widehat{\theta}) = E_\theta[(\widehat{\theta} - \theta)^2];$$

it is easy to show that $\text{MSE}_\theta(\widehat{\theta}) = \text{Var}_\theta(\widehat{\theta}) + [b_\theta(\widehat{\theta})]^2$.

The bias of $\widehat{\theta}$ gives some indication of whether the sampling distribution is centered around θ while $\text{MAE}_\theta(\widehat{\theta})$ and $\text{MSE}_\theta(\widehat{\theta})$ are measures of the dispersion of the sampling distribution of $\widehat{\theta}$ around θ. MAE and MSE are convenient measures for comparing different estimators of a parameter θ; since we would like $\widehat{\theta}$ to be close to θ, it is natural to prefer estimators with small MAE or MSE. Although MAE may seem to be a better measure for assessing the accuracy of an estimator, MSE is usually preferred to MAE. There are several reasons for preferring MSE; most of these derive from the decomposition of $\text{MSE}_\theta(\widehat{\theta})$ into variance and bias components:

$$\text{MSE}_\theta(\widehat{\theta}) = \text{Var}_\theta(\widehat{\theta}) + [b_\theta(\widehat{\theta})]^2.$$

This decomposition makes MSE much easier to work with than MAE. For example, when $\widehat{\theta}$ is a linear function of $X_1, \cdots, X_n$, the mean and variance of $\widehat{\theta}$ (and hence its MSE) are easily computed; computation of the MAE is much more difficult. Frequently, the sampling distribution of an estimator is approximately Normal; for example, it is often true that the distribution of $\widehat{\theta}$ is approximately Normal with mean θ and variance $\sigma^2(\theta)/n$. In such cases, the variance $\sigma^2(\theta)/n$ is often approximated reasonably well by $\text{MSE}_\theta(\widehat{\theta})$ and so the MSE essentially characterizes the dispersion of the sampling distribution of $\widehat{\theta}$. (Typically, the variance component of the MSE is much larger than the bias component and so $\text{MSE}_\theta(\widehat{\theta}) \approx \text{Var}_\theta(\widehat{\theta})$.) However, it is also important to note that the MSE of an estimator can be infinite even when its sampling distribution is approximately Normal.

Unbiasedness is a very controversial issue. The use of the word "biased" to describe an estimator is very loaded; it suggests that a biased estimator is somehow misleading or prejudiced. Thus, at first glance, it may seem reasonable to require an estimator to be unbiased. However, in many estimation problems, unbiased estimators do not exist; moreover, there are situations where all unbiased estimators are ridiculous. A further difficulty with unbiasedness is the fact that unbiasedness is not generally preserved by transformation; that is, if $\widehat{\theta}$ is an unbiased estimator of θ then $g(\widehat{\theta})$ is typically not an unbiased estimator of $g(\theta)$ unless g is a linear function. Thus unbiasedness is not an intrinsically desirable quality of an estimator; we should not prefer an unbiased estimator to a biased estimator based only on unbiasedness. However, this is not to say that bias should be ignored. For example, if an estimator $\widehat{\theta}$ systematically over- or under-estimates θ (in the sense that the sampling distribution of $\widehat{\theta}$ lies predominantly to the right or left of θ), steps should be taken to remove the resulting bias.

EXAMPLE 4.17: Suppose that $X_1, \cdots, X_n$ are i.i.d. Normal random variables with mean μ and variance σ^2. An unbiased estimator of σ^2 is

$$S^2 = \frac{1}{n-1}\sum_{i=1}^{n}(X_i - \bar{X})^2.$$

However, $S = \sqrt{S^2}$ is not an unbiased estimator of σ; using the fact that

$$Y = \frac{(n-1)S^2}{\sigma^2} \sim \chi^2(n-1),$$

it follows that

$$\begin{aligned} E_\sigma(S) &= \frac{\sigma}{\sqrt{n-1}} E(\sqrt{Y}) \\ &= \frac{\sigma}{\sqrt{n-1}} \frac{\sqrt{2}\Gamma(n/2)}{\Gamma((n-1)/2)} \\ &\neq \sigma. \end{aligned}$$

However, as $n \to \infty$, it can be show that $E(S) \to \sigma$. ◇

EXAMPLE 4.18: Suppose that $X_1, \cdots, X_n$ are i.i.d. random variables with a Uniform distribution on $[0, \theta]$. Let $\widehat{\theta} = X_{(n)}$, the sample maximum; the density of $\widehat{\theta}$ is

$$f(x;\theta) = \frac{n}{\theta^n} x^{n-1} \quad \text{for} \quad 0 \le x \le \theta.$$

Note that $\widehat{\theta} \le \theta$ and hence $E_\theta(\widehat{\theta}) < \theta$; in fact, it is easy to show that

$$E_\theta(\widehat{\theta}) = \frac{n}{n+1}\theta.$$

The form of $E_\theta(\widehat{\theta})$ makes it easy to construct an unbiased estimator of θ. If we define $\widetilde{\theta} = (n+1)\widehat{\theta}/n$ then clearly $\widetilde{\theta}$ is an unbiased estimator of θ. ◇

Suppose that $\widehat{\theta}_n$ is an estimator of some parameter θ based on n random variables $X_1, \cdots, X_n$. As n increases, it seems reasonable to expect that the sampling distribution of $\widehat{\theta}_n$ should become increasingly concentrated around the true parameter value θ. This property of the sequence of estimators $\{\widehat{\theta}_n\}$ is known as consistency.

DEFINITION. A sequence of estimators $\{\widehat{\theta}_n\}$ is said to be consistent for θ if $\{\widehat{\theta}_n\}$ converges in probability to θ, that is, if

$$\lim_{n\to\infty} P_\theta[|\widehat{\theta}_n - \theta| > \epsilon] = 0$$

for each $\epsilon > 0$ and each θ.

Although, strictly speaking, consistency refers to a sequence of estimators, we often say that $\widehat{\theta}_n$ is a consistent estimator of θ if it is clear that $\widehat{\theta}_n$ belongs to a well-defined sequence of estimators; an example of this occurs when $\widehat{\theta}_n$ is based on n i.i.d. random variables.

EXAMPLE 4.19: Suppose that $X_1, \cdots, X_n$ are i.i.d. random variables with $E(X_i) = \mu$ and $\text{Var}(X_i) = \sigma^2$. Define

$$S_n^2 = \frac{1}{n-1}\sum_{i=1}^n (X_i - \bar{X}_n)^2,$$

which is an unbiased estimator of σ^2. To show that S_n^2 is a consistent estimator (or more correctly $\{S_n^2\}$ is a consistent sequence of estimators), note that

$$\begin{aligned} S_n^2 &= \frac{1}{n-1}\sum_{i=1}^n (X_i - \mu + \mu - \bar{X}_n)^2 \\ &= \frac{1}{n-1}\sum_{i=1}^n (X_i - \mu)^2 + \frac{n}{n-1}(\bar{X}_n - \mu)^2 \\ &= \left(\frac{n}{n-1}\right)\frac{1}{n}\sum_{i=1}^n (X_i - \mu)^2 + \frac{n}{n-1}(\bar{X}_n - \mu)^2. \end{aligned}$$

By the WLLN, we have

$$\frac{1}{n}\sum_{i=1}^n (X_i - \mu)^2 \to_p \sigma^2 \quad \text{and} \quad \bar{X}_n \to_p \mu$$

and so by Slutsky's Theorem, it follows that

$$S_n^2 \to_p \sigma^2.$$

Note that S_n^2 will be a consistent estimator of $\sigma^2 = \text{Var}(X_i)$ for any distribution with finite variance. ◇

EXAMPLE 4.20: Suppose that $X_1, \cdots, X_n$ are independent random variables with

$$E_\beta(X_i) = \beta t_i \quad \text{and} \quad \text{Var}_\beta(X_i) = \sigma^2$$

where $t_1, \cdots, t_n$ are known constants and β, σ^2 unknown parameters. A possible estimator of β is

$$\widehat{\beta}_n = \frac{\sum_{i=1}^n t_i X_i}{\sum_{i=1}^n t_i^2}.$$

It is easy to show that $\widehat{\beta}_n$ is an unbiased estimator of β for each n and hence to show that $\widehat{\beta}_n$ is consistent, it suffices to show that $\text{Var}_\beta(\widehat{\beta}_n) \to 0$. Because of the independence of the X_i's, it follows that

$$\text{Var}_\beta(\widehat{\beta}_n) = \frac{\sigma^2}{\sum_{i=1}^n t_i^2}.$$

Thus $\widehat{\beta}_n$ is consistent provided that $\sum_{i=1}^n t_i^2 \to \infty$ as $n \to \infty$. ◇

4.5 The substitution principle

In statistics, we are frequently interested in estimating parameters that depend on the underlying distribution function of the data; we will call such parameters functional parameters (although the term statistical functional is commonly used in the statistical literature). For example, the mean of a random variable with distribution function F may be written as

$$\mu(F) = \int_{-\infty}^{\infty} x \, dF(x).$$

The value of $\mu(F)$ clearly depends on the distribution function F; thus we can think of $\mu(\cdot)$ as a real-valued function on the space of distribution functions much in the same way that $g(x) = x^2$ is a real-valued function on the real-line. Some other examples of functional parameters include

- the variance: $\sigma^2(F) = \int_{-\infty}^{\infty} (x - \mu(F))^2 \, dF(x)$.
- the median: $\text{med}(F) = F^{-1}(1/2) = \inf\{x : F(x) \geq 1/2\}$.
- the density at x_0: $\theta(F) = F'(x_0)$ ($\theta(F)$ is defined only for those distributions with a density).
- a measure of location $\theta(F)$ defined by the equation

$$\int_{-\infty}^{\infty} \psi(x - \theta(F)) \, dF(x) = 0$$

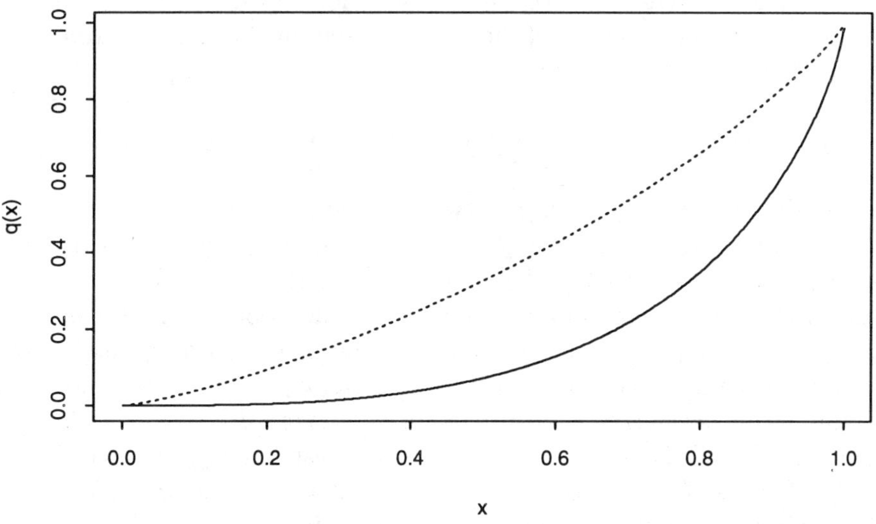

Figure 4.1 *Lorenz curves for the Gamma distribution with* $\alpha = 0.5$ *(solid curve) and* $\alpha = 5$ *(dotted curve).*

(where ψ is typically a non-decreasing, odd function).

The following example introduces a somewhat more complicated functional parameter that is of interest in economics.

EXAMPLE 4.21: Economists are often interested in the distribution of personal income in a population. More specifically, they are interested in measuring the "inequality" of this distribution. One way to do so is to consider the so-called Lorenz curve that gives the percentage of income held by the poorest $100t\%$ as a function of t. Let F be a distribution function (with $F(x) = 0$ for $x < 0$) whose expected value is $\mu(F)$. For t between 0 and 1, we define

$$q_F(t) = \frac{\int_0^t F^{-1}(s)\,ds}{\int_0^1 F^{-1}(s)\,ds} = \frac{\int_0^t F^{-1}(s)\,ds}{\mu(F)}.$$

(Note that the denominator in the definition of $q_F(t)$ is simply the expected value of the distribution F.) It is easy to verify that $q_F(t) \leq t$ with $q_F(t) = t$ (for $0 < t < 1$) if, and only if, F is concentrated at a single point (that is, all members of the population have the same income). The Lorenz curves for Gamma distributions with shape parameters 0.5 and 5 are given in Figure 4.1. (It can be shown that the Lorenz curve will not depend on the scale parameter.) One measure of inequality based on the Lorenz curve is the Gini index defined by

$$\theta(F) = 2\int_0^1 (t - q_F(t))\,dt = 1 - 2\int_0^1 q_F(t)\,dt.$$

The Gini index $\theta(F)$ is simply twice the area between the functions t and $q_F(t)$ and so $0 \leq \theta(F) \leq 1$; when perfect equality exists ($q_F(t) = t$) then $\theta(F) = 0$ while as the income gap between the richest and poorest members of the population widens, $\theta(F)$ increases. (For example, according to the World Bank (1999), estimated Gini indices for various countries range from 0.195 (Slovakia) to 0.629 (Sierra Leone); the Gini indices reported for Canada and the United States are 0.315 and 0.401, respectively.) The Gini index for the Gamma distribution with shape parameter 0.5 is 0.64 while for shape parameter 5, the Gini index is 0.25. (It can be shown that as the shape parameter tends to infinity, the Gini index tends to 0.) ◇

The substitution principle

Suppose that $X_1, \cdots, X_n$ are i.i.d. random variables with distribution function F; F may be completely unknown or may depend on a finite number of parameters. (Hence the model can be parametric or non-parametric). In this section, we will consider the problem of estimating a parameter θ that can be expressed as a functional parameter of F, that is, $\theta = \theta(F)$.

The dependence on θ of the distribution function F suggests that it may be possible to estimate θ by finding a good estimate of F and then substituting this estimate, $\widehat{F}$, for F in $\theta(F)$ to get an estimate of θ, $\widehat{\theta} = \theta(\widehat{F})$. Thus we have changed the problem from estimating θ to estimating the distribution function F. Substituting an estimator $\widehat{F}$ for F in $\theta(F)$ is known as the substitution principle. However, the substitution principle raises two basic questions: first, how do we estimate F and second, does the substitution principle always lead to good estimates of the parameter in question?

We will first discuss estimation of F. If F is the distribution function of $X_1, \cdots, X_n$ then for a given value x, $F(x)$ is the probability that any X_i is no greater than x or (according to the WLLN) the long-run proportion of X_i's that are not greater than x. Thus it seems reasonable to estimate $F(x)$ by

$$\widehat{F}(x) = \frac{1}{n}\sum_{i=1}^{n} I(X_i \le x)$$

(where $I(A)$ is 1 if condition A is true and 0 otherwise), which is simply the proportion of X_i's in the sample less than or equal to x; this estimator is called the empirical distribution function of $X_1, \cdots, X_n$. From the WLLN, it follows that $\widehat{F}(x) = \widehat{F}_n(x) \to_p F(x)$ for each value of x (as $n \to \infty$) so that $\widehat{F}_n(x)$ is a consistent estimator of $F(x)$. (In fact, the consistency of $\widehat{F}_n$ holds uniformly over the real line:

$$\sup_{-\infty<x<\infty} |\widehat{F}_n(x) - F(x)| \to_p 0.)$$

Despite these results, it is not obvious that the empirical distribution function $\widehat{F}_n$ is necessarily a good estimator of the true distribution function F. For example, $\widehat{F}_n$ is always a discrete distribution giving probability $1/n$ to each of $X_1, \cdots, X_n$; thus, if the true distribution is continuous, $\widehat{F}_n$ will not be able to capture certain features of F (because of its discontinuities). (In such cases, we could obtain a "smoother" estimator of F and apply the substitution principle with this estimator.)

Fortunately, the empirical distribution function $\widehat{F}$ can be used as described previously to estimate many functional parameters of F.

EXAMPLE 4.22: Suppose that $\theta(F) = \int_{-\infty}^{\infty} h(x)\, dF(x)$. Substituting $\widehat{F}$ for F, we get

$$\widehat{\theta} = \theta(\widehat{F}) = \int_{-\infty}^{\infty} h(x)\, d\widehat{F}(x) = \frac{1}{n}\sum_{i=1}^{n} h(X_i).$$

Thus, the substitution principle estimator is simply the sample mean of $h(X_1), \cdots, h(X_n)$. ◇

EXAMPLE 4.23: Suppose that $\theta(F) = \text{Var}(X)$ where $X \sim F$. Then

$$\theta(F) = \int_{-\infty}^{\infty} x^2\, dF(x) - \left(\int_{-\infty}^{\infty} x\, dF(x)\right)^2$$

and following Example 4.22, we get

$$\theta(\widehat{F}) = \frac{1}{n}\sum_{i=1}^{n} X_i^2 - \left(\frac{1}{n}\sum_{i=1}^{n} X_i\right)^2 = \frac{1}{n}\sum_{i=1}^{n}(X_i - \bar{X})^2.$$

Note that $\theta(\widehat{F})$ is slightly different from the unbiased estimator S^2 given in Example 4.17. ◇

EXAMPLE 4.24: Suppose $\theta(F)$ satisfies the equation

$$\int_{-\infty}^{\infty} g(x, \theta(F))\, dF(x) = 0$$

for some function $g(x, u)$. The substitution principle estimator $\widehat{\theta} = \theta(\widehat{F})$ satisfies

$$\int_{-\infty}^{\infty} g(x, \widehat{\theta})\, d\widehat{F}(x) = \frac{1}{n}\sum_{i=1}^{n} g(X_i, \widehat{\theta}) = 0.$$

In this case, $\widehat{\theta}$ is not necessarily explicitly defined. ◇

Does the substitution principle always produce reasonable estimators when the empirical distribution function is used to estimate F? The answer to this question (unfortunately) is no. For example, consider estimating a density function f corresponding to a continuous distribution function F. The density is defined by the relation

$$F(x) = \int_{-\infty}^{x} f(t)\, dt$$

and so the "natural" substitution principle estimator satisfies

$$\widehat{F}(x) = \int_{-\infty}^{x} \widehat{f}(t)\, dt.$$

However, since the empirical distribution function is a step function (with jumps at $X_1, \cdots, X_n$), no such estimator $\widehat{f}$ exists. Fortunately, it is still possible to apply the substitution principle by finding a more suitable (that is, smoother) estimator of F. One approach is to take a convolution of the empirical distribution function $\widehat{F}$ with a continuous distribution function G whose probability is concentrated closely around 0. The "smoothed" estimator of F becomes

$$\widetilde{F}(x) = \int_{-\infty}^{\infty} G(x - y)\, d\widehat{F}(y) = \frac{1}{n}\sum_{i=1}^{n} G(x - X_i).$$

Note that $\widehat{F}$ and $\widetilde{F}$ are "close" in the sense that $|\widehat{F}(x) - \widetilde{F}(x)|$ is small (provided that the probability in G is concentrated around 0) and $\widetilde{F}$ is differentiable since G is differentiable. We can then apply the substitution principle with $\widetilde{F}$ and the resulting substitution principle estimator of f is

$$\widetilde{f}(x) = \frac{1}{n}\sum_{i=1}^{n} g(x - X_i)$$

where $g = G'$ is the density function corresponding to G. (More information on density estimation can be found in the monograph by Silverman (1986).)

As the example above indicates, we cannot blindly use the substitution principle with the empirical distribution function to estimate a density function as doing so clearly results in a ridiculous estimator; however, in many other situations, the empirical distribution function may yield estimators that seem reasonable on the surface but are, in fact, inconsistent or extremely inefficient. This emphasizes the need to examine closely the properties of any estimation procedure.

Method of moments

How can the substitution principle be applied to the estimation of parameters in parametric models? That is, if $X_1, \cdots, X_n$ are i.i.d. random variables from a distribution with unknown parameters $\theta_1, \cdots, \theta_p$, can the substitution principle be used to estimate these parameters?

The simplest approach is to express $\theta_1, \cdots, \theta_p$ as functional parameters of the distribution F if possible; estimating these functional parameters by the substitution principle will then result in estimators of $\theta_1, \cdots, \theta_p$. For example, if $X_1, \cdots, X_n$ are i.i.d. Poisson random variables with parameter λ then $\lambda = \int_0^\infty x\, dF(x)$ and so an estimator of λ can be obtained by estimating $\int_0^\infty x\, dF(x)$ using the substitution principle; the resulting estimator of λ is the sample mean.

More generally, to estimate $\boldsymbol{\theta} = (\theta_1, \cdots, \theta_p)$, we can find p functional parameters of F, $\eta_1(F), \cdots, \eta_p(F)$, that depend on $\boldsymbol{\theta}$; that is,

$$\eta_k(F) = g_k(\boldsymbol{\theta}) \quad (k = 1, \cdots, p)$$

where $g_1, \cdots, g_p$ are known functions. Thus for given $\eta_1(F), \cdots,$

$\eta_p(F)$, we have a system of p equations in p unknowns $\theta_1, \cdots, \theta_p$. Using the substitution principle, we can estimate $\eta_1(F), \cdots, \eta_p(F)$ by $\eta_1(\widehat{F}), \cdots, \eta_p(\widehat{F})$. If for each possible $\eta_1(F), \cdots, \eta_p(F)$, there is a unique solution (which lies in the parameter space), we can then define $\widehat{\boldsymbol{\theta}}$ such that

$$\eta_k(\widehat{F}) = g_k(\widehat{\boldsymbol{\theta}}) \quad \text{for } k = 1, \cdots, p.$$

How do we choose $\eta_1(F), \cdots, \eta_p(F)$? Obviously, some choices will be better than others, but at this point, we are not too concerned about optimality. A common choice for $\eta_k(F)$ is $\int_{-\infty}^{\infty} x^k \, dF(x)$ (that is, the k-th moment of the X_i's) in which case $\widehat{\boldsymbol{\theta}}$ satisfies

$$\frac{1}{n}\sum_{i=1}^{n} X_i^k = g_k(\widehat{\boldsymbol{\theta}}) \quad \text{for} \quad k = 1, \cdots, p.$$

For this reason, this method of estimation is known as the method of moments. However, there is no particular reason (other than simplicity) to use moments for the $\eta_k(F)$'s.

EXAMPLE 4.25: Suppose that $X_1, \cdots, X_n$ are i.i.d. Exponential random variables with parameter λ. For any $r > 0$, we have

$$E_\lambda(X_i^r) = \frac{\Gamma(r+1)}{\lambda^r}.$$

Thus for a given $r > 0$, a method of moments estimator of λ is

$$\widehat{\lambda} = \left(\frac{1}{n\Gamma(r+1)}\sum_{i=1}^{n} X_i^r\right)^{-1/r}.$$

(If we take $r = 1$ then $\widehat{\lambda} = 1/\bar{X}$.) Since r is more-or-less arbitrary here, it is natural to ask what value of r gives the best estimator of λ; a partial answer to this question is given in Example 4.39. ◇

EXAMPLE 4.26: Suppose that $X_1, \cdots, X_n$ are i.i.d. Gamma random variables with unknown parameters α and λ. It is easy to show that

$$\eta_1(F) = E(X_i) = \frac{\alpha}{\lambda} \quad \text{and} \quad \eta_2(F) = \text{Var}(X_i) = \frac{\alpha}{\lambda^2}.$$

Thus $\widehat{\alpha}$ and $\widehat{\lambda}$ satisfy the equations

$$\bar{X} = \frac{1}{n}\sum_{i=1}^{n} X_i = \frac{\widehat{\alpha}}{\widehat{\lambda}}$$

$$\widehat{\sigma}^2 = \frac{1}{n}\sum_{i=1}^{n}(X_i - \bar{X})^2 \quad = \quad \frac{\widehat{\alpha}}{\widehat{\lambda}^2}$$

and so $\widehat{\alpha} = \bar{X}^2/\widehat{\sigma}^2$ and $\widehat{\lambda} = \bar{X}/\widehat{\sigma}^2$. ◇

EXAMPLE 4.27: Suppose that $X_1, \cdots, X_n$ are i.i.d. random variables with a "zero-inflated" Poisson distribution; the frequency function of X_i is

$$f(x;\theta,\lambda) = \begin{cases} \theta + (1-\theta)\exp(-\lambda) & \text{for } x = 0 \\ (1-\theta)\exp(-\lambda)\lambda^x/x! & \text{for } x = 1, 2, 3, \cdots \end{cases}$$

where $0 \le \theta \le 1$ and $\lambda > 0$. To estimate θ and λ via the method of moments, we will use

$$\eta_1(F) = P(X_i = 0) \quad \text{and} \quad \eta_2(F) = E(X_i);$$

it is easy to show that

$$\begin{aligned} P(X_i = 0) &= \theta + (1-\theta)\exp(-\lambda) \\ \text{and} \quad E(X_i) &= (1-\theta)\lambda. \end{aligned}$$

Thus $\widehat{\theta}$ and $\widehat{\lambda}$ satisfy the equations

$$\begin{aligned} \bar{X} &= (1-\widehat{\theta})\widehat{\lambda} \\ \frac{1}{n}\sum_{i=1}^{n} I(X_i = 0) &= \widehat{\theta} + (1-\widehat{\theta})\exp(-\widehat{\lambda}); \end{aligned}$$

however, closed form expressions for $\widehat{\theta}$ and $\widehat{\lambda}$ do not exist although they may be computed for any given sample. (For this model, the statistic $(\sum_{i=1}^n X_i, \sum_{i=1}^n I(X_i = 0))$ is sufficient for (θ, λ).) ◇

It is easy to generalize the method of moments to non-i.i.d. settings. Suppose that $(X_1, \cdots, X_n)$ has a joint distribution depending on real-valued parameters $\boldsymbol{\theta} = (\theta_1, \cdots, \theta_p)$. Suppose that $T_1, \cdots, T_p$ are statistics with

$$E_\theta(T_k) = g_k(\boldsymbol{\theta}) \quad \text{for} \quad k = 1, \cdots, p.$$

If, for all possible values of $E_\theta(T_1), \cdots, E_\theta(T_p)$, this system of equations has a unique solution then we can define the estimator $\widehat{\boldsymbol{\theta}}$ such that

$$T_k = g_k(\widehat{\boldsymbol{\theta}}) \quad \text{for} \quad k = 1, \cdots, p.$$

However, in the general (that is, non-i.i.d.) setting, greater care must be taken in choosing the statistics $T_1, \cdots, T_p$; in particular, it is important that T_k be a reasonable estimator of its mean $E_\theta(T_k)$ (for $k = 1, \cdots, p$).

4.6 Influence curves

Suppose that h is a real-valued function on the real line and that $\{x_n\}$ is a sequence of real numbers whose limit (as $n \to \infty$) is x_0. If h is continuous at x_0, then $h(x_n) \to h(x_0)$ as $n \to \infty$; thus for large enough n, $h(x_n) \approx h(x_0)$. If we assume that h is differentiable, it is possible to obtain a more accurate approximation of $h(x_n)$ by making a one term Taylor series expansion:

$$h(x_n) \approx h(x_0) + h'(x_0)(x_n - x_0).$$

This approximation can be written more precisely as

$$h(x_n) = h(x_0) + h'(x_0)(x_n - x_0) + r_n$$

where the remainder term r_n goes to 0 with n faster than $x_n - x_0$:

$$\lim_{n\to\infty} \frac{r_n}{x_n - x_0} = 0.$$

An interesting question to ask is whether notions of continuity and differentiability can be extended to functional parameters and whether similar approximations can be made for substitution principle estimators of functional parameters. Let $\theta(F)$ be a functional parameter and $\widehat{F}_n$ be the empirical distribution function of i.i.d. random variables $X_1, \cdots, X_n$. Since $\widehat{F}_n$ converges in probability to F uniformly over the real line, it is tempting to say that $\theta(\widehat{F}_n)$ converges in probability to $\theta(F)$ given the right kind of continuity of $\theta(F)$. However, continuity and differentiability of functional parameters are very difficult and abstract topics from a mathematical point of view and will not be dealt with here in any depth. In principle, defining continuity of the real-valued functional parameter $\theta(F)$ at F is not difficult; we could say that $\theta(F)$ is continuous at F if $\theta(F_n) \to \theta(F)$ whenever a sequence of distribution functions $\{F_n\}$ converges to F. However, there are several ways in which convergence of $\{F_n\}$ to F can be defined and the continuity of $\theta(F)$ may depend on which definition is chosen.

Differentiability of $\theta(F)$ is an even more difficult concept. Even if we agree on the definition of convergence of $\{F_n\}$ to F, there are several different concepts of differentiability. Thus we will not touch on differentiability in any depth. We will, however, define a type of directional derivative for $\theta(F)$ whose properties are quite useful for heuristic calculations; this derivative is commonly called the influence curve of $\theta(F)$.

The idea behind defining the influence curve is to look at the behaviour of $\theta(F)$ for distributions that are close to F in some sense. More specifically, we look at the difference between $\theta(F)$ evaluated at F and at $(1-t)F + t\Delta_x$ where Δ_x is a degenerate distribution function putting all its probability at x so that $\Delta_x(y) = 0$ for $y < x$ and $\Delta_x(y) = 1$ for $y \geq x$; for $0 \leq t \leq 1$, $(1-t)F + t\Delta_x$ is a distribution function and can be thought of as F contaminated by probability mass at x. Note that as $t \downarrow 0$, we typically have $\theta((1-t)F + t\Delta_x) \to \theta(F)$ for any x where this convergence is linear in t, that is,

$$\theta((1-t)F + t\Delta_x) - \theta(F) \approx \phi(x; F)t$$

for t close to 0.

DEFINITION. The influence curve of $\theta(F)$ at F is the function

$$\phi(x; F) = \lim_{t \downarrow 0} \frac{\theta((1-t)F + t\Delta_x) - \theta(F)}{t}$$

provided that the limit exists. The influence curve can also be evaluated as

$$\phi(x; F) = \frac{d}{dt}\theta((1-t)F + t\Delta_x)\bigg|_{t=0}$$

whenever this limit exists.

The influence curve allows for a "linear approximation" of the difference $\theta(\widehat{F}_n) - \theta(F)$ much in the same way that the derivative of a function h allows for a linear approximation of $h(x_n) - h(x_0)$; in particular, it is often possible to write

$$\theta(\widehat{F}_n) - \theta(F) = \int_{-\infty}^{\infty} \phi(x; F)d(\widehat{F}_n(x) - F(x)) + R_n$$

where R_n tends in probability to 0 at a faster rate than $\widehat{F}_n$ converges to F. This representation provides a useful heuristic method for determining the limiting distribution of $\sqrt{n}(\theta(\widehat{F}_n) - \theta(F))$. In many cases, it is possible to show that

$$\int_{-\infty}^{\infty} \phi(x; F)\, dF(x) = E[\phi(X_i; F)] = 0$$

and so

$$\theta(\widehat{F}_n) - \theta(F) = \frac{1}{n}\sum_{i=1}^{n} \phi(X_i; F) + R_n$$

where the remainder term R_n satisfies $\sqrt{n}R_n \to_p 0$. Thus by the Central Limit Theorem and Slutsky's Theorem, it follows that

$$\sqrt{n}(\theta(\widehat{F}_n) - \theta(F)) \to_d N(0, \sigma^2(F))$$

where

$$\sigma^2(F) = \int_{-\infty}^{\infty} \phi^2(x; F)\, dF(x),$$

provided that $\sigma^2(F)$ is finite. This so-called "influence curve heuristic" turns out to be very useful in practice. However, despite the fact that this heuristic approach works in many examples, we actually require a stronger notion of differentiability to make this approach rigorous; fortunately, the influence curve heuristic can typically be made rigorous using other approaches.

The influence curve is a key concept in theory of robustness, which essentially studies the sensitivity (or robustness) of estimation procedures subject to violations of the nominal model assumptions. For more information on the theory of robust estimation, see Hampel *et al* (1986).

We will now discuss some simple results that are useful for computing influence curves. To make the notation more compact, we will set $F_{t,x} = (1-t)F + t\Delta_x$.

- (Moments) Define $\theta(F) = \int_{-\infty}^{\infty} h(x)\, dF(x)$; if $X \sim F$ then $\theta(F) = E[h(X)]$. Then

$$\begin{aligned} \theta(F_{t,x}) &= (1-t)\int_{-\infty}^{\infty} h(u)\, dF(u) \\ &\quad + t\int_{-\infty}^{\infty} h(u) d\Delta_x(u) \\ &= (1-t)\int_{-\infty}^{\infty} h(u)\, dF(u) + th(x) \end{aligned}$$

  and so

$$\frac{1}{t}\left(\theta(F_{t,x}) - \theta(F)\right) = h(x) - \theta(F).$$

  Thus the influence curve is $\phi(x; F) = h(x) - \theta(F)$.

- (Sums and integrals) Suppose that $\theta(F) = \theta_1(F) + \theta_2(F)$ where

$\phi_i(x; F)$ is the influence curve of $\theta_i(F)$ (for $i = 1, 2$). Then $\phi(x; F)$, the influence curve of $\theta(F)$, is simply

$$\phi(x; F) = \phi_1(x; F) + \phi_2(x; F).$$

This result can be extend to any finite sum of functional parameters. Often we need to consider functional parameters of the form

$$\theta(F) = \int_A g(s)\theta_s(F)\, ds$$

where $\theta_s(F)$ is a functional parameter for each $s \in A$ and g is a function defined on A. Then we have

$$\begin{aligned} &\frac{1}{t}\left(\theta(F_{t,x}) - \theta(F)\right) \\ &= \int_A g(s)\frac{1}{t}\left(\theta_s(F_{t,x}) - \theta_s(F)\right) ds. \end{aligned}$$

Thus, if $\phi_s(x; F)$ is the influence curve of $\theta_s(F)$ and we can take the limit as $t \downarrow 0$ inside the integral sign, the influence curve of $\theta(F)$ is defined by

$$\phi(x; F) = \int_A g(s)\phi_s(x; F)\, ds.$$

(The trimmed mean considered in Example 4.30 is an example of such a functional parameter.)

- (The chain rule) Suppose that $\theta(F)$ has influence curve $\phi(x; F)$. What is the influence curve of $g(\theta(F))$ if g is a differentiable function? First of all, we have

$$\begin{aligned} &\frac{1}{t}\left(g(\theta(F_{t,x})) - g(\theta(F))\right) \\ &= \left(\frac{g(\theta(F_{t,x})) - g(\theta(F))}{\theta(F_{t,x}) - \theta(F)}\right)\left(\frac{\theta(F_{t,x}) - \theta(F)}{t}\right). \end{aligned}$$

As $t \to 0$, $\theta(F_{t,x}) \to \theta(F)$ (for each x) and so

$$\frac{g(\theta(F_{t,x})) - g(\theta(F))}{\theta(F_{t,x}) - \theta(F)} \to g'(\theta(F)) \quad \text{as} \quad t \to 0$$

and by definition

$$\frac{1}{t}\left(\theta(F_{t,x}) - \theta(F)\right) \to \phi(x; F).$$

Therefore the influence curve of $g(\theta(F))$ is $g'(\theta(F))\phi(x; F)$; this is a natural extension of the chain rule. For a given distribution function F, the influence curve of $g(\theta(F))$ is simply a constant multiple of the influence curve of $\theta(F)$.

- (Implicitly defined functional parameters) Functional parameters are frequently defined implicitly. For example, $\theta(F)$ may satisfy the equation
$$h(F, \theta(F)) = 0$$
where for a fixed number u, $h(F, u)$ has influence curve $\lambda(x; F, u)$ and for a fixed distribution function F, $h(F, u)$ has derivative (with respect to u), $h'(u; F)$. We then have
$$\begin{aligned} 0 &= \frac{1}{t}\left(h(F_{t,x}, \theta(F_{t,x})) - h(F, \theta(F))\right) \\ &= \frac{1}{t}\left(h(F_{t,x}, \theta(F_{t,x})) - h(F_{t,x}, \theta(F))\right) \\ &\quad + \frac{1}{t}\left(h(F_{t,x}, \theta(F)) - h(F, \theta(F))\right) \\ &\to h'(\theta(F); F)\phi(x; F) + \lambda(x; F, \theta(F)) \end{aligned}$$
as $t \to 0$ where $\phi(x; F)$ is the influence curve of $\theta(F)$. Thus
$$h'(\theta(F); F)\phi(x; F) + \lambda(x; F, \theta(F)) = 0$$
and so
$$\phi(x; F) = -\frac{\lambda(x; F, \theta(F))}{h'(\theta(F); F)}.$$

EXAMPLE 4.28: One example of an implicitly defined functional parameter is the median of a continuous distribution F, $\theta(F)$, defined by the equation
$$F(\theta(F)) = \frac{1}{2}$$
or equivalently $\theta(F) = F^{-1}(1/2)$ where F^{-1} is the inverse of F. Since
$$F(u) = \int_{-\infty}^{\infty} I(x \leq u)\, dF(x),$$
it follows that the influence curve of $F(u)$ is
$$\lambda(x; F, u) = I(x \leq u) - F(u).$$
Thus if $F(u)$ is differentiable at $u = \theta(F)$ with $F'(\theta(F)) > 0$, it follows that the influence curve of $\theta(F) = F^{-1}(1/2)$ is
$$\phi(x; F) = -\frac{I(x \leq \theta(F)) - F(\theta(F))}{F'(\theta(F))}$$

$$= \frac{\operatorname{sgn}(x - \theta(F))}{2F'(\theta(F))}$$

where sgn(u) is the "sign" of u (sgn(u) is 1 if $u > 0$, -1 if $u < 0$ and 0 if $u = 0$). Note that we require $F(u)$ to be differentiable at $u = \theta(F)$ so $\phi(x; F)$ is not defined for all F (although F does not have to be a continuous distribution function). Using the heuristic

$$\operatorname{med}(\widehat{F}_n) - \theta(F) = \frac{1}{n} \sum_{i=1}^{n} \phi(X_i; F) + R_n$$

it follows that

$$\sqrt{n}(\theta(\widehat{F}_n) - \theta(F)) \to_d N(0, [2F'(\theta(F))]^{-2})$$

since $\operatorname{Var}(\operatorname{sgn}(X_i - \theta(F))) = 1$. Indeed, the convergence indicated above can be shown to hold when the distribution function F is differentiable at its median; see Example 3.6 for a rigorous proof of the asymptotic normality of the sample median. ◇

EXAMPLE 4.29: Let $\sigma(F)$ be the standard deviation of a random variable X with distribution function F; that is,

$$\sigma(F) = \left(\theta_2(F) - \theta_1^2(F)\right)^{1/2}$$

where $\theta_1(F) = \int_{-\infty}^{\infty} y \, dF(y)$ and $\theta_2(F) = \int_{-\infty}^{\infty} y^2 \, dF(y)$. The influence curve of $\theta_2(F)$ is

$$\phi_2(x; F) = x^2 - \theta_2(F)$$

while the influence curve of $\theta_1^2(F)$ is

$$\phi_1(x; F) = 2\theta_1(F)(x - \theta_1(F))$$

by applying the chain rule for influence curves. Thus the influence curve of $\theta_2(F) - \theta_1^2(F)$ is

$$\phi_3(x; F) = x^2 - \theta_2(F) - 2\theta_1(F)(x - \theta_1(F)).$$

Since $\sigma(F) = (\theta_2(F) - \theta_1^2(F))^{1/2}$, it follows that the influence curve of $\sigma(F)$ is

$$\phi(x; F) = \frac{x^2 - \theta_2(F) - 2\theta_1(F)(x - \theta_1(F))}{2\sigma(F)}$$

by applying the chain rule. Note that $\phi(x; F) \to \infty$ as $x \to \pm\infty$ and that $\phi(x; F) = 0$ when $x = \theta_1(F) \pm \sigma(F)$. ◇

EXAMPLE 4.30: A functional parameter that includes the mean and median as limiting cases is the α-trimmed mean defined for continuous distribution functions F by

$$\mu_\alpha(F) = \frac{1}{1-2\alpha}\int_\alpha^{1-\alpha} F^{-1}(t)\,dt$$

where $0 < \alpha < 0.5$. If $f(x) = F'(x)$ is continuous and strictly positive over the interval

$$\left[F^{-1}(\alpha), F^{-1}(1-\alpha)\right]$$

as well as symmetric around some point μ then $\mu_\alpha(F) = \mu$ and the influence curve of $\mu_\alpha(F)$ is

$$\phi_\alpha(x;F) = \begin{cases} (F^{-1}(\alpha)-\mu)/(1-2\alpha) & \text{for } x < F^{-1}(\alpha) \\ (F^{-1}(1-\alpha)-\mu)/(1-2\alpha) & \text{for } x > F^{-1}(1-\alpha) \\ (x-\mu)/(1-2\alpha) & \text{otherwise.} \end{cases}$$

To find a substitution principle estimator for $\mu_\alpha(F)$ based on i.i.d. observations $X_1, \cdots, X_n$, we first find a substitution principle estimator of $F^{-1}(t)$ for $0 < t < 1$ based on the inverse of the empirical distribution function $\widehat{F}_n$:

$$\widehat{F}_n^{-1}(t) = X_{(i)} \quad \text{if } (i-1)/n < t \le i/n$$

(where $X_{(i)}$ is the i-th order statistic) and substitute this into the definition of $\mu_\alpha(F)$ yielding

$$\mu_\alpha(\widehat{F}_n) = \frac{1}{1-2\alpha}\int_\alpha^{1-\alpha} \widehat{F}_n^{-1}(t)\,dt.$$

Applying the influence curve heuristic, we have

$$\sqrt{n}(\mu_\alpha(\widehat{F}_n) - \mu_\alpha(F)) \to_d N(0, \sigma^2(F))$$

where

$$\begin{aligned} \sigma^2(F) &= \int_{-\infty}^{\infty} \phi_\alpha^2(x;F)\,dF(x) \\ &= \frac{2}{(1-2\alpha)^2}\left[\int_{F^{-1}(\alpha)}^{F^{-1}(1-\alpha)} (x-\mu)^2\,dF(x)\right] \\ &\quad + \frac{2\alpha}{(1-2\alpha)^2}\left[F^{-1}(1-\alpha)-\mu\right]^2. \end{aligned}$$

A somewhat simpler alternative that approximates $\mu_\alpha(\widehat{F}_n)$ is

$$\widetilde{\mu}_n = \frac{1}{n-2g_n} \sum_{i=g_n+1}^{n-g_n} X_{(i)}$$

where g_n is chosen so that $g_n/n \approx \alpha$. (If $g_n/n = \alpha$ then $\widetilde{\mu}_n = \mu_\alpha(\widehat{F}_n.)$ The limiting distribution of $\sqrt{n}(\widetilde{\mu}_n - \mu_\alpha(F))$ is the same as that of $\sqrt{n}(\mu_\alpha(\widehat{F}_n) - \mu_\alpha(F))$. ◇

EXAMPLE 4.31: Consider the Gini index $\theta(F)$ defined in Example 4.21. To determine the substitution principle estimator of $\theta(F)$ based on i.i.d. observations $X_1, \cdots, X_n$, we use the substitution principle estimator of $F^{-1}(t)$ from Example 4.30:

$$\begin{aligned}
\int_0^1 \int_0^t \widehat{F}_n^{-1}(s)\,ds\,dt &= \int_0^1 \int_s^1 \widehat{F}_n^{-1}(s)\,dt\,ds \\
&= \int_0^1 (1-s)\widehat{F}_n^{-1}(s)\,ds \\
&= \sum_{i=1}^n X_{(i)} \int_{(i-1)/n}^{i/n} (1-s)\,ds \\
&= \frac{1}{n}\sum_{i=1}^n \left(1 - \frac{2i-1}{2n}\right) X_{(i)}
\end{aligned}$$

and so

$$\theta(\widehat{F}_n) = \left(\sum_{i=1}^n X_i\right)^{-1} \sum_{i=1}^n \left(\frac{2i-1}{n} - 1\right) X_{(i)}.$$

As with the trimmed mean, the influence curve of the Gini index is complicated to derive. With some work, it can be shown that

$$\begin{aligned}
\phi(x;F) &= 2\left[\int_0^1 q_F(t)\,dt - q_F(F(x))\right] \\
&\quad +2\frac{x}{\mu(F)}\left(\int_0^1 q_F(t)\,dt - 1 - F(x)\right)
\end{aligned}$$

where

$$\begin{aligned}
\mu(F) &= \int_0^\infty x\,dF(x) = \int_0^1 F^{-1}(t)\,dt \\
\text{and}\quad q_F(t) &= \frac{1}{\mu(F)}\int_0^t F^{-1}(t)\,dt.
\end{aligned}$$

The influence curve heuristic suggests that

$$\sqrt{n}(\theta(\widehat{F}_n) - \theta(F)) \to_d N(0, \sigma^2(F))$$

with

$$\sigma^2(F) = \int_0^\infty \phi^2(x; F)\, dF(x).$$

Unfortunately, $\sigma^2(F)$ is difficult to evaluate (at least as a closed-form expression) for most distributions F. ◇

The influence curve has a nice finite sample interpretation. Suppose that we estimate $\theta(F)$ based on observations $x_1, \cdots, x_n$ and set $\widehat{\theta}_n = \theta(\widehat{F}_n)$. Now suppose that we obtain another observation x_{n+1} and re-estimate $\theta(F)$ by $\widehat{\theta}_{n+1} = \theta(\widehat{F}_{n+1})$ where

$$\widehat{F}_{n+1}(x) = \frac{n}{n+1}\widehat{F}_n(x) + \frac{1}{n+1}\Delta_{x_{n+1}}(x).$$

Letting $t = 1/(n+1)$ and assuming that n is sufficiently large to make t close to 0, the definition of the influence curve suggests that we can approximate $\widehat{\theta}_{n+1}$ by

$$\widehat{\theta}_{n+1} \approx \widehat{\theta}_n + \frac{1}{n+1}\phi(x_{n+1}; \widehat{F}_n).$$

(This approximation assumes that $\phi(x; \widehat{F}_n)$ is well defined; it need not be. For example, the influence curve of the median is not defined for discrete distributions such as $\widehat{F}_n$.) From this, we can see that the influence curve gives an approximation for the influence that a single observation exerts on a given estimator. For example, consider the influence curve of the standard deviation $\sigma(F)$ given in Example 4.29; based on $x_1, \cdots, x_n$, the substitution principle estimate is

$$\widehat{\sigma}_n = \left(\frac{1}{n}\sum_{i=1}^n (x_i - \bar{x}_n)^2\right)^{1/2}$$

where $\bar{x}_n$ is the sample mean. The approximation given above suggests that if the observation x_{n+1} lies between $\bar{x}_n - \widehat{\sigma}_n$ and $\bar{x}_n + \widehat{\sigma}_n$ then $\widehat{\sigma}_{n+1} < \widehat{\sigma}_n$ and otherwise $\widehat{\sigma}_{n+1} \geq \widehat{\sigma}_n$. Moreover, $\widehat{\sigma}_{n+1}$ can be made arbitrarily large by taking $|x_{n+1}|$ sufficiently large.

Suppose that $X_1, \cdots, X_n$ are i.i.d. random variables with distribution function F_θ where θ is a real-valued parameter and let $\mathcal{G} = \{\phi : \phi(F_\theta) = \theta\}$; the functional parameters ϕ in $\mathcal{G}$ are called

Fisher consistent for θ. Many statisticians consider it desirable for a functional parameter to have a bounded influence curve as this will limit the effect that a single observation can have on the value of an estimator. This would lead us to consider only those Fisher consistent ϕ's with bounded influence curves. For example, suppose that $X_1, \cdots, X_n$ are i.i.d. random variables with a distribution symmetric around θ; if $E[|X_i|] < \infty$ then we could estimate θ by the sample mean with the substitution principle estimator of $\mu(F) = \int_{-\infty}^{\infty} x\,dF(x)$. However, the influence curve of $\mu(F)$ is $\phi(x; F) = x - \mu(F)$, which is unbounded (in x) for any given F. As an alternative, we might instead estimate θ by the sample median or some trimmed mean as these are substitution principle estimators of functional parameters with bounded influence curves.

4.7 Standard errors and their estimation

The standard error of an estimator is defined to be the standard deviation of the estimator's sampling distribution. Its purpose is to convey some information about the uncertainty of the estimator.

Unfortunately, it is often very difficult to calculate the standard error of an estimator exactly. In fact, there are really only two situations where the standard error of an estimator $\widehat{\theta}$ can be computed exactly:

- the sampling distribution of $\widehat{\theta}$ is known.
- $\widehat{\theta}$ is a linear function of random variables $X_1, \cdots, X_n$ where the variances and covariances of the X_i's are known.

However, if the sampling distribution of $\widehat{\theta}$ can be approximated by a distribution whose standard deviation is known, this standard deviation can be used to give an approximate standard error for $\widehat{\theta}$. The most common example of such an approximation occurs when the sampling distribution is approximately Normal; for example, if $\sqrt{n}(\widehat{\theta} - \theta)$ is approximately Normal with mean 0 and variance σ^2 (where σ^2 may depend on θ) then $\sigma/\sqrt{n}$ can be viewed as an approximate standard error of $\widehat{\theta}$. In fact, it is not uncommon in such cases to see $\sigma/\sqrt{n}$ referred to as the standard error of $\widehat{\theta}$ despite the fact that it is merely an approximation. Moreover, approximate standard errors can be more useful than their exact counterparts. For example, $\text{Var}_\theta(\widehat{\theta})$ can be infinite despite the fact that the distribution of $\widehat{\theta}$ is approximately Normal; in this case, the approximate standard error is more informative about the uncertainty of $\widehat{\theta}$. (The variance can be distorted by small amounts of probability in the tails of the distribution; thus the variance of the approximating Normal distribution gives a better indication of the true variability.)

"Delta Method" type arguments are useful for finding approximate standard errors, especially for method of moments estimators. For example, suppose that $X_1, \cdots, X_n$ are independent random variables with $E(X_i) = \mu_i$ and $\text{Var}(X_i) = \sigma_i^2$ and

$$\widehat{\theta} = g\left(\sum_{i=1}^{n} X_i\right)$$

where

$$\theta = g\left(\sum_{i=1}^{n} \mu_i\right).$$

Then a Taylor series expansion gives

$$\begin{aligned} \widehat{\theta} - \theta &= g\left(\sum_{i=1}^{n} X_i\right) - g\left(\sum_{i=1}^{n} \mu_i\right) \\ &\approx g'\left(\sum_{i=1}^{n} \mu_i\right) \sum_{i=1}^{n} (X_i - \mu_i) \end{aligned}$$

and taking the variance of the last expression, we obtain the following approximate standard error:

$$\text{se}(\widehat{\theta}) \approx \left| g'\left(\sum_{i=1}^{n} \mu_i\right) \right| \left(\sum_{i=1}^{n} \sigma_i^2\right)^{1/2}.$$

The accuracy of this approximation depends on the closeness the distribution of $\widehat{\theta}$ to normality. When $X_1, \cdots, X_n$ are i.i.d. it is usually possible to prove directly that $\widehat{\theta}$ is approximately Normal (provided n is sufficiently large).

EXAMPLE 4.32: Suppose that $X_1, \cdots, X_n$ are i.i.d. random variables with mean μ and variance σ^2. The substitution principle estimator of μ is $\bar{X}$ whose variance is σ^2/n. Thus the standard error of $\bar{X}$ is $\sigma/\sqrt{n}$. ◇

EXAMPLE 4.33: Suppose that $X_1, \cdots, X_n$ are i.i.d. Exponential random variables with parameter λ. Since $E_\lambda(X_i) = 1/\lambda$, a method of moments estimator of λ is $\widehat{\lambda} = 1/\bar{X}$. If n is sufficiently large then $\sqrt{n}(\bar{X} - \lambda^{-1})$ is approximately Normal with mean 0 and variance λ^{-2}; applying the Delta Method, we have $\sqrt{n}(\widehat{\lambda} - \lambda)$ is approximately Normal with mean 0 and variance λ^2. Thus an approximate standard error of $\widehat{\lambda}$ is $\lambda/\sqrt{n}$. ◇

EXAMPLE 4.34: Suppose that $X_1, \cdots, X_n$ are independent Poisson random variables with $E_\beta(X_i) = \exp(\beta t_i)$ where β is an unknown parameter and $t_1, \cdots, t_n$ are known constants. Define $\widehat{\beta}$ to satisfy the equation

$$\sum_{i=1}^{n} X_i = \sum_{i=1}^{n} \exp(\widehat{\beta} t_i) = g(\widehat{\beta}).$$

To compute an approximate standard error for $\widehat{\beta}$, we will use a "Delta Method" type argument. Expanding g in a Taylor series, we get

$$\begin{aligned} g(\widehat{\beta}) - g(\beta) &= \sum_{i=1}^{n} (X_i - \exp(\beta t_i)) \\ &\approx g'(\beta)(\widehat{\beta} - \beta) \end{aligned}$$

and so

$$\begin{aligned} \widehat{\beta} - \beta &\approx \frac{g(\widehat{\beta}) - g(\beta)}{g'(\beta)} \\ &= \frac{\sum_{i=1}^{n} (X_i - \exp(\beta t_i))}{\sum_{i=1}^{n} t_i \exp(\beta t_i)}. \end{aligned}$$

Since $\text{Var}_\beta(X_i) = E_\beta(X_i) = \exp(\beta t_i)$, it follows that an approximate standard error of $\widehat{\beta}$ is

$$\text{se}(\widehat{\beta}) \approx \frac{\left(\sum_{i=1}^{n} \exp(\beta t_i)\right)^{1/2}}{\left|\sum_{i=1}^{n} t_i \exp(\beta t_i)\right|}.$$

This approximation assumes that the distribution of $\widehat{\beta}$ is approximately Normal. The standard error of $\widehat{\beta}$ can be estimated by substituting $\widehat{\beta}$ for β in the expression given above. ◇

EXAMPLE 4.35: Suppose that $X_1, \cdots, X_n$ are i.i.d. random variables with density $f(x - \theta)$ where $f(x) = f(-x)$; that is, the X_i's have distribution that is symmetric around 0. Let $\psi(x)$ be a non-decreasing odd function ($\psi(x) = -\psi(-x)$) with derivative $\psi'(x)$ and define $\widehat{\theta}$ to be the solution to the equation

$$\frac{1}{n}\sum_{i=1}^{n}\psi(x-\widehat{\theta}) = 0.$$

Note that $\widehat{\theta}$ is a substitution principle estimator of the functional parameter $\theta(F)$ defined by

$$\int_{-\infty}^{\infty}\psi(x-\theta(F))\,dF(x) = 0;$$

the influence curve of $\theta(F)$ is

$$\phi(x;F) = \frac{\psi(x-\theta(F))}{\int_{-\infty}^{\infty}\psi'(x-\theta(F))\,dF(x)}.$$

Hence for n sufficiently large, $\sqrt{n}(\widehat{\theta}-\theta)$ is approximately Normal with mean 0 and variance

$$\begin{aligned}\sigma^2 &= \int_{-\infty}^{\infty}\phi^2(x;F)\,dF(x)\\ &= \frac{\int_{-\infty}^{\infty}\psi^2(x-\theta)f(x-\theta)\,dx}{\left(\int_{-\infty}^{\infty}\psi'(x-\theta)f(x-\theta)\,dx\right)^2}\\ &= \frac{\int_{-\infty}^{\infty}\psi^2(x)f(x)\,dx}{\left(\int_{-\infty}^{\infty}\psi'(x)f(x)\,dx\right)^2}\end{aligned}$$

and so an approximate standard error of $\widehat{\theta}$ is $\sigma/\sqrt{n}$. ◇

As we noted above, standard errors (and their approximations) can and typically do depend on unknown parameters. These standard errors can themselves be estimated by substituting estimates for the unknown parameters in the expression for the standard error.

EXAMPLE 4.36: In Example 4.35, we showed that the approximate standard error of $\widehat{\theta}$ is $\sigma/\sqrt{n}$ where

$$\sigma^2 = \frac{\int_{-\infty}^{\infty}\psi^2(x-\theta)f(x-\theta)\,dx}{\left(\int_{-\infty}^{\infty}\psi'(x-\theta)f(x-\theta)\,dx\right)^2}.$$

Substituting $\widehat{\theta}$ for θ, we can obtain the following substitution principle estimator of σ^2:

$$\widehat{\sigma}^2 = \left(\frac{1}{n}\sum_{i=1}^{n}\psi'(X_i - \widehat{\theta})\right)^{-2}\left(\frac{1}{n}\sum_{i=1}^{n}\psi(X_i - \widehat{\theta})\right).$$

The estimated standard error of $\widehat{\theta}$ is simply $\widehat{\sigma}/\sqrt{n}$. ◇

Another method of estimating standard errors is given in Section 4.9.

4.8 Asymptotic relative efficiency

Suppose that $\widehat{\theta}_n$ and $\widetilde{\theta}_n$ are two possible estimators (based on $X_1, \cdots, X_n$) of a real-valued parameter θ. There are a variety of approaches to comparing two estimators. For example, we can compare the MSEs or MAEs (if they are computable) and choose the estimator whose MSE (or MAE) is smaller (although this choice may depend on the unknown value of θ). If both estimators are approximately Normal, we can use a measure called the asymptotic relative efficiency (ARE).

DEFINITION. Let $X_1, X_2, \cdots$ be a sequence of random variables and suppose that $\widehat{\theta}_n$ and $\widetilde{\theta}_n$ are estimators of θ (based on $X_1, \cdots, X_n$) such that

$$\frac{\widehat{\theta}_n - \theta}{\sigma_{1n}(\theta)} \to_d N(0,1) \quad \text{and} \quad \frac{\widetilde{\theta}_n - \theta}{\sigma_{2n}(\theta)} \to_d N(0,1)$$

for some sequences $\{\sigma_{1n}(\theta)\}$ and $\{\sigma_{2n}(\theta)\}$. Then the asymptotic relative efficiency of $\widehat{\theta}_n$ to $\widetilde{\theta}_n$ is defined to be

$$\text{ARE}_\theta(\widehat{\theta}_n, \widetilde{\theta}_n) = \lim_{n\to\infty} \frac{\sigma_{2n}^2(\theta)}{\sigma_{1n}^2(\theta)}$$

provided this limit exists.

What is the interpretation of asymptotic relative efficiency? In many applications (for example, if the X_i's are i.i.d.), we have

$$\sigma_{1n}(\theta) = \frac{\sigma_1(\theta)}{\sqrt{n}} \quad \text{and} \quad \sigma_{2n}(\theta) = \frac{\sigma_2(\theta)}{\sqrt{n}}$$

and so

$$\text{ARE}_\theta(\widehat{\theta}_n, \widetilde{\theta}_n) = \frac{\sigma_2^2(\theta)}{\sigma_1^2(\theta)}.$$

Suppose we can estimate θ using either $\widehat{\theta}_n$ or $\widetilde{\theta}_m$ where n and m are the sample sizes on which the two estimators are based. Suppose we want to choose m and n such that

$$P_\theta\left(|\widehat{\theta}_n - \theta| < \Delta\right) \approx P_\theta\left(|\widetilde{\theta}_m - \theta| < \Delta\right)$$

for any Δ. Since for m and n sufficiently large both estimators are approximately Normal, m and n satisfy

$$P\left(|Z| < \Delta\sigma_1(\theta)/\sqrt{n}\right) \approx P\left(|Z| < \Delta\sigma_2(\theta)/\sqrt{m}\right)$$

(where $Z \sim N(0,1)$), which implies that

$$\frac{\sigma_1(\theta)}{\sqrt{n}} \approx \frac{\sigma_2(\theta)}{\sqrt{m}}$$

or

$$\frac{\sigma_2^2(\theta)}{\sigma_1^2(\theta)} \approx \frac{m}{n}.$$

Thus the ratio of sample sizes needed to achieve the same precision is approximately equal to the asymptotic relative efficiency; for example, if $\text{ARE}_\theta(\widehat{\theta}_n, \widetilde{\theta}_n) = k$, we would need $m \approx kn$ so that $\widetilde{\theta}_m$ has the same precision as $\widehat{\theta}_n$ (when θ is the true value of the parameter).

In applying ARE to compare two estimators, we should keep in mind that it is a large sample measure and therefore may be misleading in small sample situations. If measures such as MSE and MAE cannot be accurately evaluated, simulation is useful for comparing estimators.

EXAMPLE 4.37: Suppose that $X_1, \cdots, X_n$ are i.i.d. Normal random variables with mean μ and variance σ^2. Let $\widehat{\mu}_n$ be the sample mean and $\widetilde{\mu}_n$ be the sample median of $X_1, \cdots, X_n$. Then we have

$$\sqrt{n}(\widehat{\mu}_n - \mu) \to_d N(0, \sigma^2) \quad \text{and} \quad \sqrt{n}(\widetilde{\mu}_n - \mu) \to_d N(0, \pi\sigma^2/2).$$

Hence

$$\text{ARE}_\mu(\widehat{\mu}_n, \widetilde{\mu}_n) = \frac{\pi}{2} = 1.571.$$

We say that $\widehat{\mu}_n$ is more efficient than $\widetilde{\mu}_n$. ◇

EXAMPLE 4.38: Suppose that $X_1, \cdots, X_n$ are i.i.d. Poisson random variables with mean λ. Suppose we want to estimate $\theta = \exp(-\lambda) = P_\lambda(X_i = 0)$. Consider the two estimators

$$\widehat{\theta}_n = \exp(-\bar{X}_n) \quad \text{and} \quad \widetilde{\theta}_n = \frac{1}{n}\sum_{i=1}^n I(X_i = 0).$$

It is easy to show (using the CLT and the Delta Method) that

$$\begin{aligned} \sqrt{n}(\widehat{\theta}_n - \theta) &\to_d N(0, \lambda \exp(-2\lambda)) \\ \text{and} \quad \sqrt{n}(\widetilde{\theta}_n - \theta) &\to_d N(0, \exp(-\lambda) - \exp(-2\lambda)). \end{aligned}$$

Hence

$$\text{ARE}_\lambda(\widehat{\theta}_n, \widetilde{\theta}_n) = \frac{\exp(\lambda) - 1}{\lambda}.$$

Using the expansion $\exp(\lambda) = 1 + \lambda + \lambda^2/2 + \cdots$, it is easy to see that this ARE is always greater than 1; however, for small values of λ, the ARE is close to 1 indicating that there is little to choose between the two estimators when λ is small. ◇

EXAMPLE 4.39: Suppose that $X_1, \cdots, X_n$ are i.i.d. Exponential random variables with parameter λ. In Example 4.25, we gave a family of method of moments estimators of λ using the fact that $E_\lambda(X_i) = \Gamma(r+1)/\lambda^r$ for $r > 0$. Define

$$\widehat{\lambda}_n^{(r)} = \left(\frac{1}{n\Gamma(r+1)} \sum_{i=1}^n X_i^r \right)^{-1/r}.$$

Using the fact that $\text{Var}_\lambda(X_i^r) = (\Gamma(2r+1) - \Gamma^2(r+1))/\lambda^{2r}$, it follows from the Central Limit Theorem and the Delta Method that

$$\sqrt{n}(\widehat{\lambda}_n^{(r)} - \lambda) \to_d N(0, \sigma^2(r))$$

where

$$\sigma^2(r) = \frac{\lambda^2}{r^2} \left(\frac{\Gamma(2r+1)}{\Gamma^2(r+1)} - 1 \right).$$

The graph of $\sigma^2(r)/\lambda^2$ is given in Figure 4.2; it is easy to see that $\sigma^2(r)$ is minimized for $r = 1$ so that $1/\bar{X}$ is the most efficient (asymptotically) estimator of λ of this form. ◇

EXAMPLE 4.40: Suppose that $X_1, \cdots, X_n$ are i.i.d. Cauchy random variables with density function

$$f(x; \theta, \sigma) = \frac{1}{\pi} \frac{\sigma}{\sigma^2 + (x-\theta)^2}.$$

This density function is symmetric around θ; however, since $E(X_i)$ is not defined for this distribution, the sample mean $\bar{X}_n$ is not a good estimator of θ. A possible estimator of θ is the α-trimmed

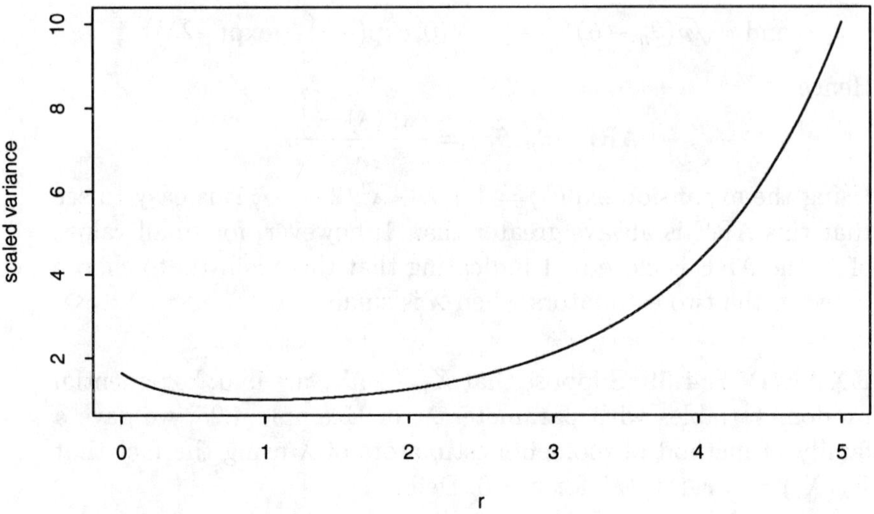

Figure 4.2 $\sigma^2(r)/\lambda^2$ *in Example 4.39 as a function of* r.

mean

$$\widehat{\theta}_n(\alpha) = \frac{1}{n - 2g_n} \sum_{i=g_n+1}^{n-g_n} X_{(i)}$$

where the $X_{(i)}$'s are the order statistics and $g_n/n \to \alpha$ as $n \to \infty$ where $0 < \alpha < 0.5$. It can be shown (for example, by using the influence curve of the trimmed mean functional parameter given in Example 4.30) that

$$\sqrt{n}(\widehat{\theta}_n(\alpha) - \theta) \to_d N(0, \gamma^2(\alpha))$$

where

$$\frac{\gamma^2(\alpha)}{\sigma^2} = \frac{2\pi^{-1}\tan(\pi(0.5-\alpha)) + 2\alpha - 1 + 2\alpha\tan^2(\pi(0.5-\alpha))}{(1-2\alpha)^2}.$$

If $\widetilde{\theta}_n$ is the sample median of $X_1, \cdots, X_n$, we have

$$\sqrt{n}(\widetilde{\theta}_n - \theta) \to_d N\left(0, \sigma^2\pi^2/4\right).$$

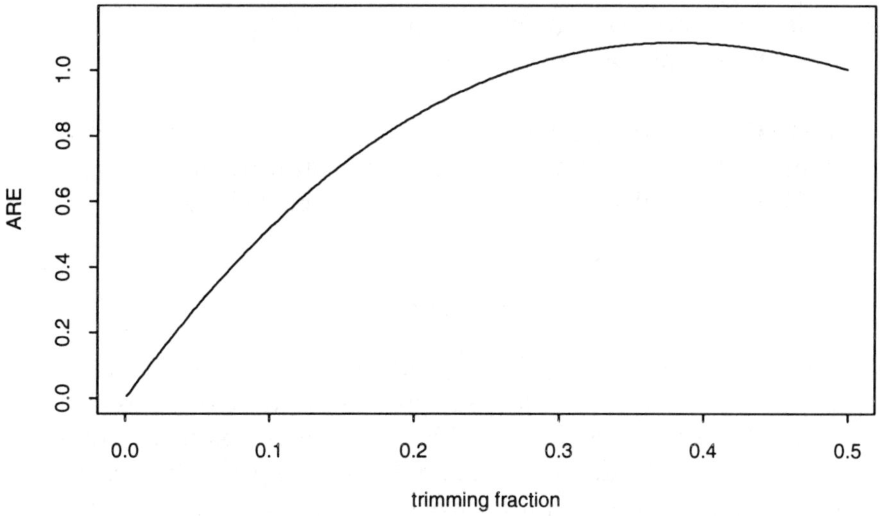

Figure 4.3 *ARE of α-trimmed means (for $0 \le \alpha \le 0.5$) with respect to the sample median in Example 4.40.*

The ARE of $\widehat{\theta}_n(\alpha)$ with respect to $\widetilde{\theta}_n$ is thus given by the formula

$$\text{ARE}_\theta(\widehat{\theta}_n(\alpha), \widetilde{\theta}_n) = \frac{\pi^2(1-2\alpha)^2}{4\left[2\pi^{-1}\tan\left(\pi(0.5-\alpha)\right) + 2\alpha - 1 + 2\alpha\tan^2\left(\pi(0.5-\alpha)\right)\right]}$$

A plot of $\text{ARE}_\theta(\widehat{\theta}_n(\alpha), \widetilde{\theta}_n)$ for α between 0 and 0.5 is given in Figure 4.3. The trimmed mean $\widehat{\theta}_n(\alpha)$ is more efficient than $\widetilde{\theta}_n$ for $\alpha > 0.269$ and the ARE is maximized at $\alpha = 0.380$. We will see in Chapter 5 that we can find even more efficient estimators of θ for this model. ◇

4.9 The jackknife

The jackknife provides a general-purpose approach to estimating the bias and variance (or standard error) of an estimator. Suppose that $\widehat{\theta}$ is an estimator of θ based on i.i.d. random variables $X_1, \cdots, X_n$; θ could be an unknown parameter from some parametric model or θ could be functional parameter of the common distribution function F of the X_i's (in which case $\theta = \theta(F)$). The jackknife is particularly useful when standard methods for computing bias and variance cannot be applied or are difficult to apply. Two such examples are given below.

EXAMPLE 4.41: Suppose that $X_1, \cdots, X_n$ are i.i.d. random variables with density $f(x-\theta)$ that is symmetric around θ ($f(x) = f(-x)$). One possible estimator of θ is the trimmed mean

$$\widehat{\theta} = \frac{1}{n-2g} \sum_{i=g+1}^{n-g} X_{(i)},$$

which averages $X_{(g+1)}, \cdots, X_{(n-g)}$, the middle $n - 2g$ order statistics. The trimmed mean is less susceptible to extreme values than the sample mean of the X_i's, and is often a useful estimator of θ. However, unless the density function f is known precisely, it is difficult to approximate the variance of $\widehat{\theta}$. (If f is known, it is possible to approximate the variance of $\widehat{\theta}$ using the influence curve given in Example 4.30; see also Example 4.40.) ◇

EXAMPLE 4.42: In survey sampling, it is necessary to estimate the ratio of two means. For example, we may be interested in estimating the unemployment rate for males aged 18 to 25. If we take a random sample of households, we can obtain both the number of males between 18 and 25 and the number of these males who are unemployed in each of the sampled households. Our estimate of the unemployment rate for males aged 18 to 25 would then be

$$\widehat{r} = \frac{\text{number of unemployed males aged 18 - 25 in sample}}{\text{number of males aged 18 - 25 in sample}}.$$

The general problem may be expressed as follows. Suppose that $(X_1, Y_1), \cdots, (X_n, Y_n)$ are independent random vectors from the same joint distribution with $E(X_i) = \mu_X$ and $E(Y_i) = \mu_Y$; we want to estimate $r = \mu_X/\mu_Y$. A method of moments estimator of r is

$$\widehat{r} = \frac{\sum_{i=1}^n X_i}{\sum_{i=1}^n Y_i} = \frac{\bar{X}}{\bar{Y}}.$$

Unfortunately, there is no easy way to evaluate either $E(\bar{X}/\bar{Y})$ or $\text{Var}(\bar{X}/\bar{Y})$ (although the Delta Method provides a reasonable approximation). ◇

The name "jackknife" was originally used by Tukey (1958) to suggest the broad usefulness of the technique as a substitute to more specialized techniques much in the way that a jackknife can be used as a substitute for a variety of more specialized tools (although, in reality, a jackknife is not a particularly versatile tool!). More complete references on the jackknife are the monographs by Efron (1982), and Efron and Tibshirani (1993).

The jackknife estimator of bias

The jackknife estimator of bias was developed by Quenouille (1949) although he did not refer to it as the jackknife. The basic idea behind the jackknife estimators of bias and variance lies in recomputing the parameter estimator using all but one of the observations.

Suppose that $\widehat{\theta}$ is an estimator of a parameter θ based on sample of i.i.d. random variables $X_1, \cdots, X_n$: $\widehat{\theta} = \widehat{\theta}(\boldsymbol{X})$. (For example, $\widehat{\theta} = \theta(\widehat{F})$ if $\theta = \theta(F)$.) Quenouille's method for estimating the bias of $\widehat{\theta}$ is based on sequentially deleting a single observation X_i and recomputing $\widehat{\theta}$ based on $n-1$ observations. Suppose that

$$E_\theta(\widehat{\theta}) = \theta + b_\theta(\widehat{\theta})$$

where $b_\theta(\widehat{\theta})$ is the bias of $\widehat{\theta}$. Let $\widehat{\theta}_{-i}$ be the estimator of θ evaluated after deleting X_i from the sample:

$$\widehat{\theta}_{-i} = \widehat{\theta}(X_1, \cdots, X_{i-1}, X_{i+1}, \cdots, X_n).$$

Now define $\widehat{\theta}_\bullet$ to be the average of $\widehat{\theta}_{-1}, \cdots, \widehat{\theta}_{-n}$:

$$\widehat{\theta}_\bullet = \frac{1}{n} \sum_{i=1}^{n} \widehat{\theta}_{-i}.$$

The jackknife estimator of bias is then

$$\widehat{b}(\widehat{\theta}) = (n-1)(\widehat{\theta}_\bullet - \widehat{\theta}).$$

A bias-corrected version of $\widehat{\theta}$ can be constructed by subtracting $\widehat{b}(\widehat{\theta})$ from $\widehat{\theta}$; we will show below that this procedure reduces the bias of $\widehat{\theta}$.

The theoretical rationale behind $\widehat{b}(\widehat{\theta})$ assumes that $E_\theta(\widehat{\theta})$ can be expressed as a series involving powers of $1/n$; for simplicity, we will first assume that for any n

$$E_\theta(\widehat{\theta}) = \theta + \frac{a_1(\theta)}{n}$$

where $a_1(\theta)$ can depend on θ or the distribution of the X_i's but not the sample size n; in this case, $b_\theta(\widehat{\theta}) = a_1(\theta)/n$. Since θ_{-i} is based on $n-1$ observations (for each i), it follows that

$$E_\theta(\widehat{\theta}_\bullet) = \frac{1}{n}\sum_{i=1}^{n} E_\theta(\widehat{\theta}_{-i}) = \theta + \frac{a_1(\theta)}{n-1}.$$

Thus

$$E_\theta(\widehat{\theta} - \widehat{\theta}_\bullet) = \frac{a_1(\theta)}{n} - \frac{a_1(\theta)}{n-1} = \frac{a_1(\theta)}{n(n-1)}$$

and so $(n-1)(\widehat{\theta} - \widehat{\theta}_\bullet)$ is an unbiased estimator of $b_\theta(\widehat{\theta})$.

In the general case, we will have

$$E_\theta(\widehat{\theta}) = \theta + \frac{a_1(\theta)}{n} + \frac{a_2(\theta)}{n^2} + \frac{a_3(\theta)}{n^3} + \cdots$$

or

$$b_\theta(\widehat{\theta}) = \frac{a_1(\theta)}{n} + \frac{a_2(\theta)}{n^2} + \frac{a_3(\theta)}{n^3} + \cdots$$

where $a_1(\theta), a_2(\theta), a_3(\theta), \cdots$ can depend on θ or the distribution of the X_i's but not on n. Again, it follows that

$$\begin{aligned} E_\theta(\widehat{\theta}_\bullet) &= \frac{1}{n}\sum_{i=1}^{n} E_\theta(\widehat{\theta}_{-i}) \\ &= \theta + \frac{a_1(\theta)}{n-1} + \frac{a_2(\theta)}{(n-1)^2} + \frac{a_3(\theta)}{(n-1)^3} + \cdots \end{aligned}$$

(since each $\widehat{\theta}_{-i}$ is based on $n-1$ observations). Thus the expected value of the jackknife estimator of bias is

$$\begin{aligned} E_\theta(\widehat{b}(\widehat{\theta})) &= (n-1)\left(E_\theta(\widehat{\theta}_\bullet) - E_\theta(\widehat{\theta})\right) \\ &= \frac{a_1(\theta)}{n} + \frac{(2n-1)a_2(\theta)}{n^2(n-1)} \\ &\quad + \frac{(3n^2-3n+1)a_3(\theta)}{n^3(n-1)^2} + \cdots. \end{aligned}$$

We can see from above that $\widehat{b}(\widehat{\theta})$ is not an unbiased estimator of $b_\theta(\widehat{\theta})$ as it was in the simple case considered earlier. However, note that the first term of $E_\theta(\widehat{b}(\widehat{\theta}))$ (namely $a_1(\theta)/n$) agrees with that of $b_\theta(\widehat{\theta})$. Thus if we define

$$\widehat{\theta}_{jack} = \widehat{\theta} - \widehat{b}(\widehat{\theta}) = n\widehat{\theta} - (n-1)\widehat{\theta}_\bullet$$

to be the bias-corrected (or jackknifed) version of $\widehat{\theta}$, it follows that

$$\begin{aligned} E_\theta(\widehat{\theta}_{jack}) &= \theta - \frac{a_2(\theta)}{n(n-1)} - \frac{(2n-1)a_3(\theta)}{n^2(n-1)^2} + \cdots \\ &\approx \theta - \frac{a_2(\theta)}{n^2} - \frac{2a_3(\theta)}{n^3} + \cdots \end{aligned}$$

for large n. Since $1/n^2$, $1/n^3, \cdots$ go to 0 faster than $1/n$ goes to 0 (as n gets large), it follows that the bias of $\widehat{\theta}_{jack}$ is smaller than the bias of $\widehat{\theta}$ for n sufficiently large. In the case where

$$E_\theta(\widehat{\theta}) = \theta + \frac{a_1(\theta)}{n}$$

(so that $a_2(\theta) = a_3(\theta) = \cdots = 0$), $\widehat{\theta}_{jack}$ will be unbiased.

EXAMPLE 4.43: Suppose that $X_1, \cdots, X_n$ are i.i.d. random variables from a distribution with mean μ and variance σ^2, both unknown. The estimator

$$\widehat{\sigma}^2 = \frac{1}{n}\sum_{i=1}^n (X_i - \bar{X})^2$$

is a biased estimator with $b(\widehat{\sigma}^2) = -\sigma^2/n$. Thus the bias in $\widehat{\sigma}^2$ can be removed by using the jackknife. An educated guess for the resulting unbiased estimator is

$$S^2 = \frac{1}{n-1}\sum_{i=1}^n (X_i - \bar{X})^2.$$

To find the unbiased estimator using the jackknife, we first note that

$$\bar{X}_{-i} = \frac{1}{n-1}\sum_{j \neq i} X_j = \frac{1}{n-1}(n\bar{X} - X_i)$$

and so

$$\begin{aligned} \widehat{\sigma}^2_{-i} &= \frac{1}{n-1}\sum_{j \neq i}(X_j - \bar{X}_{-i})^2 \\ &= \frac{1}{n-1}\sum_{j \neq i}\left(X_j - \frac{n}{n-1}\bar{X} + \frac{X_i}{n-1}\right) \\ &= \frac{1}{n-1}\sum_{j=1}^n \left(X_j - \bar{X} + \frac{1}{n-1}(X_i - \bar{X})\right) \end{aligned}$$

$$
\begin{aligned}
& -\frac{n^2}{(n-1)^3}(X_i - \bar{X})^2 \\
= & \frac{1}{n-1}\sum_{j=1}^{n}(X_j - \bar{X})^2 + \frac{n}{(n-1)^3}(X_i - \bar{X})^2 \\
& -\frac{n^2}{(n-1)^3}(X_i - \bar{X})^2 \\
= & \frac{1}{n-1}\sum_{j=1}^{n}(X_j - \bar{X})^2 - \frac{n}{(n-1)^2}(X_i - \bar{X})^2.
\end{aligned}
$$

Now $\widehat{\sigma}^2_\bullet$ is just the average of the $\widehat{\sigma}^2_{-i}$'s so that

$$
\widehat{\sigma}^2_\bullet = \frac{1}{n-1}\sum_{i=1}^{n}(X_i - \bar{X})^2 - \frac{1}{(n-1)^2}\sum_{i=1}^{n}(X_i - \bar{X})^2
$$

and the unbiased estimator of σ^2 is

$$
n\widehat{\sigma}^2 - (n-1)\widehat{\sigma}^2_\bullet = \frac{1}{n-1}\sum_{i=1}^{n}(X_i - \bar{X})^2 = S^2
$$

as was guessed above. ◇

EXAMPLE 4.44: Suppose that $X_1, \cdots, X_n$ are i.i.d. random variables with probability density function

$$
f(x;\theta) = \frac{1}{\theta} \quad \text{for} \quad 0 \le x \le \theta
$$

where θ is an unknown parameter. Since θ is the maximum possible value of the X_i's, a natural estimator of θ is

$$
\widehat{\theta} = X_{(n)} = \max(X_1, \cdots, X_n).
$$

However, since the X_i's cannot exceed θ, it follows that their maximum cannot exceed θ and so $\widehat{\theta}$ is biased; in fact,

$$
\begin{aligned}
E(\widehat{\theta}) &= \frac{n}{n+1}\theta \\
&= \theta\frac{1}{1+1/n} \\
&= \theta\left(1 - \frac{1}{n} + \frac{1}{n^2} - \frac{1}{n^3} + \cdots\right).
\end{aligned}
$$

Since

$$
\widehat{\theta}_{-i} = \max(X_1, \cdots, X_{i-1}, X_{i+1} \cdots, X_n),
$$

it follows that $\widehat{\theta}_{-i} = X_{(n)}$ for $n-1$ values of i and $\widehat{\theta}_{-i} = X_{(n-1)}$ for the other value of i. Thus, we obtain

$$\widehat{\theta}_\bullet = \frac{n-1}{n} X_{(n)} + \frac{1}{n} X_{(n-1)}$$

and so the jackknifed estimator of θ is

$$\widehat{\theta}_{jack} = X_{(n)} + \frac{n-1}{n}(X_{(n)} - X_{(n-1)}).$$

The bias of $\widehat{\theta}_{jack}$ will be smaller than that of $\widehat{\theta}$; nonetheless, we can easily modify $\widehat{\theta}$ to make it unbiased without resorting to the jackknife by simply multiplying it by $(n+1)/n$. ◇

The latter example points out one of the drawbacks in using any general purpose method (such as the jackknife), namely that in specific situations, it is often possible to improve upon that method with one that is tailored specifically to the situation at hand. Removing the bias in $\widehat{\theta} = X_{(n)}$ by multiplying $X_{(n)}$ by $(n+1)/n$ relies on the fact that the form of the density is known. Suppose instead that the range of the X_i's was still $[0, \theta]$ but that the density $f(x)$ was unknown for $0 \le x \le \theta$. Then $X_{(n)}$ is still a reasonable estimator of θ and still always underestimates θ. However, $(n+1)X_{(n)}/n$ need not be unbiased and, in fact, may be more severely biased than $X_{(n)}$. However, the jackknifed estimator

$$\widehat{\theta}_{jack} = X_{(n)} + \frac{n-1}{n}(X_{(n)} - X_{(n-1)})$$

will have a smaller bias than $X_{(n)}$ and may be preferable to $X_{(n)}$ in this situation.

The jackknife estimator of variance

The jackknife estimator of bias uses the estimators $\widehat{\theta}_{-1}, \cdots, \widehat{\theta}_{-n}$ (which use all the observations but one in their computation) to construct an estimator of bias of an estimator $\widehat{\theta}$. Tukey (1958) suggested a method of estimating $\text{Var}(\widehat{\theta})$ that uses $\widehat{\theta}_{-1}, \cdots, \widehat{\theta}_{-n}$. Tukey's jackknife estimator of $\text{Var}(\widehat{\theta})$ is

$$\widehat{\text{Var}}(\widehat{\theta}) = \frac{n-1}{n} \sum_{i=1}^{n} (\widehat{\theta}_{-i} - \widehat{\theta}_\bullet)^2$$

where as before $\widehat{\theta}_{-i}$ is the estimator evaluated using all the observations except X_i and

$$\widehat{\theta}_{\bullet} = \frac{1}{n}\sum_{i=1}^{n} \widehat{\theta}_{-i}.$$

The formula for the jackknife estimator of variance is somewhat unintuitive. In deriving the formula, Tukey assumed that the estimator $\widehat{\theta}$ can be approximated well by an average of independent random variables; this assumption is valid for a wide variety of estimators but is not true for some estimators (for example, sample maxima or minima). More precisely, Tukey assumed that

$$\widehat{\theta} \approx \frac{1}{n}\sum_{i=1}^{n} \phi(X_i),$$

which suggests that

$$\text{Var}(\widehat{\theta}) \approx \frac{\text{Var}(\phi(X_1))}{n}.$$

(In the case where the parameter of interest θ is a functional parameter of the distribution function F (that is, $\theta = \theta(F)$), the function $\phi(\cdot) - \theta(F)$ is typically the influence curve of $\theta(F)$.)

In general, we do not know the function $\phi(x)$ so we cannot use the above formula directly. However, it is possible to find reasonable surrogates for $\phi(X_1), \cdots, \phi(X_n)$. Using the estimators $\widehat{\theta}_{-i}$ ($i = 1, \cdots, n$) and $\widehat{\theta}$, we define pseudo-values

$$\Phi_i = \widehat{\theta} + (n-1)(\widehat{\theta} - \widehat{\theta}_{-i})$$

(for $i = 1, \cdots, n$) that essentially play the same role as the $\phi(X_i)$'s above; in the case where $\theta = \theta(F)$, $(n-1)(\widehat{\theta} - \widehat{\theta}_{-i})$ is an attempt to estimate the influence curve of $\theta(F)$ at $x = X_i$. (In the case where $\widehat{\theta}$ is exactly a sample mean

$$\widehat{\theta} = \frac{1}{n}\sum_{i=1}^{n} \phi(X_i),$$

it easy to show that $\Phi_i = \phi(X_i)$ and so the connection between Φ_i and $\phi(X_i)$ is clear in this simple case.) We can then take the sample variance of the pseudo-values Φ_i to be an estimate of the variance of $\phi(X_1)$ and use it to estimate the variance of $\widehat{\theta}$. Note

Table 4.1 *Pre-tax incomes for Example 4.45.*

3841	7084	7254	15228	18042	19089
22588	23972	25694	27592	27927	31576
32528	32921	33724	36887	37776	37992
39464	40506	44516	46538	51088	51955
54339	57935	75137	82612	83381	84741

that

$$\begin{aligned}\frac{1}{n}\sum_{i=1}^{n}\Phi_i &= n\widehat{\theta} - \frac{n-1}{n}\sum_{i=1}^{n}\widehat{\theta}_{-i}\\ &= n\widehat{\theta} - (n-1)\widehat{\theta}_{\bullet}\\ &= \widehat{\theta}_{jack}\end{aligned}$$

where $\widehat{\theta}_{jack}$ is the bias-corrected version of $\widehat{\theta}$. The sample variance of the Φ_i's is

$$\begin{aligned}\frac{1}{n-1}\sum_{i=1}^{n}(\Phi_i - \bar{\Phi})^2 &= \frac{1}{n-1}\sum_{i=1}^{n}[(n-1)(\widehat{\theta}_{\bullet} - \widehat{\theta}_{-i})]^2\\ &= (n-1)\sum_{i=1}^{n}(\widehat{\theta}_{-i} - \widehat{\theta}_{\bullet})^2.\end{aligned}$$

We now get the jackknife estimator of variance by dividing the sample variance of the Φ_i's by n:

$$\widehat{\mathrm{Var}}(\widehat{\theta}) = \frac{n-1}{n}\sum_{i=1}^{n}(\widehat{\theta}_{-i} - \widehat{\theta}_{\bullet})^2.$$

It should be noted that the jackknife estimator of variance does not work in all situations. One such situation is the sample median; the problem here seems to be the fact that the influence curve of the median is defined only for continuous distributions and so is difficult to approximate adequately from finite samples.

EXAMPLE 4.45: The data in Table 4.1 represent a sample of 30 pre-tax incomes. We will assume that these data are outcomes of i.i.d. random variables $X_1, \cdots, X_{30}$ from a distribution function

Table 4.2 *Values of $\widehat{\theta}_{-i}$ obtained by leaving out the corresponding entry in Table 4.1.*

0.2912	0.2948	0.2950	0.3028	0.3055	0.3064
0.3092	0.3103	0.3115	0.3127	0.3129	0.3148
0.3153	0.3154	0.3157	0.3166	0.3168	0.3168
0.3170	0.3170	0.3169	0.3167	0.3161	0.3159
0.3152	0.3140	0.3069	0.3033	0.3028	0.3020

F; we will use the data to estimate the Gini index

$$\theta(F) = 1 - 2\int_0^1 q_F(t)\,dt$$

where

$$q_F(t) = \frac{\int_0^t F^{-1}(s)\,ds}{\int_0^1 F^{-1}(s)\,ds}$$

is the Lorenz curve. The substitution principle estimator of $\theta(F)$ is

$$\widehat{\theta} = \theta(\widehat{F}) = \left(\sum_{i=1}^{30} X_i\right)^{-1} \sum_{i=1}^{30} \left(\frac{2i-1}{30} - 1\right) X_{(i)}$$

where $X_{(1)} \le X_{(2)} \le \cdots \le X_{(30)}$ are the order statistics of $X_1, \cdots, X_{30}$.

For these data, the estimate of $\theta(F)$ is 0.311. The standard error of this estimate can be estimated using the jackknife. The leave-out-estimates $\widehat{\theta}_{-i}$ of $\theta(F)$ are given in Table 4.2.

The jackknife estimate of the standard error of $\widehat{\theta}$ is

$$\widehat{\text{se}}(\widehat{\theta}) = \frac{29}{30}\sum_{i=1}^{30}(\widehat{\theta}_{-i} - \widehat{\theta}_{\bullet})^2 = 0.0398$$

where $\widehat{\theta}_{\bullet} = 0.310$ is the average of $\widehat{\theta}_{-1}, \cdots, \widehat{\theta}_{-30}$. ◇

Comparing the jackknife and Delta Method estimators

How does the jackknife estimator of variance compare to the Delta Method estimator? We will consider the simple case of estimating the variance of $g(\bar{X})$ where $\bar{X}$ is the sample mean of i.i.d. random variables $X_1, \cdots, X_n$. The Delta Method estimator is

$$\widehat{\text{Var}}_d(g(\bar{X})) = [g'(\bar{X})]^2 \frac{1}{n(n-1)} \sum_{i=1}^n (X_i - \bar{X})^2$$

while the jackknife estimator is

$$\widehat{\text{Var}}_j(g(\bar{X})) = \frac{n-1}{n} \sum_{i=1}^n (g(\bar{X}_{-i}) - g_\bullet)^2$$

where

$$g_\bullet = \frac{1}{n} \sum_{i=1}^n g(\bar{X}_{-i}).$$

Recalling that

$$\begin{aligned} \bar{X}_{-i} &= \frac{1}{n-1}(n\bar{X} - X_i) \\ &= \bar{X} - \frac{1}{n-1}(X_i - \bar{X}), \end{aligned}$$

it follows from a Taylor series expansion that

$$\begin{aligned} g(\bar{X}_{-i}) &\approx g(\bar{X}) + (\bar{X}_{-i} - \bar{X})g'(\bar{X}) \\ &= g(\bar{X}) - \frac{1}{n-1}(X_i - \bar{X})g'(\bar{X}) \end{aligned}$$

and hence

$$\begin{aligned} g_\bullet &= \frac{1}{n} \sum_{i=1}^n g(\bar{X}_{-i}) \\ &\approx g(\bar{X}). \end{aligned}$$

Substituting these approximations into $\widehat{\text{Var}}_j(g(\bar{X}))$, we get

$$\begin{aligned} \widehat{\text{Var}}_j(g(\bar{X})) &\approx [g'(\bar{X})]^2 \frac{1}{n(n-1)} \sum_{i=1}^n (X_i - \bar{X})^2 \\ &= \widehat{\text{Var}}_d(g(\bar{X})). \end{aligned}$$

Thus the jackknife and Delta Method estimators are approximately equal when $\widehat{\theta} = g(\bar{X})$.

4.10 Problems and complements

4.1: Suppose that $\boldsymbol{X} = (X_1, \cdots, X_n)$ has a one-parameter exponential family distribution with joint density or frequency function

$$f(\boldsymbol{x};\theta) = \exp\left[\theta T(\boldsymbol{x}) - d(\theta) + S(\boldsymbol{x})\right]$$

where the parameter space Θ is an open subset of R. Show that

$$E_\theta[\exp(sT(\boldsymbol{X}))] = d(\theta+s) - d(\theta)$$

if s is sufficiently small. (Hint: Since Θ is open, $f(\boldsymbol{x};\theta+s)$ is a density or frequency function for s sufficiently small and hence integrates or sums to 1.)

4.2: Suppose that $\boldsymbol{X} = (X_1, \cdots, X_n)$ has a k-parameter exponential family distribution with joint density or frequency function

$$f(\boldsymbol{x};\boldsymbol{\theta}) = \exp\left[\sum_{i=1}^{p} \theta_i T_i(\boldsymbol{x}) - d(\boldsymbol{\theta}) + S(\boldsymbol{x})\right]$$

where the parameter space Θ is an open subset of R^k.

(a) Show that

$$E_\theta[T_i(\boldsymbol{X})] = \frac{\partial}{\partial\theta_i} d(\boldsymbol{\theta})$$

for $i = 1, \cdots, k$.

(b) Show that

$$\mathrm{Cov}_\theta[T_i(\boldsymbol{X}), T_j(\boldsymbol{X})] = \frac{\partial^2}{\partial\theta_i \partial\theta_j} d(\boldsymbol{\theta})$$

for $i, j = 1, \cdots, k$.

4.3: Suppose that $X_1, \cdots, X_n$ are i.i.d. random variables with density

$$f(x;\theta_1,\theta_2) = \begin{cases} a(\theta_1,\theta_2)h(x) & \text{for } \theta_1 \le x \le \theta_2 \\ 0 & \text{otherwise} \end{cases}$$

where $h(x)$ is a known function defined on the real line.

(a) Show that

$$a(\theta_1,\theta_2) = \left(\int_{\theta_1}^{\theta_2} h(x)\,dx\right)^{-1}.$$

(b) Show that $(X_{(1)}, X_{(n)})$ is sufficient for (θ_1, θ_2).

4.4: Suppose that $\boldsymbol{X} = (X_1, \cdots, X_n)$ has joint density or frequency function $f(\boldsymbol{x};\theta_1,\theta_2)$ where θ_1 and θ_2 vary independently (that is, $\Theta = \Theta_1 \times \Theta_2$) and the set

$$S = \{\boldsymbol{x} : f(\boldsymbol{x};\theta_1,\theta_2) > 0\}$$

does not depend on (θ_1, θ_2). Suppose that T_1 is sufficient for θ_1 when θ_2 is known and T_2 is sufficient for θ_2 when θ_1 is known. Show that (T_1, T_2) is sufficient for (θ_1, θ_2) if T_1 and T_2 do not depend on θ_2 and θ_1 respectively. (Hint: Use the Factorization Criterion.)

4.5: Suppose that the lifetime of an electrical component is known to depend on some stress variable that varies over time; specifically, if U is the lifetime of the component, we have

$$\lim_{\Delta \downarrow 0} \frac{1}{\Delta} P(x \le U \le x + \Delta | U \ge x) = \lambda \exp(\beta \phi(x))$$

where $\phi(x)$ is the stress at time x. Assuming that we can measure $\phi(x)$ over time, we can conduct an experiment to estimate λ and β by replacing the component when it fails and observing the failure times of the components. Because $\phi(x)$ is not constant, the inter-failure times will not be i.i.d. random variables.

Define nonnegative random variables $X_1 < \cdots < X_n$ such that X_1 has hazard function

$$\lambda_1(x) = \lambda \exp(\beta \phi(x))$$

and conditional on $X_i = x_i$, X_{i+1} has hazard function

$$\lambda_{i+1}(x) = \begin{cases} 0 & \text{if } x < x_i \\ \lambda \exp(\beta \phi(x)) & \text{if } x \ge x_i \end{cases}$$

where λ, β are unknown parameters and $\phi(x)$ is a known function.

(a) Find the joint density of $(X_1, \cdots, X_n)$.

(b) Find sufficient statistics for (λ, β).

4.6: Let $X_1, \cdots, X_n$ be i.i.d. Exponential random variables with parameter λ. Suppose that we observe only the smallest r values of $X_1, \cdots, X_n$, that is, the order statistics $X_{(1)}, \cdots, X_{(r)}$. (This is called type II censoring in reliability.)

(a) Find the joint density of $X_{(1)}, \cdots, X_{(r)}$.

(b) Show that

$$V = X_{(1)} + \cdots + X_{(r-1)} + (n - r + 1) X_{(r)}$$

is sufficient for λ.

4.7: Suppose that $X_1, \cdots, X_n$ are i.i.d. Uniform random variables on $[0, \theta]$:

$$f(x;\theta) = \frac{1}{\theta} \quad \text{for } 0 \le x \le \theta.$$

Let $X_{(1)} = \min(X_1, \cdots, X_n)$ and $X_{(n)} = \max(X_1, \cdots, X_n)$.

(a) Define $T = X_{(n)}/X_{(1)}$. Is T ancillary for θ?

(b) Find the joint distribution of T and $X_{(n)}$. Are T and $X_{(n)}$ independent?

4.8: Suppose that $X_1, \cdots, X_n$ are i.i.d. random variables with density function

$$f(x;\theta) = \theta(1+x)^{-(\theta+1)} \quad \text{for } x \ge 0$$

where $\theta > 0$ is an unknown parameter.

(a) Show that $T = \sum_{i=1}^n \ln(1 + X_i)$ is sufficient for θ.

(b) Find the mean and variance of T.

4.9: Consider the Gini index $\theta(F)$ as defined in Example 4.21.

(a) Suppose that $X \sim F$ and let G be the distribution function of $Y = aX$ for some $a > 0$. Show that $\theta(G) = \theta(F)$.

(b) Suppose that F_p is a discrete distribution with probability p at 0 and probability $1 - p$ at $x > 0$. Show that $\theta(F_p) \to 0$ as $p \to 0$ and $\theta(F_p) \to 1$ as $p \to 1$.

(c) Suppose that F is a Pareto distribution whose density is

$$f(x;\alpha) = \frac{\alpha}{x_0}\left(\frac{x}{x_0}\right)^{-\alpha-1} \quad \text{for } x > x_0 > 0$$

$\alpha > 0$. (This is sometimes used as a model for incomes exceeding a threshold x_0.) Show that $\theta(F) = (2\alpha - 1)^{-1}$ for $\alpha > 1$. ($f(x;\alpha)$ is a density for $\alpha > 0$ but for $\alpha \le 1$, the expected value is infinite.)

4.10: An alternative to the Gini index as a measure of inequality is the Theil index. Given a distribution function F whose probability is concentrated on nonnegative values, the Theil index is defined to be the functional parameter

$$\theta(F) = \int_0^\infty \frac{x}{\mu(F)} \ln\left(\frac{x}{\mu(F)}\right) dF(x)$$

where $\mu(F) = \int_0^\infty x\, dF(x)$.

(a) Suppose that $X \sim F$ and let G be the distribution function of $Y = aX$ for some $a > 0$. Show that $\theta(G) = \theta(F)$.

(b) Find the influence curve of $\theta(F)$.

(c) Suppose that $X_1, \cdots, X_n$ are i.i.d. random variables with distribution function F. Show that

$$\widehat{\theta}_n = \frac{1}{n}\sum_{i=1}^{n} \frac{X_i}{\bar{X}_n} \ln\left(\frac{X_i}{\bar{X}_n}\right)$$

is the substitution principle estimator of $\theta(F)$.

(d) Find the limiting distribution of $\sqrt{n}(\widehat{\theta}_n - \theta(F))$.

4.11: The influence curve heuristic can be used to obtain the joint limiting distribution of a finite number of substitution principle estimators. Suppose that $\theta_1(F), \cdots, \theta_k(F)$ are functional parameters with influence curves $\phi_1(x; F), \cdots, \phi_k(x; F)$. Then if $X_1, \cdots, X_n$ is an i.i.d. sample from F, we typically have

$$\begin{aligned}
\sqrt{n}(\theta_1(\widehat{F}_n) - \theta_1(F)) &= \frac{1}{\sqrt{n}}\sum_{i=1}^{n} \phi_1(X_i; F) + R_{n1} \\
&\vdots \\
\sqrt{n}(\theta_k(\widehat{F}_n) - \theta_k(F)) &= \frac{1}{\sqrt{n}}\sum_{i=1}^{n} \phi_k(X_i; F) + R_{nk}
\end{aligned}$$

where $R_{n1}, \cdots, R_{nk} \to_p 0$.

(a) Suppose that $X_1, \cdots, X_n$ are i.i.d. random variables from a distribution F with mean μ and median θ; assume that $\text{Var}(X_i) = \sigma^2$ and $F'(\theta) > 0$. If $\widehat{\mu}_n$ is the sample mean and $\widehat{\theta}_n$ is the sample median, use the influence curve heuristic to show that

$$\sqrt{n}\begin{pmatrix} \widehat{\mu}_n - \mu \\ \widehat{\theta}_n - \theta \end{pmatrix} \to_d N_2(\mathbf{0}, C)$$

and give the elements of the variance-covariance matrix C.

(b) Now assume that the X_i's are i.i.d. with density

$$f(x;\theta) = \frac{p}{2\Gamma(1/p)} \exp(-|x - \theta|^p)$$

where θ is the mean and median of the distribution and $p > 0$ is another parameter (that may be known or unknown). Show that the matrix C in part (a) is

$$C = \begin{pmatrix} \Gamma(3/p)/\Gamma(1/p) & \Gamma(2/p)/p \\ \Gamma(2/p)/p & [\Gamma(1/p)/p]^2. \end{pmatrix}$$

(c) Consider estimators of θ of the form $\widetilde{\theta}_n = s\widehat{\mu}_n + (1-s)\widehat{\theta}_n$. For a given s, find the limiting distribution of $\sqrt{n}(\widetilde{\theta}_n - \theta)$.

(d) For a given value of $p > 0$, find the value of s that minimizes the variance of this limiting distribution. For which value(s) of p is this optimal value equal to 0; for which value(s) is it equal to 1?

4.12: (a) Suppose that F is a continuous distribution function with density $f = F'$. Find the influence curve of the functional parameter $\theta_p(F)$ defined by $F(\theta_p(F)) = p$ for some $p \in (0,1)$. ($\theta_p(F)$ is the p-quantile of F.)

(b) Let $\widehat{F}_n(x)$ be the empirical distribution function of i.i.d. random variables $X_1, \cdots, X_n$ (with continuous distribution F and density $f = F'$) and for $0 < t < 1$ define

$$\widehat{F}_n^{-1}(t) = \inf\{x : \widehat{F}_n(x) \geq t\}.$$

Define $\widehat{\tau}_n = \widehat{F}_n^{-1}(0.75) - \widehat{F}_n^{-1}(0.25)$ to be the interquartile range of $X_1, \cdots, X_n$. Find the limiting distribution of

$$\sqrt{n}(\widehat{\tau}_n - \tau(F))$$

where $\tau(F) = \theta_{3/4}(F) - \theta_{1/4}(F)$. (Hint: Find the influence curve of $\tau(F)$; a rigorous derivation of the limiting distribution can be obtained by mimicking Examples 3.5 and 3.6.)

4.13: Suppose that $X_1, X_2, \cdots$ are i.i.d. nonnegative random variables with distribution function F and define the functional parameter

$$\theta(F) = \frac{\left(\int_0^\infty x\, dF(x)\right)^2}{\int_0^\infty x^2\, dF(x)}.$$

(Note that $\theta(F) = (E(X))^2/E(X^2)$ where $X \sim F$.)

(a) Find the influence curve of $\theta(F)$.

(b) Using $X_1, \cdots, X_n$, find a substitution principle estimator, $\widehat{\theta}_n$, of $\theta(F)$ and find the limiting distribution of $\sqrt{n}(\widehat{\theta}_n - \theta)$. (You can use either the influence curve or the Delta Method to do this.)

4.14: Size-biased (or length-biased) sampling occurs when the size or length of a certain object affects its probability of being sampled. For example, suppose we are interested in estimating the mean number of people in a household. We could take a random sample of households, in which case the natural estimate would be the sample mean (which is an unbiased estimator). Alternatively, we could take a random sample of individuals and record the number of people in each individual's household; in this case, the sample mean is typically not a good estimator since the sampling scheme is more likely to include individuals from large households than would be the case if households were sampled. In many cases, it is possible to correct for the biased sampling if the nature of the biased sampling is known. (Another example of biased sampling is given in Example 2.21.)

(a) Suppose we observe i.i.d. random variables $X_1, \cdots, X_n$ from the distribution

$$G(x) = \left(\int_0^\infty w(t)\, dF(t)\right)^{-1} \int_0^x w(t)\, dF(t)$$

where $w(t)$ is a known (nonnegative) function and F is an unknown distribution function. Define

$$\widehat{F}_n(x) = \left(\sum_{i=1}^n [w(X_i)]^{-1}\right)^{-1} \sum_{i=1}^n [w(X_i)]^{-1} I(X_i \le x).$$

Show that for each x, $\widehat{F}_n(x)$ is a consistent estimator of $F(x)$ provided that $E[1/w(X_i)] < \infty$.

(b) Using the estimator in part (a), give a substitution principle estimator of $\theta(F) = \int g(x)\, dF(x)$. What is the estimator of $\int x\, dF(x)$ when $w(x) = x$? Find the limiting distribution of this estimator when $E[1/w^2(X_i)] < \infty$.

(c) Suppose that we have the option of sampling from F or from the biased version G where $w(x) = x$. Show that the estimator of $\int x\, dF(x)$ based on the biased sample is asymptotically more efficient than that based on the sample from F if

$$\left(\int x\, dF(x)\right)^3 \left(\int x^{-1}\, dF(x)\right) < \int x^2\, dF(x).$$

4.15: Suppose that $X_1, \cdots, X_n$ are i.i.d. Normal random variables with mean 0 and unknown variance σ^2.

(a) Show that $E(|X_i|) = \sigma\sqrt{2/\pi}$.

(b) Use the result of (a) to construct a method of moments estimator, $\widehat{\sigma}_n$, of σ. Find the limiting distribution of $\sqrt{n}(\widehat{\sigma}_n - \sigma)$.

(c) Another method of moments estimator of σ is

$$\widetilde{\sigma}_n = \left(\frac{1}{n}\sum_{i=1}^{n} X_i^2\right)^{1/2}.$$

Find the limiting distribution of $\sqrt{n}(\widetilde{\sigma}_n - \sigma)$ and compare the results of parts (b) and (c).

4.16: Suppose that $X_1, \cdots, X_n$ are i.i.d. random variables with density function

$$f(x;\mu) = \exp[-(x-\mu)] \quad \text{for } x \geq \mu$$

(a) Show that $Z_n = \min(X_1, \cdots, X_n)$ is a sufficient statistic for μ.

(b) Show that $Z_n \to_p \mu$ as $n \to \infty$.

4.17: Let $U_1, \cdots, U_n$ be i.i.d. Uniform random variables on $[0, \theta]$. Suppose that only the smallest r values are actually observed, that is, the order statistics $U_{(1)} < U_{(2)} < \cdots < U_{(r)}$.

(a) Find the joint density of $U_{(1)}, U_{(2)}, \cdots, U_{(r)}$ and and find a one-dimensional sufficient statistic for θ. (Hint: The joint density of $(U_{(1)}, \cdots, U_{(n)})$ is $f(u_1, \cdots, u_n) = n!\theta^{-n}$ for $0 < u_1 < \cdots < u_n < 1$.)

(b) Find a unbiased estimator of θ based on the sufficient statistic found in (a).

4.18: Suppose that $\Lambda_1, \cdots, \Lambda_n$ are i.i.d. Gamma random variables with (reparametrized) density function

$$g(x) = \frac{(\alpha/\mu)^\alpha x^{\alpha-1}\exp(-\alpha x/\mu)}{\Gamma(\alpha)} \quad \text{for } x > 0$$

so that $E(\Lambda_i) = \mu$. Given Λ_i, let X_i and Y_i be independent Poisson random variables with $E(X_i|\Lambda_i) = \Lambda_i$ and $E(Y_i|\Lambda_i) = \theta\Lambda_i$. We will observe i.i.d. pairs of (dependent) random variables $(X_1, Y_1), \cdots, (X_n, Y_n)$ (that is, the Λ_i's are unobservable). (See Lee (1996) for an application of such a model.)

(a) Show that the joint frequency function of (X_i, Y_i) is

$$f(x,y) = \frac{\theta^y}{x!y!}\frac{\Gamma(x+y+\alpha)}{\Gamma(\alpha)(\mu/\alpha)^\alpha}\left(1+\theta+\frac{\alpha}{\mu}\right)^{-(x+y+\alpha)}$$

for $x, y = 0, 1, 2, 3, \cdots$. (Hint: $P(X_i = x, Y_i = y) = E[P(X_i = x, Y_i = y|\Lambda_i)]$.)

(b) Find the expected values and variances of X_i and Y_i as well as $\text{Cov}(X_i, Y_i)$.

(c) Show that $\widehat{\theta}_n = \bar{Y}_n/\bar{X}_n$ is a consistent estimator of θ.

(d) Find the asymptotic distribution of $\sqrt{n}(\widehat{\theta}_n - \theta)$.

4.19: Suppose that $X_1, \cdots, X_n$ are i.i.d. random variables with a continuous distribution function F. It can be shown that $g(t) = E(|X_i - t|)$ (or $g(t) = E[|X_i - t| - |X_i|]$) is minimized at $t = \theta$ where $F(\theta) = 1/2$ (see Problem 1.25). This suggests that the median θ can be estimated by choosing $\widehat{\theta}_n$ to minimize

$$g_n(t) = \sum_{i=1}^{n} |X_i - t|.$$

(a) Let $X_{(1)} \leq X_{(2)} \leq \cdots \leq X_{(n)}$ be the order statistics. Show that if n is even then $g_n(t)$ is minimized for $X_{(n/2)} \leq t \leq X_{(1+n/2)}$ while if n is odd then $g_n(t)$ is minimized at $t = X_{((n+1)/2)}$. (Hint: Evaluate the derivative of $g_n(t)$ for $X_{(i)} < t < X_{(i+1)}$ $(i = 1, \cdots, n-1)$; determine for which values of t $g_n(t)$ is decreasing and for which it is increasing.)

(b) Let $\widehat{F}_n(x)$ be the empirical distribution function. Show that $\widehat{F}_n^{-1}(1/2) = X_{(n/2)}$ if n is even and $\widehat{F}_n^{-1}(1/2) = X_{((n+1)/2)}$ if n is odd.

4.20: Suppose that $X_1, X_2, \cdots$ are i.i.d. random variables with distribution function

$$F(x) = (1-\theta)\Phi\left(\frac{x-\mu}{\sigma}\right) + \theta\Phi\left(\frac{x-\mu}{5\sigma}\right)$$

where $0 < \theta < 1$, μ and σ are unknown parameters. (Φ is the $N(0,1)$ distribution function.) Define $\widehat{\mu}_n$ to be the sample mean and $\widetilde{\mu}_n$ to be the sample median of $X_1, \cdots, X_n$. (This is an example of a contaminated Normal model that is sometimes used to study the robustness of estimators.)

(a) Find the limiting distributions of $\sqrt{n}(\widehat{\mu}_n - \mu)$ and $\sqrt{n}(\widetilde{\mu}_n - \mu)$. (These will depend on θ and σ^2.)

(b) For which values (if any) of θ is the sample median more efficient than the sample mean?

4.21: Suppose that $X_1, \cdots, X_n$ are i.i.d. random variables with distribution function. The substitution principle can be extended to estimating functional parameters of the form

$$\theta(F) = E[h(X_1, \cdots, X_k)]$$

where h is some specified function. (We assume that this expected value is finite.) If $n \geq k$, a substitution principle estimator of $\theta(F)$ is

$$\widehat{\theta} = \binom{n}{k}^{-1} \sum_{i_1 < \cdots < i_k} h(X_{i_1}, \cdots, X_{i_k})$$

where the summation extends over all combinations of k integers drawn from the integers 1 through n. The estimator $\widehat{\theta}$ is called a U-statistic.

(a) Show that $\widehat{\theta}$ is a unbiased estimator of $\theta(F)$.

(b) Suppose $\text{Var}(X_i) < \infty$. Show that

$$\text{Var}(X_i) = E[(X_1 - X_2)^2]/2.$$

How does the "U-statistic" substitution principle estimator differ from the the substitution principle estimator in Example 4.23?

4.22: Suppose that $(X_1, Y_1), \cdots, (X_n, Y_n)$ are independent pairs of correlated Bernoulli random variables. The correlation between X_i and Y_i is introduced using a mixing distribution. Take $\theta_1, \cdots, \theta_n$ to be i.i.d. random variables with a non-degenerate distribution function G. Then given θ_i, X_i and Y_i are independent random variables with

$$P(X_i = 1|\theta_i) = \theta_i \quad \text{and} \quad P(Y_i = 1|\theta_i) = h(\theta_i, \phi)$$

where h is such that

$$\frac{h(\theta_i, \phi)}{1 - h(\theta_i, \phi)} = \phi \frac{\theta_i}{1 - \theta_i}.$$

Thus, for example, we have

$$P(X_i = 1) = \int_0^1 \theta \, dG(\theta)$$

and

$$P(Y_i = 1) = \int_0^1 h(\theta, \phi) \, dG(\theta).$$

We would like to estimate ϕ, the so-called odds ratio, given $(X_1, Y_1), \cdots, (X_n, Y_n)$. (Both ϕ and the distribution function

G are unknown in this problem; however, we will not have to explicitly estimate G.)

(a) Show that $E[Y_i(1 - X_i)] = \phi E[X_i(1 - Y_i)]$. (Hint: Look at the conditional expectations given θ_i.)

(b) Show that the estimator

$$\widehat{\phi}_n = \frac{\sum_{i=1}^n Y_i(1 - X_i)}{\sum_{i=1}^n X_i(1 - Y_i)}$$

is a consistent estimator of ϕ. ($\widehat{\phi}_n$ is a special case of the Mantel-Haenszel estimator of the odds ratio.)

(c) Find the limiting distribution of $\sqrt{n}(\widehat{\phi}_n - \phi)$ and suggest an estimator of the standard error of $\widehat{\phi}_n$.

(d) Find the limiting distribution of $\sqrt{n}(\ln(\widehat{\phi}_n) - \ln(\phi))$ and suggest an estimator of the standard error of $\ln(\widehat{\phi}_n)$.

4.23: Suppose that $X_1, \cdots, X_n$ are i.i.d. random variables and define an estimator $\widehat{\theta}$ by

$$\sum_{i=1}^n \psi(X_i - \widehat{\theta}) = 0$$

where ψ is an odd function ($\psi(x) = -\psi(-x)$) with derivative ψ'.

(a) Let $\widehat{\theta}_{-j}$ be the estimator computed from all the X_i's except X_j. Show that

$$\sum_{i=1}^n \psi(X_i - \widehat{\theta}_{-j}) = \psi(X_j - \widehat{\theta}_{-j}).$$

(b) Use the approximation

$$\psi(X_i - \widehat{\theta}_{-j}) \approx \psi(X_i - \widehat{\theta}) + (\widehat{\theta} - \widehat{\theta}_{-j})\psi'(X_i - \widehat{\theta})$$

to show that

$$\widehat{\theta}_{-j} \approx \widehat{\theta} - \frac{\psi(X_j - \widehat{\theta})}{\sum_{i=1}^n \psi'(X_i - \widehat{\theta})}.$$

(c) Show that the jackknife estimator of $\text{Var}(\widehat{\theta})$ can be approximated by

$$\frac{n-1}{n} \frac{\sum_{i=1}^n \psi^2(X_i - \widehat{\theta})}{\left(\sum_{i=1}^n \psi'(X_i - \widehat{\theta})\right)^2}.$$

4.24: Consider the jackknife estimator of the variance of the sample median. Suppose that $X_1, \cdots, X_n$ are i.i.d. random variables with distribution function F with median θ and $F'(\theta) = \lambda > 0$; for simplicity, we will assume that n is even ($n = 2m$). Define $\widehat{\theta}_n$ to be the sample median.

(a) Let $X_{(1)} \leq X_{(2)} \leq \cdots \leq X_{(n)}$ be the order statistics. Show that the jackknife estimator of $\widehat{\theta}_n$ is

$$\widehat{\mathrm{Var}}(\widehat{\theta}_n) = \frac{n-1}{4}(X_{(m+1)} - X_{(m)})^2.$$

(b) Ideally, we would like $n\widehat{\mathrm{Var}}(\widehat{\theta}_n)$ to have the same limit as $n\mathrm{Var}(\widehat{\theta}_n)$. However, this does not work for the sample median. To see this, start by assuming that the X_i's are i.i.d. Exponential random variables with mean 1. Show that (see Problem 2.26) $m(X_{(m+1)} - X_{(m)})$ has an Exponential distribution with mean 1.

(c) In general, show that for $n = 2m$,

$$m(X_{(m+1)} - X_{(m)}) \to_d \frac{Z}{\lambda}$$

as $n \to \infty$ where Z has an Exponential distribution with mean 1. (Hint: Find a function that transforms the Exponential random variables in part (b) to Uniform random variables on $[0, 1]$ and then to random variables with distribution function F.)

(d) Show that

$$\widehat{\mathrm{Var}}(\widehat{\theta}_n) \to_d \frac{1}{4\lambda^2}Z^2 \neq \frac{1}{4\lambda^2}.$$

Note that the mean of the limiting distribution is $1/(2\lambda^2)$.

CHAPTER 5

Likelihood-Based Estimation

5.1 Introduction

We saw in the last chapter that the substitution principle (or the method of moments) provides an approach for finding reasonable estimators of parameters in statistical models. However, the substitution principle does not prescribe any particular estimator and, in fact, the quality of different substitution principle estimators can vary greatly. Moreover, the substitution principle is tailored to i.i.d. data and can be difficult to apply for non-i.i.d. data. These problems are somewhat unsettling since we would like to find a general-purpose algorithm for generating good (if not "optimal") estimators. We will see in this chapter that one such algorithm is provided by maximum likelihood estimation. The maximum likelihood estimator is defined to be the maximizing value of a certain function called the likelihood function. We will see in Chapter 6 that the maximum likelihood estimator has some very nice optimality properties.

The likelihood function (defined in the next section) has a much wider significance in statistical theory. An important principle in statistics (called the likelihood principle) essentially states that the likelihood function contains all of the information about an unknown parameter in the data. The likelihood function also plays an integral role in Bayesian inference in which we deal with the uncertainty in the value of parameters via probability distributions on the parameter space.

5.2 The likelihood function

Suppose that $\boldsymbol{X} = (X_1, \cdots, X_n)$ are random variables with joint density or frequency function $f(\boldsymbol{x}; \theta)$ where $\theta \in \Theta$. Given outcomes $\boldsymbol{X} = \boldsymbol{x}$, we define the likelihood function

$$\mathcal{L}(\theta) = f(\boldsymbol{x}; \theta);$$

for each possible sample $\boldsymbol{x} = (x_1, \cdots, x_n)$, the likelihood function $\mathcal{L}(\theta)$ is a real-valued function defined on the parameter space Θ. Note that we do not need to assume that $X_1, \cdots, X_n$ are i.i.d. random variables.

DEFINITION. Suppose that for a sample $\boldsymbol{x} = (x_1, \cdots, x_n)$, $L(\theta)$ is maximized (over Θ) at $\theta = S(\boldsymbol{x})$:

$$\sup_{\theta \in \Theta} \mathcal{L}(\theta) = \mathcal{L}(S(\boldsymbol{x}))$$

(with $S(\boldsymbol{x}) \in \Theta$). Then the statistic $\widehat{\theta} = S(\boldsymbol{X})$ is called the maximum likelihood estimator (MLE) of θ. ($S(\boldsymbol{x})$ is sometimes called the maximum likelihood estimate based on $\boldsymbol{x}$.)

For continuous random variables, the likelihood function is not uniquely defined since the joint density is not uniquely defined. In practice, we usually choose a form for the likelihood function that guarantees (if possible) the existence of a MLE for all possible values of $X_1, \cdots, X_n$. For discrete random variables, such difficulties do not occur since the joint frequency function (and hence the likelihood function) is uniquely defined.

If $T = T(\boldsymbol{X})$ is a sufficient statistic for θ then from the Factorization Criterion, it follows that

$$\mathcal{L}(\theta) \propto g(T(\boldsymbol{x}); \theta).$$

From this, it follows that if the MLE $\widehat{\theta}$ is unique then $\widehat{\theta}$ is a function of the sufficient statistic T.

Another attractive property of MLEs is invariance. For example, if $\phi = g(\theta)$ where g is a monotone function (or, more generally, one-to-one) and $\widehat{\theta}$ is the MLE of θ then $g(\widehat{\theta})$ is the MLE of ϕ. It is conventional to extend this invariance property to arbitrary functions; thus if $\phi = g(\theta)$ then we typically say that $\widehat{\phi} = g(\widehat{\theta})$ is the MLE of ϕ.

There are essentially two distinct methods for finding MLEs:

- Direct maximization: Examine $\mathcal{L}(\theta)$ directly to determine which value of θ maximizes $\mathcal{L}(\theta)$ for a given sample $x_1, \cdots, x_n$. This method is particularly useful when the range (or support) of the data depends on the parameter.
- Likelihood equations: If the range of the data does not depend on the data, the parameter space Θ is an open set, and the likelihood function is differentiable with respect to $\boldsymbol{\theta} = (\theta_1, \cdots, \theta_p)$

over Θ, then the maximum likelihood estimate $\widehat{\boldsymbol{\theta}}$ satisfies the equations

$$\frac{\partial}{\partial \theta_k} \ln \mathcal{L}(\widehat{\boldsymbol{\theta}}) = 0 \quad \text{for } k = 1, \cdots, p.$$

These equations are called the likelihood equations and $\ln \mathcal{L}(\boldsymbol{\theta})$ is called the log-likelihood function.

For the vast majority of statistical models, we can use the likelihood equations to determine the MLE. We use the log-likelihood function for convenience. If $\widehat{\theta}$ maximizes $\mathcal{L}(\theta)$, it also maximizes $\ln \mathcal{L}(\theta)$. In addition, $\mathcal{L}(\theta)$ is often expressed as a product so that $\ln \mathcal{L}(\theta)$ becomes a sum, which is easier to differentiate. The likelihood equations can have multiple solutions, so it is important to check that a given solution indeed maximizes the likelihood function. If the parameter space Θ is not an open set then the likelihood equations can be used provided that we verify that the maximum does not occur on the boundary of the parameter space.

EXAMPLE 5.1: Suppose that $X_1, \cdots, X_n$ are i.i.d. Uniform random variables on the interval $[0, \theta]$ for some $\theta > 0$. The likelihood function is

$$\mathcal{L}(\theta) = \theta^{-n} I(0 \le x_1, \cdots, x_n \le \theta) = \theta^{-n} I(\theta \ge \max(x_1, \cdots, x_n)).$$

Thus if $\theta < \max(x_1, \cdots, x_n)$, $\mathcal{L}(\theta) = 0$ while $\mathcal{L}(\theta)$ is a decreasing function of θ for $\theta \ge \max(x_1, \cdots, x_n)$. Hence, $\mathcal{L}(\theta)$ attains its maximum at $\theta = \max(x_1, \cdots, x_n)$ and so

$$\widehat{\theta} = X_{(n)} = \max(X_1, \cdots, X_n)$$

is the MLE of θ. ◇

Note that in Example 5.1, we could have defined the likelihood function to be

$$\mathcal{L}(\theta) = \theta^{-n} I(\theta < \max(x_1, \cdots, x_n))$$

(by defining the density of X_i to be $f(x; \theta) = 1/\theta$ for $0 < x < \theta$). In this case, the MLE does not exist. We have

$$\sup_{\theta > 0} \mathcal{L}(\theta) = [\max(x_1, \cdots, x_n)]^{-n};$$

however, there exists no $S(\boldsymbol{x})$ such that

$$\mathcal{L}(S(\boldsymbol{x})) = \sup_{\theta > 0} \mathcal{L}(\theta).$$

EXAMPLE 5.2: Suppose that $X_1, \cdots, X_n$ are i.i.d. Poisson random variables with mean $\lambda > 0$. The likelihood function is

$$\mathcal{L}(\lambda) = \prod_{i=1}^{n} \left\{ \frac{\exp(-\lambda)\lambda^{x_i}}{x_i!} \right\}$$

and the log-likelihood is

$$\ln \mathcal{L}(\lambda) = -n\lambda + \ln(\lambda) \sum_{i=1}^{n} x_i - \sum_{i=1}^{n} \ln(x_i!).$$

Assuming that $\sum_{i=1}^{n} x_i > 0$ and taking the derivative with respect to λ, we find that

$$\frac{d}{d\lambda} \ln \mathcal{L}(\lambda) = -n + \frac{1}{\lambda} \sum_{i=1}^{n} x_i$$

and setting the derivative to 0 suggests that $\widehat{\lambda} = \bar{x}$. To verify that this is indeed a maximum, note that

$$\frac{d^2}{d\lambda^2} \ln \mathcal{L}(\lambda) = -\frac{1}{\lambda^2} \sum_{i=1}^{n} x_i,$$

which is always negative. Thus $\bar{x}$ maximizes the likelihood function (for a given sample $x_1, \cdots, x_n$) and the MLE of λ is $\widehat{\lambda} = \bar{X}$ provided that $\sum_{i=1}^{n} X_i > 0$. If $\sum_{i=1}^{n} X_i = 0$ then strictly speaking no MLE exists since the log-likelihood function $\ln \mathcal{L}(\lambda) = -n\lambda$ has no maximum on the interval $(0, \infty)$. ◇

EXAMPLE 5.3: Suppose that the joint density or frequency function of $\boldsymbol{X} = (X_1, \cdots, X_n)$ is a one-parameter exponential family; the log-likelihood function is then

$$\ln \mathcal{L}(\theta) = c(\theta)T(\boldsymbol{x}) - d(\theta) + S(\boldsymbol{x}).$$

Differentiating with respect to θ and setting this derivative to 0, we get the following equation for the MLE $\widehat{\theta}$:

$$\frac{c'(\widehat{\theta})}{d'(\widehat{\theta})} = T(\boldsymbol{X}).$$

However, since $T(\boldsymbol{X})$ is the sufficient statistic, we know that

$$E_\theta[T(\boldsymbol{X})] = \frac{d'(\theta)}{c'(\theta)}$$

and so it follows that the MLE is simply a method of moments estimator for one-parameter exponential families. Also note that $\widehat{\theta}$ does maximize the likelihood function; the second derivative of the log-likelihood function is

$$\frac{d^2}{d\theta^2} \ln \mathcal{L}(\theta) = c''(\theta) T(\boldsymbol{x}) - d''(\theta)$$

and substituting $\widehat{\theta}$ for θ, we get

$$\frac{d^2}{d\theta^2} \ln \mathcal{L}(\widehat{\theta}) = c''(\widehat{\theta}) \frac{d'(\widehat{\theta})}{c'(\widehat{\theta})} - d''(\widehat{\theta}) < 0$$

since

$$\mathrm{Var}_\theta[T(\boldsymbol{X})] = d''(\theta) - c''(\theta) \frac{d'(\theta)}{c'(\theta)} > 0.$$

A similar result also holds for k-parameter exponential families. ◇

Why does maximum likelihood estimation make sense? For simplicity, assume that $X_1, \cdots, X_n$ are i.i.d. random variables with density or frequency function $f_0(x)$ and distribution function $F_0(x)$. For any other density (frequency) function $f(x)$, we can define the Kullback-Leibler information number

$$K(f : f_0) = E_0 \left[\ln \left(f_0(X_i) / f(X_i) \right) \right]$$

where the expected value E_0 is computed assuming that $f_0(x)$ is the true density (frequency) function of X_i. $K(f : f_0)$ can be interpreted as measuring the distance to the "true" density (frequency) function f_0 of some other density (frequency) function f; it is easy to see that

$$K(f_0 : f_0) = 0$$

and since $-\ln(x)$ is a convex function, it follows (from Jensen's inequality) that

$$\begin{aligned} K(f : f_0) &= E_0 \left[-\ln \left(f(X_i) / f_0(X_i) \right) \right] \\ &\geq -\ln E_0 \left[\frac{f(X_i)}{f_0(X_i)} \right] \\ &= 0. \end{aligned}$$

Thus over all density (frequency) functions f, $K(f : f_0)$ is minimized (for a given f_0) at $f = f_0$. Moreover, unless $f(x) = f_0(x)$ for all x, $K(f : f_0) > 0$. $K(f : f_0)$ can also be interpreted as the inefficiency in assuming that the density (frequency) function is f when the true density (frequency) function is f_0.

We will now use these facts to rationalize maximum likelihood estimation in the case of i.i.d. observations. Since

$$K(f : f_0) = -E_0\left[\ln(f(X_i))\right] + E_0\left[\ln(f_0(X_i))\right],$$

we can also see that for fixed f_0,

$$L(f : f_0) = E_0\left[\ln(f(X_i))\right]$$

is maximized over all f at $f = f_0$. This suggests the following method for estimating the density (frequency) function f_0 of i.i.d. observations $X_1, \cdots, X_n$:

- For each f in some family F, estimate $L(f : f_0)$ using the substitution principle

$$\widehat{L}(f : f_0) = \frac{1}{n}\sum_{i=1}^{n} \ln f(X_i);$$

- Find f to maximize $\widehat{L}(f : f_0)$.

If $\mathcal{F} = \{f(x;\theta) : \theta \in \Theta\}$, then $n\widehat{L}(f : f_0)$ is simply the log-likelihood function. Thus maximum likelihood estimation can be viewed as a sort of substitution principle estimation.

Maximum likelihood estimation has been described as the original jackknife in the sense that it is an estimation procedure that is applicable in a wide variety of problems. As will be shown in Chapter 6, maximum likelihood estimation has some very attractive optimality properties and for this reason, it is often viewed as the "gold standard" of estimation procedures. However, as with any methodology, maximum likelihood estimation should be viewed with the appropriate degree of scepticism as there may be estimation procedures that are better behaved in terms of robustness or efficiency when the model is slightly (or grossly) misspecified.

5.3 The likelihood principle

The likelihood function has a much greater significance in statistical inference; we will discuss this briefly in this section. Suppose we are given a choice between two experiments for estimating a parameter θ. From the first experiment, we obtain data $\boldsymbol{x}$ while $\boldsymbol{x}^*$ is obtained from the second experiment. The likelihood principle provides a simple criterion for when identical inferences for θ should be drawn from both $\boldsymbol{x}$ and $\boldsymbol{x}^*$; it is stated as follows:

Likelihood principle. Let $\mathcal{L}(\theta)$ be the likelihood function for θ based on observing $\boldsymbol{X} = \boldsymbol{x}$ and $\mathcal{L}^*(\theta)$ be the likelihood function for θ based on observations $\boldsymbol{X}^* = \boldsymbol{x}^*$. If $\mathcal{L}(\theta) = k\mathcal{L}^*(\theta)$ (for some k that does not depend on θ) then the same inference for θ should be drawn from both samples.

In point estimation, the likelihood principle implies that if $\boldsymbol{x}$ and $\boldsymbol{x}^*$ are two samples with likelihood functions $\mathcal{L}(\theta)$ and $\mathcal{L}^*(\theta)$ where $\mathcal{L}(\theta) = k\mathcal{L}^*(\theta)$ (for all θ) then the two point estimates $T(\boldsymbol{x})$ and $T^*(\boldsymbol{x}^*)$ should be equal. Clearly, maximum likelihood estimation satisfies this condition but many of the substitution principle and other estimators discussed in Chapter 4 do not. For example, the biased-reduced jackknife estimator typically violates the likelihood principle. However, certain inferential procedures based on maximum likelihood estimation may violate the likelihood principle.

It is also worth noting at this point that the likelihood principle refers only to the information about θ contained in the sample; we may also have additional information about θ available to us from outside the sample. For example, we can express our beliefs about the true value of θ (prior to observing the sample) by specifying a probability distribution over the parameter space Θ; this distribution is called a prior distribution. This approach (called the Bayesian approach) is often cited as the "correct" approach to implementing the likelihood principle. More details on the Bayesian approach are given in section 5.7.

Does the likelihood principle make sense? This is a very controversial issue in theoretical statistics and there is really no clear answer. The likelihood principle can be shown to follow by assuming two other somewhat less controversial principles, the (weak) sufficiency principle and the (weak) conditionality principle. The sufficiency principle states that the sufficient statistic contains as much information about the value of a parameter as the data themselves while the conditionality principle essentially states that no information is lost by conditioning on an ancillary statistic. For some compelling arguments in favour of the likelihood principle, see the monograph by Berger and Wolpert (1988). The book by Lee (1989) also contains some interesting discussion and examples on the likelihood principle.

The likelihood principle is not universally accepted; in fact, much of classical statistical practice violates it. Of course, the likelihood principle assumes that the true parametric model for the data is known; we have stated previously that statistical models are almost inevitably used to approximate reality and are seldom exactly true. This, of course, does not immediately invalidate the likelihood principle but rather emphasizes the point that any principle or philosophy should not be accepted without closely examining its tenets.

5.4 Asymptotic theory for MLEs

Under what conditions is the MLE a consistent and asymptotically Normal estimator of a parameter? We will show in this section that, under fairly mild regularity conditions, it is possible to prove consistency and asymptotic normality for the MLE of a real-valued parameter based on i.i.d. observations. However, in many cases, it is possible to find the asymptotic distribution of a sequence of MLEs using standard techniques. This is common when the MLE corresponds to a method of moments estimator.

EXAMPLE 5.4: Suppose that $X_1, \cdots, X_n$ are i.i.d. Geometric random variables with frequency function

$$f(x; \theta) = \theta(1-\theta)^x \quad \text{for } x = 0, 1, 2, \cdots$$

The MLE of θ based on $X_1, \cdots, X_n$ is

$$\widehat{\theta}_n = \frac{1}{\bar{X}_n + 1}.$$

By the Central Limit Theorem, we have that

$$\sqrt{n}(\bar{X}_n - (\theta^{-1} - 1)) \to_d N(0, \theta^{-2}(1-\theta)).$$

Thus we obtain

$$\begin{aligned} \sqrt{n}(\widehat{\theta}_n - \theta) &= \sqrt{n}(g(\bar{X}_n) - g(\theta^{-1} - 1)) \\ &\to_d N(0, \theta^2(1-\theta)) \end{aligned}$$

by applying the Delta Method with $g(x) = 1/(1+x)$ and $g'(x) = -1/(1+x)^2$. ◇

EXAMPLE 5.5: Suppose that $X_1, \cdots, X_n$ are i.i.d. Uniform random variables on $[0, \theta]$. The MLE of θ is

$$\widehat{\theta}_n = X_{(n)} = \max(X_1, \cdots, X_n)$$

whose distribution function is given by

$$P(\widehat{\theta}_n \leq x) = \left(\frac{x}{\theta}\right)^n \quad \text{for } 0 \leq x \leq \theta.$$

Thus for $\epsilon > 0$,

$$\begin{aligned} P(|\widehat{\theta}_n - \theta| > \epsilon) &= P(\widehat{\theta}_n < \theta - \epsilon) \\ &= \left(\frac{\theta - \epsilon}{\theta}\right)^n \\ &\to 0 \quad \text{as } n \to \infty \end{aligned}$$

since $(\theta - \epsilon)/\theta < 1$. We also have that

$$\begin{aligned} P\left[n(\theta - \widehat{\theta}_n) \leq x\right] &= P\left[\widehat{\theta}_n \geq \theta - \frac{x}{n}\right] \\ &= 1 - \left(1 - \frac{x}{\theta n}\right)^n \\ &\to 1 - \exp(-x/\theta) \quad \text{for } x \geq 0 \end{aligned}$$

and so $n(\theta - \widehat{\theta}_n)$ converges in distribution to an Exponential random variable. ◇

For the remainder of this section, we will assume that $X_1, X_2, \cdots, X_n$ are i.i.d. random variables with common density or frequency function $f(x;\theta)$ where θ is a real-valued parameter.

Define $\ell(x;\theta) = \ln f(x;\theta)$ and let $\ell'(x;\theta)$, $\ell''(x;\theta)$, and $\ell'''(x;\theta)$ be the first three partial derivatives of $\ell(x;\theta)$ with respect to θ. We will make the following assumptions about $f(x;\theta)$:

(A1) The parameter space Θ is an open subset of the real-line.
(A2) The set $A = \{x : f(x;\theta) > 0\}$ does not depend on θ.
(A3) $f(x;\theta)$ is three times continuously differentiable with respect to θ for all x in A.
(A4) $E_\theta[\ell'(X_i;\theta)] = 0$ for all θ and $\text{Var}_\theta[\ell'(X_i;\theta)] = I(\theta)$ where $0 < I(\theta) < \infty$ for all θ.
(A5) $E_\theta[\ell''(X_i;\theta)] = -J(\theta)$ where $0 < J(\theta) < \infty$ for all θ.
(A6) For each θ and $\delta > 0$, $|\ell'''(x;t)| \leq M(x)$ for $|\theta - t| \leq \delta$ where $E_\theta[M(X_i)] < \infty$.

Suppose that $f(x;\theta)$ is a density function. If condition (A2) holds then

$$\int_A f(x;\theta)\,dx = 1 \quad \text{for all } \theta \in \Theta$$

and so

$$\frac{d}{d\theta}\int_A f(x;\theta)\,dx = 0.$$

If the derivative can be taken inside the integral we then have

$$\begin{aligned} 0 &= \int_A \frac{\partial}{\partial\theta} f(x;\theta)\,dx \\ &= \int_A \ell'(x;\theta) f(x;\theta)\,dx \\ &= E_\theta[\ell'(X_i;\theta)] \end{aligned}$$

and so the assumption that $E_\theta[\ell'(X_i;\theta)] = 0$ is, in many cases, a natural consequence of condition (A2). Moreover, if we can differentiate $\int_A f(x;\theta)\,dx$ twice inside the integral sign, we have

$$\begin{aligned} 0 &= \int_A \frac{\partial}{\partial\theta}\left(\ell'(x;\theta) f(x;\theta)\right)\,dx \\ &= \int_A \ell''(x;\theta) f(x;\theta)\,dx + \int_A \left(\ell'(x;\theta)\right)^2 f(x;\theta)\,dx \\ &= -J(\theta) + I(\theta) \end{aligned}$$

and so $I(\theta) = J(\theta)$. (Similar results apply if $f(x;\theta)$ is a frequency function.) $I(\theta)$ is called the Fisher information.

EXAMPLE 5.6: Suppose that $X_1, X_2, \cdots$ are i.i.d. random variables with the one-parameter exponential family density or frequency function

$$f(x;\theta) = \exp[c(\theta)T(x) - d(\theta) + S(x)] \quad \text{for } x \in A.$$

In this case,

$$\begin{aligned} \ell'(x;\theta) &= c'(\theta)T(x) - d'(\theta) \\ \text{and} \quad \ell''(x;\theta) &= c''(\theta)T(x) - d''(\theta). \end{aligned}$$

Since we have

$$\begin{aligned} E_\theta[T(X_i)] &= \frac{d'(\theta)}{c'(\theta)} \\ \text{and} \quad \text{Var}_\theta[T(X_i)] &= \frac{1}{[c'(\theta)]^2}\left(d''(\theta) - c''(\theta)\frac{d'(\theta)}{c'(\theta)}\right), \end{aligned}$$

it follows that

$$\begin{aligned}
E_\theta[\ell'(X_i;\theta)] &= c'(\theta)E_\theta[T(X_i)] - d'(\theta) \\
&= 0, \\
I(\theta) &= [c'(\theta)]^2 \mathrm{Var}_\theta[T(X_i)] \\
&= d''(\theta) - c''(\theta)\frac{d'(\theta)}{c'(\theta)} \\
\text{and} \quad J(\theta) &= d''(\theta) - c''(\theta)E_\theta[T(X_i)] \\
&= d''(\theta) - c''(\theta)\frac{d'(\theta)}{c'(\theta)}
\end{aligned}$$

and so $I(\theta) = J(\theta)$. ◇

EXAMPLE 5.7: Suppose that $X_1, \cdots, X_n$ are i.i.d. random variables with a Logistic distribution whose density function is

$$f(x;\theta) = \frac{\exp(x-\theta)}{[1+\exp(x-\theta)]^2}.$$

The derivatives of $\ell(x;\theta) = \ln f(x;\theta)$ with respect to θ are

$$\ell'(x;\theta) = \frac{\exp(x-\theta) - 1}{1+\exp(x-\theta)}$$

and

$$\ell''(x;\theta) = -2\frac{\exp(x-\theta)}{[1+\exp(x-\theta)]^2}.$$

It follows then that

$$E_\theta[\ell'(X_i;\theta)] = \int_{-\infty}^{\infty} \frac{\exp(2(x-\theta)) - \exp(x-\theta)}{[1+\exp(x-\theta)]^3}\,dx = 0,$$

$$\begin{aligned}
I(\theta) &= \mathrm{Var}_\theta[\ell'(X_i;\theta)] \\
&= \int_{-\infty}^{\infty} (\exp(x-\theta)-1)^2 \frac{\exp(x-\theta)}{[1+\exp(x-\theta)]^4}\,dx \\
&= \frac{1}{3} \\
\text{and} \quad J(\theta) &= -E_\theta[\ell''(X_i;\theta)] \\
&= 2\int_{-\infty}^{\infty} \frac{\exp(2(x-\theta))}{[1+\exp(x-\theta)]^4}\,dx \\
&= \frac{1}{3}
\end{aligned}$$

and so $I(\theta) = J(\theta)$. ◇

Under conditions (A1) to (A3), if $\widehat{\theta}_n$ maximizes the likelihood, we have

$$\sum_{i=1}^{n} \ell'(X_i; \widehat{\theta}_n) = 0$$

and expanding this equation in a Taylor series expansion, we get

$$\begin{aligned} 0 = \sum_{i=1}^{n} \ell'(X_i; \widehat{\theta}_n) &= \sum_{i=1}^{n} \ell'(X_i; \theta) \\ &\quad + (\widehat{\theta}_n - \theta) \sum_{i=1}^{n} \ell''(X_i; \theta) \\ &\quad + \frac{1}{2}(\widehat{\theta}_n - \theta)^2 \sum_{i=1}^{n} \ell'''(X_i; \theta_n^*) \end{aligned}$$

where θ_n^* lies between θ and $\widehat{\theta}_n$. Dividing both sides by $\sqrt{n}$, we get

$$\begin{aligned} 0 &= \frac{1}{\sqrt{n}} \sum_{i=1}^{n} \ell'(X_i; \theta) + \sqrt{n}(\widehat{\theta}_n - \theta) \frac{1}{n} \sum_{i=1}^{n} \ell''(X_i; \theta) \\ &\quad + \frac{1}{2}\sqrt{n}(\widehat{\theta}_n - \theta)^2 \frac{1}{n} \sum_{i=1}^{n} \ell'''(X_i; \theta_n^*), \end{aligned}$$

which suggests that

$$\begin{aligned} &\sqrt{n}(\widehat{\theta}_n - \theta) \\ &= \frac{-n^{-1/2} \sum_{i=1}^{n} \ell'(X_i; \theta)}{n^{-1} \sum_{i=1}^{n} \ell''(X_i; \theta) + (\widehat{\theta}_n - \theta)(2n)^{-1} \sum_{i=1}^{n} \ell'''(X_i; \theta_n^*)}. \end{aligned}$$

From the Central Limit Theorem (and condition (A4)), it follows that

$$\frac{1}{\sqrt{n}} \sum_{i=1}^{n} \ell'(X_i; \theta) \to_d Z \sim N(0, I(\theta))$$

and from the WLLN (and condition (A5)), we have that

$$\frac{1}{n} \sum_{i=1}^{n} \ell''(X_i; \theta) \to_p -J(\theta).$$

Thus it follows from Slutsky's Theorem that

$$\sqrt{n}(\widehat{\theta}_n - \theta) \to_d \frac{Z}{J(\theta)} \sim N(0, I(\theta)/J^2(\theta))$$

provided that

$$(\widehat{\theta}_n - \theta)\frac{1}{n}\sum_{i=1}^{n}\psi'''(X_i;\theta_n^*) \to_p 0;$$

we will show later that this latter statement holds provided that condition (A6) holds and $\widehat{\theta}_n \to_p \theta$.

Proving consistency of the MLE $\widehat{\theta}_n$ is somewhat subtle. Consider the (random) function

$$\phi_n(t) = \frac{1}{n}\sum_{i=1}^{n}\left[\ln f(X_i;t) - \ln f(X_i;\theta)\right],$$

which is maximized at $t = \widehat{\theta}_n$. By the WLLN, for each fixed $t \in \Theta$,

$$\phi_n(t) \to_p \phi(t) = E_\theta\left[\ln\left(\frac{f(X_i;t)}{f(X_i;\theta)}\right)\right].$$

Now note that $-\phi(t)$ is simply a Kullback-Leibler information number, which is minimized when $t = \theta$ and so $\phi(t)$ is maximized at $t = \theta$ ($\phi(\theta) = 0$). Moreover, unless $f(x;t) = f(x;\theta)$ for all $x \in A$ then $\phi(t) < 0$; since we are assuming (implicitly) identifiability of the model, it follows that $\phi(t)$ is uniquely maximized at $t = \theta$.

Does the fact that $\phi_n(t) \to_p \phi(t)$ for each t (where $\phi(t)$ is maximized at $t = \theta$) imply that $\widehat{\theta}_n \to_p \theta$? Unfortunately, the answer to this question is, in general, "no" (unless we make more assumptions about the ϕ_n's). To keep things simple, we will consider some examples where $\{\phi_n(t)\}$ and $\phi(t)$ are non-random functions.

EXAMPLE 5.8: Let $\{\phi_n(t)\}$ be a sequence of functions with

$$\phi_n(t) = \begin{cases} 1 - n\left|t - \frac{1}{n}\right| & \text{for } 0 \le t \le 2/n \\ \frac{1}{2} - |t-2| & \text{for } 3/2 \le t \le 5/2 \\ 0 & \text{otherwise.} \end{cases}$$

Note that $\phi_n(t)$ is maximized at $t_n = 1/n$ and $\phi_n(1/n) = 1$. It is easy to see that for each t, $\phi_n(t) \to \phi(t)$ where

$$\phi(t) = \begin{cases} \frac{1}{2} - |t-2| & \text{for } 3/2 \le t \le 5/2 \\ 0 & \text{otherwise.} \end{cases}$$

Thus $\phi(t)$ is maximized at $t_0 = 2$; clearly, $t_n \to 0 \neq t_0$. ◇

What goes wrong in the previous example? The main problem

seems to be that although $\phi_n(t) \to \phi(t)$ for each t, the convergence is not uniform; that is, for any $M > 0$,

$$\sup_{|t|\leq M} |\phi_n(t) - \phi(t)| = 1 \quad \text{(for } n \text{ sufficiently large)}$$

and so $\phi_n(t)$ does not converge to $\phi(t)$ uniformly for $|t| \leq M$. However, this uniform convergence is not by itself sufficient to guarantee the convergence of the maximizers of $\phi_n(t)$ to the maximizer of $\phi(t)$ as evidenced by the following example.

EXAMPLE 5.9: Let $\{\phi_n(t)\}$ be a sequence of functions with

$$\phi_n(t) = \begin{cases} \frac{1}{2}(1 - 2|t|) & \text{for } |t| \leq 1/2 \\ 1 - 2|t-n| & \text{for } n - 1/2 \leq t \leq t + 1/2 \\ 0 & \text{otherwise.} \end{cases}$$

It is easy to see that for any $M > 0$,

$$\sup_{|t|\leq M} |\phi_n(t) - \phi(t)| \to 0$$

where

$$\phi(t) = \begin{cases} \frac{1}{2}(1 - 2|t|) & \text{for } |t| \leq 1/2 \\ 0 & \text{otherwise.} \end{cases}$$

However, $\phi_n(t)$ is maximized at $t_n = n$ while $\phi(t)$ is maximized at $t_0 = 0$; again, $t_n \to \infty \neq t_0$. ◇

Even though uniform convergence of $\phi_n(t)$ to $\phi(t)$ holds over closed and bounded sets in this example, the sequence of maximizers $\{t_n\}$ cannot be contained within a closed and bounded set and so the sequence does not converge to the maximizer of $\phi(t)$. The following result shows that adding the condition that the sequence of maximizers is bounded is sufficient for convergence of t_n to t_0 where t_0 maximizes $\phi(t)$; moreover, this result covers the case where $\{\phi_n(t)\}$ and $\phi(t)$ are random.

THEOREM 5.1 *Suppose that $\{\phi_n(t)\}$ and $\phi(t)$ are real-valued random functions defined on the real line. Suppose that*
(a) for each $M > 0$,

$$\sup_{|t|\leq M} |\phi_n(t) - \phi(t)| \to_p 0;$$

(b) T_n maximizes $\phi_n(t)$ and T_0 is the unique maximizer of $\phi(t)$;
(c) for each $\epsilon > 0$, there exists M_ϵ such that $P(|T_n| > M_\epsilon) < \epsilon$ for

all n.
Then $T_n \to_p T_0$.

As appealing as Theorem 5.1 seems to be, it is very difficult to use in practice. For example, although it is not too difficult to establish the uniform convergence, it can be extremely difficult to show that $P(|T_n| > M_\epsilon) < \epsilon$. However, if we assume that $\phi_n(t)$ is a concave function for each n, we can weaken the conditions of the previous theorem considerably.

THEOREM 5.2 *Suppose that* $\{\phi_n(t)\}$ *and* $\phi(t)$ *are random concave functions. If*
(a) for each t, $\phi_n(t) \to_p \phi(t)$, *and*
(b) T_n *maximizes* $\phi_n(t)$ *and* T_0 *is the unique maximizer of* $\phi(t)$
then $T_n \to_p T_0$.

In most applications of the preceding theorem, $\phi_n(t)$ will be an average of n random variables (depending on t) and so we can use the WLLN (or a similar result) to show that $\phi_n(t) \to_p \phi(t)$; in such cases, the limiting function $\phi(t)$ will be non-random and so its maximizer T_0 will also be non-random. Theorem 5.2 also holds if the functions $\{\phi_n\}$ are defined on R^p; the same is true for Theorem 5.1.

Theorem 5.2 can also be used if the functions $\phi_n(t)$ are convex. Since the negative of a convex function is a concave function, we can use this theorem to establish convergence in probability of minimizers of convex functions.

EXAMPLE 5.10: Suppose that $X_1, X_2, \cdots, X_n$ are i.i.d. random variables with continuous distribution function $F(x)$ and assume that μ is the unique median of the distribution so that $F(\mu) = 0.5$. The sample median $\widehat{\mu}_n$ of $X_1, \cdots, X_n$ can be defined as a minimizer of

$$\sum_{i=1}^{n} |X_i - t|$$

or equivalently as the minimizer of

$$\phi_n(t) = \frac{1}{n}\sum_{i=1}^{n} \left[|X_i - t| - |X_i|\right].$$

It is easy to see that $\phi_n(t)$ is a convex function since $|a - t|$ is convex in t for any a. By the WLLN,

$$\phi_n(t) \to_p E\left[|X_1 - t| - |X_1|\right] = \phi(t)$$

and it can be easily shown that $\phi(t)$ is minimized at $t = \mu$. Thus $\widehat{\mu}_n \to_p \mu$. (Note that this result is valid even if $E[|X_i|] = \infty$ since $E[|X_i - t| - |X_i|] < \infty$ for all t.) It is interesting to compare this proof of consistency to the proof given in Example 3.3. ◇

EXAMPLE 5.11: Suppose that $X_1, X_2, \cdots, X_n$ are i.i.d. random variables from a one-parameter exponential family density or frequency function

$$f(x;\theta) = \exp\left[c(\theta)T(x) - d(\theta) + S(x)\right] \quad \text{for } x \in A.$$

The MLE of θ maximizes

$$\phi_n(t) = \frac{1}{n}\sum_{i=1}^{n}\left[c(t)T(X_i) - d(t)\right];$$

however, $\phi_n(t)$ is not necessarily a concave function. Nonetheless, if $c(\cdot)$ is a one-to-one continuous function with inverse $c^{-1}(\cdot)$, we can define $u = c(t)$ and consider

$$\phi_n^*(u) = \frac{1}{n}\sum_{i=1}^{n}\left[uT(X_i) - d_0(u)\right] = \phi_n(c(t))$$

where $d_0(u) = d(c^{-1}(u))$. It follows that $\phi_n^*(u)$ is a concave function since its second derivative is $-d_0''(u)$, which is negative. By the WLLN, for each u, we have

$$\phi_n^*(u) \to_p uE_\theta[T(X_1)] - d_0(u) = \phi^*(u)$$

and $\phi^*(u)$ is maximized when $d_0'(u) = E_\theta[T(X_1)]$. Since

$$E_\theta[T(X_1)] = d_0'(c(\theta)),$$

it follows that $\phi^*(u)$ is maximized at $c(\theta)$. Since $u = c(t)$, it follows that $c(\widehat{\theta}_n) \to_p c(\theta)$ and since $c(\cdot)$ is one-to-one and continuous, it follows that $\widehat{\theta}_n \to_p \theta$. ◇

We will now state a result concerning the asymptotic normality of MLEs for i.i.d. sequences. We will assume that consistency of the estimators has been proved.

THEOREM 5.3 (Asymptotic normality of MLEs) *Suppose that $X_1, X_2, \cdots, X_n$ are i.i.d. random variables with density or frequency function $f(x;\theta)$ that satisfies conditions (A1)-(A6) and* *suppose that the MLEs satisfy* $\widehat{\theta}_n \to_p \theta$ *where*

$$\sum_{i=1}^{n} \ell'(X_i; \widehat{\theta}_n) = 0.$$

Then

$$\sqrt{n}(\widehat{\theta}_n - \theta) \to_d N(0, I(\theta)/J^2(\theta)).$$

When $I(\theta) = J(\theta)$, *we have* $\sqrt{n}(\widehat{\theta}_n - \theta) \to_d N(0, 1/I(\theta))$.

Proof. From above we have that

$$\sqrt{n}(\widehat{\theta}_n - \theta) = \frac{-n^{-1/2}\sum_{i=1}^{n} \ell'(X_i; \theta)}{n^{-1}\sum_{i=1}^{n} \ell''(X_i; \theta) + (\widehat{\theta}_n - \theta)(2n)^{-1}\sum_{i=1}^{n} \ell'''(X_i; \theta_n^*)}.$$

Given our previous development, we need only show that

$$R_n = (\widehat{\theta}_n - \theta)\frac{1}{2n}\sum_{i=1}^{n} \ell'''(X_i; \theta_n^*) \to_p 0.$$

We have that for any $\epsilon > 0$,

$$P(|R_n| > \epsilon) = P(|R_n| > \epsilon, |\widehat{\theta}_n - \theta| > \delta) + P(|R_n| > \epsilon, |\widehat{\theta}_n - \theta| \le \delta)$$

and

$$P(|R_n| > \epsilon, |\widehat{\theta}_n - \theta| > \delta) \le P(|\widehat{\theta}_n - \theta| > \delta) \to 0$$

as $n \to \infty$. If $|\widehat{\theta}_n - \theta| \le \delta$, we have (by condition (A6)),

$$|R_n| \le \frac{\delta}{2n}\sum_{i=1}^{n} M(X_i)$$

and since

$$\frac{1}{n}\sum_{i=1}^{n} M(X_i) \to_p E_\theta[M(X_1)] < \infty$$

(by the WLLN), it follows that

$$P(|R_n| > \epsilon, |\widehat{\theta}_n - \theta| \le \delta)$$

can be made arbitrarily small (for large n) by taking δ sufficiently small. Thus $R_n \to_p 0$ and so

$$\sqrt{n}(\widehat{\theta}_n - \theta) \to_d N(0, I(\theta)/J^2(\theta))$$

applying Slutsky's Theorem. □

The regularity conditions (A1) through (A6) are by no means minimal conditions. In particular, it is possible to weaken the differentiability conditions at the cost of increasing the technical difficulty of the proof. For example, we could replace conditions (A5) and (A6) by a somewhat weaker condition that

$$\frac{1}{n}\sum_{i=1}^{n} \ell''(X_i; T_n) \to_p -J(\theta)$$

for any sequence of random variables $\{T_n\}$ with $T_n \to_p \theta$; it is easy to see that this latter condition is implied by (A5) and (A6). However, conditions (A1) through (A6) are satisfied for a wide variety of one-parameter models and are relatively simple to verify. An alternative condition to (A6) is considered in Problem 5.11.

Estimating standard errors

In all but rare cases, we have $I(\theta) = J(\theta)$ and so the result of Theorem 5.3 suggests that for sufficiently large n, the MLE $\widehat{\theta}_n$ is approximately Normal with mean θ and variance $1/(nI(\theta))$. This result can be used to approximate the standard error of $\widehat{\theta}_n$ by $[nI(\theta)]^{-1/2}$. Since $I(\theta)$ typically depends on θ, it is necessary to estimate $I(\theta)$ to estimate the standard error of $\widehat{\theta}_n$. There are two approaches to estimating $I(\theta)$ and hence the standard error of $\widehat{\theta}_n$.

- If $I(\theta)$ has a closed-form, we can substitute $\widehat{\theta}_n$ for θ; our standard error estimator becomes

  $$\widehat{\text{se}}(\widehat{\theta}_n) = \frac{1}{\sqrt{nI(\widehat{\theta}_n)}}.$$

  $nI(\widehat{\theta}_n)$ is called the expected Fisher information for θ.
- Since $I(\theta) = -E_\theta[\ell''(X_i; \theta)]$, we can estimate $I(\theta)$ by

  $$\widehat{I(\theta)} = -\frac{1}{n}\sum_{i=1}^{n} \ell''(X_i; \widehat{\theta}_n),$$

  which leads to the standard error estimator

  $$\widehat{\text{se}}(\widehat{\theta}_n) = \frac{1}{\sqrt{n\widehat{I(\theta)}}} = \left(-\sum_{i=1}^{n} \ell''(X_i; \widehat{\theta}_n)\right)^{-1/2}.$$

  $n\widehat{I(\theta)}$ is called the observed Fisher information for θ.

EXAMPLE 5.12: Suppose $X_1, \cdots, X_n$ have a one-parameter exponential family density or frequency function

$$f(x;\theta) = \exp\left[c(\theta)T(x) - d(\theta) + S(x)\right] \quad \text{for } x \in A$$

Using the facts that

$$\frac{1}{n}\sum_{i=1}^{n} T(X_i) = \frac{d'(\widehat{\theta}_n)}{c'(\widehat{\theta}_n)}$$

and

$$I(\theta) = d''(\theta) - c''(\theta)\frac{d'(\theta)}{c'(\theta)},$$

it follows that the expected and observed Fisher information for θ are the same. ◇

EXAMPLE 5.13: Suppose that $X_1, \cdots, X_n$ are i.i.d. Cauchy random variables with density function

$$f(x;\theta) = \frac{1}{\pi}\frac{1}{1+(x-\theta)^2}.$$

Then

$$\ell'(x;\theta) = \frac{2(x-\theta)}{1+(x-\theta)^2}$$

and

$$\ell''(x;\theta) = -\frac{2(1-(x-\theta)^2)}{(1+(x-\theta)^2)^2}.$$

It possible to show that

$$I(\theta) = \frac{1}{2}$$

and so the expected Fisher information for θ is $n/2$ and the corresponding standard error estimate is

$$\widehat{\text{se}}(\widehat{\theta}_n) = \sqrt{\frac{2}{n}}.$$

On the other hand, the observed Fisher information for θ is

$$\sum_{i=1}^{n} \frac{2(1-(X_i-\widehat{\theta}_n)^2)}{(1+(X_i-\widehat{\theta}_n)^2)^2}$$

(which is not equal to the expected Fisher information) and so the corresponding standard error estimator is

$$\widehat{\text{se}}(\widehat{\theta}_n) = \left(\sum_{i=1}^{n} \frac{2(1-(X_i-\widehat{\theta}_n)^2)}{(1+(X_i-\widehat{\theta}_n)^2)^2}\right)^{-1/2}.$$

It is interesting to compare the limiting variance of the MLE with that of the trimmed means in Example 4.40; we can compute the minimum limiting variance for all trimmed means to 2.278, which implies that the MLE is more efficient since its limiting variance is 2. ◇

In Example 5.13, we see that different estimates of the standard error can result by using the observed and expected Fisher information. This raises the question of whether either of the two estimators can be shown to be superior to the other. While there is some debate, results of Efron and Hinkley (1978) indicate that there are reasons to prefer the estimator based on the observed Fisher information. Their rationale is that this estimator actually estimates $\text{Var}_\theta^{1/2}(\widehat{\theta}_n | S = s)$ where $S = S(\boldsymbol{X})$ is an ancillary statistic for θ and $s = S(\boldsymbol{x})$.

Multiparameter models

Extending the consistency and asymptotic normality results from the single- to multi-parameter cases is simple, if somewhat notationally messy. Assume that $X_1, X_2, \cdots$ are i.i.d. random variables with density or frequency function $f(x; \boldsymbol{\theta})$ where $\boldsymbol{\theta} = (\theta_1, \cdots, \theta_p)$. The MLE $\widehat{\boldsymbol{\theta}}_n$ based on $X_1, \cdots, X_n$ satisfies the likelihood equations

$$\sum_{i=1}^{n} \ell'(X_i; \widehat{\boldsymbol{\theta}}_n) = 0$$

where now $\ell'(x; \boldsymbol{\theta})$ is the vector of partial derivatives of $\ell(x; \boldsymbol{\theta}) = \ln f(x; \boldsymbol{\theta})$ (with respect to the components of $\boldsymbol{\theta}$). The idea behind proving asymptotic normality is really exactly the same as that used in the single parameter case: we make a Taylor series expansion of the likelihood equations around the true parameter value. Doing this we get

$$0 = \frac{1}{\sqrt{n}} \sum_{i=1}^{n} \ell'(X_i; \boldsymbol{\theta}) + \left(\frac{1}{n} \sum_{i=1}^{n} \ell''(X_i; \boldsymbol{\theta}_n^*) \right) \sqrt{n}(\widehat{\boldsymbol{\theta}}_n - \boldsymbol{\theta})$$

where
(a) $\ell''(x; \boldsymbol{\theta})$ is the matrix of second partial derivatives (with respect to $\theta_1, \cdots, \theta_p$); the (j, k) element of this matrix is given by

$$\ell''_{jk}(x; \boldsymbol{\theta}) = \frac{\partial^2}{\partial\theta_j \partial\theta_k} \ell(x; \boldsymbol{\theta}).$$

(b) $\boldsymbol{\theta}_n^*$ is on the line segment joining $\boldsymbol{\theta}$ and $\widehat{\boldsymbol{\theta}}_n$.

Now solving for $\sqrt{n}(\widehat{\boldsymbol{\theta}}_n - \boldsymbol{\theta})$, we get

$$\sqrt{n}(\widehat{\boldsymbol{\theta}}_n - \boldsymbol{\theta}) = \left(-\frac{1}{n}\sum_{i=1}^n \ell''(X_i;\boldsymbol{\theta}_n^*)\right)^{-1} \frac{1}{\sqrt{n}}\sum_{i=1}^n \ell'(X_i;\boldsymbol{\theta})$$

and so, under appropriate regularity conditions, we should have

$$\frac{1}{\sqrt{n}}\sum_{i=1}^n \ell'(X_i;\boldsymbol{\theta}) \to_d N_p(\mathbf{0}, I(\boldsymbol{\theta}))$$

(with $I(\boldsymbol{\theta}) = \mathrm{Cov}_\theta[\ell'(X_i;\boldsymbol{\theta})]$) and

$$\frac{1}{n}\sum_{i=1}^n \ell''(X_i;\boldsymbol{\theta}_n^*) \to_p -J(\boldsymbol{\theta}) = E_\theta\left[\ell''(X_1;\boldsymbol{\theta})\right]$$

where $I(\boldsymbol{\theta})$ and $J(\boldsymbol{\theta})$ are $p \times p$ matrices. Now provided $J(\boldsymbol{\theta})$ is invertible, we have

$$\sqrt{n}(\widehat{\boldsymbol{\theta}}_n - \boldsymbol{\theta}) \to_d N_p(\mathbf{0}, J(\boldsymbol{\theta})^{-1} I(\boldsymbol{\theta}) J(\boldsymbol{\theta})^{-1}).$$

As in the single parameter case, for many models (including exponential families) we have $I(\boldsymbol{\theta}) = J(\boldsymbol{\theta})$ in which case the limiting variance-covariance matrix $J(\boldsymbol{\theta})^{-1} I(\boldsymbol{\theta}) J(\boldsymbol{\theta})^{-1}$ above becomes $I(\boldsymbol{\theta})^{-1}$. $I(\boldsymbol{\theta}) = J(\boldsymbol{\theta})$ if we are able to differentiate twice inside the integral (or summation) sign with respect to all p components of $\boldsymbol{\theta}$. ($I(\boldsymbol{\theta})$ is called the Fisher information matrix.)

We now state the regularity conditions that are sufficient to "rigorize" the previous argument. The conditions are simply analogues of conditions (A1)-(A6) used previously.

(B1) The parameter space Θ is an open subset of R^p.
(B2) The set $A = \{x : f(x;\boldsymbol{\theta}) > 0\}$ does not depend on $\boldsymbol{\theta}$.
(B3) $f(x;\boldsymbol{\theta})$ is three times continuously differentiable with respect to $\boldsymbol{\theta}$ for all x in A.
(B4) $E_\theta[\ell'(X_i;\boldsymbol{\theta})] = 0$ for all $\boldsymbol{\theta}$ and $\mathrm{Cov}_\theta[\ell'(X_i;\boldsymbol{\theta})] = I(\boldsymbol{\theta})$ where $I(\boldsymbol{\theta})$ is positive definite for all $\boldsymbol{\theta}$.
(B5) $E_\theta[\ell''(X_i;\boldsymbol{\theta})] = -J(\boldsymbol{\theta})$ where $J(\boldsymbol{\theta})$ is positive definite for all $\boldsymbol{\theta}$.
(B6) Let $\ell'''_{jkl}(x;\boldsymbol{\theta})$ be the mixed partial derivative of $\ell(x;\boldsymbol{\theta})$ with respect to $\theta_j, \theta_k, \theta_l$. For each $\boldsymbol{\theta}$, $\delta > 0$ and $1 \le j, k, l \le p$,

$$|\ell'''_{jkl}(x;\mathbf{t})| \le M_{jkl}(x)$$

for $||\boldsymbol{\theta} - \mathbf{t}|| \le \delta$ where $E_\theta[M_{jkl}(X_i)] < \infty$.

THEOREM 5.4 *Suppose that $X_1, X_2, \cdots, X_n$ are i.i.d. random variables with density or frequency function $f(x; \boldsymbol{\theta})$ that satisfies conditions (B1)-(B6) and suppose that the MLEs satisfy $\widehat{\boldsymbol{\theta}}_n \to_p \boldsymbol{\theta}$ where*

$$\sum_{i=1}^{n} \ell'(X_i; \widehat{\boldsymbol{\theta}}_n) = 0.$$

Then

$$\sqrt{n}(\widehat{\boldsymbol{\theta}}_n - \boldsymbol{\theta}) \to_d N_p(0, J(\boldsymbol{\theta})^{-1} I(\boldsymbol{\theta}) J(\boldsymbol{\theta})^{-1}).$$

When $I(\boldsymbol{\theta}) = J(\boldsymbol{\theta})$, we have $\sqrt{n}(\widehat{\boldsymbol{\theta}}_n - \boldsymbol{\theta}) \to_d N_p(0, I(\boldsymbol{\theta})^{-1})$.

The proof of Theorem 5.4 parallels that of Theorem 5.3 and will not be given. Note that Theorem 5.4 assumes consistency of $\widehat{\boldsymbol{\theta}}_n$. This can be proved in a variety of ways; for example, in the case of p-parameter exponential families, Theorem 5.2 can be used as in Example 5.11.

It should also be noted that both Theorems 5.3 and 5.4 hold even if $X_1, X_2, \cdots, X_n$ are i.i.d. random vectors.

EXAMPLE 5.14: Suppose that $X_1, \cdots, X_n$ are i.i.d. Normal random variables with mean μ and variance σ^2. The MLEs of μ and σ^2 are

$$\widehat{\mu}_n = \frac{1}{n} \sum_{i=1}^{n} X_i \quad \text{and} \quad \widehat{\sigma}_n^2 = \frac{1}{n} \sum_{i=1}^{n} (X_i - \widehat{\mu}_n)^2.$$

The joint limiting distribution of $\sqrt{n}(\widehat{\mu}_n - \mu)$ and $\sqrt{n}(\widehat{\sigma}_n - \sigma)$ can be derived quite easily using Theorem 5.4. (Of course, $\sqrt{n}(\widehat{\mu}_n - \mu)$ is exactly Normal for any n.) Writing

$$\ell(x; \mu, \sigma) = -\frac{1}{2\sigma^2}(x - \mu)^2 - \ln(\sigma) - \frac{1}{2}\ln(2\pi)$$

and taking partial derivatives of ℓ with respect to μ and σ, it is easy to show that

$$I(\mu, \sigma) = J(\mu, \sigma) = \begin{pmatrix} \sigma^{-2} & 0 \\ 0 & 2\sigma^{-2} \end{pmatrix}.$$

Thus (after verifying conditions (B1)-(B6) of Theorem 5.4 hold), we have

$$\sqrt{n} \begin{pmatrix} \widehat{\mu}_n - \mu \\ \widehat{\sigma}_n - \sigma \end{pmatrix} \to_d N_2(0, I^{-1}(\mu, \sigma));$$

note that the variance-covariance matrix $I^{-1}(\mu, \sigma)$ is diagonal with diagonal entries σ^2 and $\sigma^2/2$. It is also worth noting that we could have determined the joint limiting distribution via the Multivariate CLT more or less directly. ◇

EXAMPLE 5.15: Suppose that $X_1, \cdots, X_n$ are i.i.d. Gamma random variables with shape parameter α and scale parameter λ. The MLEs of α and λ satisfy the equations

$$\begin{aligned}\frac{1}{n}\sum_{i=1}^{n} X_i &= \frac{\widehat{\alpha}_n}{\widehat{\lambda}_n}\\ \frac{1}{n}\sum_{i=1}^{n} \ln(X_i) &= \psi(\widehat{\alpha}_n) - \ln(\widehat{\lambda}_n)\end{aligned}$$

where $\psi(\alpha)$ is the derivative of $\ln \Gamma(\alpha)$. ($\psi(\alpha)$ is called the digamma function and its derivative $\psi'(\alpha)$ the trigamma function.) Since this is a two-parameter exponential family model, we can compute the information matrix by looking at the second partial derivatives of the log-density; we have

$$\begin{aligned}\frac{\partial^2}{\partial\alpha^2}\ln f(x;\alpha,\lambda) &= -\psi'(\alpha)\\ \frac{\partial^2}{\partial\lambda^2}\ln f(x;\alpha,\lambda) &= -\frac{\alpha}{\lambda^2}\\ \text{and}\quad \frac{\partial^2}{\partial\alpha\partial\lambda}\ln f(x;\alpha,\lambda) &= \frac{1}{\lambda}.\end{aligned}$$

Since none of these derivatives depends on x, we can easily compute

$$I(\alpha,\lambda) = \begin{pmatrix} \psi'(\alpha) & -1/\lambda \\ -1/\lambda & \alpha/\lambda^2 \end{pmatrix}.$$

Inverting $I(\alpha,\lambda)$, we obtain the limiting variance-covariance matrix

$$I^{-1}(\alpha,\lambda) = \frac{\lambda^2}{\alpha\psi'(\alpha)-1}\begin{pmatrix} \alpha/\lambda^2 & 1/\lambda \\ 1/\lambda & \psi'(\alpha) \end{pmatrix}.$$

In particular, this implies that

$$\sqrt{n}(\widehat{\alpha}_n - \alpha) \to_d N\left(0, \alpha/(\alpha\psi'(\alpha)-1)\right)$$

and

$$\sqrt{n}(\widehat{\lambda}_n - \lambda) \to_d N\left(0, \lambda^2\psi'(\alpha)/(\alpha\psi'(\alpha)-1)\right).$$

It is interesting to compare the limiting distribution of $\sqrt{n}(\widehat{\lambda}_n - \lambda)$ (assuming α is unknown) to the limiting distribution of the MLE

Table 5.1 *Comparison of* α^{-1} *and* $g(\alpha) = \psi'(\alpha)/(\alpha\psi'(\alpha)-1)$ *in Example 5.15 for various* α.

α	α^{-1}	$g(\alpha)$
0.1	10	11.09
0.5	2	3.36
1	1	2.55
2	0.5	2.22
5	0.2	2.08
10	0.1	2.04
100	0.01	2.00

of λ when α is known. In this latter case, the MLE is

$$\widetilde{\lambda}_n = \frac{\alpha}{\bar{X}_n}$$

and we have

$$\sqrt{n}(\widetilde{\lambda}_n - \lambda) \to_d N(0, \lambda^2/\alpha).$$

We should expect $\widetilde{\lambda}_n$ to be the more efficient estimator of λ since we are able to incorporate our knowledge of α into the estimation of λ. In fact, it can be shown that $\alpha^{-1} < \psi'(\alpha)/(\alpha\psi'(\alpha)-1)$ for any $\alpha > 0$; Table 5.1 compares α^{-1} and $g(\alpha) = \psi'(\alpha)/(\alpha\psi'(\alpha) - 1)$ for several values of α. Note that as α becomes larger, the difference in efficiency is more substantial. In practice, of course, one would rarely "know" α and so we would have no choice but to use $\widehat{\lambda}_n$. ◇

Example 5.15 illustrates a more general point. Given two equally valid statistical models, we can obtain more efficient estimators in the model with fewer parameters (or more precisely, we will not do worse with the lower parameter model). However, the penalties we pay in assuming a "too small" model can be quite severe. In practice, model selection is often the most difficult and important step in the analysis of data. Some of the consequences of misspecifying a model are examined in the next section.

5.5 Misspecified models

It is important to remember that statistical models are typically merely approximations to reality and so the wrong model is, more often than not, fit to the observed data. As troubling as this observation may seem, it may not be a problem from a practical point of view. First, the assumed model may be "close enough" to the true model so that very little is lost by assuming the wrong model. Second, the parameters estimated for a given model can often be interpreted usefully even if the assumed model is wrong. The following two examples illustrate these points.

EXAMPLE 5.16: Suppose that $X_1, \cdots, X_n$ are i.i.d. Exponential random variables with parameter λ. However, suppose that we decide that the appropriate model for the data is given by the following two-parameter family of densities for the X_i's:

$$f(x; \alpha, \theta) = \frac{\alpha}{\theta}\left(1 + \frac{x}{\theta}\right)^{-(\alpha+1)} \quad \text{for } x > 0.$$

$\alpha > 0$ and $\theta > 0$ are the unknown parameters. Even though the Exponential distribution is not a member of this family of distributions, it is easy to see that by letting α and θ tend to infinity such that α/θ tends to λ, we have (for $x > 0$),

$$f(x; \alpha, \theta) \to \lambda \exp(-\lambda x) \quad (\text{as } \alpha, \theta \to \infty \text{ and } \alpha/\theta \to \lambda).$$

Thus given reasonable estimators $\widehat{\alpha}$ and $\widehat{\theta}$, the estimated density $f(x; \widehat{\alpha}, \widehat{\theta})$ will be close to true density of the X_i's. ◇

EXAMPLE 5.17: Suppose that $X_1, \cdots, X_n$ are independent random variables with

$$E(X_i) = \alpha + \beta t_i \quad \text{and} \quad \text{Var}(X_i) = \sigma^2$$

where α, β, and σ^2 are unknown, and $t_1, \cdots, t_n$ are known constants. In estimating α, β and σ^2, it is often assumed that the X_i's are Normal and the parameters estimated by maximum likelihood; we will see in Chapter 8 that these estimators remain valid even when the X_i's are non-normal. ◇

Suppose that $X_1, X_2, \cdots, X_n$ are i.i.d. random variables with distribution function F. We assume, however, that the X_i's have a common density or frequency function $f(x; \theta)$ for some $\theta \in \Theta$ where the true distribution function F does not necessarily correspond to any $f(x; \theta)$. Suppose that an estimator $\widehat{\theta}_n$ satisfies the likelihood equation

$$\sum_{i=1}^{n} \ell'(X_i; \widehat{\theta}_n) = 0$$

where $\ell'(x;\theta) = \ell'(x;\theta)$. (In general, this type of relation that defines an estimator is called an estimating equation and includes the likelihood equation as a special case.) What exactly is $\widehat{\theta}_n$ estimating? What is the behaviour of the sequence of estimators $\{\widehat{\theta}_n\}$ for large n?

Consider the functional parameter $\theta(F)$ defined by

$$\int_{-\infty}^{\infty} \ell'(x;\theta(F))\,dF(x) = 0.$$

The substitution principle estimator of $\theta(F)$ is simply the solution to the likelihood equation given above. The influence curve of $\theta(F)$ is

$$\phi(x;F) = -\frac{\ell'(x;\theta(F))}{\int_{-\infty}^{\infty} \ell''(x;\theta(F))},$$

which suggests that

$$\sqrt{n}(\widehat{\theta}_n - \theta(F)) \to_d N(0,\sigma^2)$$

where

$$\sigma^2 = \frac{\int_{-\infty}^{\infty} [\ell'(x;\theta(F))]^2\,dF(x)}{\left(\int_{-\infty}^{\infty} \ell''(x;\theta(F))\,dF(x)\right)^2}.$$

The following theorem gives precise conditions under which the preceding statement is true; these conditions parallel the differentiability conditions (A4) through (A6).

THEOREM 5.5 *Suppose that $X_1, X_2, \cdots, X_n$ are i.i.d. random variables with distribution function F and that the estimator $\widehat{\theta}_n$ satisfies the estimating equation*

$$\sum_{i=1}^{n} \ell'(X_i;\widehat{\theta}_n) = 0$$

for some $\widehat{\theta}_n$ in an open set Θ. If
(a) $\ell'(x;\theta)$ is a strictly decreasing (or strictly decreasing) function of θ (over the open set Θ) for each x,
(b) $\int_{-\infty}^{\infty} \ell'(x;\theta)\,dF(x) = 0$ has a unique solution $\theta = \theta(F)$ where $\theta(F) \in \Theta$,
(c) $I(F) = \int_{-\infty}^{\infty} [\ell'(x;\theta(F))]^2\,dF(x) < \infty$,
(d) $J(F) = -\int_{-\infty}^{\infty} \ell''(x;\theta(F))\,dF(x) < \infty$,
(e) $|\ell'''(x;t)| \le M(x)$ for $\theta(F) - \delta \le t \le \theta(F) + \delta$ and some $\delta > 0$ where $\int_{-\infty}^{\infty} M(x)\,dF(x) < \infty$,
then $\widehat{\theta}_n \to_p \theta(F)$ and $\sqrt{n}(\widehat{\theta}_n - \theta(F)) \to_d N(0, I(F)/J^2(F))$.

Proof. Since $\ell'(x;\theta)$ is strictly decreasing in θ and $\theta(F)$ is the unique solution of the equation

$$\int_{-\infty}^{\infty} \ell'(x;\theta)\, dF(x) = 0,$$

we have for any $\epsilon > 0$,

$$P\left[\frac{1}{n}\sum_{i=1}^{n} \ell'(X_i;\theta(F)+\epsilon) > 0\right] \to 0$$

and

$$P\left[\frac{1}{n}\sum_{i=1}^{n} \ell'(X_i;\theta(F)-\epsilon) < 0\right] \to 0$$

as $n \to \infty$. From this it follows that

$$P\left[|\widehat{\theta}_n - \theta(F)| > \epsilon\right] \to 0$$

and so $\widehat{\theta}_n \to_p \theta(F)$. Now expanding the estimating equation in a Taylor series, we have

$$\begin{aligned} 0 &= \frac{1}{\sqrt{n}}\sum_{i=1}^{n} \ell'(X_i;\theta(F)) + \sqrt{n}(\widehat{\theta}_n - \theta(F))\frac{1}{n}\sum_{i=1}^{n} \ell''(X_i;\theta(F)) \\ &\quad + \sqrt{n}(\widehat{\theta}_n - \theta(F))^2 \frac{1}{2n}\sum_{i=1}^{n} \ell'''(X_i;\theta_n^*) \end{aligned}$$

where θ_n^* lies between $\widehat{\theta}_n$ and $\theta(F)$. The remainder of the proof is identical to the proof of Theorem 5.3. □

We can remove the assumption that $\ell'(x;\theta)$ is strictly monotone in θ for each x by adding the assumption that $\widehat{\theta}_n \to_p \theta(F)$. In certain cases, we will not have

$$\int \ell'(x;\theta)\, dF(x) = 0$$

for any $\theta \in \Theta$ but may instead have

$$\lim_{\theta \to a} \int \ell'(x;\theta)\, dF(x) = 0$$

for some a lying at the boundary of the set Θ (but not in Θ); for example, a can be $\pm\infty$. In this case, it is usually possible to show that the sequence of estimators $\{\widehat{\theta}_n\}$ converges in probability to a.

It is also possible to extend the result to the multiparameter case.

Suppose $X_1, \cdots, X_n$ are i.i.d. random variables with distribution F and let $\widehat{\boldsymbol{\theta}}_n$ satisfy

$$\sum_{i=1}^{n} \ell'(X_i; \widehat{\boldsymbol{\theta}}_n) = 0$$

where $\ell'(x; \boldsymbol{\theta})$ is the vector of partial derivatives (with respect to the components of $\boldsymbol{\theta}$) of $\ln f(x; \boldsymbol{\theta})$. Then under appropriate regularity conditions, we have

$$\sqrt{n}(\widehat{\boldsymbol{\theta}}_n - \boldsymbol{\theta}(F)) \to_d N_p(0, J(F)^{-1} I(F) J(F)^{-1})$$

where $\boldsymbol{\theta}(F) = (\theta_1(F), \cdots, \theta_p(F))$ satisfies

$$\int_{-\infty}^{\infty} \ell'(x; \boldsymbol{\theta}(F))\, dF(x) = 0$$

and the matrices $I(F)$, $J(F)$ are defined by

$$\begin{aligned} I(F) &= \int_{-\infty}^{\infty} [\ell'(x; \boldsymbol{\theta}(F))][\ell'(x; \boldsymbol{\theta})]^T\, dF(x) \\ J(F) &= -\int_{-\infty}^{\infty} \ell''(x; \boldsymbol{\theta}(F))\, dF(x). \end{aligned}$$

(As before, $\ell''(x; \boldsymbol{\theta})$ is the matrix of second partial derivatives of $\ell(x; \boldsymbol{\theta})$ with respect to $\theta_1, \cdots, \theta_p$; the integrals defining $I(F)$ and $J(F)$ are defined element by element.)

EXAMPLE 5.18: Suppose the $X_1, X_2, \cdots, X_n$ are i.i.d. Normal random variables with (true) mean θ_0 and variance σ^2. However, we (erroneously) assume that the density of the X_i's is

$$f(x; \theta) = \frac{\exp(x - \theta)}{[1 + \exp(x - \theta)]^2}$$

for some θ. The MLE of θ for this model based on $X_1, \cdots, X_n$ is the solution to the equation

$$\sum_{i=1}^{n} \frac{\exp(X_i - \widehat{\theta}_n) - 1}{\exp(X_i - \widehat{\theta}_n) + 1} = 0.$$

It is possible to show that $\widehat{\theta}_n \to_p \theta_0$ since

$$\int_{-\infty}^{\infty} \left(\frac{\exp(x - \theta_0) - 1}{\exp(x - \theta_0) + 1} \right) \frac{1}{\sigma\sqrt{2\pi}} \exp\left(-\frac{(x - \theta_0)^2}{2\sigma^2} \right) dx = 0.$$

We also have that $\sqrt{n}(\widehat{\theta}_n - \theta_0) \to_d N(0, \gamma^2(\sigma^2))$. The values of

Table 5.2 *Asymptotic variances of $\sqrt{n}(\widehat{\theta}_n - \theta)$ in Example 5.18.*

σ	0.5	1	2	5	10	20	50
$\gamma^2(\sigma^2)/\sigma^2$	1.00	1.02	1.08	1.24	1.37	1.46	1.53

$\gamma^2(\sigma^2)/\sigma^2$ (as a function of σ) are given in Table 5.2. (The quantity γ^2/σ^2 is the asymptotic relative efficiency of the sample mean to the estimator $\widehat{\theta}_n$.) ◇

EXAMPLE 5.19: Suppose that $X_1, \cdots, X_n$ are i.i.d. Uniform random variables on the interval $[0, b]$. We erroneously assume that the X_i's have the Gamma density

$$f(x; \alpha, \lambda) = \frac{\lambda^\alpha}{\Gamma(\alpha)} x^{\alpha-1} \exp(-\lambda x) \quad \text{for } x > 0.$$

The estimators $\widehat{\alpha}_n$ and $\widehat{\lambda}_n$ satisfy the equations

$$\begin{aligned} n \ln(\widehat{\lambda}_n) + \sum_{i=1}^n \ln(X_i) - n\psi(\widehat{\alpha}_n) &= 0 \\ n\frac{\widehat{\alpha}_n}{\widehat{\lambda}_n} - \sum_{i=1}^n X_i &= 0 \end{aligned}$$

where $\psi(\alpha)$ is the derivative of $\ln \Gamma(\alpha)$. It follows that $\widehat{\alpha}_n \to_p \alpha(b)$ and $\widehat{\lambda}_n \to_p \lambda(b)$ where $\alpha(b)$ and $\lambda(b)$ satisfy

$$\begin{aligned} \ln(\lambda(b)) + \frac{1}{b}\int_0^b \ln(x)\,dx - \psi(\alpha(b)) &= 0 \\ \frac{\alpha(b)}{\lambda(b)} - \frac{1}{b}\int_0^b x\,dx &= 0. \end{aligned}$$

From the second equation above, it is easy to see that $\alpha(b) = b\lambda(b)/2$ and so $\lambda(b)$ satisfies

$$\ln(\lambda(b)) + \ln(b) - 1 - \psi\left(b\lambda(b)/2\right) = 0.$$

Numerically solving this equation, we get

$$\lambda(b) = \frac{3.55585}{b}$$

and so $\alpha(b) = b\lambda(b)/2 = 1.77793$. Now using $\alpha(b)$ and $\lambda(b)$, we can compute the matrices $I(F)$ and $J(F)$:

$$I(F) = \begin{pmatrix} 1 & -b/4 \\ -b/4 & b^2/12 \end{pmatrix}$$
$$J(F) = \begin{pmatrix} 0.74872 & -b/3.55585 \\ -b/3.55585 & b^2/7.11171 \end{pmatrix}.$$

From this it follows that

$$\sqrt{n}\begin{pmatrix} \widehat{\alpha}_n - \alpha(b) \\ \widehat{\lambda}_n - \lambda(b) \end{pmatrix} \to_d N_2(0, J(F)^{-1}I(F)J(F)^{-1})$$

with

$$J(F)^{-1}I(F)J(F)^{-1} = \begin{pmatrix} 9.61 & 16.03/b \\ 16.03/b & 29.92/b^2 \end{pmatrix}. \qquad \diamond$$

EXAMPLE 5.20: Suppose that $X_1, \cdots, X_n$ are i.i.d. Gamma random variables with shape parameter α and scale parameter λ that represent incomes sampled from some population. Using $X_1, \cdots, X_n$, we wish to estimate the Gini index (see Examples 4.21 and 4.31) of the distribution; for a Gamma distribution, this depends only on α and is given by the expression

$$\pi^{-1/2}\frac{\Gamma(\alpha + 1/2)}{\sqrt{\pi}\Gamma(\alpha+1)}.$$

However, we erroneously assume that the observations are log-Normal with density

$$f(x;\mu,\sigma) = \frac{1}{\sqrt{2\pi}\sigma x}\exp\left(-\frac{(\ln(x)-\mu)^2}{2\sigma^2}\right) \quad \text{for } x > 0$$

($\ln(X_i)$ will have a Normal distribution with mean μ and variance σ^2). The Gini index for the log-Normal depends only on σ and is given by $\gamma(\sigma) = 2\Phi(\sigma/\sqrt{2}) - 1$ where $\Phi(x)$ is the standard Normal distribution function. The MLE of σ is

$$\widehat{\sigma}_n = \left(\frac{1}{n}\sum_{i=1}^n (Y_i - \bar{Y}_n)^2\right)^{1/2} \quad \text{where } Y_i = \ln(X_i)$$

and the corresponding estimator of the Gini index (assuming the log-Normal model) is $\gamma(\widehat{\sigma}_n)$. By the WLLN,

$$\frac{1}{n}\sum_{i=1}^n (Y_i - \bar{Y}_n)^2 \to_p \text{Var}(\ln(X_i)) = \psi'(\alpha)$$

Table 5.3 *Comparison of true Gini index (from a Gamma model) with its misspecified version based on a log-Normal model in Example 5.20.*

α	True	Misspecified
0.1	0.883	1.000
0.2	0.798	1.000
0.5	0.637	0.884
1	0.500	0.636
2	0.375	0.430
5	0.246	0.261
10	0.176	0.181
20	0.125	0.127
50	0.080	0.080

where ψ' is the derivative of the function ψ in Example 5.19 (that is, the second derivative of the logarithm of the Gamma function). Thus applying the Continuous Mapping Theorem, it follows that

$$\gamma(\widehat{\sigma}_n) \to_p 2\Phi\left((\psi'(\alpha)/2)^{1/2}\right) - 1.$$

The limit can now be compared to the true value of the Gini index in order to assess the asymptotic bias involved in assuming the wrong model; these values are given in Table 5.3 for various α.

Table 5.3 shows that the estimator $\gamma(\widehat{\sigma}_n)$ is quite badly biased for small α but also that the bias gradually disappears as α increases. Note that we could also use the substitution principle estimator discussed in Example 4.31; this estimator does not require us to know the form of the distribution of the X_i's but could be extremely inefficient compared to an estimator that assumes the data comes from a particular parametric model. ◇

The result of Theorem 5.5 suggests that an estimator of the standard error of $\widehat{\theta}_n$ (when a single parameter is estimated) is given by

$$\widehat{\text{se}}(\widehat{\theta}_n) = \left(-\sum_{i=1}^{n} \ell''(X_i; \widehat{\theta}_n)\right)^{-1} \left(\sum_{i=1}^{n} [\ell'(X_i; \widehat{\theta}_n)]^2\right)^{1/2}.$$

In the multiparameter case, the variance-covariance matrix of $\widehat{\boldsymbol{\theta}}_n$

can be estimated by

$$\widehat{\text{Cov}}(\widehat{\boldsymbol{\theta}}_n) = \widehat{J}(\widehat{\boldsymbol{\theta}}_n)^{-1}\widehat{I}(\widehat{\boldsymbol{\theta}}_n)\widehat{J}(\widehat{\boldsymbol{\theta}}_n)^{-1}$$

where

$$\begin{aligned}
\widehat{J}(\widehat{\boldsymbol{\theta}}_n) &= -\frac{1}{n}\sum_{i=1}^{n} \ell''(X_i; \widehat{\boldsymbol{\theta}}_n) \\
\widehat{I}(\widehat{\boldsymbol{\theta}}_n) &= \frac{1}{n}\sum_{i=1}^{n} [\ell'(X_i; \widehat{\boldsymbol{\theta}}_n)][\ell'(X_i; \widehat{\boldsymbol{\theta}}_n)]^T.
\end{aligned}$$

This estimator has come to be known as the sandwich estimator.

5.6 Non-parametric maximum likelihood estimation

In discussing maximum likelihood estimation, we have assumed that the observations have a joint density or frequency function (depending on a real- or vector-valued parameter), from which we obtain the likelihood function. This formulation effectively rules out maximum likelihood estimation for non-parametric models as for these models we typically do not make sufficiently strong assumptions to define a likelihood function in the usual sense.

However, it is possible to define a notion of non-parametric maximum likelihood estimation although its formulation is somewhat tenuous. Suppose that $X_1, \cdots, X_n$ are i.i.d. random variables with unknown distribution function F; we want to define a (non-parametric) MLE of F. In order to make the estimation problem well-defined, we will consider only distributions putting positive probability mass only at the points $X_1, \cdots, X_n$. For simplicity, we will assume here that the X_i's are distinct (as would be the case if the X_i's were sampled from a continuous distribution). If p_i is the probability mass at X_i then the non-parametric log-likelihood function is

$$\ln \mathcal{L}(p_1, \cdots, p_n) = \sum_{i=1}^{n} \ln(p_i)$$

where $p_i \geq 0$ $(i = 1, \cdots, n)$ and $p_1 + \cdots + p_n = 1$. Maximizing the non-parametric log-likelihood, we obtain $\widehat{p}_i = 1/n$ $(i = 1, \cdots, n)$; thus the non-parametric MLE of F is the empirical distribution function $\widehat{F}$ with

$$\widehat{F}(x) = \frac{1}{n}\sum_{i=1}^{n} I(X_i \leq x).$$

See Scholz (1980) and Vardi (1985) for more discussion of non-parametric maximum likelihood estimation.

Given our non-parametric MLE of F, we can determine the non-parametric MLE of an arbitrary functional parameter $\theta(F)$ to be $\widehat{\theta} = \theta(\widehat{F})$, provided that this latter estimator is well-defined. Thus the non-parametric MLE of $\theta(F)$ is simply the substitution principle estimator as described in section 4.5.

More discussion of non-parametric maximum likelihood estimation will be given in section 7.5 where we discuss using the non-parametric likelihood function to find confidence intervals for certain functional parameters $\theta(F)$.

5.7 Numerical computation of MLEs

In many estimation problems, it is difficult to obtain closed-form analytical expressions for maximum likelihood estimates. In these situations, it is usually necessary to calculate maximum likelihood estimates numerically. Many numerical methods are available for calculating maximum likelihood (or other) estimates; a good survey of such methods is available in the monograph by Thisted (1988). We will describe just two methods here: the Newton-Raphson and EM algorithms.

The Newton-Raphson algorithm

The Newton-Raphson algorithm is a general purpose algorithm for finding the solution of a non-linear equation; it can also be generalized to finding the solution of a system of non-linear equations. The Newton-Raphson algorithm is natural in the context of computing maximum likelihood estimates as these estimates are often the solution of a system of equations (the likelihood equations).

Suppose that we want to find the solution to the equation $g(x_0) = 0$ where g is a differentiable function. Given a number x that is close to x_0, it follows from a Taylor series expansion around x that

$$0 = g(x_0) \approx g(x) + g'(x)(x_0 - x)$$

and solving for x_0, we get

$$x_0 \approx x - \frac{g(x)}{g'(x)}.$$

Thus given an estimate x_k, we can obtain a new estimate x_{k+1} by

$$x_{k+1} = x_k - \frac{g(x_k)}{g'(x_k)}$$

and this procedure can be iterated for $k = 1, 2, 3, \cdots$ until $|g(x_k)|$ (or $|g(x_k)/g'(x_k)|$) is sufficiently small.

The Newton-Raphson algorithm can be applied to the computation of maximum likelihood estimates (as well as other estimates defined as the solution to a single equation or system of equations). Suppose that the joint density or joint frequency function of $\boldsymbol{X} = (X_1, \cdots, X_n)$ is $f(\boldsymbol{x};\theta)$ and let $\mathcal{L}(\theta) = f(\boldsymbol{x};\theta)$ be the likelihood function based on $\boldsymbol{X} = \boldsymbol{x}$. Suppose that the maximum likelihood estimate $\widehat{\theta}$ satisfies $S(\widehat{\theta}) = 0$ where $S(\theta)$ is the derivative of the log-likelihood function, $\ln \mathcal{L}(\theta)$. ($S(\theta)$ is often called the score function.) Let $\widehat{\theta}^{(k)}$ be the estimate of θ after k iterations of the algorithm; then

$$\widehat{\theta}^{(k+1)} = \widehat{\theta}^{(k)} + \frac{S(\widehat{\theta}^{(k)})}{H(\widehat{\theta}^{(k)})}$$

where

$$H(\theta) = -\frac{\partial^2}{\partial\theta^2} \ln \mathcal{L}(\theta).$$

The procedure is then iterated until convergence (that is, when $|S(\widehat{\theta}^{(k)})|$ or the absolute difference between $\widehat{\theta}^{(k)}$ and $\widehat{\theta}^{(k+1)}$ is sufficiently small).

In order to use the Newton-Raphson algorithm, one needs an initial estimate of θ, $\widehat{\theta}^{(0)}$. In fact, in some cases, this initial estimate is critical as the algorithm will not always converge for a given $\widehat{\theta}^{(}0)$. It is also possible that there may be several solutions to the equation $S(\widehat{\theta}) = 0$, each solution corresponding to either a local maximum, local minimum, or "saddle-point" of the log-likelihood function; thus it is possible that the sequence of estimates $\{\widehat{\theta}^{(k)}\}$ will converge to the "wrong" solution of $S(\widehat{\theta}) = 0$. (The convergence question is much more important when estimating three or more parameters; with one or two unknown parameters, it is possible to plot the log-likelihood function and determine appropriate initial estimates from this plot.) If it is not clear that the algorithm will converge to the maximizer of the likelihood function then several different initial estimates can be tried. Alternatively, one can use another estimate (for example, a method of moments estimate) as an initial estimate for the algorithm. In fact, if $\widehat{\theta}^{(0)}$ is a sufficiently good estimator of θ then the "one-step" estimator

$$\widehat{\theta}^{(1)} = \widehat{\theta}^{(0)} + \frac{S(\widehat{\theta}^{(0)})}{H(\widehat{\theta}^{(0)})}$$

has virtually the same large sample properties as the MLE. More precisely, if $\widehat{\theta}_n^{(0)}$ is based on n observations and $\sqrt{n}(\widehat{\theta}_n^{(0)} - \theta)$ converges in distribution then typically $\sqrt{n}(\widehat{\theta}_n^{(1)} - \widehat{\theta}_n) \to_p 0$ where $\widehat{\theta}_n$ is the MLE and $\widehat{\theta}_n^{(1)}$ is the "one-step" estimator using $\widehat{\theta}_n^{(0)}$ as a starting value. Thus we could take $\widehat{\theta}_n^{(0)}$ to be a method of moments estimator (or some other substitution principle estimator) of θ.

EXAMPLE 5.21: Suppose that $X_1, \cdots, X_n$ are i.i.d. Cauchy random variables with density function

$$f(x; \theta) = \frac{1}{\pi(1 + (x - \theta)^2}.$$

Given outcomes $x_1, \cdots, x_n$, the log-likelihood function is

$$\ln \mathcal{L}(\theta) = -\sum_{i=1}^{n} \ln[1 + (x_i - \theta)^2] - n \ln(\pi).$$

The maximum likelihood estimate $\widehat{\theta}$ satisfies the equation

$$S(\widehat{\theta}) = \sum_{i=1}^{n} \frac{2(x_i - \widehat{\theta})}{1 + (x_i - \widehat{\theta})^2} = 0.$$

Note that $S(\theta)$ is not monotone in θ and hence the equation $S(\widehat{\theta}) = 0$ may have more than one solution for a given sample $x_1, \cdots, x_n$. To illustrate the Newton-Raphson algorithm, we take a sample of 100 observations with $\theta = 10$. Plots of the log-likelihood $\ln \mathcal{L}(\theta)$ and the score function $S(\theta)$ are given in Figures 5.1 and 5.2.

To find $\widehat{\theta}^{(0)}$, we must first find a reasonable initial estimate of θ. Since the density of the X_i's is symmetric around θ, it makes sense to consider either the sample mean or sample median as an initial estimate; however, $E(X_i)$ is not well-defined so that the sample mean need not be a good estimate of θ. Thus we will use the sample median as an initial estimate. Successive values of $\widehat{\theta}^{(k)}$ are defined by

$$\widehat{\theta}^{(k+1)} = \widehat{\theta}^{(k)} + \frac{S(\widehat{\theta}^{(k)})}{H(\widehat{\theta}^{(k)})}$$

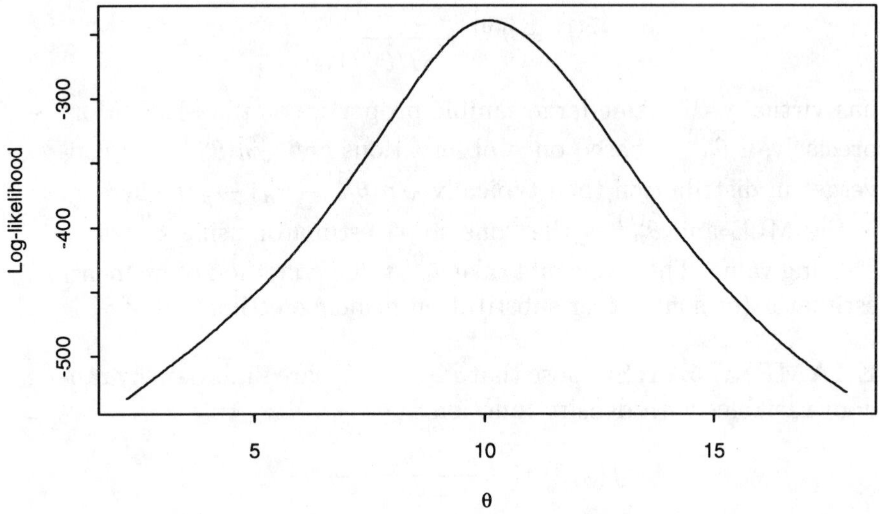

Figure 5.1 *Log-likelihood function for the Cauchy data in Example 5.21.*

Table 5.4 *Iterates of the Newton-Raphson algorithm for Cauchy data in Example 5.21.*

k	$\widehat{\theta}^{(k)}$	$\ln \mathcal{L}(\widehat{\theta}^{(k)})$
0	10.04490	-239.6569
1	10.06934	-239.6433
2	10.06947	-239.6433
3	10.06947	-239.6433

where

$$H(\theta) = 2\sum_{i=1}^{n} \frac{1 - (x_i - \theta)^2}{(1 + (x_i - \theta)^2)^2}.$$

The values of $\widehat{\theta}^{(k)}$ and $\ln \mathcal{L}(\widehat{\theta}^{(k)})$ are given in Table 5.4. The choice of $\widehat{\theta}^{(0)}$ is crucial here; for example, if $\widehat{\theta}^{(0)}$ is taken to be less than 8.74 or greater than 11.86 then the sequence $\{\widehat{\theta}^{(k)}\}$ will not con-

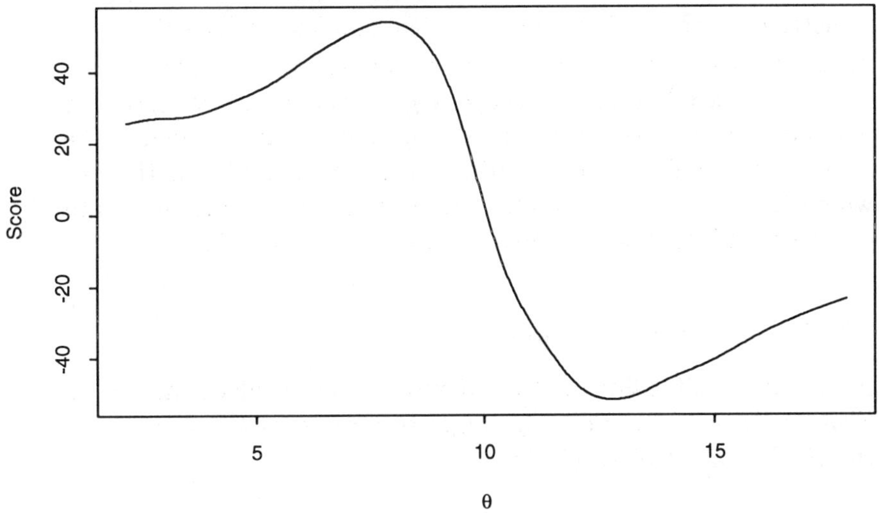

Figure 5.2 *Score function for the Cauchy data in Example 5.21.*

verge (at least to the maximum likelihood estimate). ◇

The Newton-Raphson algorithm can also be extended to the multi-parameter case. Let $\boldsymbol{\theta} = (\theta_1, \cdots, \theta_p)^T$ and suppose that the maximum likelihood estimate of $\boldsymbol{\theta}$ is given by the equation $S(\widehat{\boldsymbol{\theta}}) = 0$ where

$$S(\boldsymbol{\theta}) = \left(\frac{\partial}{\partial\theta_1}\ln\mathcal{L}(\boldsymbol{\theta}), \cdots, \frac{\partial}{\partial\theta_p}\ln\mathcal{L}(\boldsymbol{\theta})\right)^T.$$

Then given $\widehat{\boldsymbol{\theta}}^{(k)}$, we define $\widehat{\boldsymbol{\theta}}^{(k+1)}$ by

$$\widehat{\boldsymbol{\theta}}^{(k+1)} = \widehat{\boldsymbol{\theta}}^{(k)} + \left[H\left(\widehat{\boldsymbol{\theta}}^{(k)}\right)\right]^{-1} S\left(\widehat{\boldsymbol{\theta}}^{(k)}\right)$$

where $H(\boldsymbol{\theta})$ is the matrix of negative second partial derivatives of $\ln\mathcal{L}(\boldsymbol{\theta})$; the (i,j) element of H is given by

$$H''_{ij}(\boldsymbol{\theta}) = -\frac{\partial^2}{\partial\theta_i\partial\theta_j}\ln\mathcal{L}(\boldsymbol{\theta}).$$

($H(\widehat{\boldsymbol{\theta}})$ is sometimes called the observed Fisher information matrix.)

Estimates of standard error for the parameter estimates can be obtained from the Newton-Raphson algorithm. In the single parameter case, the variance of $\widehat{\theta}$ can be estimated by $[H(\widehat{\theta})]^{-1}$ while in the multi-parameter case, the variance-covariance matrix of $\widehat{\theta}$ can be estimated by the matrix $[H(\widehat{\boldsymbol{\theta}})]^{-1}$. These variance estimates assume that the model is correctly specified (that is, the data belong to the family of distributions being estimated). In the case where the model is incorrectly specified, these variance estimates can be modified to make them valid as in Section 4.5.

The Fisher scoring algorithm

A simple modification of the Newton-Raphson algorithm is the Fisher scoring algorithm. This algorithm replaces H by H^* where the (i, j) element of $H^*(\boldsymbol{\theta})$ is

$$H_{ij}^*(\boldsymbol{\theta}) = E_\theta[H_{ij}(\boldsymbol{\theta})] = -E_\theta\left[\frac{\partial^2}{\partial\theta_i\partial\theta_j}\ln f(\boldsymbol{X};\boldsymbol{\theta})\right];$$

the expected value above is computed assuming that $\boldsymbol{\theta}$ is the true value of the parameter. (H is the observed Fisher information matrix while H^* is the expected Fisher information matrix.) Now if $\widehat{\boldsymbol{\theta}}^{(k)}$ is the estimate of $\boldsymbol{\theta}$ after k iterations, we define $\widehat{\boldsymbol{\theta}}^{(k+1)}$ by

$$\widehat{\boldsymbol{\theta}}^{(k+1)} = \widehat{\boldsymbol{\theta}}^{(k)} + \left[H\left(\widehat{\boldsymbol{\theta}}^{(k)}\right)\right]^{-1} S\left(\widehat{\boldsymbol{\theta}}^{(k)}\right).$$

The important distinction between the Newton-Raphson and Fisher scoring algorithms is the fact that $H^*(\boldsymbol{\theta})$ depends on the observed value of $\boldsymbol{X}$, $\boldsymbol{x}$, only through the value of $\boldsymbol{\theta}$ while $H(\boldsymbol{\theta})$ depends, in general, on both $\boldsymbol{\theta}$ and $\boldsymbol{x}$.

EXAMPLE 5.22: As in Example 5.21, let $X_1, \cdots, X_n$ be i.i.d. Cauchy random variables. From before, we have

$$H(\theta) = 2\sum_{i=1}^{n}\frac{1-(x_i-\theta)^2}{(1+(x_i-\theta)^2)^2}$$

and so

$$H^*(\theta) = \frac{n}{\pi}\int_{-\infty}^{\infty}\frac{1-(x-\theta)^2}{(1+(x-\theta)^2)^3}\,dx = \frac{n}{2}.$$

Table 5.5 *Iterates of the Fisher scoring algorithm for the Cauchy data in Example 5.21.*

k	$\widehat{\theta}^{(k)}$	$\ln \mathcal{L}(\widehat{\theta}^{(k)})$
0	10.04490	-239.6569
1	10.06710	-239.6434
2	10.06923	-239.6433
3	10.06945	-239.6433
4	10.06947	-239.6433

Hence the Fisher scoring algorithm is

$$\widehat{\theta}^{(k+1)} = \widehat{\theta}^{(k)} + \frac{4}{n} \sum_{i=1}^{n} \frac{x_i - \widehat{\theta}^{(k)}}{1 + (x_i - \widehat{\theta}^{(k)})^2}.$$

Values for $\widehat{\theta}^{(k)}$ and $\ln \mathcal{L}(\widehat{\theta}^{(k)})$ are given in Table 5.5; as before, $\widehat{\theta}^{(0)}$ is taken to be the sample median. One advantage of the Fisher scoring algorithm in this particular example is the fact that $\widehat{\theta}^{(k)}$ converges for a much wider interval of starting values $\widehat{\theta}^{(0)}$. ◇

The differences between the Newton-Raphson and Fisher scoring algorithms are subtle but still important. Although it is difficult to generalize too much, we can make the following observations:

- The convergence of the Newton-Raphson algorithm is often faster when both algorithms converge.
- The radius of convergence for the Fisher scoring algorithm is often larger; this suggests that the choice of an initial estimate is less important for the Fisher scoring algorithm.

In the case of exponential family models, the Newton-Raphson and Fisher scoring algorithms are almost equivalent; if $(\theta_1, \cdots, \theta_p)$ are the natural parameters then $H(\boldsymbol{\theta}) = H^*(\boldsymbol{\theta})$ and so the two algorithms are identical.

The EM algorithm

The EM algorithm provides a useful framework for computing maximum likelihood estimates in so-called incomplete data situations, for example, when data are missing or are not observed exactly. It was formalized in a paper by Dempster, Laird and Rubin (1977) although special cases of the algorithm had been used in specific problems for many years before 1977. Today, the EM algorithm is widely used in a variety of problems. The following example illustrates an incomplete data problem where a continuous random variable is not observed exactly but rather in a given interval.

EXAMPLE 5.23: Suppose that $X_1, \cdots, X_n$ are i.i.d. Exponential random variables with parameter λ. Given outcomes $x_1, \cdots, x_n$ of $X_1, \cdots, X_n$, the maximum likelihood estimate of θ is $1/\bar{x}$, where $\bar{x}$ is the average of the x_i's. However, suppose that rather than observing X_i exactly, we observe lower and upper bounds, u_i and v_i, for X_i so that $u_i < X_i < v_i$. Given $(u_1, v_1), \cdots, (u_n, v_n)$, the likelihood function of λ is

$$\begin{aligned} \mathcal{L}(\lambda) &= \prod_{i=1}^{n} P_\lambda(u_i \le X_i \le v_i) \\ &= \prod_{i=1}^{n} (\exp(-\lambda u_i) - \exp(-\lambda v_i)). \end{aligned}$$

This is simple example of interval censored data. ◇

The main idea behind the EM algorithm is to construct an estimate of the complete data likelihood, which can then be maximized using traditional numerical methods (such as the Newton-Raphson algorithm). Thus the EM algorithm is not really a numerical algorithm but instead a general purpose procedure for computing parameter estimates from incomplete data by iteratively computing parameter maximum likelihood estimates based on the complete data likelihood function; implicit in the use of the EM algorithm is the assumption that complete data maximum likelihood estimates can be readily computed and the incomplete data likelihood function is difficult to work with. The EM algorithm iterates 2 steps (called the E and M steps) to convergence. Before discussing the algorithm in any depth, we will illustrate it using the example given above.

EXAMPLE 5.24: In Example 5.23, the log-likelihood function of λ based on the exact outcomes $x_1, \cdots, x_n$ is

$$\ln \mathcal{L}_c(\lambda) = n \ln(\lambda) - \lambda \sum_{i=1}^{n} x_i.$$

The first step of the EM algorithm involves finding an estimate of $\ln \mathcal{L}_c(\lambda)$ given the incomplete data $(u_1, v_1), \cdots, (u_n, v_n)$ and assuming that the true parameter value is $\widehat{\lambda}^{(k)}$ where $\widehat{\lambda}^{(k)}$ is the estimate of λ after k iterations of the algorithm; this is done by finding the expected value of

$$n \ln(\lambda) - \lambda \sum_{i=1}^{n} X_i$$

given $u_1 < X_1 < v_1, \cdots, u_n < X_n < v_n$ and assuming that the value of the parameter is $\widehat{\lambda}^{(k)}$. By simple integration, it follows that

$$E(X_i | u_i < X_i < v_i; \lambda) = \frac{1}{\lambda} + \frac{u_i \exp(-\lambda u_i) - v_i \exp(-\lambda v_i)}{\exp(-\lambda u_i) - \exp(-\lambda v_i)}$$

and so to complete the E step of the algorithm, we substitute

$$\widehat{x}_i^{(k)} = \frac{1}{\widehat{\lambda}^{(k)}} + \frac{u_i \exp(-\widehat{\lambda}^{(k)} u_i) - v_i \exp(-\widehat{\lambda}^{(k)} v_i)}{\exp(-\widehat{\lambda}^{(k)} u_i) - \exp(-\widehat{\lambda}^{(k)} v_i)}$$

for x_i in $\ln \mathcal{L}_c$. The M step of the algorithm involves maximizing this new log-likelihood; it is maximized at

$$\widehat{\lambda}^{(k+1)} = \left(\frac{1}{n} \sum_{i=1}^{n} \widehat{x}_i^{(k)} \right)^{-1}.$$

The E and M steps are then iterated until convergence. ◇

We can now discuss the EM algorithm in more or less complete generality. Suppose that $\boldsymbol{X} = (X_1, \cdots, X_n)$ are continuous random variables with joint density function $f_X(\boldsymbol{x}; \theta)$ (where θ can be real- or vector-valued) and let $\boldsymbol{Y} = (Y_1, \cdots, Y_m)$ be random variables such that $Y_i = g_i(\boldsymbol{X})$ where $g_1, \cdots, g_m$ are known functions. The mapping that produces the Y_i's from the X_i's is typically not one-to-one; this means that any given outcome of the Y_i's, $y_1, \cdots, y_m$ can be produced by more than one outcome of the X_i's. We thus think of the X_i's as being the complete data and the Y_i's as being the incomplete data. The joint density of $\boldsymbol{Y}$ is

$$f_Y(\boldsymbol{y}; \theta) = \int_{A(y)} f_X(\boldsymbol{x}; \theta)\, d\boldsymbol{x}$$

where

$$A(y) = \{\boldsymbol{x} : y_i = g_i(\boldsymbol{x}) \text{ for } i = 1, \cdots m\}.$$

If $X_1, \cdots, X_n$ are discrete, the integral sign is replaced by a summation over $A(y)$.

If we observed $\boldsymbol{X} = \boldsymbol{x}$, we could estimate θ by maximizing the complete data likelihood function $\mathcal{L}_c(\theta) = f_X(\boldsymbol{x};\theta)$. Instead we observe $\boldsymbol{Y} = \boldsymbol{y}$. In general, the joint density function of $\boldsymbol{Y}$ (as given above) will be quite difficult to compute directly; hence, it will be difficult to compute maximum likelihood estimates from incomplete data using standard algorithms. However, it is often possible to impute (or "estimate") the values of the complete data using the observed (incomplete) data. Given a preliminary estimate of the parameter, the EM algorithm first constructs an estimate of the complete data likelihood function and then maximizes this likelihood to obtain a new parameter estimate; this two step procedure is then iterated until convergence.

- (E step) The E (for expectation) step of the EM algorithm involves finding an estimate of the likelihood function of θ for the complete data given the observed (or incomplete) data. Given an estimate of θ, $\widehat{\theta}^{(k)}$, after k iterations of the algorithm, we define

$$\ln \mathcal{L}^{(k)}(\theta) = E[\ln f_X(\boldsymbol{X};\theta)|\boldsymbol{Y} = \boldsymbol{y}; \widehat{\theta}^{(k)}]$$

where the expectation is taken assuming that the true parameter value is $\widehat{\theta}^{(k)}$.

- (M step) An updated estimate of θ is obtained in the M (maximization) step of the EM algorithm. The updated estimate, $\widehat{\theta}^{(k+1)}$, is chosen to maximize the estimate of the complete data log-likelihood function, $\ln \mathcal{L}^{(k)}(\theta)$, that was obtained in the E step. Computation of $\widehat{\theta}^{(k+1)}$ will often involve the use of some numerical method such as the Newton-Raphson algorithm; it is often useful to use $\widehat{\theta}^{(k)}$ as an initial estimate of $\widehat{\theta}^{(k+1)}$ in this case. Once $\widehat{\theta}^{(k+1)}$ is computed, we return to the E step.

The E and M steps are repeated until convergence.

In many situations (for example, exponential family models), $f_X(\boldsymbol{x};\theta)$ is a product of two non-constant factors, one of which does not depend on θ. We can then write

$$\ln f_X(\boldsymbol{x};\theta) = h_1(\boldsymbol{x};\theta) + h_2(\boldsymbol{x}).$$

In such cases, we can redefine $\ln \mathcal{L}^{(k)}(\theta)$ in the E step by

$$\ln \mathcal{L}^{(k)}(\theta) = E[h_1(\boldsymbol{X};\theta)|\boldsymbol{Y} = \boldsymbol{y}; \widehat{\theta}^{(k)}].$$

This modification of the algorithm can considerably simplify computations. The M step of the algorithm remains the same; as before, $\widehat{\theta}^{(k+1)}$ maximizes $\ln \mathcal{L}^{(k)}(\theta)$.

What can be said about the convergence properties of the EM algorithm? If $\mathcal{L}(\theta)$ is the likelihood function based on $y_1, \cdots, y_m$, it can be shown that

$$\ln \mathcal{L}(\widehat{\theta}^{(k+1)}) \geq \ln \mathcal{L}(\widehat{\theta}^{(k)})$$

for any initial estimate $\widehat{\theta}^{(0)}$. Unfortunately, this does not necessarily imply that the sequence $\{\widehat{\theta}^{(k)}\}$ converges to the maximum likelihood estimate (namely the maximizer of $\mathcal{L}(\theta)$). This suggests that the choice of initial estimate can be very important as it is for the Newton-Raphson algorithm. In practice, it is a good idea to use a variety of initial estimates. If $\widehat{\theta}^{(k+1)}$ is computed numerically in the M step of the algorithm, it is important to ensure that an appropriate initial estimate is chosen to guarantee that $\widehat{\theta}^{(k+1)}$ will maximize $\mathcal{L}^{(k)}(\theta)$. A rigorous treatment of the convergence properties of the EM algorithm is given by Wu (1983).

EXAMPLE 5.25: Suppose that $X_1, \cdots, X_n$ are i.i.d. random variables whose distribution is a mixture of two Poisson distributions:

$$f_X(x; \lambda, \theta) = \theta \frac{\exp(-\lambda)\lambda^x}{x!} + (1-\theta)\frac{\exp(-\mu)\mu^x}{x!}$$

for $x = 1, 2, \cdots$. The frequency function given above arises if we observe a Poisson random variable with mean λ with probability θ and a Poisson distribution with mean μ with probability $1-\theta$. (Note that when $\mu = \lambda$, the model becomes a i.i.d. Poisson model and the parametrization is not identifiable.

The log-likelihood function for θ, λ, and μ is given by

$$\ln \mathcal{L}(\theta, \lambda, \mu) = \sum_{i=1}^{n} \ln \left(\theta \frac{\exp(-\lambda)\lambda^{x_i}}{x_i!} + (1-\theta)\frac{\exp(-\mu)\mu^{x_i}}{x_i!} \right).$$

Given $x_1, \cdots, x_n$, it is possible to estimate the parameters using the Newton-Raphson algorithm; however, the implementation of the Newton-Raphson algorithm in this example is somewhat difficult and so the EM algorithm is a natural alternative.

To use the EM algorithm, we must find a suitable "complete" data problem. Assume that $(X_1, Y_1), \cdots, (X_n, Y_n)$ are i.i.d. pairs of random variables with

$$\begin{aligned} P(Y_i = y) &= \theta^y (1-\theta)^{1-y} \quad \text{for } y = 0, 1 \\ P(X_i = x | Y = 0) &= \frac{\exp(-\mu)\mu^x}{x!} \quad \text{for } x = 0, 1, 2, \cdots \\ P(X_i = x | Y = 1) &= \frac{\exp(-\lambda)\lambda^x}{x!} \quad \text{for } x = 0, 1, 2, \cdots . \end{aligned}$$

The complete data likelihood based on $(x_1, y_1), \cdots, (x_n, y_n)$ is then

$$\begin{aligned} \ln \mathcal{L}_c(\theta, \lambda, \mu) &= \sum_{i=1}^n y_i \left[\ln(\theta) + x_i \ln(\lambda) - \lambda\right] \\ &\quad + \sum_{i=1}^n (1 - y_i) \left[\ln(1-\theta) + x_i \ln(\mu) - \mu\right] \\ &\quad - \sum_{i=1}^n \ln(x_i!). \end{aligned}$$

We must now find the expected value of the complete data log-likelihood given the observations $x_1, \cdots, x_n$; to do this, it is sufficient to find the expected value of Y_i given $X_i = x_i$ for any values of the parameters θ, λ, and μ; an easy computation using Bayes' Theorem yields

$$E(Y_i | X_i = x; \theta, \lambda, \mu) = \frac{\theta \exp(-\lambda)\lambda^x}{\theta \exp(-\lambda)\lambda^x + (1-\theta)\exp(-\mu)\mu^x}.$$

Thus given estimates $\widehat{\theta}^{(k)}$, $\widehat{\lambda}^{(k)}$, and $\widehat{\mu}^{(k)}$, we obtain $\widehat{\theta}^{(k+1)}$, $\widehat{\lambda}^{(k+1)}$, and $\widehat{\mu}^{(k+1)}$ by maximizing

$$\begin{aligned} \ln \mathcal{L}^{(k)}(\theta, \lambda, \mu) &= \sum_{i=1}^n \widehat{y}_i^{(k)} \left[\ln(\theta) + x_i \ln(\lambda) - \lambda\right] \\ &\quad + \sum_{i=1}^n (1 - \widehat{y}_i^{(k)}) \left[\ln(1-\theta) + x_i \ln(\mu) - \mu\right] \end{aligned}$$

where

$$\widehat{y}_i^{(k)} = E\left(Y_i | X_i = x; \widehat{\theta}^{(k)}, \widehat{\lambda}^{(k)}, \widehat{\mu}^{(k)}\right).$$

It is easy to see that

$$\widehat{\theta}^{(k+1)} = \frac{1}{n} \sum_{i=1}^n \widehat{y}_i^{(k)}$$

Table 5.6 *Frequency distribution of goals in First Division matches.*

Goals	0	1	2	3	4	5	6	7
Frequency	252	344	180	104	28	11	2	3

$$\widehat{\lambda}^{(k+1)} = \frac{\sum_{i=1}^n x_i \widehat{y}_i^{(k)}}{\sum_{i=1}^n \widehat{y}_i^{(k)}}$$

$$\widehat{\mu}^{(k+1)} = \frac{\sum_{i=1}^n x_i (1 - \widehat{y}_i^{(k)})}{\sum_{i=1}^n (1 - \widehat{y}_i^{(k)})}.$$

To illustrate the convergence EM algorithm in this example, we consider the number of goals scored during a game by teams in the First Division of the English Football League during the 1978-79 season; at that time, there were 22 teams with each team playing 42 games for a total of 924 observations. The data are summarized in Table 5.6. The EM algorithm is very slow to converge in this example even when the initial estimates are taken to be close to the maximizing values. For example, starting from initial estimates $\widehat{\theta}^{(0)} = 0.95$, $\widehat{\lambda}^{(0)} = 1.23$, and $\widehat{\mu}^{(k)} = 3.04$, the EM algorithm described above takes about 300 iterations to converge (to three decimal places) to the maximum likelihood estimates $\widehat{\theta} = 0.954$, $\widehat{\lambda} = 1.232$, and $\widehat{\mu} = 3.043$. The Newton-Raphson algorithm can, of course, also be applied; its performance in this example is very erratic. For example, using the same starting values as above, the Newton-Raphson algorithm actually diverges. ◇

Comparing the Newton-Raphson and EM algorithms

Frequently, it is feasible to use both the Newton-Raphson and EM algorithms to compute maximum likelihood estimates. This raises the question of which algorithm is best to use. There are two compelling theoretical reasons to prefer the Newton-Raphson algorithm:

- standard error estimates are a by-product of the Newton-Raphson algorithm and not of the EM algorithm;
- the Newton-Raphson algorithm has a faster rate of convergence

in the sense that $\widehat{\theta}^{(k)} \to \widehat{\theta}$ more quickly when $\widehat{\theta}^{(k)}$ is in a neighbourhood of $\widehat{\theta}$.

Nonetheless, there are also many reasons to prefer the EM algorithm, perhaps the most important being ease of implementation. For example, the EM algorithm often leads to a considerable savings in programming time that far outweigh the savings in computer time incurred by using the Newton-Raphson or other more sophisticated algorithm. Moreover, in many problems, the number of iterations to convergence for the EM algorithm can be comparable to the Newton-Raphson algorithm; much depends on how close the complete data problem is to the incomplete data problem. For example, in Example 5.24, there is virtually no difference in the convergence rate between the EM and Newton-Raphson algorithm.

5.8 Bayesian estimation

To this point, we have regarded a parameter θ as a fixed quantity whose value is unknown and used the data to estimate (or make inference about) its value. In essence, we treat the observable quantities (the data) as outcomes of random variables and the unobservable quantity (the parameter) as a constant. The Bayesian approach describes the uncertainty in θ with a probability distribution on the parameter space Θ. This approach works by specifying a distribution for θ prior to observing the data and considering $f(\boldsymbol{x};\theta)$ as a "conditional" density or frequency function given θ; the observed data is then used to update this distribution. The probabilities placed on different subsets of Θ are generally not interpreted as "long-run frequencies" but instead are usually subjective probabilities that reflect "degrees of belief" about θ lying in the various subsets of Θ.

Before we go into the specifics of the Bayesian approach, we will discuss some of its pros and cons. First, if we accept the notion of describing uncertainty in the parameter by means of a probability distribution, the Bayesian approach yields a unified procedure for solving practically any problem in statistical inference. Second, mathematical arguments can be used to show that the only coherent systems for describing uncertainty are those based on probability distributions. A common complaint with Bayesian inference is that it is not "objective" in the sense that estimates depend on the prior distribution that is specified by the scientist or investigator (perhaps in consultation with someone else). Therefore, the possibility exists for the unscrupulous scientist to influence the results by choosing the appropriate prior. The counter-argument to this is that most scientific inference is very subjective; for example, scientists constantly make subjective judgements about how to carry out experiments. Moreover, much of non-Bayesian inference is also to some extent subjective. If a scientist or investigator is in a good position to specify a prior distribution then the Bayesian approach to estimation (and inference) can be extremely useful. However, in many cases, there is no obvious, realistic prior distribution available; in such cases, the usefulness of the Bayesian approach is questionable although some remedies do exist.

We will now elaborate on the implementation of the Bayesian approach. Suppose that $\boldsymbol{X} = (X_1, \cdots, X_n)$ are random variables with joint density or frequency function $f(\boldsymbol{x}; \theta)$ where θ is a real-valued parameter lying in Θ, which we will assume to an open set. Let $\Pi(\theta)$ be a distribution function defined on Θ and suppose that $\pi(\theta)$ is the corresponding density function; $\pi(\theta)$ is called the prior density for θ and describes the uncertainty in the value of the parameter θ prior to observing $\boldsymbol{X}$. We define the posterior density of θ given $\boldsymbol{X} = \boldsymbol{x}$ as follows:

$$\pi(\theta|\boldsymbol{x}) = \frac{f(\boldsymbol{x}; \theta)\pi(\theta)}{\int_\Theta f(\boldsymbol{x}; t)\pi(t)\,dt}.$$

Note that the denominator of the right-hand-side above does not depend on θ and so

$$\pi(\theta|\boldsymbol{x}) \propto f(\boldsymbol{x}; \theta)\pi(\theta) = \mathcal{L}(\theta)\pi(\theta)$$

where $\mathcal{L}(\theta)$ is the likelihood function based on $\boldsymbol{x}$.

It is very easy to generalize posterior distributions to the cases where θ is vector-valued or is discrete-valued. If $\boldsymbol{\theta} = (\theta_1, \cdots, \theta_p)$ is vector-valued, the (joint) posterior density of θ is

$$\pi(\boldsymbol{\theta}|\boldsymbol{x}) = \frac{f(\boldsymbol{x}; \boldsymbol{\theta})\pi(\boldsymbol{\theta})}{\int \cdots \int_\Theta f(\boldsymbol{x}; \boldsymbol{t})\pi(\boldsymbol{t})\,d\boldsymbol{t}}$$

where $\pi(\boldsymbol{\theta})$ is the joint prior density of $\boldsymbol{\theta}$. The marginal posterior density of any single parameter (or the joint posterior density of a collection of parameters) in $\boldsymbol{\theta}$ can be obtained by integrating the joint posterior density over all the other parameters. If θ is discrete-valued and $\pi(\theta)$ is the prior frequency function for θ then the posterior frequency function is given by

$$\pi(\theta|\boldsymbol{x}) = \frac{f(\boldsymbol{x};\theta)\pi(\theta)}{\sum_t f(\boldsymbol{x};t)\pi(t)}.$$

Again in both these cases, we have that

$$\pi(\theta|\boldsymbol{x}) \propto \pi(\theta)\mathcal{L}(\theta)$$

where $\mathcal{L}(\theta)$ is the likelihood function based on $\boldsymbol{x}$.

The posterior distribution describes the uncertainty in the value of θ after observing the data, taking into account the prior distribution. This brings up an important distinction between Bayesian and Frequentist statistics. In the Bayesian approach, all of the information about the uncertainty about θ is contained in the posterior distribution. More importantly, inference for θ depends only on the observed data (through the likelihood function) as well as the prior distribution. In contrast, the Frequentist approach often measures uncertainty (for example, by the standard error of a point estimator) by averaging over all possible (but unobserved) samples. Since the posterior distribution depends on the data only through the likelihood function, it follows that all Bayesian procedures satisfy the likelihood principle.

EXAMPLE 5.26: Suppose that $X_1, \cdots, X_n$ are i.i.d. Bernoulli random variables with unknown parameter θ and assume the following Beta prior density for θ

$$\pi(\theta) = \frac{\Gamma(\alpha+\beta)}{\Gamma(\alpha)\Gamma(\beta)}\theta^{\alpha-1}(1-\theta)^{\beta-1} \quad \text{for } 0 < \theta < 1$$

where α and β are specified. Then given $X_1 = x_1, \cdots, X_n = x_n$, the posterior density for θ is

$$\begin{aligned}
\pi(\theta|\boldsymbol{x}) &= \frac{f(x_1,\cdots,x_n;\theta)\pi(\theta)}{\int_0^1 f(\boldsymbol{x};t)\pi(t)\,dt} \\
&= \frac{\theta^{y+\alpha-1}(1-\theta)^{n-y+\beta-1}}{\int_0^1 t^{y+\alpha-1}(1-t)^{n-y+\beta-1}\,dt} \\
&\qquad \text{(where } y = x_1 + \cdots + x_n\text{)} \\
&= \frac{\Gamma(n+\alpha+\beta)}{\Gamma(y+\alpha)\Gamma(n-y+\beta)}\theta^{y+\alpha-1}(1-\theta)^{n-y+\beta-1}.
\end{aligned}$$

Note that the posterior distribution is also a Beta distribution with data dependent parameters. ◇

EXAMPLE 5.27: Suppose that $X_1, \cdots, X_n$ are i.i.d. Exponential random variables with parameter λ where the prior density for λ is also an Exponential distribution with known parameter α. To complicate matters somewhat, we will assume that only the r smallest of $X_1, \cdots, X_n$ are observed. Letting $Y_i = X_{(i)}$ for $i = 1, \cdots, r$, the joint density of $Y_1, \cdots, Y_r$ is

$$f(\boldsymbol{y}; \lambda) = \lambda^r \exp\left[-\lambda\left(\sum_{i=1}^{r} y_i + (n-r)y_r\right)\right].$$

Given $Y_1 = y_1, \cdots, Y_r = y_r$, the posterior density of λ is

$$\pi(\lambda|\boldsymbol{y}) = [K(\boldsymbol{y}, \alpha)]^{-1} \alpha\lambda^r \exp\left[-\lambda\left(\sum_{i=1}^{r} y_i + (n-r)y_r + \alpha\right)\right]$$

where

$$\begin{aligned} K(\boldsymbol{y}, \alpha) &= \int_0^\infty \alpha\lambda^r \exp\left[-\lambda\left(\sum_{i=1}^{r} y_i + (n-r)y_r + \alpha\right)\right] d\lambda \\ &= \frac{\alpha r!}{\left(\sum_{i=1}^{r} y_i + (n-r)y_r + \alpha\right)^{r+1}}. \end{aligned}$$

It is easy to see that the posterior distribution is a Gamma distribution with shape parameter $r+1$ and scale parameter $\sum_{i=1}^{r} y_i + (n-r)y_r + \alpha$. ◇

In the previous two examples, the prior distributions themselves depend on parameters (α and β in Example 5.26, α in Example 5.27); these parameters are often called hyperparameters. Orthodox Bayesian statisticians maintain that the values of hyperparameters must be specified independently of the observed data so the prior distribution reflects true *a priori* beliefs. However, if we are not willing *a priori* to specify the values of the hyperparameters, it is possible to use the so-called empirical Bayes approach in which hyperparameter values are estimated from the observed data and then substituted into the posterior distribution. More precisely, if $\pi(\theta; \alpha)$ is a prior density for θ depending on a hyperparameter α (possibly vector-valued), we can define the "marginal" joint density or frequency function

$$g(\boldsymbol{x}; \alpha) = \int_\Theta f(\boldsymbol{x}; \theta)\pi(\theta; \alpha)\, d\theta.$$

The hyperparameter α can then be estimated by maximum likelihood or substitution principle estimation assuming that the joint density or frequency function of the data is $g(\boldsymbol{x}; \alpha)$. Empirical Bayes methods are often very useful although despite their name, they are essentially non-Bayesian since the prior distribution is estimated from the data.

EXAMPLE 5.28: Suppose that $X_1, \cdots, X_n$ are i.i.d. Poisson random variables with mean λ where the prior distribution for λ is given by

$$\pi(\lambda; \alpha, \beta) = \frac{\alpha^\beta}{\Gamma(\beta)} \lambda^{\beta-1} \exp(-\alpha\lambda) \quad \text{for } \lambda > 0$$

with α and β the hyperparameters. Setting $y = x_1 + \cdots + x_n$, the "marginal" joint frequency function is given by

$$\begin{aligned} g(\boldsymbol{x}; \alpha, \beta) &= \int_0^\infty \frac{\exp(-n\lambda)\lambda^y}{x_1! \cdots x_n!} \pi(\theta; \alpha, \beta)\, d\lambda \\ &= \left(\frac{\alpha}{n+\alpha}\right)^\beta \frac{\Gamma(y+\beta)}{\Gamma(\alpha) x_1! \cdots x_n!} \left(\frac{1}{n+\alpha}\right)^y. \end{aligned}$$

Estimates for α and β can be obtained by maximum likelihood using $g(\boldsymbol{x}; \alpha, \beta)$ as the likelihood function. Alternatively, method of moments estimates can be used. For example, if we think of λ as a random variable (with density $\pi(\lambda; \alpha, \beta)$, we have that $E(X_i|\lambda) = \lambda$ and $\text{Var}(X_i|\lambda) = \lambda$. Then

$$\begin{aligned} E(X_i) &= E[E(X_i|\lambda)] = \frac{\alpha}{\beta} \\ \text{and } \text{Var}(X_i) &= E[\text{Var}(X_i|\lambda)] + \text{Var}[E(X_i|\lambda)] \\ &= \frac{\alpha}{\beta} + \frac{\alpha}{\beta^2} \end{aligned}$$

using the fact that the mean and variance of the Gamma prior are α/β and α/β^2, respectively. Estimates $\widehat{\alpha}$ and $\widehat{\beta}$ can be obtained by setting the sample mean and variance of the x_i's equal to $\widehat{\alpha}/\widehat{\beta}$ and $\widehat{\alpha}/\widehat{\beta} + \widehat{\alpha}/\widehat{\beta}^2$ respectively. ◇

How do we go from the posterior density for θ to a point estimator for θ? A common practice is to take the mean of the posterior distribution (or the median) to be the Bayesian point estimator. Another commonly used estimator is the posterior mode, that is, the value of θ that maximizes $\pi(\theta|\boldsymbol{x})$. In some sense, Bayesian point estimators are really just descriptive measures of the posterior distribution as it is the posterior distribution itself that is of primary importance. However, one can also study Bayesian point estimators from a Frequentist point of view; in fact, Bayesian point estimators have very attractive Frequentist properties and are often superior to estimators obtained by Frequentist principles.

EXAMPLE 5.29: In Example 5.26, the posterior distribution for θ is

$$\pi(\theta|\boldsymbol{x}) = \frac{\Gamma(n+\alpha+\beta)}{\Gamma(y+\alpha)\Gamma(n-y+\beta)}\theta^{y+\alpha-1}(1-\theta)^{n-y+\beta-1}$$

where $y = x_1 + \cdots + x_n$. The posterior mean is

$$\widehat{\theta} = \int_0^1 \theta\,\pi(\theta|\boldsymbol{x})\,d\theta = \frac{y+\alpha}{n+\alpha+\beta}$$

while the posterior mode is

$$\widetilde{\theta} = \frac{y+\alpha-1}{n+\alpha+\beta-2}.$$

It is easy to see that when both y and n are large compared to α and β then both $\widehat{\theta}$ and $\widetilde{\theta}$ are approximately y/n (which is the maximum likelihood estimate of θ). ◇

EXAMPLE 5.30: In Example 5.27, the posterior distribution for λ is a Gamma distribution with shape parameter $r+1$ and scale parameter $\sum_{i=1}^r y_i + (n-r)y_r + \alpha$. The posterior mean is

$$\widehat{\theta} = \frac{r+1}{\sum_{i=1}^r y_i + (n-r)y_r + \alpha}$$

while the posterior mode is

$$\widetilde{\theta} = \frac{r}{\sum_{i=1}^r y_i + (n-r)y_r + \alpha}.$$

Note that the difference between the posterior mean and mode becomes smaller as n increases. ◇

Conjugate and ignorance priors

Although prior distributions are essentially arbitrary, it is often convenient to choose the prior distribution so that the posterior distribution is easily derivable. The classical examples of such prior distributions are conjugate families, which are typically parametric families of priors such that posterior distribution belongs the same parametric family. For example, suppose that the model for $(X_1, \cdots, X_n)$ is a one-parameter exponential family so that

$$f(\boldsymbol{x}; \theta) = \exp\left[c(\theta)T(\boldsymbol{x}) - d(\theta) + S(\boldsymbol{x})\right]$$

and suppose that the prior density for θ is of the form

$$\pi(\theta) = K(\alpha, \beta) \exp\left[\alpha c(\theta) - \beta d(\theta)\right]$$

for some α and β (where the $K(\alpha, \beta)$ is chosen so that $\pi(\theta)$ is a density). Then it is easy to see that

$$\begin{aligned} \pi(\theta|\boldsymbol{x}) &\propto \pi(\theta) f(\boldsymbol{x}; \theta) \\ &\propto \exp\left[(T(\boldsymbol{x}) + \alpha)c(\theta) - (\beta + 1)d(\theta)\right]. \end{aligned}$$

From this, we can generally deduce that

$$\pi(\theta|\boldsymbol{x}) = K(T(\boldsymbol{x}) + \alpha, \beta + 1) \exp\left[(T(\boldsymbol{x}) + \alpha)c(\theta) - (\beta + 1)d(\theta)\right],$$

which has exactly the same form as $\pi(\theta)$. The prior distributions $\{\pi(\theta)\}$ indexed by the hyperparameters α and β is a conjugate family of priors for the one-parameter exponential family.

EXAMPLE 5.31: Suppose that $X_1, \cdots, X_n$ are i.i.d. Geometric random variables with frequency function

$$f(x; \theta) = \theta(1 - \theta)^x$$

for $x = 0, 1, 2, \cdots$ where $0 < \theta < 1$. The joint frequency function of $X_1, \cdots, X_n$ is a one-parameter exponential family:

$$f(\boldsymbol{x}; \theta) = \exp\left[\ln(1 - \theta) \sum_{i=1}^{n} x_i + n \ln(\theta)\right].$$

This suggests that a conjugate family of priors for θ is given by

$$\pi(\theta) = K(\alpha, \beta) \exp\left(\alpha \ln(1 - \theta) + \beta \ln(\theta)\right) = K(\alpha, \beta)(1 - \theta)^\alpha \theta^\beta$$

for $0 < \theta < 1$. For $\alpha > 0$ and $\beta > 0$, $\pi(\theta)$ is simply a Beta density with

$$K(\alpha, \beta) = \frac{\Gamma(\alpha, \beta)}{\Gamma(\alpha)\Gamma(\beta))}.$$

Note that the same family is conjugate in Example 5.26. ◇

EXAMPLE 5.32: Suppose that $X_1, \cdots, X_n$ are i.i.d. Exponential random variables with parameter λ. Again we have a one-parameter exponential family whose joint density function is

$$f(\boldsymbol{x};\theta) = \exp\left[-\lambda \sum_{i=1}^{n} x_i + n\ln(\lambda)\right].$$

This suggests that a conjugate family of prior densities for λ is given by

$$\pi(\lambda) = K(\alpha,\beta)\exp\left(-\alpha\lambda + \beta\ln(\lambda)\right) = K(\alpha,\beta)\lambda^{\beta}\exp\left(-\alpha\lambda\right)$$

for $\lambda > 0$. For $\alpha > 0$ and $\beta > -1$, this is a Gamma density and $K(\alpha,\beta) = \alpha^{\beta+1}/\Gamma(\beta+1)$. Note that the Exponential prior density given in Example 5.27 is a special case. ◇

Conjugate prior distributions are often used simply for convenience rather than for their ability to accurately describe *a priori* beliefs. Basically, assuming a conjugate prior for θ greatly facilitates evaluation of the integral

$$\int_{\Theta} \pi(\theta) f(\boldsymbol{x};\theta)\, d\theta.$$

However, with the rapid increase in computational power over the past 20 years, the use of conjugate priors to simplify computation has become less important; numerical integration techniques can be used to evaluate posterior distributions with minimal difficulty. These numerical techniques are particularly important in multiparameter problems where useful conjugate families are not generally available. Monte Carlo integration (see Chapter 3) is often useful in this context; in recent years, techniques such as Gibbs sampling and related Markov chain Monte Carlo methods have been used effectively in problems with large numbers of parameters. A good reference is the monograph by Gilks *et al* (1996).

EXAMPLE 5.33: Suppose we want to evaluate the integral

$$K(\boldsymbol{x}) = \int_{\Theta} \pi(\theta) f(\boldsymbol{x};\theta)\, d\theta.$$

Let g be a density function on Θ; then

$$K(\boldsymbol{x}) = \int_{\Theta} \frac{\pi(\theta) f(\boldsymbol{x};\theta)}{g(\theta)} g(\theta)\, d\theta.$$

To estimate $K(\boldsymbol{x})$, let $T_1, \cdots, T_m$ be i.i.d. random variables with density g; for m sufficiently large, we will have (by the WLLN)

$$\widehat{K}(\boldsymbol{x}) = \frac{1}{m} \sum_{i=1}^{m} \frac{\pi(T_i) f(\boldsymbol{x}; T_i)}{g(T_i)} \approx K(\boldsymbol{x})$$

and $\widehat{K}(\boldsymbol{x})$ serves as estimate of $K(\boldsymbol{x})$. This is simply the importance sampling estimate of the integral considered in section 3.6. ◇

As stated above, Bayesian inference is often perceived as not objective. While this perception is perhaps undeserved, it is nonetheless desirable to find prior distributions that express an indifference among all values of the parameter space. In the case where the parameter space is discrete and finite, the solution is clear; if there are m elements in the parameter space, we can specify a prior distribution putting probability $1/m$ on each element. However, when the parameter space is not finite then some difficulties arise.

As above, we will concentrate on the case where the parameter space Θ is a subset of the real-line. Suppose first of all that Θ is the interval (a, b). Then a natural prior distribution is a Uniform distribution with density

$$\pi(\theta) = (b - a)^{-1} \quad \text{for } a < \theta < b.$$

However, expression of indifference by Uniform distributions is not invariant under transformations: the prior density of $g(\theta)$ will not itself be Uniform if $g(\cdot)$ is a non-linear function. If Θ is a infinite interval (for example, the real-line) then "proper" Uniform prior densities do not exist in the sense that $\int_\Theta k \, d\theta = \infty$ for any $k > 0$. Nonetheless, these "improper" prior densities will often yield valid posterior densities; the posterior density is

$$\pi(\theta|\boldsymbol{x}) = \frac{f(\boldsymbol{x}; \theta)}{\int_\Theta f(\boldsymbol{x}; t) \, dt}.$$

Uniform improper priors share the same problem as Uniform proper priors, namely, the lack of invariance under non-linear transformation.

EXAMPLE 5.34: Suppose that $X_1, \cdots, X_n$ are i.i.d. Normal random variables with mean μ and variance 1. We will assume a uniform prior on the real-line for μ. It follows that the posterior for μ is

$$\pi(\mu|\boldsymbol{x}) = k(\boldsymbol{x}) \exp\left[-\frac{1}{2}\sum_{i=1}^{n}(x_i - \mu)^2\right]$$

where

$$k(\boldsymbol{x}) = \left(\int_{-\infty}^{\infty} \exp\left[-\frac{1}{2}\sum_{i=1}^{n}(x_i - t)^2\right] dt\right)^{-1}$$
$$= \sqrt{\frac{n}{2\pi}} \exp\left[\frac{1}{2}\sum_{i=1}^{n}(x_i - \bar{x})^2\right].$$

Thus

$$\pi(\theta|\boldsymbol{x}) = \sqrt{\frac{n}{2\pi}} \exp\left[-\frac{n}{2}(\mu - \bar{x})^2\right]$$

and so the posterior distribution is Normal with mean $\bar{x}$ and variance $1/n$. ◇

One approach to defining ignorance priors has been proposed by Jeffreys (1961). Let g be a monotone function on Θ and define $\nu = g(\theta)$. The information for ν can be defined in the usual way:

$$I(\nu) = \text{Var}_\nu\left[\frac{\partial}{\partial\nu}\ln f(\boldsymbol{X}; g^{-1}(\nu))\right]$$
$$= \frac{1}{(g'(\theta))^2}\text{Var}_\theta\left[\frac{\partial}{\partial\theta}\ln f(\boldsymbol{X};\theta)\right].$$

Now choose the function g so that $I(\nu)$ is constant and put a uniform prior distribution on $g(\Theta)$; this implies that the prior distribution on Θ is

$$\pi(\theta) = k|g'(\theta)| \propto \text{Var}_\theta\left[\frac{\partial}{\partial\theta}\ln f(\boldsymbol{X};\theta)\right]^{1/2} = I^{1/2}(\theta)$$

where k is some positive constant. Note that this prior distribution could be improper. These prior distributions are called Jeffreys priors.

EXAMPLE 5.35: Suppose that $X_1, \cdots, X_n$ are i.i.d. Poisson random variables with mean λ. The information for λ is $I(\lambda) = n/\lambda$. It is easy to verify that the transformation g must satisfy

$$\lambda(g'(\lambda))^2 = \text{constant}.$$

Thus the Jeffreys prior for λ is

$$\pi(\lambda) = \frac{k}{\sqrt{\lambda}} \quad \text{for } \lambda > 0.$$

Since $\int_0^\infty \lambda^{-1/2}\, d\lambda = \infty$, the prior is improper and so k can be chosen arbitrarily. Using the Jeffreys prior, the posterior density of λ given $x_1, \cdots, x_n$ is the Gamma density

$$\pi(\lambda|\boldsymbol{x}) = \frac{n^{y+1/2}\lambda^{y-1/2}\exp(-n\lambda)}{\Gamma(y+1/2)}$$

where $y = x_1 + \cdots + x_n$. This posterior density is always proper (even though the prior density is improper). ◇

5.9 Problems and complements

5.1: Suppose that $X_1, \cdots, X_n$ are i.i.d. random variables with density

$$f(x;\theta_1,\theta_2) = \begin{cases} a(\theta_1,\theta_2)h(x) & \text{for } \theta_1 \le x \le \theta_2 \\ 0 & \text{otherwise} \end{cases}$$

where $h(x) > 0$ is a known continuous function defined on the real line.

(a) Show that the MLEs of θ_1 and θ_2 are $X_{(1)}$ and $X_{(n)}$ respectively.

(b) Let $\widehat{\theta}_{1n}$ and $\widehat{\theta}_{2n}$ be the MLEs of θ_1 and θ_2 and suppose that $h(\theta_1) = \lambda_1 > 0$ and $h(\theta_2) = \lambda_2 > 0$. Show that

$$n\begin{pmatrix} \widehat{\theta}_{1n} - \theta_1 \\ \theta_2 - \widehat{\theta}_{2n} \end{pmatrix} \to_d \begin{pmatrix} Z_1 \\ Z_2 \end{pmatrix}$$

where Z_1 and Z_2 are independent Exponential random variables with parameters $\lambda_1 a(\theta_1,\theta_2)$ and $\lambda_2 a(\theta_1,\theta_2)$ respectively.

5.2: Suppose that $(X_1,Y_1), \cdots, (X_n,Y_n)$ are i.i.d. pairs of Normal random variables where X_i and Y_i are independent $N(\mu_i,\sigma^2)$ random variables.

(a) Find the MLEs of $\mu_1, \cdots, \mu_n$ and σ^2.

(b) Show that the MLE of σ^2 is not consistent. Does this result contradict the theory we have established regarding the consistency of MLEs? Why or why not?

(c) Suppose we observe only $Z_1, \cdots, Z_n$ where $Z_i = X_i - Y_i$. Find the MLE of σ^2 based on $Z_1, \cdots, Z_n$ and show that it is consistent.

5.3: Suppose that $X_1, \cdots, X_n, Y_1, \cdots Y_n$ are independent Expo-

Table 5.7 *Data for Problem 5.3.*

x_i	y_i	x_i	y_i	x_i	y_i	x_i	y_i
0.7	3.8	20.2	2.8	1.1	2.8	15.2	8.8
11.3	4.6	0.3	1.9	1.9	3.2	0.2	7.6
2.1	2.1	0.9	1.4	0.5	8.5	0.7	1.3
30.7	5.6	0.7	0.4	0.8	14.5	0.4	2.2
4.6	10.3	2.3	0.9	1.2	14.4	2.3	4.0

nential random variables where the density of X_i is

$$f_i(x) = \lambda_i \theta \exp(-\lambda_i \theta x) \quad \text{for } x \geq 0$$

and the density of Y_i is

$$g_i(x) = \lambda_i \exp(-\lambda_i x) \quad \text{for } x \geq 0$$

where $\lambda_1, \cdots, \lambda_n$ and θ are unknown parameters.

(a) Show that the MLE of θ (based on $X_1, \cdots, X_n, Y_1, \cdots Y_n$) satisfies the equation

$$\frac{n}{\widehat{\theta}} - 2\sum_{i=1}^{n} \frac{R_i}{1 + \widehat{\theta} R_i} = 0$$

where $R_i = X_i / Y_i$.

(b) Show that the density of R_i is

$$f_R(x; \theta) = \theta \left(1 + \theta x\right)^{-2} \quad \text{for } x \geq 0.$$

and show that the MLE for θ based on $R_1, \cdots, R_n$ is the same as that given in part (a).

(c) Let $\widehat{\theta}_n$ be the MLE in part (c). Find the limiting distribution of $\sqrt{n}(\widehat{\theta}_n - \theta)$.

(d) Use the data for $(X_i, Y_i), i = 1, \cdots, 20$ given in Table 5.7 to compute the maximum likelihood estimate of θ using either the Newton-Raphson or Fisher scoring algorithm. Find an appropriate starting value for the iterations and justify your choice.

(e) Give an estimate of the standard error for the maximum likelihood estimate computed in part (c).

5.4: Suppose that $X_1, \cdots, X_n$ are i.i.d. nonnegative random variables whose hazard function is given by

$$\lambda(x) = \begin{cases} \lambda_1 & \text{for } x \le x_0 \\ \lambda_2 & \text{for } x > x_0 \end{cases}$$

where λ_1, λ_2 are unknown parameters and x_0 is a known constant.

(a) Show that the density of X_i is

$$f(x; \lambda_1, \lambda_2) = \begin{cases} \lambda_1 \exp(-\lambda_1 x) & \text{for } x \le x_0 \\ \lambda_2 \exp(-\lambda_2(x - x_0) - \lambda_1 x_0) & \text{for } x > x_0 \end{cases}$$

(b) Find the MLEs of λ_1 and λ_2 as well as their joint limiting distribution.

5.5: Suppose that $X_1, \cdots, X_n$ are i.i.d. discrete random variables with frequency function

$$f(x; \theta) = \begin{cases} \theta & \text{for } x = -1 \\ (1-\theta)^2 \theta^x & \text{for } x = 0, 1, 2, \cdots \end{cases}$$

where $0 < \theta < 1$.

(a) Show that the MLE of θ based on $X_1, \cdots, X_n$ is

$$\widehat{\theta}_n = \frac{2\sum_{i=1}^n I(X_i = -1) + \sum_{i=1}^n X_i}{2n + \sum_{i=1}^n X_i}$$

and show that $\{\widehat{\theta}_n\}$ is consistent for θ.

(b) Show that $\sqrt{n}(\widehat{\theta}_n - \theta) \to_d N(0, \sigma^2(\theta)$ and find the value of $\sigma^2(\theta)$.

5.6: Suppose that $U_1, \cdots, U_n$ are i.i.d. Uniform random variables on $[0, \theta]$. Suppose that only the smallest r values are actually observed, that is, the order statistics $U_{(1)} < U_{(2)} < \cdots < U_{(r)}$.

(a) Find the MLE of θ based on $U_{(1)}$, $U_{(2)}$, $\cdots$, $U_{(r)}$.

(b) If $r = r_n = n - k$ where k is fixed, find the limiting distribution of $n(\theta - \widehat{\theta}_n)$ as $n \to \infty$ where $\widehat{\theta}_n$ is the MLE.

5.7: Suppose that $\boldsymbol{X} = (X_1, \cdots, X_n)$ has a k-parameter exponential family distribution with joint density or frequency function

$$f(\boldsymbol{x}; \boldsymbol{\theta}) = \exp\left[\sum_{i=1}^k c_i(\boldsymbol{\theta}) T_i(\boldsymbol{x}) - d(\boldsymbol{\theta}) + S(\boldsymbol{x})\right]$$

where the parameter space Θ is an open subset of R^k and the function $\boldsymbol{c} = (c_1, \cdots, c_k)$ is one-to-one on Θ.

(a) Suppose that $E_\theta[T_i(\boldsymbol{X})] = b_i(\boldsymbol{\theta})$ $(i = 1, \cdots, k)$. Show that the MLE $\widehat{\boldsymbol{\theta}}$ satisfies the equations

$$T_i(\boldsymbol{X}) = b_i(\widehat{\boldsymbol{\theta}}) \quad (i = 1, \cdots, k).$$

(b) Suppose that the X_i's are also i.i.d. so that $T_i(\boldsymbol{X})$ can be taken to be an average of i.i.d. random variables. If $\widehat{\boldsymbol{\theta}}_n$ is the MLE, use the Delta Method to show that $\sqrt{n}(\widehat{\boldsymbol{\theta}}_n - \boldsymbol{\theta})$ has the limiting distribution given in Theorem 5.4.

5.8: Suppose that $X_1, \cdots, X_n$ are i.i.d. continuous random variables with density $f(x;\theta)$ where θ is real-valued.

(a) We are often not able to observe the X_i's exactly rather only if they belong to some region B_k $(k = 1, \cdots, m)$; an example of this is interval censoring in survival analysis. Intuitively, we should be able to estimate θ more efficiently with the actual values of the X_i's; in this problem, we will show that this is true (at least) for MLEs.

Assume that $B_1, \cdots, B_m$ are disjoint sets such that

$$P_\theta\left(X_i \in \bigcup_{k=1}^{m} B_k\right) = 1.$$

Define i.i.d. discrete random variables $Y_1, \cdots, Y_n$ where $Y_i = k$ if $X_i \in B_k$; the frequency function of Y_i is

$$p(k;\theta) = P_\theta(X_i \in B_k) = \int_{B_k} f(x;\theta)\,dx \quad \text{for } k = 1, \cdots, m.$$

Also define

$$\begin{aligned} I_X(\theta) &= \mathrm{Var}_\theta\left[\frac{\partial}{\partial\theta}\ln f(X_i;\theta)\right] \\ \text{and} \quad I_Y(\theta) &= \mathrm{Var}_\theta\left[\frac{\partial}{\partial\theta}\ln p(Y_i;\theta)\right]. \end{aligned}$$

Assume the usual regularity conditions for $f(x;\theta)$, in particular, that $f(x;\theta)$ can be differentiated with respect to θ inside integral signs with impunity! Show that $I_X(\theta) \ge I_Y(\theta)$ and indicate under what conditions there will be strict inequality.

(Hint: Note that (i) $f(x;\theta)/p(k;\theta)$ is a density function on B_k, and (ii) for any random variable U, $E(U^2) \geq [E(U)]^2$ with strict inequality unless U is constant.)

(b) Suppose that we observe $Y_i = g(X_i)$ $(i = 1, \cdots, n)$ for some differentiable function g. Show that $I_X(\theta) \geq I_Y(\theta)$ with equality if g is a monotone function.

5.9: Let $X_1, \cdots, X_n$ be i.i.d. Exponential random variables with parameter λ. Suppose that the X_i's are not observed exactly but rather we observe random variables $Y_1, \cdots, Y_n$ where

$$Y_i = k\delta \quad \text{if } k\delta \leq X_i < (k+1)\delta$$

for $k = 0, 1, 2, \cdots$ where $\delta > 0$ is known.

(a) Give the joint frequency function of $\boldsymbol{Y} = (Y_1, \cdots, Y_n)$ and show that $\sum_{i=1}^n Y_i$ is sufficient for λ.

(b) Find the MLE of λ based on $Y_1, \cdots, Y_n$.

(c) Let $\widehat{\lambda}_n$ be the MLE of λ in part (b). Show that

$$\sqrt{n}(\widehat{\lambda}_n - \lambda) \to_d N(0, \sigma^2(\lambda, \delta))$$

where $\sigma^2(\lambda, \delta) \to \lambda^2$ as $\delta \to 0$.

5.10: Let $X_1, \cdots, X_n$ be i.i.d. random variables with density or frequency function $f(x; \theta)$ satisfying conditions (A1)-(A6) with $I(\theta) = J(\theta)$. Suppose that $\widetilde{\theta}_n$ is such that $\sqrt{n}(\widetilde{\theta}_n - \theta) \to_d Z$ for some random variable Z (not necessarily Normal); $\widetilde{\theta}_n$ is said to be $\sqrt{n}$-consistent. Define

$$\widehat{\theta}_n = \widetilde{\theta}_n - \left(\sum_{i=1}^n \ell''(X_i; \widetilde{\theta}_n)\right)^{-1} \sum_{i=1}^n \ell'(X_i; \widetilde{\theta}_n)$$

(a) Show that $\sqrt{n}(\widehat{\theta}_n - \theta) \to_d N(0, 1/I(\theta))$.

(b) Suppose that $X_1, \cdots, X_n$ are i.i.d. random variables with density function

$$f(x; \theta) = \frac{\theta}{(1 + \theta x)^2}$$

for $x \geq 0$ and define $\widehat{\mu}_n$ to be the sample median of $X_1, \cdots, X_n$. Use $\widehat{\mu}_n$ to construct a $\sqrt{n}$-consistent estimator of θ, $\widetilde{\theta}_n$. What is the asymptotic distribution of $\sqrt{n}(\widetilde{\theta}_n - \theta)$?

(c) Using $\widetilde{\theta}_n$ from part (b), show how to construct an estimator with the same limiting distribution as the MLE in Problem 5.3(b).

5.11: The key condition in Theorem 5.3 is (A6) as this allows us to approximate the likelihood equation by a linear equation in $\sqrt{n}(\widehat{\theta}_n - \theta)$. However, condition (A6) can be replaced by other similar conditions, some of which may be weaker than (A6).

Assume that $\widehat{\theta}_n \to_p \theta$ and that conditions (A1)-(A5) hold. Suppose that for some $\delta > 0$, there exists a function $K_\delta(x)$ and a constant $\alpha > 0$ such that

$$|\ell'(x;t) - \ell'(x;\theta)| \le K(x)|t-\theta|^\alpha$$

for $|t-\theta| \le \delta$ where $E_\theta\left[K_\delta(X_i)\right] < \infty$. Show that the conclusion of Theorem 5.3 holds.

5.12: In section 5.4, we noted that the consistency of the MLE is straightforward when the log-likelihood function is concave. Similarly, it is possible to exploit the concavity or convexity of objective functions to derive the limiting distributions of estimators. In this problem, we will derive the limiting distribution of the sample median using the fact that the median minimizes a convex objective function; see Problem 4.19.

Suppose that $\{Z_n(u)\}$ is a sequence of random convex functions and for any $(u_1, u_2, \cdots, u_k)$, we have

$$(Z_n(u_1), \cdots, Z_n(u_k)) \to_d (Z(u_1), \cdots, Z(u_k))$$

where $Z(u)$ is a random convex function that is uniquely minimized at U (which will be a random variable). Then it can be shown that if U_n minimizes Z_n then $U_n \to_d U$ (Davis *et al*, 1992).

This result can be used to rederive the limiting distribution of the sample median (see Examples 3.5 and 3.6). Suppose that $X_1, \cdots, X_n$ are i.i.d. random variables with distribution function F where $F(\mu) = 1/2$ and $F'(\mu) = \lambda > 0$. If $\widehat{\mu}_n$ is the sample median of $X_1, \cdots, X_n$ then we know that

$$\sqrt{n}(\widehat{\mu}_n - \mu) \to_d N(0, 1/(4\lambda^2)).$$

(a) Show that $U_n = \sqrt{n}(\widehat{\mu}_n - \mu)$ minimizes the objective function

$$Z_n(u) = \sum_{i=1}^n \left[|X_i - \mu - u/\sqrt{n}| - |X_i - \mu|\right]$$

and that Z_n is a convex function of u.

(b) Show that

$$|x - y| - |x| = -y\,\mathrm{sgn}(x) + 2\int_0^y [I(x \le s) - I(x \le 0)]\,ds$$

for $x \neq 0$ where $\mathrm{sgn}(x) = I(x > 0) - I(x < 0)$ is the "sign" of x.

(c) Show that

$$\begin{aligned} Z_n(u) &= -\frac{u}{\sqrt{n}}\sum_{i=1}^n \mathrm{sgn}(X_i - \mu) \\ &\quad + \frac{2}{\sqrt{n}}\sum_{i=1}^n \int_0^u \left[I\left(X_i \le \mu + \frac{s}{\sqrt{n}} \right) - I(X_i \le \mu) \right] ds \end{aligned}$$

using the formula in part (b).

(d) Show that

$$(Z_n(u_1), \cdots, Z_n(u_k)) \to_d (Z(u_1), \cdots, Z(u_k))$$

where $Z(u) = -uW + 2\lambda u^2$ and $W \sim N(0,1)$. (Hint: Note that

$$E[I(X_i \le \mu + s/\sqrt{n}) - I(X_i \le \mu)] = F(\mu + s/\sqrt{n}) - F(\mu)$$

and

$$\mathrm{Var}[I(X_i \le \mu + s/\sqrt{n}) - I(X_i \le \mu)] \le F(\mu + s/\sqrt{n}) - F(\mu)$$

for each s, with $F(\mu + s/\sqrt{n}) - F(\mu) \approx \lambda s/\sqrt{n}$.)

(e) Show that $\sqrt{n}(\widehat{\mu}_n - \mu) \to_d W/(2\lambda)$.

5.13: The same approach used in Problem 5.12 can be used to determine the limiting distribution of the sample median under more general conditions. Again let $X_1, \cdots, X_n$ be i.i.d. with distribution function F and median μ where now

$$\lim_{n\to\infty} \sqrt{n}[F(\mu + s/a_n) - F(\mu)] = \psi(s)$$

for some increasing function ψ and sequence of constants $a_n \to \infty$. The asymptotic distribution of $a_n(\widehat{\mu}_n - \mu)$ will be determined by considering the objective function

$$Z_n(u) = \frac{a_n}{\sqrt{n}}\sum_{i=1}^n [|X_i - \mu - u/a_n| - |X_i - \mu|].$$

(a) Show that $U_n = a_n(\widehat{\mu}_n - \mu)$ minimizes Z_n.

(b) Repeat the steps used in Problem 5.12 to show that

$$(Z_n(u_1), \cdots, Z_n(u_k)) \to_d (Z(u_1), \cdots, Z(u_k))$$

where $Z(u) = -uW + 2\int_0^u \psi(s)\, ds$ and $W \sim N(0, 1)$.

(c) Show that $a_n(\widehat{\mu}_n - \mu) \to_d \psi^{-1}(W/2)$.

5.14: The conditions assumed in Theorems 5.3 and 5.4 effectively imply that the log-likelihood function is approximately quadratic in a neighbourhood of the true parameter value. These quadratic approximations can be used to give "heuristic" proofs of Theorems 5.3 and 5.4.

Suppose that $X_1, \cdots, X_n$ are i.i.d. random variables with density or frequency function $f(x; \boldsymbol{\theta})$ satisfying conditions (B1)-(B6). Define

$$Z_n(\boldsymbol{u}) = \sum_{i=1}^{n} \ln\left[f(X_i; \boldsymbol{\theta} + \boldsymbol{u}/\sqrt{n})/f(X_i; \boldsymbol{\theta})\right]$$

and note that Z_n is maximized at $\boldsymbol{u} = \sqrt{n}(\widehat{\boldsymbol{\theta}}_n - \boldsymbol{\theta})$ where $\widehat{\boldsymbol{\theta}}_n$ is the MLE of $\boldsymbol{\theta}$.

(a) Show that

$$Z_n(\boldsymbol{u}) = \boldsymbol{u}^T \boldsymbol{W}_n - \frac{1}{2}\boldsymbol{u}^T J(\boldsymbol{\theta})\boldsymbol{u} + R_n(\boldsymbol{u})$$

where $\sup_K |R_n(\boldsymbol{u})| \to_p 0$ for any compact set K and $\boldsymbol{W}_n \to_d N_p(\boldsymbol{0}, I(\boldsymbol{\theta}))$.

(b) Part (a) suggests that the limit of Z_n is

$$Z(\boldsymbol{u}) = \boldsymbol{u}^T \boldsymbol{W} - \frac{1}{2}\boldsymbol{u}^T J(\boldsymbol{\theta})\boldsymbol{u}$$

where $\boldsymbol{W} \sim N_p(\boldsymbol{0}, I(\boldsymbol{\theta}))$. Show that Z is maximized at $\boldsymbol{u} = J^{-1}(\boldsymbol{\theta})\boldsymbol{W}$, which suggests that $\sqrt{n}(\widehat{\boldsymbol{\theta}}_n - \boldsymbol{\theta}) \to_d J^{-1}(\boldsymbol{\theta})\boldsymbol{W}$. (If $Z_n(\boldsymbol{u})$ is a concave function of $\boldsymbol{u}$ for each n then this argument is a rigorous proof of Theorem 5.4.)

5.15: In Theorems 5.3 and 5.4, we assume that the parameter space Θ is an open subset of R^p. However, in many situations, this assumption is not valid; for example, the model may impose constraints on the parameter $\boldsymbol{\theta}$, which effectively makes Θ a closed set. If Θ is not an open set then the MLE of $\boldsymbol{\theta}$ need not satisfy the likelihood equations as the MLE $\widehat{\boldsymbol{\theta}}_n$ may lie on the boundary of Θ. In determining the asymptotic distribution of $\widehat{\boldsymbol{\theta}}_n$ the main concern is whether or not the true value of the parameter lies on the boundary of the parameter space. If $\boldsymbol{\theta}$ lies in the interior of Θ then eventually (for sufficiently large n) $\widehat{\boldsymbol{\theta}}_n$ will satisfy the likelihood equations and so Theorems 5.3 and 5.4 will still hold; however, the situation becomes more complicated if $\boldsymbol{\theta}$ lies on the boundary of Θ.

Suppose that $X_1, \cdots, X_n$ are i.i.d. random variables with density or frequency function $f(x; \boldsymbol{\theta})$ (satisfying conditions (B2)-(B6)) where $\boldsymbol{\theta}$ lies on the boundary of Θ. Define (as in Problem 5.14) the function

$$Z_n(\boldsymbol{u}) = \sum_{i=1}^{n} \ln \left[f(X_i; \boldsymbol{\theta} + \boldsymbol{u}/\sqrt{n}) / f(X_i; \boldsymbol{\theta}) \right]$$

and the set

$$C_n = \{\boldsymbol{u} : \boldsymbol{\theta} + \boldsymbol{u}/\sqrt{n} \in \Theta\}.$$

The limiting distribution of the MLE can be determined by the limiting behaviour of Z_n and C_n; see Geyer (1994) for details.

(a) Show that $\sqrt{n}(\widehat{\boldsymbol{\theta}}_n - \boldsymbol{\theta})$ maximizes $Z_n(\boldsymbol{u})$ subject to the constraint $\boldsymbol{u} \in C_n$.

(b) Suppose that $\{C_n\}$ is a decreasing sequence of sets whose limit is C. Show that C is non-empty.

(c) Parts (a) and (b) (together with Problem 5.14) suggest that $\sqrt{n}(\widehat{\boldsymbol{\theta}}_n - \theta)$ converges in distribution to the minimizer of

$$Z(\boldsymbol{u}) = \boldsymbol{u}^T \boldsymbol{W} - \frac{1}{2} \boldsymbol{u}^T J(\boldsymbol{\theta}) \boldsymbol{u}$$

(where $\boldsymbol{W} \sim N_p(\boldsymbol{0}, I(\boldsymbol{\theta}))$) subject to $\boldsymbol{u} \in C$. Suppose that $X_1, \cdots, X_n$ are i.i.d. Gamma random variables with shape parameter α and scale parameter λ where the parameter space is restricted so that $\alpha \geq \lambda > 0$ (that is, $E(X_i) \geq 1$). If $\alpha = \lambda$, describe the limiting distribution of the MLEs. (Hint: Show that $C = \{(u_1, u_2) : u_1 \geq u_2\}$.)

5.16: Suppose that $X_1, \cdots, X_n$ are i.i.d. random variables with density or frequency function $f(x; \theta)$ where θ is real-valued. In many cases, the MLE of θ satisfies the likelihood equation

$$\sum_{i=1}^{n} \ell'(X_i; \widehat{\theta}_n) = 0.$$

If $\ell(x;\theta)$ is an unbounded function in x then a single observation can potentially have an arbitrarily large influence on the value of the MLE. For this reason, it is often desirable to robustify maximum likelihood estimation by replacing $\ell'(x;\theta)$ by a function $\psi_c(x;\theta)$ that is bounded in x. In order to have Fisher consistency, we choose ψ_c so that $E_\theta[\psi_c(X_i;\theta)] = 0$ where the expected value is taken with respect to the density or frequency function $f(x;\theta)$.

Suppose that $X_1, \cdots, X_n$ are i.i.d. Exponential random variables with parameter λ. The MLE of λ satisfies the likelihood equation

$$\sum_{i=1}^{n} \left(\frac{1}{\widehat{\lambda}_n} - X_i \right) = 0,$$

which can be solved to yield $\widehat{\lambda}_n = 1/\bar{X}_n$. Consider replacing $\ell'(x;\lambda) = 1/\lambda - x$ (which is unbounded in x) by

$$\psi_c(x;\lambda) = \begin{cases} 1/\lambda + g_c(\lambda) - x & \text{if } x \le c/\lambda + 1/\lambda + g_c(\lambda) \\ c/\lambda & \text{otherwise} \end{cases}$$

where $c > 0$ is a tuning parameter and $g_c(\lambda)$ is determined so that $E_\lambda[\psi_c(X_i;\lambda)] = 0$.

(a) Show that $g_c(\lambda)$ satisfies the equation

$$\exp(-c - 1 - g_c(\lambda)\lambda) + g_c(\lambda) = 0$$

(b) Define $\widetilde{\lambda}_n$ to satisfy the equation

$$\sum_{i=1}^{n} \psi_c(X_i;\widetilde{\lambda}_n) = 0.$$

Assuming that the standard regularity conditions hold, find the limiting distribution of $\sqrt{n}(\widetilde{\lambda}_n - \lambda)$.

(c) Find the asymptotic relative efficiency of $\widetilde{\lambda}_n$ with respect to the MLE $\widehat{\lambda}_n$. For what value of c is the asymptotic relative efficiency equal to 0.95?

5.17: Let $X_1, \cdots, X_n$ be i.i.d. random variables with density or frequency function $f(x;\theta)$ where θ is a real-valued parameter. Suppose that MLE of θ, $\widehat{\theta}$, satisfies the likelihood equation

$$\sum_{i=1}^{n} \ell'(X_i;\widehat{\theta}) = 0$$

where $\ell'(x;\theta)$ is the derivative with respect to θ of $\ln f(x;\theta)$.

(a) Let $\widehat{\theta}_{-j}$ be MLE of θ based on all the X_i's except X_j. Show that

$$\widehat{\theta}_{-j} \approx \widehat{\theta} + \frac{\ell'(X_j;\widehat{\theta})}{\sum_{i=1}^{n} \ell''(X_i;\widehat{\theta})}$$

(if n is reasonably large).

(b) Show that the jackknife estimator of $\widehat{\theta}$ satisfies

$$\widehat{\mathrm{Var}}(\widehat{\theta}) \approx \frac{n-1}{n} \frac{\sum_{j=1}^{n} [\ell'(X_j;\widehat{\theta})]^2}{\left(\sum_{i=1}^{n} \ell''(X_i;\widehat{\theta})\right)^2}.$$

(c) The result of part (b) suggests that the jackknife estimator of $\mathrm{Var}(\widehat{\theta})$ is essentially the "sandwich" estimator; the latter estimator is valid when the model is misspecified. Explain the apparent equivalence between these two estimators of $\mathrm{Var}(\widehat{\theta})$. (Hint: Think of the MLE as a substitution principle estimator of some functional parameter.)

5.18: Millar (1987) considers a statistical model for determining the composition of a mixed stock fishery. The statistical model can be described as follows: We have a sample of N fish that can be classified into one of G genotypes (where typically $G >> N$). If Y_i is the number of fish in the sample with genotype i then $\boldsymbol{Y} = (Y_1, \cdots, Y_G)$ is a Multinomial random vector with

$$P(\boldsymbol{Y} = \boldsymbol{y}) = \frac{N!}{y_1! \times \cdots \times y_G!} \prod_{i=1}^{G} \lambda_i$$

where

$$\lambda_i = \sum_{j=1}^{S} x_{ij} \theta_j$$

where $\theta_1, \cdots, \theta_S$ (unknown) are the proportion of fish belonging to the S sub-populations or stocks, and x_{ij} (known) is the (conditional) probability of belonging to genotype i given membership in stock j.

To estimate $\boldsymbol{\theta} = (\theta_1, \cdots, \theta_S)$, we maximize the log-likelihood function

$$\ln \mathcal{L}(\boldsymbol{\theta}) = \sum_{i=1}^{G} y_i \ln \left(\sum_{j=1}^{S} x_{ij} \theta_j \right) + k(\boldsymbol{y})$$

where $k(\boldsymbol{y})$ does not depend on $\boldsymbol{\theta}$. (We also must assume that $\theta_j \geq 0$ for all j and $\theta_1 + \cdots + \theta_S = 1$.)

(a) Show that $\ln \mathcal{L}(\boldsymbol{\theta})$ can be written as

$$\ln \mathcal{L}(\boldsymbol{\theta}) = \sum_{r=1}^{N} \ln\left(\sum_{j=1}^{S} x_{i_r j}\theta_j\right)$$

where i_r is the genotype of the r-th fish in the sample.

(b) Show that the MLEs of $\theta_1, \cdots, \theta_S$ satisfy the equations

$$\widehat{\theta}_j = \frac{1}{N}\sum_{r=1}^{N} \frac{x_{i_r j}\widehat{\theta}_j}{\sum_{k=1}^{S} x_{i_r k}\widehat{\theta}_k} \quad \text{for } j = 1, \cdots, S.$$

(c) Assume that $S = 2$. In this case, we need only estimate a single parameter θ (equal, say, to the proportion belonging to stock 1). Assuming that the standard regularity conditions for maximum likelihood estimators hold, find an estimator of the standard error of $\widehat{\theta}$.

5.19: Suppose that $\boldsymbol{X} = (X_1, \cdots, X_n)$ has a joint density or frequency function $f(\boldsymbol{x}; \theta)$ where θ has prior density $\pi(\theta)$. If $T = T(\boldsymbol{X})$ is sufficient for θ, show that the posterior density of θ given $\boldsymbol{X} = \boldsymbol{x}$ is the same as the posterior density of θ given $T = T(\boldsymbol{x})$.

5.20: Suppose that $X_1, \cdots, X_n$ are i.i.d. Uniform random variables on $[0, \theta]$ where θ has a Pareto prior density function:

$$\pi(\theta) = \frac{\alpha}{\theta_0}\left(\frac{\theta}{\theta_0}\right)^{-\alpha-1}$$

for $\theta > \theta_0 > 0$ and $\alpha > 0$ where θ_0 and α are hyperparameters.

(a) Show that the posterior distribution of θ is also Pareto.

(b) Suppose that θ^* is the true value of θ. Under i.i.d. sampling, what happens to the posterior density of θ as $n \to \infty$? (Hint: There are two cases to consider: $\theta^* \geq \theta_0$ and $\theta^* < \theta_0$.)

5.21: The Zeta distribution is sometimes used in insurance as a model for the number of policies held by a single person in an insurance portfolio. The frequency function for this distribution is

$$f(x; \alpha) = \frac{x^{-(\alpha+1)}}{\zeta(\alpha+1)}$$

Table 5.8 *Data for Problem 5.21.*

Observation	1	2	3	4	5
Frequency	63	14	5	1	2

for $x = 1, 2, 3, \cdots$ where $\alpha > 0$ and

$$\zeta(p) = \sum_{k=1}^{\infty} k^{-p}.$$

(The function $\zeta(p)$ is called the Riemann zeta function.)

(a) Suppose that $X_1, \cdots, X_n$ are i.i.d. Zeta random variables. Show that the MLE of α satisfies the equation

$$\frac{1}{n}\sum_{i=1}^{n} \ln(X_i) = -\frac{\zeta'(\widehat{\alpha}_n + 1)}{\zeta(\widehat{\alpha}_n + 1)}$$

and find the limiting distribution of $\sqrt{n}(\widehat{\alpha}_n - \alpha)$.

(b) Assume the following prior density for α:

$$\pi(\alpha) = \frac{1}{2}\alpha^2 \exp(-\alpha) \quad \text{for } \alpha > 0$$

A sample of 85 observations is collected; its frequency distribution is given in Table 5.8.

Find the posterior distribution of α. What is the mode (approximately) of this posterior distribution?

(c) Repeat part (b) using the improper prior density

$$\pi(\alpha) = \frac{1}{\alpha} \quad \text{for } \alpha > 0.$$

Compare the posterior densities in part (b) and (c).

5.22: Suppose that X has a Binomial distribution with parameters n and θ where θ is unknown.

(a) Find the Jeffreys prior for θ. Is this prior density proper?

(b) Find the posterior density for θ given $X = x$ using the Jeffreys prior.

5.23: The concept of Jeffreys priors can be extended to derive "noninformative" priors for multiple parameters. Suppose that

$\boldsymbol{X}$ has joint density or frequency function $f(\boldsymbol{x};\boldsymbol{\theta})$ and define the matrix

$$I(\boldsymbol{\theta}) = E_\theta[\boldsymbol{S}(\boldsymbol{X};\boldsymbol{\theta})\boldsymbol{S}^T(\boldsymbol{X};\boldsymbol{\theta})]$$

where $\boldsymbol{S}(\boldsymbol{x};\boldsymbol{\theta})$ is the gradient (vector of partial derivatives) of $\ln f(\boldsymbol{x};\boldsymbol{\theta})$ with respect to $\boldsymbol{\theta}$. The Jeffreys prior for $\boldsymbol{\theta}$ is proportional to $\det(I(\boldsymbol{\theta}))^{1/2}$.

(a) Show that the Jeffreys prior can be derived using the same considerations made in the single parameter case. That is, if $\boldsymbol{\phi} = \boldsymbol{g}(\boldsymbol{\theta})$ for some one-to-one function $\boldsymbol{g}$ such that $I(\boldsymbol{\phi})$ is constant then the Jeffreys prior for θ corresponds to a uniform prior for $\boldsymbol{\phi}$.

(b) Suppose that $X_1, \cdots, X_n$ are i.i.d. Normal random variables with mean μ and variance σ^2. Find the Jeffreys prior for (μ, σ).

CHAPTER 6

Optimality in Estimation

6.1 Introduction

To this point, we have discussed a number of approaches for obtaining point estimators. We have also discussed approaches for determining various properties of these estimators, such as limiting distributions and approximate standard errors. However, we have not (so far) attempted to determine if a given estimator is optimal in any sense.

As it turns out, "optimal" is a poorly defined term in the context of estimation as there are a number of criteria for optimality. For example, we can approach optimality from a large sample (or asymptotic) point of view. Suppose that $X_1, X_2, \cdots$ is a sequence of random variables and $\widehat{\theta}_n$ is an estimator of some parameter θ based on $\boldsymbol{X} = (X_1, \cdots, X_n)$ such that

$$\sqrt{n}(\widehat{\theta}_n - \theta) \to_d N(0, \sigma^2(\theta)).$$

A natural question to ask is whether a lower bound for $\sigma^2(\theta)$ exists; if such a lower bound exists and is attained by some sequence of estimators $\{\widehat{\theta}_n\}$ then we can say that this sequence is optimal.

In this chapter, we will discuss optimal estimation from three perspectives. First, we will look at estimation from a decision theoretic point of view, comparing estimators based on their risk for a given loss function. Next, we will narrow our focus to unbiased estimation and attempt to find unbiased estimators with uniformly minimum variance over the parameter space. Finally, we will take an asymptotic point of view and consider estimators with asymptotic Normal distributions.

6.2 Decision theory

While we have discussed a number of different approaches to point estimation, we have so far avoided the issue of which estimator is "best" or optimal in a given situation. One approach to finding and evaluating estimators is based on decision theory. We will discuss only the most basic elements of decision theory here; a more detailed treatment is given in Ferguson (1967).

In decision theory, in addition to specifying a model for $\boldsymbol{X}$ (depending on some unknown parameter θ), we also specify a loss function $L(\widehat{\theta}, \theta)$ that describes the "loss" incurred by making an estimate $\widehat{\theta}$ when the true value of the parameter is θ. Typically, the loss function is chosen so that for fixed θ, $L(a, \theta)$ increases as a moves away from θ and $L(\theta, \theta) = 0$.

DEFINITION. A loss function $L(\cdot, \cdot)$ is a nonnegative-valued function defined on $\Theta \times \Theta$ so that $L(\widehat{\theta}, \theta)$ indicates the loss incurred in estimating a parameter θ by $\widehat{\theta}$.

Perhaps the most commonly used loss functions for real-valued parameters are squared error loss ($L(a, b) = (a - b)^2$) and absolute error loss ($L(a, b) = |a - b|$). However, many other loss functions are possible. Strictly speaking, the loss function used should be dictated by the particular problem; however, almost invariably squared error loss is used in the vast majority of estimation problems.

DEFINITION. Given a loss function L and an estimator $\widehat{\theta}$, the risk function $R_\theta(\widehat{\theta})$ is defined to be the expected value of $L(\widehat{\theta}, \theta)$:

$$R_\theta(\widehat{\theta}) = E_\theta[L(\widehat{\theta}, \theta)].$$

If squared error loss is used, then $R_\theta(\widehat{\theta}) = \mathrm{MSE}_\theta(\widehat{\theta})$, the mean square error of $\widehat{\theta}$. Likewise, if absolute error loss is used then $R_\theta(\widehat{\theta}) = \mathrm{MAE}_\theta(\widehat{\theta})$, the mean absolute error.

Given a particular loss function L, we can try to find an estimator $\widehat{\theta}$ such that $R_\theta(\widehat{\theta})$ is minimized for all θ. Unfortunately, this is impossible to do except in trivial cases; typically, the estimator that minimizes $R_\theta(\cdot)$ for a fixed θ will not minimize $R_\theta(\cdot)$ uniformly over all θ. However, if we accept risk (for example, mean square error) as a measure of the quality of an estimator, we can rule out estimators whose risk is uniformly higher than another estimator.

DEFINITION. For a given loss function L, an estimator $\widehat{\theta}$ is an inadmissible estimator of θ if there exists an estimator $\widetilde{\theta}$ such that

$$R_\theta(\widetilde{\theta}) \leq R_\theta(\widehat{\theta}) \quad \text{for all } \theta \in \Theta$$

and

$$R_{\theta_0}(\widetilde{\theta}) < R_{\theta_0}(\widehat{\theta}) \quad \text{for some } \theta_0 \in \Theta.$$

If no such estimator $\widetilde{\theta}$ exists then $\widehat{\theta}$ is admissible.

EXAMPLE 6.1: Suppose that $X_1, \cdots, X_n$ are i.i.d. Exponential random variables with parameter λ. The MLE of λ is

$$\widehat{\lambda} = \frac{1}{\bar{X}}$$

where $\bar{X}$ is the sample mean. It can be shown that $E_\lambda(\widehat{\lambda}) = n\lambda/(n-1)$ when $n \geq 2$. An unbiased estimator of λ is $\widetilde{\lambda} = (n-1)\widehat{\lambda}/n$. Clearly,

$$\text{MSE}_\lambda(\widetilde{\lambda}) < \text{MSE}_\lambda(\widehat{\lambda})$$

since $\widetilde{\lambda}$ is unbiased and $\text{Var}_\lambda(\widetilde{\lambda}) < \text{Var}_\lambda(\widehat{\lambda})$. Since $\text{MSE}_\lambda(\widehat{\lambda}) = R_\lambda(\widehat{\lambda})$ when the loss function is squared error loss, $\widehat{\lambda}$ is an inadmissible estimator of λ under squared error loss. However, in the case of estimating positive parameters, mean square error tends to penalize over-estimation more heavily than under-estimation (since the maximum possible under-estimation is bounded) and thus squared error loss may not be the best loss function to consider in this example. Instead, we might define the loss function

$$L(a, b) = \frac{b}{a} - 1 - \ln(b/a)$$

for which we have for each fixed $b > 0$,

$$\lim_{a \to 0} L(a, b) = \lim_{a \to \infty} L(a, b) = \infty.$$

Evaluating the risk function of $\widetilde{\lambda}$, we get

$$\begin{aligned}
R_\lambda(\widetilde{\lambda}) &= E_\lambda\left[\frac{n\lambda\bar{X}}{n-1} - 1 - \ln\left(\frac{n\lambda\bar{X}}{n-1}\right)\right] \\
&= E_\lambda\left[\lambda\bar{X} - 1 - \ln\left(\lambda\bar{X}\right)\right] \\
&\quad + \frac{1}{n-1}E_\lambda(\lambda\bar{X}) - \ln\left(\frac{n}{n-1}\right) \\
&= E_\lambda\left[\lambda\bar{X} - 1 - \ln\left(\lambda\bar{X}\right)\right] + \frac{1}{n-1} - \ln\left(\frac{n}{n-1}\right) \\
&> E_\lambda\left[\lambda\bar{X} - 1 + \ln\left(\lambda\bar{X}\right)\right] \\
&= R_\lambda(\widehat{\lambda})
\end{aligned}$$

(since $1/(n-1) - \ln(n/(n-1)) > 0$ for $n \geq 2$) and so $\widetilde{\lambda}$ is inadmissible under this loss function. ◇

Note that we cannot make any claims about the admissibility of the estimators in Example 6.1. More precisely, to prove admissibility, we need to show that there exists no other estimator with uniformly lower risk. At first glance, this may seem a formidable task; however, there are technical devices available that facilitate it. One of these devices arises by putting a prior distribution on the parameter space.

Suppose that we put a (proper) prior density function $\pi(\theta)$ on the parameter space Θ. Then given a loss function $L(\cdot, \theta)$ and risk function $R_\theta(\cdot)$, we can define the Bayes risk of an estimator $\widehat{\theta}$ by

$$R_B(\widehat{\theta}) = \int_\Theta R_\theta(\widehat{\theta})\, \pi(\theta)\, d\theta.$$

(Note that $R_B(\widehat{\theta})$ depends on the loss function, the distribution of $\boldsymbol{X}$ as well as the prior distribution on Θ.) Then a Bayes estimator of θ is an estimator that minimizes the Bayes risk.

Bayes estimators can usually be determined from the posterior distribution of θ; more precisely, if the expected posterior loss function

$$\int_\Theta L(T(\boldsymbol{x}), \theta)\pi(\theta|\boldsymbol{x})\, d\theta$$

is minimized at $T(\boldsymbol{x}) = T^*(\boldsymbol{x})$ (where $\pi(\theta|\boldsymbol{x})$ is the posterior density function of θ) then $T^*(\boldsymbol{X})$ is a Bayes estimator. For example, for squared error loss, the Bayes estimator is simply the mean of the posterior distribution while for absolute error loss, the Bayes estimator is any median of the posterior distribution.

Bayes estimators are also admissible estimators provided that they are unique. For example, suppose that $\widehat{\theta}$ is a unique Bayes estimator and suppose that $R_\theta(\widetilde{\theta}, \theta) \leq R_\theta(\widehat{\theta}, \theta)$ for some other estimator $\widetilde{\theta}$. Then

$$\begin{aligned} R_B(\widetilde{\theta}) &= \int_\Theta R_\theta(\widetilde{\theta})\, \pi(\theta)\, d\theta \\ &\leq \int_\Theta R_\theta(\widehat{\theta})\, \pi(\theta)\, d\theta \\ &= R_B(\widehat{\theta}), \end{aligned}$$

which contradicts the uniqueness of $\widehat{\theta}$ as a Bayes estimator; thus $\widehat{\theta}$ must be admissible.

An important optimality concept in decision theory is minimax estimation.

DEFINITION. An estimator $\widehat{\theta}$ is said to be minimax (with respect to a loss function L) if

$$\sup_{\theta \in \Theta} R_\theta(\widehat{\theta}) \leq \sup_{\theta \in \Theta} R_\theta(\widetilde{\theta})$$

for any estimator $\widetilde{\theta}$.

It is easy to see that a minimax estimator is admissible. In general, proving that an estimator is minimax is not easy although this can be facilitated if it can be shown that the estimator is a Bayes estimator (or a limit of Bayes estimators) and has constant risk over the parameter space; see Problem 6.5 for more details.

It should be noted that there are pitfalls involved in using any criterion for evaluating estimators based on risk functions (such as admissibility or minimaxity), particularly when the loss function is not appropriate for the problem. In some problems, estimators are often inadmissible only because they are dominated (in terms of risk) by estimators whose practical use is somewhat dubious; in such cases, it may be the loss function rather than the estimator that is suspect. However, even if we are confident that our loss function is appropriate, it may be worthwhile to limit the class of estimators under consideration. For example, if θ is a parameter describing the center of a distribution, it may be desirable to consider only those estimators $\widehat{\theta} = S(X_1, \cdots, X_n)$ for which

$$S(X_1 + c, \cdots, X_n + c) = S(X_1, \cdots, X_n) + c.$$

More generally, we might consider only those estimators that satisfy a certain invariance or equivariance property, or only unbiased estimators. In the next section, we will consider unbiased estimation under squared error loss.

6.3 Minimum variance unbiased estimation

As mentioned above, one approach to finding estimators is to find an estimator having uniformly smallest risk over some restricted class of estimators. If we take the loss function to be squared error loss and consider only unbiased estimators of a parameter then we reduce the problem to finding the unbiased estimator with the minimum variance (since the mean square error and variance are the same for unbiased estimators); if such an estimator has minimum variance over the parameter space then this estimator is called a uniformly minimum variance unbiased (UMVU) estimator.

Suppose that $X_1, \cdots, X_n$ are random variables whose joint distribution depends on some parameter θ, which may be real- or vector-valued. In this section, we will consider unbiased estimation of $g(\theta)$ where g is a real-valued function defined on the parameter space Θ.

As discussed previously in Chapter 4, it is not clear that unbiased estimators are always desirable. For example, if we believe in the likelihood principle then it is easily shown that the requirement of unbiasedness is a violation of this principle. For the time being, however, we will ignore the possible shortcomings of unbiased estimators and focus our energy on the theory of optimal unbiased estimator under squared error loss. Nonetheless, it should be noted that unbiased estimators do not always exist or, if they do exist, can be nonsensical estimators; the following examples illustrate these points.

EXAMPLE 6.2: Suppose that X is a Binomial random variable with parameters n and θ, where $0 < \theta < 1$ is unknown. We wish to find an unbiased estimator of $g(\theta) = 1/\theta$; that is, we need to find a statistic $T(X)$ such that $E_\theta[T(X)] = 1/\theta$ for all $0 < \theta < 1$. Thus we need to find $T(0), T(1), \cdots, T(n)$ such that

$$\sum_{x=0}^{n} T(x) \binom{n}{x} \theta^x (1-\theta)^{n-x} = \frac{1}{\theta}.$$

Multiplying both sides by θ, we get

$$\begin{aligned}\sum_{x=0}^{n} T(x) \binom{n}{x} \theta^{x+1} (1-\theta)^{n-x} &= \sum_{k=1}^{n+1} a(k)\theta^k \\ &= 1\end{aligned}$$

where $a(1), \cdots, a(n+1)$ depend on $T(0), T(1), \cdots, T(n)$. It follows that whatever the choice of $a(1), \cdots, a(n+1)$, the equality

$$a(1)\theta + \cdots + a(n+1)\theta^{n+1} = 1$$

is satisfied for at most $n+1$ values of θ between 0 and 1 and cannot be satisfied for all θ. Thus there exists no unbiased estimator of $1/\theta$. $\diamond$

EXAMPLE 6.3: Suppose that X is a Poisson random variable with mean λ. We wish to find an unbiased estimator of $\exp(-2\lambda)$ based on X. If $T(X)$ is this estimator, we must have

$$E_\lambda[T(X)] = \sum_{x=0}^{\infty} T(x)\frac{\exp(-\lambda)\lambda^x}{x!} = \exp(-2\lambda)$$

for all $\lambda > 0$. Multiplying both sides of the equation by $\exp(\lambda)$, we get

$$\begin{aligned}\sum_{x=0}^{\infty} T(x)\frac{\lambda^x}{x!} &= \exp(-\lambda) \\ &= \sum_{x=0}^{\infty}(-1)^x\frac{\lambda^x}{x!},\end{aligned}$$

which implies that $T(x) = (-1)^x$. Thus the only unbiased estimator of $\exp(-2\lambda)$ is $(-1)^X$; since $0 < \exp(-2\lambda) < 1$ for $\lambda > 0$, this is clearly a ridiculous estimator. It should be noted, however, that if $X_1, \cdots, X_n$ are i.i.d. Poisson random variables with parameter λ then $(1 - 2/n)^T$ (with $T = \sum_{i=1}^n X_i$) is unbiased; this estimator is somewhat more sensible especially for larger values of n. ◇

Examples 6.2 and 6.3 notwithstanding, in many cases the class of unbiased estimators is non-trivial.

EXAMPLE 6.4: Suppose that $X_1, \cdots, X_n$ are i.i.d. random variables with density

$$f(x;\mu) = \exp(-(x-\mu)) \quad \text{for } x \geq \mu$$

(where $-\infty < \mu < \infty$). Two possible unbiased estimators of μ are

$$\widehat{\mu}_1 = X_{(1)} - \frac{1}{n} \quad \text{and} \quad \widehat{\mu}_2 = \bar{X} - 1;$$

note also that $t\widehat{\mu}_1 + (1-t)\widehat{\mu}_2$ is also unbiased for any t. Simple calculations reveal that

$$\text{Var}(\widehat{\mu}_1) = \frac{1}{n^2} \quad \text{and} \quad \text{Var}(\widehat{\mu}_2) = \frac{1}{n}$$

so that $\widehat{\mu}_1$ has the smaller variance. It is interesting to note that $\widehat{\mu}_1$ depends only on the one-dimensional sufficient statistic $X_{(1)}$. ◇

Does the estimator $\widehat{\mu}_1$ have the minimum variance among all unbiased estimators of μ in the previous example? The following result indicates that any candidate for the minimum variance unbiased estimator should be a function of the sufficient statistic.

THEOREM 6.1 (Rao-Blackwell Theorem) *Suppose that $\boldsymbol{X}$ has a joint distribution depending on some unknown parameter θ and that $T = T(\boldsymbol{X})$ is a sufficient statistic for θ. (Both θ and T can be vector-valued.) Let $S = S(\boldsymbol{X})$ be a statistic with $E_\theta(S) = g(\theta)$ and $Var_\theta(S) < \infty$ for all θ.*
If $S^* = E(S|T)$ then
(a) S^* is an unbiased estimator of $g(\theta)$, and
(b) $Var_\theta(S^*) \leq Var_\theta(S)$ for all θ.
Moreover, $Var_\theta(S^*) < Var_\theta(S)$ unless $P_\theta(S = S^*) = 1$.

Proof. Since T is sufficient for θ, $h(t) = E(S|T = t)$ does not depend on θ and so $S^* = h(T)$ is a statistic with $E_\theta(S^*) = E_\theta[E(S|T)] = E_\theta(S) = g(\theta)$. Also

$$\begin{aligned} \text{Var}_\theta(S) &= \text{Var}_\theta[E(S|T)] + E_\theta[\text{Var}(S|T)] \\ &\geq \text{Var}_\theta[E(S|T)] \\ &= \text{Var}_\theta(S^*) \end{aligned}$$

and so $\text{Var}_\theta(S^*) \leq \text{Var}_\theta(S)$. Also

$$\text{Var}(S|T) = E[(S - S^*)^2|T] > 0$$

unless $P(S = S^*|T) = 1$. Thus $E_\theta[\text{Var}(S|T)] > 0$ unless $S = S^*$ with probability 1. □

The Rao-Blackwell Theorem says that any unbiased estimator should be a function of a sufficient statistic; if not, we can construct an estimator with smaller variance merely by taking the conditional expectation given a sufficient statistic. However, this raises the question of which sufficient statistic to use to compute the conditional expectation. For example, suppose that S is an unbiased estimator of $g(\theta)$ (with finite variance) and T_1 and T_2 are both sufficient statistics for θ with $T_2 = h(T_1)$ for some function h. (Both T_1 and T_2 can be vector-valued.) We define $S_1^* = E(S|T_1)$ and $S_2^* = E(S|T_2)$. By the Rao-Blackwell Theorem, the variances of S_1^* and S_2^* cannot exceed Var(S); however, it is not obvious which "Rao-Blackwellized" estimator will have the smaller variance although intuition suggests that $\text{Var}_\theta(S_2^*)$ should be smaller since T_2 is the "simpler" statistic (as $T_2 = h(T_1)$).

PROPOSITION 6.2 *Suppose that S is an unbiased estimator* *of $g(\theta)$ with finite variance and define $S_1^* = E(S|T_1)$ and $S_2^* = E(S|T_2)$ for sufficient statistics T_1 and T_2. If $T_2 = h(T_1)$,*

$$\mathit{Var}_\theta(S_2^*) \le \mathit{Var}_\theta(S_1^*).$$

Proof. We will use the fact that $E_\theta[(S-\phi(T_1))^2]$ is minimized over all functions ϕ by $E(S|T_1)$. Since T_2 is a function of T_1, it follows that

$$\begin{aligned} E(S_1^*|T_2) &= E\left[E(S|T_1)|T_2\right] \\ &= E\left[E(S|T_2)|T_1\right] \\ &= E(S_2^*|T_1) \\ &= S_2^* \end{aligned}$$

where the last equality holds since S_2^* is a function of T_1. The conclusion now follows from the Rao-Blackwell Theorem. □

Proposition 6.2 essentially says that for any unbiased estimator S of $g(\theta)$ the best "Rao-Blackwellization" of S is achieved by conditioning on a minimal sufficient statistic. (Recall that a statistic T is minimal sufficient if for any other sufficient statistic T^*, $T = h(T^*)$ for some function h.) However, even if T is minimal sufficient, the estimator $S^* = E(S|T)$ will not necessarily have the minimum variance among all unbiased estimators since there may exist another unbiased estimator S_1 such that $S_1^* = E(S_1|T)$ has a smaller variance than S^*. In the next section, we will show that in certain problems, it is possible to find a sufficient statistic T such that $S^* = E(S|T)$ is independent of S. The following simple (and rather silly) example illustrates the potential problem.

EXAMPLE 6.5: Suppose X is a discrete random variable with frequency function

$$f(x;\theta) = \begin{cases} \theta & \text{for } x = -1 \\ (1-\theta)^2\theta^x & \text{for } x = 0, 1, 2, \cdots \end{cases}$$

where $0 < \theta < 1$. The statistic X is sufficient for θ and can be shown to be minimal sufficient. Two unbiased estimators of θ are $S_1 = I(X = -1)$ and $S_2 = I(X = -1) + X$ (since $E_\theta(X) = 0$ for all θ). Since both S_1 and S_2 are functions of the minimal sufficient statistic X, we have $S_1^* = E(S_1|X) = S_1$ and $S_2^* = E(S_2|X) = S_2$. But

$$\text{Var}_\theta(S_1) = \theta(1-\theta)$$

while

$$\begin{aligned}
\text{Var}_\theta(S_2) &= \text{Var}(S_1) + \text{Var}(X) + 2\text{Cov}(S_1, X) \\
&= \theta(1-\theta) + \frac{2\theta}{1-\theta} - 2\theta \\
&= \frac{\theta(\theta^2+1)}{1-\theta} \\
&> \text{Var}_\theta(S_1)
\end{aligned}$$

thus illustrating that dependence on the minimal sufficient statistic does not guarantee that an unbiased estimator will have minimum variance. ◇

Complete and Sufficient Statistics

Suppose that S_1 and S_2 are two unbiased estimators of $g(\theta)$ (with finite variance) and suppose that T is sufficient for θ. We can define $S_1^* = E(S_1|T)$ and $S_2^* = E(S_2|T)$. Although $\text{Var}_\theta(S_i^*) \leq \text{Var}_\theta(S_i)$ (for $i = 1, 2$), there is no way of knowing *a priori* whether S_1^* or S_2^* will have a smaller variance. However, for an appropriate choice of T, we would like to have

- $\text{Var}_\theta(S_1^*) = \text{Var}_\theta(S_2^*)$ for all θ, or
- $P_\theta(S_1^* = S_2^*) = 1$ for all θ.

More precisely, for this particular choice of T, there will be only one unbiased estimator that is a function of T and, if T is minimal sufficient, this unbiased estimator will have minimum variance.

EXAMPLE 6.6: Suppose that $X_1, \cdots, X_n$ are i.i.d. Poisson random variables with mean λ and consider unbiased estimators of $g(\lambda) = \exp(-\lambda) = P_\lambda(X_i = 0)$. Two unbiased estimators of $g(\lambda)$ are

$$S_1 = I(X_1 = 0) \quad \text{and} \quad S_2 = \frac{1}{n}\sum_{i=1}^{n} I(X_i = 0).$$

A minimal sufficient statistic for λ is $T = \sum_{i=1}^{n} X_i$. To find the "Rao-Blackwellized" estimators, we must find the conditional distributions of S_1 and S_2 given $T = t$. It is easy to see that

$$\begin{aligned}
P(S_1 = 1|T = t) &= \frac{P_\lambda(S_1 = s, T = t)}{P_\lambda(T = t)} \\
&= \left(1 - \frac{1}{n}\right)^t
\end{aligned}$$

and hence

$$S_1^* = E(S_1|T) = \left(1 - \frac{1}{n}\right)^T.$$

For S_2, define $U = \sum_{i=1}^n I(X_i = 0)$; we get

$$P(U = u|T = t) = \binom{n}{u}\left(1 - \frac{u}{n}\right)^t$$

for $u = \max(n-t, 0), \cdots, \max(n-1, n-t)$. It follows then that

$$E(U|T = t) = \sum_u u\binom{n}{u}\left(1 - \frac{u}{n}\right)^t = n\left(1 - \frac{1}{n}\right)^t$$

and so $S_2^* = (1 - 1/n)^T = S_1^*$. Thus, in this example, "Rao-Blackwellizing" S_1 and S_2 leads to the same estimator. ◇

DEFINITION. Suppose that $X_1, \cdots, X_n$ are random variables whose joint distribution depends on some unknown parameter θ. A statistic $T = T(X_1, \cdots, X_n)$ (possibly vector-valued) is said to be complete for θ if for any function g, $E_\theta[g(T)] = 0$ for all θ implies that $P_\theta[g(T) = 0] = 1$ for all θ.

Completeness of a statistic T essentially means that T contains no "ancillary" (meaningless) information about θ. For example, if T is complete for θ then $g(T)$ is an ancillary statistic for θ if, and only if, $g(T)$ is constant over the range of T. (More precisely, for some constant k, $P_\theta[g(T) = k] = 1$ for all θ.) It can be shown that if a statistic T is sufficient and complete then T is also minimal sufficient. However, a minimal sufficient statistic need not be complete.

EXAMPLE 6.7: Suppose that $X_1, \cdots, X_n$ are i.i.d. Poisson random variables with mean λ and define $T = \sum_{i=1}^n X_i$. T, of course, is sufficient for λ; to see if T is complete, we need to see if there exists a function g such that $E_\lambda[g(T)] = 0$ for all $\lambda > 0$. Since T has a Poisson distribution with mean $n\lambda$, we have

$$E_\lambda[g(T)] = \sum_{x=0}^{\infty} g(x)\frac{\exp(-n\lambda)(n\lambda)^x}{x!}$$

and so $E_\lambda[g(T)] = 0$ for all λ if, and only if,

$$\sum_{x=0}^{\infty} g(x)\frac{(n\lambda)^x}{x!} = 0 \quad \text{for all } \lambda > 0.$$

Since $\sum_{k=0}^{\infty} c_k \lambda^k = 0$ for all $a < \lambda < b$ if, and only if, $c_k = 0$ for all $k \geq 0$, it follows that $E_\lambda[g(T)] = 0$ for all $\lambda > 0$ if, and only if, $g(x)n^x/x! = 0$ for $x = 0, 1, \cdots$. Hence $E_\lambda[g(T)] = 0$ for all $\lambda > 0$ implies that $P_\lambda(g(T) = 0) = 1$ for all $\lambda > 0$. ◇

EXAMPLE 6.8: Suppose that $X_1, \cdots, X_n$ are i.i.d. discrete random variables with frequency function

$$f(x;\theta) = \begin{cases} \theta & \text{for } x = -1 \\ (1-\theta)^2\theta^x & \text{for } x = 0, 1, 2, \cdots \end{cases}$$

where $0 < \theta < 1$. The joint frequency function of $\boldsymbol{X} = (X_1, \cdots, X_n)$ can be written as a two-parameter exponential family:

$$f(\boldsymbol{x};\theta) = \exp\left[c_1(\theta)T_1(\boldsymbol{x}) + c_2(\theta)T_2(\boldsymbol{x}) + 2\ln(1-\theta)\right]$$

where $c_1(\theta) = \ln(\theta) - 2\ln(1-\theta)$, $c_2(\theta) = \ln(\theta)$ and

$$\begin{aligned} T_1(\boldsymbol{x}) &= \sum_{i=1}^{n} I(x_i = -1) \\ T_2(\boldsymbol{x}) &= \sum_{i=1}^{n} x_i I(x_i \geq 0). \end{aligned}$$

Thus $(T_1(\boldsymbol{X}), T_2(\boldsymbol{X})$ is sufficient (in fact, minimal sufficient) for θ. However, this sufficient statistic is not complete. To see this, note that

$$E_\theta(X_i) = -\theta + \sum_{x=0}^{\infty} x(1-\theta)^2\theta^x = 0.$$

Since

$$\begin{aligned} \sum_{i=1}^{n} X_i &= \sum_{i=1}^{n} X_i I(X_i \geq 0) - \sum_{i=1}^{n} I(X_i = -1) \\ &= T_2 - T_1, \end{aligned}$$

it follows that $E_\theta(T_2 - T_1) = 0$ for all θ. Since $P_\theta(T_2 = T_1) < 1$ for all θ, it follows that the sufficient statistic (T_1, T_2) is not complete. (This explains what happens in Example 6.5.) ◇

The following result gives a condition for a sufficient statistic to be complete in an exponential family.

THEOREM 6.3 *Suppose that $\boldsymbol{X} = (X_1, \cdots, X_n)$ have joint density or joint frequency function that is a k-parameter exponential* *family:*

$$f(\boldsymbol{x};\theta) = \exp\left[\sum_{i=1}^{k} c_i(\theta)T_i(\boldsymbol{x}) - d(\theta) + S(\boldsymbol{x})\right] \quad \textit{for } \boldsymbol{x} \in A$$

Define $C = \{(c_1(\theta), \cdots, c_k(\theta)) : \theta \in \Theta\}$. *If the set* C *contains an open set (rectangle) of the form* $(a_1, b_1) \times \cdots \times (a_k, b_k)$ *then the statistic* $(T_1(\boldsymbol{X}), \cdots, T_k(\boldsymbol{X}))$ *is complete as well as sufficient for* θ.

The proof of Theorem 6.3 is beyond the scope of this book; a proof can be found in Lehmann (1991). Essentially this result is a consequence of the uniqueness of characteristic functions. It says, roughly speaking, that a k-dimensional sufficient statistic in a k-parameter exponential family will also be complete provided that the dimension of the parameter space is k. Note that in Example 6.8, the parameter space is one-dimensional while the model is a two-parameter exponential family.

EXAMPLE 6.9: Suppose $X_1, \cdots, X_n$ are i.i.d. Normal random variables with mean μ and variance σ^2. The joint density can be written as a two-parameter exponential family:

$$f(\boldsymbol{x};\mu,\sigma) = \exp\left[-\frac{1}{2\sigma^2}\sum_{i=1}^{n} x_i^2 + \frac{\mu}{\sigma^2}\sum_{i=1}^{n} x_i - d(\mu,\sigma)\right]$$

with $d(\mu, \sigma) = n\mu^2/(2\sigma^2) + n\ln(\sigma) + n\ln(2\pi)/2$. Clearly range of the function

$$c(\mu,\sigma) = \left(-\frac{1}{2\sigma}, \frac{\mu}{\sigma^2}\right)$$

for $-\infty < \mu < \infty$ and $\sigma > 0$ contains an open rectangle in R^2 so the sufficient statistic

$$\left(\sum_{i=1}^{n} X_i, \sum_{i=1}^{n} X_i^2\right)$$

is complete as well as sufficient for (μ, σ). Moreover, as with sufficiency, any one-to-one function of a complete statistic will also be complete; thus, for example, $\left(\bar{X}, \sum_{i=1}^{n}(X_i - \bar{X})^2\right)$ is also complete. ◇

The following result gives a simple criterion for the existence of a UMVU estimator when a complete and sufficient statistic exists.

THEOREM 6.4 (Lehmann-Scheffé Theorem) *Suppose that T is a sufficient and complete statistic for θ and that S is a statistic with $E_\theta(S) = g(\theta)$ and $Var_\theta(S) < \infty$ for all θ. If $S^* = E(S|T)$ and V is any other unbiased estimator of $g(\theta)$ then*
(a) $Var_\theta(S^*) \leq Var_\theta(V)$
(b) $Var_\theta(S^*) = Var_\theta(V)$ implies that $P_\theta(S^* = V) = 1$.
(Thus S^* is the unique UMVU estimator of $g(\theta)$.)

Proof. Take V to be an arbitrary statistic with $E_\theta(V) = g(\theta)$ and $\text{Var}_\theta(V) < \infty$. Define $V^* = E(V|T)$ and note that $\text{Var}_\theta(V^*) \leq \text{Var}_\theta(V)$ by the Rao-Blackwell Theorem. It suffices to show that $P_\theta(S^* = V^*) = 1$ for all θ. Since both S^* and V^* are unbiased estimators of $g(\theta)$, we have

$$0 = E_\theta(S^* - V^*) = E_\theta\left[E(S|T) - E(V|T)\right] \quad \text{for all } \theta$$

or $E_\theta[h(T)] = 0$ for all θ where $h(T) = E(S|T) - E(V|T) = S^* - V^*$. Since T is complete, it follows that $P_\theta(S^* - V^* = 0) = 1$ for all θ, which completes the proof. □

The Lehmann-Scheffé Theorem states that if a complete and sufficient statistic T exists, then the UMVU estimator of $g(\theta)$ (if it exists) must be a function of T; moreover, if the UMVU estimator exists then it is unique. The Lehmann-Scheffé Theorem also simplifies the search for unbiased estimators considerably: if a complete and sufficient statistic T exists and there exists no function h such that $E_\theta[h(T)] = g(\theta)$ then no unbiased estimator of $g(\theta)$ exists.

Taken together, the Rao-Blackwell and Lehmann-Scheffé Theorems also suggest two approaches to finding UMVU estimators when a complete and sufficient statistic T exists.

- Find a function h such that $E_\theta[h(T)] = g(\theta)$. If $\text{Var}_\theta[h(T)] < \infty$ for all θ then $h(T)$ is the unique UMVU estimator of $g(\theta)$. The function h can be determined by solving the equation $E_\theta[h(T)] = g(\theta)$ or by making an educated guess.
- Given an unbiased estimator S of $g(\theta)$, define the "Rao-Blackwellized" estimator $S^* = E(S|T)$. Then S^* is the unique UMVU estimator of $g(\theta)$.

EXAMPLE 6.10: Suppose that $X_1, \cdots, X_n$ are i.i.d. Bernoulli random variables with parameter θ. By the Neyman Factorization Criterion, $T = X_1 + \cdots + X_n$ is sufficient for θ and since the distribution of $(X_1, \cdots, X_n)$ is a one-parameter exponential family,

T is also complete. Suppose we want to find the UMVU estimator of θ^2.

First suppose that $n = 2$. If a UMVU estimator exists, it must be of the form $h(T)$ where the function h satisfies

$$\theta^2 = \sum_{k=0}^{2} h(k) \binom{2}{k} \theta^k (1-\theta)^{2-k}.$$

It is easy to see that $h(0) = h(1) = 0$ while $h(2) = 1$. Thus $h(T) = T(T-1)/2$ is the unique UMVU estimator of θ^2 when $n = 2$.

For $n > 2$, we set $S = I(X_1 + X_2 = 2)$ and note that this is an unbiased estimator of θ^2. By the Lehman-Scheffé Theorem, $S^* = E(S|T)$ is the unique UMVU estimator of θ^2. We have

$$\begin{aligned} E(S|T=t) &= P(X_1 + X_2 = 2|T=t) \\ &= \frac{P_\theta(X_1 + X_2 = 2, X_3 + \cdots + X_n = t-2)}{P_\theta(T=t)} \\ &= \begin{cases} 0 & \text{if } t \le 1 \\ \binom{n-2}{t-2} \Big/ \binom{n}{t} & \text{if } t \ge 2 \end{cases} \\ &= \frac{t(t-1)}{n(n-1)}. \end{aligned}$$

Thus $S^* = T(T-1)/[n(n-1)]$ is the UMVU estimator of θ^2. ◇

EXAMPLE 6.11: Suppose that $X_1, \cdots, X_n$ are i.i.d. Normal random variables with unknown mean and variance, μ and σ^2. It was shown earlier that the statistic

$$(T_1, T_2) = \left(\sum_{i=1}^{n} X_i, \sum_{i=1}^{n} (X_i - \bar{X})^2 \right)$$

is sufficient and complete for (μ, σ). Consider unbiased estimators of μ/σ. It is not obvious that an unbiased estimator exists; however, T_1 is independent of T_2 with $T_2/\sigma^2 \sim \chi^2_{n-1}$, which suggests that

$$E\left(T_1/\sqrt{T_2}\right) = E(T_1) E\left(T_2^{-1/2}\right) = k_n \frac{\mu}{\sigma}$$

where the constant k_n depends only on n and not on μ or σ. It follows that $E(T_1) = n\mu$ while

$$E\left(T_2^{-1/2}\right) = \frac{\Gamma(n/2 - 1)}{\sqrt{2}\Gamma\left((n-1)/2\right)} \sigma^{-1} \quad \text{(for } n \ge 3\text{)}$$

and so

$$k_n = \frac{n\Gamma(n/2-1)}{\sqrt{2}\Gamma\left((n-1)/2\right)}.$$

Thus $k_n^{-1}T_1/\sqrt{T_2}$ is an unbiased estimator of μ/σ when $n \geq 3$ and has finite variance when $n \geq 4$; since the estimator is a function of the sufficient and complete statistic, it is the unique UMVU estimator. ◇

6.4 The Cramér-Rao lower bound

In the previous section, we saw that UMVU estimators of $g(\theta)$ could be found if a complete and sufficient statistic existed. However, in many problems, the minimal sufficient statistic is not complete and so we cannot appeal to the Lehmann-Scheffé Theorem to find UMVU estimators. In this section, we will derive a lower bound for the variance of an unbiased estimator of $g(\theta)$; if the variance of some unbiased estimator achieves this lower bound, then the estimator will be UMVU.

Suppose that $X_1, \cdots, X_n$ are random variables with joint density or frequency function depending on a real-valued parameter θ and consider unbiased estimators of $g(\theta)$ (if they exist). Suppose that $S = S(X_1, \cdots, X_n)$ is an unbiased estimator of $g(\theta)$. Under fairly weak regularity conditions, we would like to find a function $\phi(\theta)$ such that

$$\mathrm{Var}_\theta(S) \geq \phi(\theta) \quad \text{(for all } \theta)$$

for any statistic S with $E_\theta(S) = g(\theta)$. Moreover, we would like to be able to find unbiased estimators such that the lower bound is achieved or comes close to being achieved. (For example, 0 is always a lower bound for $\mathrm{Var}_\theta(S)$ but is typically unattainable.)

The Cauchy-Schwarz inequality states that if U and V are random variables with $E(U^2) < \infty$ and $E(V^2) < \infty$ then

$$[\mathrm{Cov}(U,V)]^2 \leq \mathrm{Var}(U)\mathrm{Var}(V).$$

Using this result, we obtain the following lower bound for $\mathrm{Var}_\theta(S)$,

$$\mathrm{Var}_\theta(S) \geq \frac{[\mathrm{Cov}_\theta(S,U)]^2}{\mathrm{Var}_\theta(U)},$$

which is valid for any random variable U. However, as it stands, this lower bound is not particularly useful since $\mathrm{Cov}_\theta(S,U)$ will generally depend on S and not merely on θ. Fortunately, it is possible to find a random variable U such that $\text{Cov}_\theta(S, U)$ depends only on $g(\theta) = E_\theta(S)$.

We will assume that θ is real-valued and make the following assumptions about the joint density or frequency function of $\boldsymbol{X} = (X_1, \cdots, X_n)$:

(1) The set $A = \{\boldsymbol{x} : f(\boldsymbol{x}; \theta) > 0\}$ does not depend on θ.

(2) For all $\boldsymbol{x} \in A$, $f(\boldsymbol{x}; \theta)$ is differentiable with respect to θ.

(3) $E_\theta[U_\theta(X_1, \cdots, X_n)] = 0$ where

$$U_\theta(\boldsymbol{x}) = \frac{\partial}{\partial \theta} \ln f(\boldsymbol{x}; \theta).$$

(4) For a statistic $T = T(\boldsymbol{X})$ with $E_\theta(|T|) < \infty$ (for all θ) and $g(\theta) = E_\theta(T)$ differentiable,

$$g'(\theta) = E_\theta[T U_\theta(\boldsymbol{X})]$$

for all θ.

At first glance, assumptions (3) and (4) may not appear to make sense. However, suppose that $f(\boldsymbol{x}; \theta)$ is a density function. Then for some statistic S,

$$\begin{aligned}
\frac{d}{d\theta} E_\theta(S) &= \frac{d}{d\theta} \int \cdots \int S(\boldsymbol{x}) f(\boldsymbol{x}; \theta)\, d\boldsymbol{x} \\
&= \int \cdots \int S(\boldsymbol{x}) \frac{\partial}{\partial \theta} f(\boldsymbol{x}; \theta)\, d\boldsymbol{x} \\
&= \int \cdots \int S(\boldsymbol{x}) U_\theta(\boldsymbol{x})\, d\boldsymbol{x} \\
&= E_\theta[S U_\theta]
\end{aligned}$$

provided that the derivative may be taken inside the integral sign. Setting $S = 1$ and $S = T$, it is easy to see that assumptions (3) and (4) hold.

THEOREM 6.5 (Cramér-Rao lower bound)
Suppose that $\boldsymbol{X} = (X_1, \cdots, X_n)$ has a joint density (frequency) function $f(\boldsymbol{x}; \theta)$ satisfying assumptions (1), (2), and (3). If the statistic T satisfies assumption (4) then

$$Var_\theta(T) \geq \frac{[g'(\theta)]^2}{I(\theta)}$$

where $I(\theta) = E_\theta(U_\theta^2)$.

Proof. By the Cauchy-Schwarz inequality,

$$\mathrm{Var}_\theta(T) \geq \frac{\mathrm{Cov}_\theta^2(T, U_\theta)}{\mathrm{Var}_\theta(U_\theta)}.$$

Since $E_\theta(U_\theta) = 0$, it follows that $\mathrm{Var}_\theta(U_\theta) = I(\theta)$. Also

$$\begin{aligned}\mathrm{Cov}_\theta(T, U_\theta) &= E_\theta(TU_\theta) - E_\theta(T)E_\theta(U_\theta) \\ &= E_\theta(TU_\theta) \\ &= \frac{d}{d\theta} E_\theta(T) = g'(\theta),\end{aligned}$$

which completes the proof. $\square$

When is the Cramér-Rao lower bound attained? If

$$\mathrm{Var}_\theta(T) = \frac{[g'(\theta)]^2}{I(\theta)}$$

then

$$\mathrm{Var}_\theta(T) = \frac{\mathrm{Cov}_\theta^2(T, U_\theta)}{\mathrm{Var}_\theta(U_\theta)},$$

which occurs if, and only if, U_θ is a linear function of T; that is, with probability 1,

$$U_\theta = A(\theta)T + B(\theta)$$

or

$$\frac{\partial}{\partial\theta} \ln f(\boldsymbol{x}; \theta) = A(\theta)T(\boldsymbol{x}) + B(\theta)$$

for all $\boldsymbol{x} \in A$. Hence

$$\ln f(\boldsymbol{x}; \theta) = A^*(\theta)T(\boldsymbol{x}) + B^*(\theta) + S(\boldsymbol{x})$$

and so $\mathrm{Var}_\theta(T)$ attains the Cramér-Rao lower bound if, and only if, the density or frequency function of $(X_1, \cdots, X_n)$ has the one-parameter exponential family form given above. In particular, the estimator T must be a sufficient statistic.

If $X_1, \cdots, X_n$ are i.i.d. random variables with common density $f(x; \theta)$ then

$$U_\theta(\boldsymbol{x}) = \sum_{i=1}^{n} \frac{\partial}{\partial\theta} \ln f(x_i; \theta)$$

and so

$$I(\theta) = \mathrm{Var}_\theta(U_\theta^2) = n\mathrm{Var}_\theta\left(\frac{\partial}{\partial\theta} \ln f(X_1; \theta)\right).$$

An alternative method for computing $I(\theta)$ can frequently be used. If $E_\theta(T)$ can be differentiated twice with respect to θ under the integral or sum signs then

$$I(\theta) = E_\theta[H_\theta(\boldsymbol{X})]$$

where

$$H_\theta(\boldsymbol{x}) = -\frac{\partial^2}{\partial\theta^2} \cdot \ln f(\boldsymbol{x};\theta)$$

A similar result was proved in Chapter 5 in the case of finding the asymptotic variance of the MLE. This alternative method works in many models including one-parameter exponential families.

EXAMPLE 6.12: Suppose that $X_1, \cdots, X_n$ are i.i.d. Poisson random variables with mean λ. A simple calculation yields

$$U_\lambda(\boldsymbol{x}) = \sum_{i=1}^{n} \frac{\partial}{\partial\lambda} \ln f(x_i;\lambda) = -n + \frac{1}{\lambda}\sum_{i=1}^{n} x_i$$

and so

$$I(\lambda) = \text{Var}_\lambda(U_\lambda) = \frac{1}{\lambda^2}\sum_{i=1}^{n} \text{Var}_\lambda(X_i) = \frac{n}{\lambda}.$$

Also note that

$$H_\lambda(\boldsymbol{x}) = -\sum_{i=1}^{n} \frac{\partial^2}{\partial\lambda^2} \ln f(x_i;\lambda) = \frac{1}{\lambda^2}\sum_{i=1}^{n} x_i$$

and so $E_\lambda(H_\lambda) = n/\lambda = I(\lambda)$.

The Cramér-Rao lower bound for unbiased estimators of λ is simply λ/n and this lower bound is attained by the estimator $T = \bar{X}$. In fact, the lower bound can only be attained by estimators of the form $aT + b$ for constants a and b.

Now consider unbiased estimators of λ^2; the Cramér-Rao lower bound in this case is

$$\frac{(2\lambda)^2}{I(\lambda)} = \frac{4\lambda^3}{n}.$$

From above, we know that no unbiased estimator attains this lower bound. However, we would like to see how close different estimators come to attaining the lower bound. Using the fact that

$$E_\lambda[X_i(X_i - 1)] = \lambda^2,$$

it follows that

$$T_1 = \frac{1}{n}\sum_{i=1}^{n} X_i(X_i - 1)$$

is an unbiased estimator of λ^2; a tedious calculation yields

$$\text{Var}_\lambda(T_1) = \frac{4\lambda^3}{n} + \frac{2\lambda^2}{n}.$$

Since $\bar{X}$ is sufficient and complete for λ, the UMVU estimator of λ^2 is $T_2 = E(T_1|\bar{X})$ with

$$\text{Var}_\lambda(T_2) = \frac{4\lambda^3}{n} + \frac{2\lambda^2}{n^2}.$$

Finally consider the biased estimator $T_3 = (\bar{X})^2$, which is the MLE of λ^2. Since $n\bar{X}$ has a Poisson distribution with mean $n\lambda$, another tedious calculation yields

$$\begin{aligned} E_\lambda(T_3) &= \lambda^2 + \frac{\lambda}{n} \\ \text{and} \quad \text{Var}_\lambda(T_3) &= \frac{4\lambda^3}{n} + \frac{5\lambda^2}{n^2} + \frac{\lambda}{n^3}. \end{aligned}$$

The difference between the UMVE estimator and the MLE of λ^2 becomes negligible for large n. ◇

EXAMPLE 6.13: Suppose that $X_1, \cdots, X_n$ are i.i.d. Logistic random variables with density function

$$f(x;\theta) = \frac{\exp(x-\theta)}{[1+\exp(x-\theta)]^2}.$$

It is easy to verify that $I(\theta) = n/3$ and so if $\widehat{\theta}$ is an unbiased estimator of θ based on $X_1, \cdots, X_n$, we have

$$\text{Var}_\theta(\widehat{\theta}) \geq \frac{3}{n}.$$

Since the model is not a one-parameter exponential family, no unbiased estimator attains the lower bound. For example, $\bar{X}$ is an unbiased estimator of θ with $\text{Var}_\theta(\bar{X}) = \pi^2/(3n) \approx 3.29/n$. The sample median is also unbiased; asymptotic theory predicts that its variance is approximately $4/n$. It follows from standard asymptotic theory for MLEs that $\sqrt{n}(\widehat{\theta}_n - \theta) \to_d N(0,3)$ for the MLE, which suggests that for large n, $\text{Var}_\theta(\widehat{\theta}_n) \approx 3/n$. ◇

6.5 Asymptotic efficiency

Suppose that $X_1, \cdots, X_n$ are i.i.d. random variables with density or frequency function $f(x;\theta)$ where θ is a real-valued parameter. In

Chapter 4, we showed that, subject to some regularity conditions, if $\widehat{\theta}_n$ is the MLE of θ then

$$\sqrt{n}(\widehat{\theta}_n - \theta) \to_d N(0, 1/I(\theta))$$

where

$$I(\theta) = \text{Var}_\theta \left[\frac{\partial}{\partial \theta} \ln f(X_1; \theta) \right].$$

This result suggests that for sufficiently large n,

$$E_\theta(\widehat{\theta}_n) \approx \theta \quad \text{and} \quad \text{Var}_\theta(\widehat{\theta}_n) \approx \frac{1}{nI(\theta)}.$$

On the other hand if T is any statistic (based on $X_1, \cdots, X_n$) with $E_\theta(T) = \theta$) then

$$\text{Var}_\theta(T) \geq \frac{1}{nI(\theta)}.$$

Juxtaposing the Cramér-Rao lower bound with the asymptotic theory for MLEs (as developed in Chapter 5) raises the following question: If $\widetilde{\theta}_n$ is a sequence of estimators with

$$\sqrt{n}(\widetilde{\theta}_n - \theta) \to_d N(0, \sigma^2(\theta))$$

then is $\sigma^2(\theta) \geq 1/I(\theta)$ for all θ? The answer to the question is a qualified "yes" although we will show it is possible to find estimators for which $\sigma^2(\theta) < 1/I(\theta)$ for some θ.

What qualifications must be made in order to conclude that the asymptotic variance $\sigma^2(\theta) \geq 1/I(\theta)$? It will follow from results given below that if $\sigma^2(\theta)$ is a continuous function of θ then $\sigma^2(\theta) \geq 1/I(\theta)$. The following example shows that $\sigma^2(\theta) < 1/I(\theta)$ for some θ if $\sigma^2(\theta)$ is not continuous.

EXAMPLE 6.14: Suppose $X_1, X_2, \cdots, X_n$ are i.i.d. Normal random variables with mean θ and variance 1; for this model, $I(\theta) = 1$ for all θ. Consider the estimator

$$\widetilde{\theta}_n = \begin{cases} \bar{X}_n & \text{if } |\bar{X}_n| \geq n^{-1/4} \\ a\bar{X}_n & \text{if } |\bar{X}_n| < n^{-1/4} \end{cases}$$

where a is a constant with $|a| < 1$. We will show that $\sqrt{n}(\widetilde{\theta}_n - \theta) \to_d N(0, \sigma^2(\theta))$ where

$$\sigma^2(\theta) = \begin{cases} 1 & \text{if } \theta \neq 0 \\ a^2 & \text{if } \theta = 0 \end{cases};$$

thus, $\sigma^2(0) < 1/I(0)$. To prove this, note that $\sqrt{n}(\bar{X}_n - \theta) =_d Z$ where $Z \sim N(0,1)$. It follows then that

$$\begin{aligned}\sqrt{n}(\widetilde{\theta}_n - \theta) &= \sqrt{n}(\bar{X}_n - \theta)I(|\bar{X}_n| \geq n^{-1/4}) \\ &\quad + \sqrt{n}(a\bar{X}_n - \theta)I(|\bar{X}_n| < n^{-1/4}) \\ &= \sqrt{n}(\bar{X}_n - \theta)I(\sqrt{n}|\bar{X}_n - \theta + \theta| \geq n^{1/4}) \\ &\quad + \sqrt{n}(a\bar{X}_n - \theta)I(\sqrt{n}|\bar{X}_n - \theta + \theta| < n^{1/4}) \\ &=_d ZI(|Z + \sqrt{n}\theta| \geq n^{1/4}) \\ &\quad + [aZ + \sqrt{n}\theta(a-1)]I(|Z + \sqrt{n}\theta| < n^{1/4}).\end{aligned}$$

Now note that $Z + \sqrt{n}\theta \sim N(\sqrt{n}\theta, 1)$ and so

$$I(|Z + \sqrt{n}\theta| \geq n^{1/4}) \to_p \begin{cases} 0 & \text{if } \theta = 0 \\ 1 & \text{if } \theta \neq 0. \end{cases}$$

Thus we have

$$ZI(|Z + \sqrt{n}\theta| \geq n^{1/4}) \to_p \begin{cases} Z & \text{if } \theta \neq 0 \\ 0 & \text{if } \theta = 0 \end{cases}$$

and also

$$[aZ + \sqrt{n}\theta(a-1)]I(|Z + \sqrt{n}\theta| < n^{1/4}) \to_p \begin{cases} 0 & \text{if } \theta \neq 0 \\ aZ & \text{if } \theta = 0. \end{cases}$$

Thus

$$\sqrt{n}(\widetilde{\theta}_n - \theta) \to_d \begin{cases} Z & \text{if } \theta \neq 0 \\ aZ & \text{if } \theta = 0. \end{cases}$$

This example was first given by J.L. Hodges in 1952. ◇

Asymptotically Normal estimators whose limiting variance $\sigma^2(\theta)$ satisfies

$$\sigma^2(\theta_0) < \frac{1}{I(\theta_0)}$$

for some θ_0 are often called superefficient estimators. Bahadur (1964) showed that (subject to some weak regularity conditions) the set of θ for which $\sigma^2(\theta) < 1/I(\theta)$ is at most countable. We can also define the notion of "regularity" of a sequence of estimators, an idea dating back to Hájek (1970). We say that a sequence of estimators $\{\widehat{\theta}_n\}$ is regular at θ if, for $\theta_n = \theta + c/\sqrt{n}$,

$$\lim_{n\to\infty} P_{\theta_n}\left(\sqrt{n}(\widehat{\theta}_n - \theta_n) \leq x\right) = G_\theta(x)$$

where the limiting distribution function $G_\theta(x)$ can depend on θ but not c. It is possible to verify that the sequence of estimators

$\{\widehat{\theta}_n\}$ given in Example 6.14 is not regular at $\theta = 0$. Virtually all "standard" estimators such as maximum likelihood and substitution principle estimators are regular.

EXAMPLE 6.15: Suppose that $X_1, X_2, \cdots, X_n$ are i.i.d. Normal random variables with mean θ and variance 1; define $\widehat{\theta}_n = \bar{X}_n$. If $\theta_n = \theta + c/\sqrt{n}$ is the true parameter value then $\widehat{\theta}_n$ has a Normal distribution with mean θ_n and variance $1/n$; hence $\sqrt{n}(\widehat{\theta}_n - \theta_n)$ has a standard Normal distribution for all n and so $\{\widehat{\theta}_n\}$ is regular (at any θ). ◇

EXAMPLE 6.16: Suppose that $X_1, X_2, \cdots, X_n$ are i.i.d. Exponential random variables with parameter λ and define $\widehat{\lambda}_n = 1/\bar{X}_n$. If $\lambda_n = \lambda + c/\sqrt{n}$, it is possible to show (using, for example, the Lyapunov Central Limit Theorem) that

$$P_{\lambda_n}\left(\sqrt{n}(\bar{X}_n - \lambda_n^{-1}) \le x\right) \to \Phi(\lambda x)$$

where $\Phi(\cdot)$ is the standard Normal distribution function. Then using a "Delta Method"-type argument, we get

$$P_{\lambda_n}\left(\sqrt{n}(\widehat{\lambda}_n - \lambda_n) \le x\right) \to \Phi(x/\lambda)$$

and so $\{\widehat{\lambda}_n\}$ is a regular sequence of estimators. ◇

The following theorem gives a representation for regular estimators based on i.i.d. random variables in the case where the log-likelihood function can be approximated by a quadratic function in a neighbourhood of the true parameter θ.

THEOREM 6.6 *Let $X_1, X_2, \cdots, X_n$ be i.i.d. random variables with density or frequency function $f(x;\theta)$ and suppose that $\{\widehat{\theta}_n\}$ is a sequence of estimators that is regular at θ. If*

$$\sum_{i=1}^{n}\left[\ln f(X_i;\theta + c/\sqrt{n}) - \ln f(X_i;\theta)\right] = cS_n(\theta) - \frac{1}{2}c^2 I(\theta) + R_n(c,\theta)$$

where $S_n(\theta) \to_d N(0, I(\theta))$ and $R_n(c,\theta) \to_p 0$ for all c then

$$\sqrt{n}(\widehat{\theta}_n - \theta) \to_d Z_1 + Z_2$$

where $Z_1 \sim N(0, 1/I(\theta))$ and Z_2 is independent of Z_1.

In most cases, $\sqrt{n}(\widehat{\theta}_n - \theta) \to_d N(0, \sigma^2(\theta))$ so that the random variable Z_2 in Theorem 6.6 has a Normal distribution with mean

0 and variance $\sigma^2(\theta) - 1/I(\theta)$. If $\sigma^2(\theta) = 1/I(\theta)$ then the sequence $\{\widehat{\theta}_n\}$ is said to be asymptotically efficient. Theorem 6.6 can also be extended to handle an estimator of a vector parameter $\boldsymbol{\theta}$ in which case Z_1 becomes a multivariate Normal random vector with variance-covariance matrix $[I(\boldsymbol{\theta})]^{-1}$.

It is important to note that, under the conditions of Theorem 6.6, the sequence of MLEs $\{\widehat{\theta}_n\}$ typically satisfies $\sqrt{n}(\widehat{\theta}_n - \theta) \to_d N(0, 1/I(\theta))$, which establishes the MLE as the most efficient of all regular estimators. However, it is also important to recognize that there may be other regular estimators that, while not MLEs, have the same asymptotic variance as the MLE and, at the same time, have superior finite sample properties.

The proof of Theorem 6.6 is quite technical and will not be given here. We will however sketch a proof of the result under somewhat stronger assumptions. In particular, we will assume that

$$E_\theta\left[\exp\left(t_1\sqrt{n}(\widehat{\theta}_n - \theta) + t_2 S_n(\theta)\right)\right] \to m(t_1, t_2)$$

and that as $n \to \infty$

$$E_{\theta_n}\left[\exp(\left(t_1\sqrt{n}(\widehat{\theta}_n - \theta_n)\right)\right] \to m(t_1, 0)$$

(where $\theta_n = \theta + c/\sqrt{n}$) for $|t_1| \le b$, $|t_2| \le b$ where $b > 0$. We need to show that $m(t_1, 0)$ is the product of two moment generating functions, one of which is the moment generating function of a $N(0, 1/I(\theta))$ random variable. First of all, for $\theta_n = \theta + c/\sqrt{n}$, we have that

$$\begin{aligned} & E_{\theta_n}\left[\exp\left(t_1\sqrt{n}(\widehat{\theta}_n - \theta)\right)\right] \\ &= \exp(-t_1 c) E_{\theta_n}\left[\exp\left(t_1\sqrt{n}(\widehat{\theta}_n - \theta_n)\right)\right] \\ &\to \exp(-t_1 c) m(t_1, 0). \end{aligned}$$

On the other hand, if we set

$$W_n(\theta, c) = \sum_{i=1}^n \left[\ln f(X_i; \theta + c/\sqrt{n}) - \ln f(X_i; \theta)\right],$$

we also have

$$\begin{aligned} & E_{\theta_n}\left[\exp\left(t_1\sqrt{n}(\widehat{\theta}_n - \theta)\right)\right] \\ &= E_\theta\left[\exp\left(t_1\sqrt{n}(\widehat{\theta}_n - \theta) + W_n(\theta, c)\right)\right] \end{aligned}$$

$$\to \quad m(t_1, c) \exp\left(-\frac{1}{2}c^2 I(\theta)\right),$$

which follows by substituting the approximately quadratic function (in c) for $W_n(\theta, c)$. Equating the two limits above, it follows that

$$m(t_1, 0) = m(t_1, c) \exp\left(-t_1 c - \frac{1}{2}c^2 I(\theta)\right)$$

and setting $c = -t_1/I(\theta)$, we get

$$m(t_1, 0) = m(t_1, -t_1/I(\theta)) \exp\left(\frac{t_1^2}{2I(\theta)}\right).$$

It is easy to verify that $m(t_1, -t_1/I(\theta))$ is a moment generating function and $\exp\left(t_1^2/(2I(\theta))\right)$ is the moment generating function of a $N(0, 1/I(\theta))$ random variable.

It should be noted that the rigorous proof of Theorem 6.6 is very similar to that given above except that characteristic functions are used and a few other technical difficulties must be addressed. Theorem 6.6 also holds in the multi-parameter case where Z_1 and Z_2 are random vectors and $I(\theta)$ is a matrix.

Verifying the regularity of a sequence of estimators is generally a tedious process. The assumption of regularity in Theorem 6.6 can be replaced by a more natural condition (Tierney, 1987), namely that

$$\lim_{n\to\infty} P_\theta\left(\sqrt{n}(\widehat{\theta}_n - \theta) \le x\right) = G_\theta(x)$$

where the limiting distribution function $G_\theta(x)$ is "continuous" in θ in the sense that

$$\int_{-\infty}^{\infty} h(x)\, dG_\theta(x)$$

is a continuous function of θ for all bounded, continuous functions $h(x)$. For example, if $G_\theta(x)$ is the $N(0, \sigma^2(\theta))$ distribution function and $\sigma^2(\theta)$ is a continuous function of θ then this condition is satisfied. Sequences of estimators satisfying this "continuous limit" condition are typically regular and vice versa although exceptions in both directions can be found.

It is important to view Theorem 6.6 in its proper context, that is, as an asymptotic optimality result for MLEs within a particular class of estimators, namely regular estimators. In particular, it is somewhat tempting to dismiss non-regular estimators as contrivances that would never be used in practice. While this is probably true of the estimator in Example 6.14, there are estimators used in practice (particularly in multi-parameter problems) that are similar in spirit to the Hodges estimator of Example 6.14; for example, in regression models, it is quite common to set certain parameter estimators to 0 if some preliminary estimator of that parameter falls below a specified threshold.

6.6 Problems and complements

6.1: Suppose that $\boldsymbol{X} = (X_1, \cdots, X_n)$ have joint density or frequency function $f(\boldsymbol{x}; \theta)$ where θ is a real-valued parameter with a proper prior density function $\pi(\theta)$. For squared error loss, define the Bayes risk of an estimator $\widehat{\theta} = S(\boldsymbol{X})$:

$$R_B(\widehat{\theta}, \theta) = \int_{\Theta} E_\theta[(\widehat{\theta} - \theta)^2]\pi(\theta)\, d\theta.$$

The Bayes estimator of minimizes the Bayes risk.

(a) Show that the Bayes estimator is the mean of the posterior distribution of θ.

(b) Suppose that the Bayes estimator in (a) is also an unbiased estimator. Show that the Bayes risk of this estimator must be 0. (This result implies that Bayes estimators and unbiased estimators agree only in pathological examples.)

6.2: Suppose that $X \sim \text{Bin}(n, \theta)$ where θ has a Beta prior:

$$\pi(\theta)\frac{\Gamma(\alpha + \beta)}{\Gamma(\alpha)\Gamma(\beta)}\theta^{\alpha-1}(1 - \theta)^{\beta-1}$$

for $0 < \theta < 1$.

(a) Show that the Bayes estimator of θ under squared error loss is $(X + \alpha)/(\alpha + \beta + n)$.

(b) Find the mean square error of the Bayes estimator in (a). Compare the mean square error to that of the unbiased estimator X/n.

6.3: Suppose that $X_1, \cdots, X_n$ are i.i.d. Poisson random variables with mean θ where θ has a Gamma (α , β) prior distribution.

(a) Show that

$$\widehat{\theta} = \frac{\alpha + \sum_{i=1}^n X_i}{\beta + n}$$

is the Bayes estimator of θ under squared error loss.

(b) Use the result of (a) to show that any estimator of the form

$a\bar{X} + b$ for $0 < a < 1$ and $b > 0$ is an admissible estimator of θ under squared error loss.

6.4: Suppose that $X_1, \cdots, X_n$ are i.i.d. Normal random variables with mean θ and variance 1 where θ has the improper prior density $\pi(\theta) = \exp(\theta)$ for $-\infty < \theta < \infty$.

(a) Find the posterior density of θ given $X_1, \cdots, X_n$.

(b) Find the posterior mean of θ. Is this estimator admissible under squared error loss?

6.5: Given a loss function L, we want to find a minimax estimator of a parameter θ.

(a) Suppose that $\widehat{\theta}$ is a Bayes estimator of θ for some prior distribution $\pi(\theta)$ with

$$R_B(\widehat{\theta}) = \sup_{\theta \in \Theta} R_\theta(\widehat{\theta}).$$

Show that $\widehat{\theta}$ is a minimax estimator. (The prior distribution π is called a least favourable distribution.)

(b) Let $\{\pi_n(\theta)\}$ be a sequence of prior density functions on Θ and suppose that $\{\widehat{\theta}_n\}$ are the corresponding Bayes estimators. If $\widehat{\theta}_0$ is an estimator with

$$\sup_{\theta \in \Theta} R_\theta(\widehat{\theta}_0) = \lim_{n \to \infty} \int_\Theta R_\theta(\widehat{\theta}_n) \pi(\theta)\, d\theta,$$

show that $\widehat{\theta}_0$ is a minimax estimator.

(c) Suppose that $X \sim \text{Bin}(n, \theta)$. Assuming squared error loss, find a minimax estimator of θ. (Hint: Use a Beta prior as in Problem 6.2.)

6.6: The Rao-Blackwell Theorem can be extended to convex loss functions. Let $L(a, b)$ be a loss function that is convex in a for each fixed b. Let $S = S(\boldsymbol{X})$ be some estimator and suppose that $T = T(\boldsymbol{X})$ is sufficient for θ. Show that

$$E_\theta[L(E(S|T), \theta)] \leq E_\theta[L(S, \theta)].$$

(Hint: Write $E_\theta[L(S,\theta)] = E_\theta[E[L(S,\theta)|T]]$. Then apply Jensen's inequality to the conditional expected value.)

6.7: Suppose that $\boldsymbol{X} = (X_1, \cdots, X_n)$ are random variables with joint density or frequency function $f(\boldsymbol{x}; \theta)$ and suppose that $T = T(\boldsymbol{X})$ is sufficient for θ.

(a) Suppose that there exists no function $\phi(t)$ such that $\phi(T)$ is an unbiased estimator of $g(\theta)$. Show that no unbiased estimator of $g(\theta)$ (based on $\boldsymbol{X}$) exists.

6.8: Suppose that X has a Binomial distribution with parameters n and θ where θ is unknown. Consider unbiased estimators of $g(\theta) = \theta^2(1-\theta)^2$.

(a) Show that no unbiased estimator of $g(\theta)$ exists if $n \leq 3$.

(b) Find the UMVU estimator of $g(\theta)$ when $n \geq 4$. (Hint: Consider the case $n = 4$ first.)

6.9: Suppose that $\boldsymbol{X} = (X_1, \cdots, X_n)$ have a joint distribution depending on a parameter θ where $T = T(\boldsymbol{X})$ is sufficient for θ.

(a) Prove Basu's Theorem: If $S = S(\boldsymbol{X})$ is an ancillary statistic and the sufficient statistic T is complete then T and S are independent. (Hint: It suffices to show that $P(S \in A|T) = P(S \in A)$ for any set A; note that neither $P(S \in A|T)$ nor $P(S \in A)$ depends on θ. Use the completeness of T to argue that $P(S \in A|T) = P(S \in A)$.)

(b) Suppose that X and Y are independent Exponential random variables with parameter λ. Use Basu's Theorem to show that $X + Y$ and $X/(X+Y)$ are independent.

(c) Suppose that $X_1, \cdots, X_n$ are i.i.d. Normal random variables with mean μ and variance σ^2. Let $T = T(X_1, \cdots, X_n)$ be a statistic such that

$$T(X_1 + a, X_2 + a, \cdots, X_n + a) = T(X_1, \cdots, X_n) + a$$

and $E(T) = \mu$. Show that

$$\text{Var}(T) = \text{Var}(\bar{X}) + E[(T - \bar{X})^2].$$

(Hint: Show that $T - \bar{X}$ is ancillary for μ with σ^2 known.)

6.10: Suppose that $X_1, \cdots, X_n$ are i.i.d. Normal random variables with mean μ and variance σ^2, both unknown. We want to find the UMVU estimator of

$$g_c(\mu, \sigma) = \Phi\left(\frac{c - \mu}{\sigma}\right) = P_{\mu,\sigma}(X_i \leq c)$$

for some specified c.

(a) State why $(\bar{X}, \sum_{i=1}^n (X_i - \bar{X})^2)$ is sufficient and complete for (μ, σ).

(b) Show that $X_i - \bar{X}$ is independent of $\bar{X}$ for any i. (Hint: It suffices to show (why?) that the covariance between the two r.v.'s is 0; alternatively, we could use Basu's Theorem.)

(c) Using the preliminary estimator $S = I(X_1 \le c)$, find the UMVU estimator of $g_c(\mu, \sigma)$ by "Rao-Blackwellizing" S.

6.11: Suppose that $X_1, \cdots, X_n$ are i.i.d. Poisson random variables with mean λ.

(a) Use the fact that

$$\sum_{k=0}^{\infty} c_k x^k = 0 \quad \text{for all } a < x < b$$

if, and only if, $c_0 = c_1 = c_2 = \cdots = 0$ to show that $T = \sum_{i=1}^{n} X_i$ is complete for λ.

(b) Find the unique UMVU estimator of λ^2. (Hint: Find $g(0)$, $g(1)$, $g(2), \cdots$ to solve

$$\sum_{k=0}^{\infty} g(k) \frac{\exp(-n\lambda)(n\lambda)^k}{k!} = \lambda^2$$

by multiplying both sides by $\exp(n\lambda)$ and matching the coefficients of λ^k.)

(c) Find the unique UMVU estimator of λ^r for any integer $r > 2$.

6.12: Suppose that $X_1, \cdots, X_n$ are i.i.d. Exponential random variables with parameter λ. Define $g(\lambda) = P_\lambda(X_i > t)$ for some specified $t > 0$.

(a) Show that $T = X_1 + \cdots + X_n$ is independent of X_1/T.

(b) Find the UMVU estimator of $g(\lambda)$. (Hint: "Rao-Blackwellize" the unbiased estimator $S = I(X_1 > t)$ using the result of part (a).)

6.13: Suppose that $\boldsymbol{X} = (X_1, \cdots, X_n)$ has a joint distribution that depends on an unknown parameter θ and define

$$\mathcal{U} = \{U : E_\theta(U) = 0, E_\theta(U^2) < \infty\}$$

to be the space of all statistics $U = U(\boldsymbol{X})$ that are unbiased estimators of 0 with finite variance.

(a) Suppose that $T = T(\boldsymbol{X})$ is an unbiased estimator of $g(\theta)$ with $\text{Var}_\theta(T) < \infty$. Show that any unbiased estimator S of

$g(\theta)$ with $\text{Var}_\theta(S) < \infty$ can be written as

$$S = T + U$$

for some $U \in \mathcal{U}$.

(b) Let T be an unbiased estimator of $g(\theta)$ with $\text{Var}_\theta(T) < \infty$. Suppose that $\text{Cov}_\theta(T, U) = 0$ for all $U \in \mathcal{U}$ (and all θ). Show that T is a UMVU estimator of $g(\theta)$. (Hint: Use the result of part (a).)

(c) Suppose that T is a UMVU estimator of $g(\theta)$. Show that $\text{Cov}_\theta(T, U) = 0$ for all $U \in \mathcal{U}$. (Hint: Let $S_\lambda = T + \lambda U$ for some $U \in \mathcal{U}$ and find the minimum value of $\text{Var}_\theta(S_\lambda)$ for $-\infty < \lambda < \infty$.)

6.14: Suppose that $X_1, \cdots, X_n$ are i.i.d. Normal random variables with mean $\theta > 0$ and variance θ^2. A (minimal) sufficient statistic for θ is

$$T = \left(\sum_{i=1}^n X_i, \sum_{i=1}^n X_i^2 \right).$$

Show that T is not complete.

6.15: Suppose that $X_1, \cdots, X_n$ are i.i.d. Normal random variables with mean θ and variance θ^2 where $\theta > 0$. Define

$$\widehat{\theta}_n = \bar{X}_n \left(1 + \frac{\sum_{i=1}^n (X_i - \bar{X}_n)^2 - n\bar{X}_n^2}{3\sum_{i=1}^n (X_i - \bar{X}_n)^2} \right)$$

where $\bar{X}_n$ is the sample mean of $X_1, \cdots, X_n$.

(a) Show that $\widehat{\theta}_n \to_p \theta$ as $n \to \infty$.

(b) Find the asymptotic distribution of $\sqrt{n}(\widehat{\theta}_n - \theta)$. Is $\widehat{\theta}_n$ asymptotically efficient?

(c) Find the Cramér-Rao lower bound for unbiased estimators of θ. (Assume all regularity conditions are satisfied.)

(d) Does there exist an unbiased estimator of θ that achieves the lower bound in (a)? Why or why not?

6.16: Suppose that $X_1, \cdots, X_n$ are independent random variables where the density function of X_i is

$$f_i(x; \beta) = \frac{1}{\beta t_i} \exp(-x/(\beta t_i)) \quad \text{for } x \ge 0$$

where $t_1, \cdots, t_n$ are known constants. (Note that each X_i has an Exponential distribution.)

(a) Show that

$$\widehat{\beta} = \frac{1}{n}\sum_{i=1}^{n} X_i/t_i$$

is an unbiased estimator of β.

(b) Compute the Cramér-Rao lower bound for the variance of unbiased estimators of β. Does the estimator in (a) achieve this lower bound? (Hint: Write the joint density as a one-parameter exponential family.)

6.17: Suppose that $X_1, \cdots, X_n$ are i.i.d. random variables with frequency function

$$f(x;\theta) = \begin{cases} \theta & \text{for } x = -1 \\ (1-\theta)^2\theta^x & \text{for } x = 0, 1, 2, \cdots \end{cases}$$

where $0 < \theta < 1$.

(a) Find the Cramér-Rao lower bound for unbiased estimators of θ based on $X_1, \cdots, X_n$.

(b) Show that the maximum likelihood estimator of θ based on $X_1, \cdots, X_n$ is

$$\widehat{\theta}_n = \frac{2\sum_{i=1}^n I(X_i = -1) + \sum_{i=1}^n X_i}{2n + \sum_{i=1}^n X_i}$$

and show that $\{\widehat{\theta}_n\}$ is consistent for θ.

(c) Show that $\sqrt{n}(\widehat{\theta}_n - \theta) \to_d N(0, \sigma^2(\theta))$ and find the value of $\sigma^2(\theta)$. Compare $\sigma^2(\theta)$ to the Cramér-Rao lower bound found in part (a).

6.18: Suppose that $\boldsymbol{X} = (X_1, \cdots, X_n)$ are random variables with joint density or frequency function $f(\boldsymbol{x};\theta)$ where θ is a one-dimensional parameter. Let $T = T(\boldsymbol{X})$ be some statistic with $\text{Var}_\theta(T) < \infty$ for all θ and suppose that

(i) $A = \{\boldsymbol{x} : f(\boldsymbol{x};\theta) > 0\}$ does not depend on θ

(ii) $E_\theta(T) = g(\theta)$

Show that

$$\text{Var}_\theta(T) \ge \frac{[g(\theta+\Delta) - g(\theta)]^2}{\text{Var}_\theta(\psi(\boldsymbol{X};\theta))}$$

(provided that $\theta + \Delta$ lies in the parameter space) where

$$\psi(\boldsymbol{x};\theta) = \frac{f(\boldsymbol{x};\theta+\Delta) - f(\boldsymbol{x};\theta)}{f(\boldsymbol{x};\theta)}.$$

(This lower bound for $\text{Var}_\theta(T)$ is called the Chapman-Robbins lower bound.)

6.19: Suppose that $X_1, \cdots, X_n$ be i.i.d. Bernoulli random variables with parameter θ.

(a) Indicate why $S = X_1 + \cdots + X_n$ is a sufficient and complete statistic for θ.

(b) Find the UMVU estimator of $\theta(1-\theta)$. (Hint: $I(X_1 = 0, X_2 = 1)$ is an unbiased estimator of $\theta(1-\theta)$.)

6.20: Suppose that $X_1, \cdots, X_n$ are i.i.d. Normal random variables with mean μ and variance 1. Given $\lambda_n > 0$, define $\widehat{\mu}_n$ to minimize

$$g_n(t) = \sum_{i=1}^{n} (X_i - t)^2 + \lambda_n |t|$$

(a) Show that

$$\widehat{\mu}_n = \begin{cases} 0 & \text{if } |\bar{X}_n| \le \lambda_n/(2n) \\ \bar{X}_n - \lambda_n \text{sgn}(\bar{X}_n)/(2n) & \text{if } |\bar{X}_n| > \lambda_n/(2n). \end{cases}$$

(b) Suppose that $\lambda_n/\sqrt{n} \to \lambda_0 > 0$. Find the limiting distribution of $\sqrt{n}(\widehat{\mu}_n - \mu)$ for $\mu = 0$ and $\mu \ne 0$. Is $\{\widehat{\mu}_n\}$ a sequence of regular estimators in either the Hájek or Tierney sense?

6.21: Suppose that $X_1, \cdots, X_n$ are i.i.d. random variables with density or frequency function $f(x;\theta)$ satisfying the conditions of Theorem 6.6. Let $\widehat{\theta}_n$ be the MLE of θ and $\widetilde{\theta}_n$ be another (regular) estimator of θ such that

$$\sqrt{n} \begin{pmatrix} \widehat{\theta}_n - \theta \\ \widetilde{\theta}_n - \theta \end{pmatrix} \to_d N_2(\mathbf{0}, C(\theta)).$$

Show that $C(\theta)$ must have the form

$$C(\theta) = \begin{pmatrix} I^{-1}(\theta) & I^{-1}(\theta) \\ I^{-1}(\theta) & \sigma^2(\theta) \end{pmatrix}.$$

(Hint: Consider estimators of the form $t\widehat{\theta}_n + (1-t)\widetilde{\theta}_n$; by Theorem 6.6, the minimum asymptotic variance must occur at $t = 1$.)

CHAPTER 7

Interval Estimation and Hypothesis Testing

7.1 Confidence intervals and regions

Suppose $X_1, \cdots, X_n$ are random variables with some joint distribution depending on a parameter θ that may be real- or vector-valued. To this point, we have dealt exclusively with the problem of finding point estimators of θ.

The obvious problem with point estimation is the fact that typically $P_\theta(\widehat{\theta} = \theta)$ is small (if not 0) for a given point estimator $\widehat{\theta}$. Of course, in practice, we usually attach to any point estimator an estimator of its variability (for example, its standard error); however, this raises the question of exactly how to interpret such an estimate of variability.

An alternative approach to estimation is interval estimation. Rather than estimating θ by a single statistic, we instead give a range of values for θ that we feel are consistent with observed values of $X_1, \cdots, X_n$, in the sense, that these parameter values could have produced (with some degree of plausibility) the observed data.

We will start by considering interval estimation for a single (that is, real-valued) parameter.

DEFINITION. Let $\boldsymbol{X} = (X_1, \cdots, X_n)$ be random variables with joint distribution depending on a real-valued parameter θ and let $L(\boldsymbol{X}) < U(\boldsymbol{X})$ be two statistics. Then the (random) interval $[L(\boldsymbol{X}), U(\boldsymbol{X})]$ is called a $100p\%$ confidence interval for θ if

$$P_\theta\left[L(\boldsymbol{X}) \le \theta \le U(\boldsymbol{X})\right] \ge p$$

for all θ with equality for at least one value of θ.

The number p is called the coverage probability (or simply coverage) or confidence level of the confidence interval. In many cases, we will be able to find an interval $[L(\boldsymbol{X}), U(\boldsymbol{X})]$ with

$$P_\theta\left[L(\boldsymbol{X}) \le \theta \le U(\boldsymbol{X})\right] = p$$

for all θ. We can also define upper and lower confidence bounds for θ. For example, suppose that

$$P_\theta\left[\theta \geq L(\boldsymbol{X})\right] = p$$

for some statistic $L(\boldsymbol{X})$ and for all θ; then $L(\boldsymbol{X})$ is called a $100p\%$ lower confidence bound for θ. Likewise, if

$$P_\theta\left[\theta \leq U(\boldsymbol{X})\right] = p$$

for some statistic $U(\boldsymbol{X})$ and all θ then $U(\boldsymbol{X})$ is called a $100p\%$ upper confidence bound for θ. It is easy to see that if $L(\boldsymbol{X})$ is a $100p_1\%$ lower confidence bound and $U(\boldsymbol{X})$ a $100p_2\%$ upper confidence bound for θ then the interval

$$[L(\boldsymbol{X}), U(\boldsymbol{X})]$$

is a $100p\%$ confidence interval for θ where $p = p_1 + p_2 - 1$ (provided that $L(\boldsymbol{X}) < U(\boldsymbol{X})$).

The interpretation of confidence intervals is frequently misunderstood. Much of the confusion stems from the fact that confidence intervals are defined in terms of the distribution of $\boldsymbol{X} = (X_1, \cdots, X_n)$ but, in practice, are stated in terms of the observed values of these random variables leaving the impression that a probability statement is being made about θ rather than about the random interval. However, given data $\boldsymbol{X} = \boldsymbol{x}$, the interval $[L(\boldsymbol{x}), U(\boldsymbol{x})]$ will either contain the true value of θ or not contain the true value of θ; under repeated sampling, $100p\%$ of these intervals will contain the true value of θ. This distinction is important but poorly understood by many non-statisticians.

In many problems, it is difficult or impossible to find an exact confidence interval; this is particularly true if a model is not completely specified. However, it may be possible to find an interval $[L(\boldsymbol{X}), U(\boldsymbol{X})]$ for which

$$P_\theta\left[L(\boldsymbol{X}) \leq \theta \leq U(\boldsymbol{X})\right] \approx p,$$

in which case the resulting interval is called an approximate $100p\%$ confidence interval for θ.

EXAMPLE 7.1: Suppose that $X_1, \cdots, X_n$ are i.i.d. Normal random variables with mean μ and variance 1. Then $\sqrt{n}(\bar{X} - \mu) \sim N(0,1)$ and so

$$P_\mu\left[-1.96 \leq \sqrt{n}(\bar{X} - \mu) \leq 1.96\right] = 0.95.$$

The event $[-1.96 \le \sqrt{n}(\bar{X} - \mu) \le 1.96]$ is clearly the same as the event $[\bar{X} - 1.96/\sqrt{n} \le \mu \le \bar{X} + 1.96/\sqrt{n}]$ and so we have

$$P_\mu \left[\bar{X} - \frac{1.96}{\sqrt{n}} \le \mu \le \bar{X} + \frac{1.96}{\sqrt{n}} \right] = 0.95.$$

Thus the interval whose endpoints are $\bar{X} \pm 1.96/\sqrt{n}$ is a 95% confidence interval for μ.

Note in this example, if we assume only that $X_1, \cdots, X_n$ are i.i.d. with mean μ and variance 1 (not necessarily normally distributed), we have (by the CLT),

$$P_\mu \left[-1.96 \le \sqrt{n}(\bar{X} - \mu) \le 1.96 \right] \approx 0.95$$

if n is sufficiently large. Using the same argument used above, it follows that the interval whose endpoints are $\bar{X} \pm 1.96/\sqrt{n}$ is an approximate 95% confidence interval for μ. ◇

Pivotal method

Example 7.1 illustrates a simple but useful approach to finding confidence intervals; this approach is called the pivotal method. Suppose $\boldsymbol{X} = (X_1, \cdots, X_n)$ have a joint distribution depending on a real-valued parameter θ and let $g(\boldsymbol{X}; \theta)$ be a random variable whose distribution does not depend on θ; that is, the distribution function $P_\theta[g(\boldsymbol{X}; \theta) \le x] = G(x)$ is independent of θ. Thus we can find constants a and b such that

$$p = P_\theta[a \le g(\boldsymbol{X}; \theta) \le b]$$

for all θ. The event $[a \le g(\boldsymbol{X}; \theta) \le b]$ can (hopefully) be manipulated to yield

$$p = P_\theta[L(\boldsymbol{X}) \le \theta \le U(\boldsymbol{X})]$$

and so the interval $[L(\boldsymbol{X}), U(\boldsymbol{X})]$ is a $100p$% confidence interval for θ. The random variable $g(\boldsymbol{X}; \theta)$ is called a pivot for the parameter θ.

EXAMPLE 7.2: Suppose that $X_1, \cdots, X_n$ are i.i.d. Uniform random variables on $[0, \theta]$. The MLE of θ is $X_{(n)}$, the sample maximum, and the distribution function of $X_{(n)}/\theta$ is

$$G(x) = x^n \quad \text{for } 0 \le x \le 1.$$

Thus $X_{(n)}/\theta$ is a pivot for θ. To find a $100p$% confidence interval for θ, we need to find a and b such that

$$P_\theta\left[a \le \frac{X_{(n)}}{\theta} \le b\right] = p.$$

There are obviously infinitely many choices for a and b; however, it can be shown that setting $b = 1$ and $a = (1-p)^{1/n}$ results in the shortest possible confidence interval using the pivot $X_{(n)}/\theta$, namely $[X_{(n)}, X_{(n)}/(1-p)^{1/n}]$. ◇

EXAMPLE 7.3: Suppose that $X_1, \cdots, X_{10}$ are i.i.d. Exponential random variables with parameter λ. Then the random variable $\lambda \sum_{i=1}^{10} X_i$ is a pivot for λ having a Gamma distribution with shape parameter 10 and scale parameter 1. (Alternatively, we can use $2\lambda \sum_{i=1}^{10} X_i$, which has a χ^2 distribution with 20 degrees of freedom, as our pivot.) To find a 90% confidence interval for λ, we need to find a and b such that

$$P_\lambda\left[a \le \lambda \sum_{i=1}^{10} X_i \le b\right] = 0.90.$$

Again there are infinitely many choices for a and b; one approach is to choose a and b so that

$$\begin{aligned} P_\lambda\left[\lambda \sum_{i=1}^{10} X_i < a\right] &= 0.05 \quad \text{and} \\ P_\lambda\left[\lambda \sum_{i=1}^{10} X_i > b\right] &= 0.05; \end{aligned}$$

this yields $a = 5.425$ and $b = 15.705$. We thus get

$$[L(\boldsymbol{X}), U(\boldsymbol{X})] = \left[\frac{5.425}{\sum_{i=1}^{10} X_i}, \frac{15.705}{\sum_{i=1}^{10} X_i}\right]$$

as a 90% confidence interval for λ. As one might expect, this confidence interval is not the shortest possible based on the pivot used here; in fact, by using $a = 4.893$ and $b = 14.938$ we obtain the shortest possible 90% confidence interval based on the pivot $\lambda \sum_{i=1}^{10} X_i$. ◇

It is easy to extend the pivotal method to allow us to find confidence intervals for a single real-valued parameter when there are other unknown parameters. Suppose that $\theta = (\theta_1, \cdots, \theta_p)$ and consider finding a confidence interval for θ_1. Let $g(\boldsymbol{X}; \theta_1)$ be a random variable that depends on θ_1 but not on $\theta_2, \cdots, \theta_p$ and suppose that

$$P_\theta \left[g(\boldsymbol{X}; \theta_1) \le x \right] = G(x)$$

where $G(x)$ is independent of θ; the random variable $g(\boldsymbol{X}; \theta_1)$ is then a pivot for θ_1 and can be used to obtain a confidence interval for θ_1 in exactly the same way as before.

EXAMPLE 7.4: Suppose that $X_1, \cdots, X_n$ are i.i.d. Normal random variables with unknown mean and variance μ and σ^2, respectively. To find a confidence interval for μ, define

$$S^2 = \frac{1}{n-1} \sum_{i=1}^{n} (X_i - \bar{X})^2$$

and note that the random variable

$$\frac{\sqrt{n}(\bar{X} - \mu)}{S}$$

has Student's t distribution with $n - 1$ degrees of freedom; this distribution is independent of both μ and σ^2 and hence

$$\sqrt{n}(\bar{X} - \mu)/S$$

is a pivot for μ.

To find confidence intervals for σ^2, note that

$$\frac{(n-1)S^2}{\sigma^2} = \frac{1}{\sigma^2} \sum_{i=1}^{n} (X_i - \bar{X})^2 \sim \chi^2(n-1)$$

and is therefore a pivot for σ^2. ◇

In many problems, it is difficult to find exact pivots or to determine the distribution of an exact pivot if it does exist. However, in these cases, it is often possible to find an approximate pivot, that is, a random variable $g(\boldsymbol{X}; \theta)$ for which

$$P_\theta \left[g(\boldsymbol{X}; \theta) \le x \right] \approx G(x)$$

where $G(x)$ is independent of θ; almost inevitably approximate pivots are justified via asymptotic arguments and so we assume that n is large enough to justify the approximation. In such cases, the approximate pivot can be used to find approximate confidence intervals for θ. The classic example of this occurs when we have a point estimator $\widehat{\theta}$ whose distribution is approximately Normal with mean θ and variance $\sigma^2(\theta)$; in this case, we have

$$\frac{\widehat{\theta} - \theta}{\sigma(\theta)} \sim N(0, 1)$$

and so $(\widehat{\theta} - \theta)/\sigma(\theta)$ is an approximate pivot for θ. (Note that $\sigma(\theta)$ is essentially the standard error of $\widehat{\theta}$.) If $\sigma(\theta)$ depends on θ, it may be desirable to substitute $\widehat{\theta}$ for θ and use $(\widehat{\theta} - \theta)/\sigma(\widehat{\theta})$ as the approximate pivot. This approximate pivot is particularly easy to use; if z_p satisfies $\Phi(z_p) - \Phi(-z_p) = p$ then

$$P_\theta \left[-z_p \leq \frac{\widehat{\theta} - \theta}{\sigma(\widehat{\theta})} \leq z_p \right] \approx p,$$

which yields an approximate $100p\%$ confidence interval whose endpoints are $\widehat{\theta} \pm z_p \sigma(\widehat{\theta})$.

EXAMPLE 7.5: Suppose that $X_1, \cdots, X_n$ are i.i.d. Exponential random variables with unknown parameter λ. The MLE of λ is $\widehat{\lambda} = 1/\bar{X}$ and, if n is sufficiently large, $\sqrt{n}(\widehat{\lambda} - \lambda)$ is approximately Normal with mean 0 and variance λ^2 (from asymptotic theory for MLEs). Thus the random variable

$$\frac{\sqrt{n}(\widehat{\lambda} - \lambda)}{\lambda}$$

is an approximate pivot for λ and has a standard Normal distribution. To find an approximate 95% confidence interval for λ, we note that

$$P_\lambda \left[-1.96 \leq \frac{\sqrt{n}(\widehat{\lambda} - \lambda)}{\lambda} \leq 1.96 \right] \approx 0.95$$

and so the interval

$$\left[\widehat{\lambda}(1 + 1.96/\sqrt{n})^{-1}, \widehat{\lambda}(1 - 1.96/\sqrt{n})^{-1} \right]$$

is an approximate 95% confidence interval for λ. We can also use $\sqrt{n}(\widehat{\lambda} - \lambda)/\widehat{\lambda}$ as an approximate pivot; using the same argument as before, we obtain an approximate 95% confidence interval whose endpoints are $\widehat{\lambda} \pm 1.96\widehat{\lambda}/\sqrt{n}$. The two confidence intervals are quite similar when n is large since

$$(1 - 1.96/\sqrt{n})^{-1} = 1 + 1.96/\sqrt{n} + (1.96)^2/n + \cdots$$

$$\text{and} \quad (1 + 1.96/\sqrt{n})^{-1} = 1 - 1.96/\sqrt{n} + (1.96)^2/n + \cdots;$$

a more careful analysis is needed in order to determine which interval gives a coverage closer to 95%. ◇

EXAMPLE 7.6: Suppose that X has a Binomial distribution with parameters n and θ where θ is unknown. The MLE of θ is $\widehat{\theta} = X/n$ and if n is sufficiently large, $\sqrt{n}(\widehat{\theta} - \theta)/[\theta(1-\theta)]^{1/2}$ has approximately a standard Normal distribution and is an approximate pivot for θ. To find an approximate 95% confidence interval for θ, note that

$$\begin{aligned} P_\theta\left[-1.96 \le \frac{\sqrt{n}(\widehat{\theta} - \theta)}{[\theta(1-\theta)]^{1/2}} \le 1.96\right] &= P_\theta\left[\frac{n(\widehat{\theta} - \theta)^2}{\theta(1-\theta)} \le 1.96^2\right] \\ &\approx 0.95. \end{aligned}$$

Thus an approximate 95% confidence interval for θ will consist of all values of t for which

$$g(t) = n(\widehat{\theta} - t)^2 - 1.96^2 t(1-t) \le 0.$$

Note that $g(t)$ is a quadratic function and will have zeros at

$$t = \frac{\widehat{\theta} + 1.96^2/(2n) \pm 1.96\left(\widehat{\theta}(1-\widehat{\theta}) + 1.96^2/(4n^2)\right)^{1/2}}{1 + 1.96^2/n}$$

and so the region between these two zeros becomes the approximate 95% confidence interval for θ. Figure 7.1 shows the function $g(t)$ and the confidence interval for θ when $n = 20$ and $X = 13$. Alternatively, we can use

$$\frac{\sqrt{n}(\widehat{\theta} - \theta)}{[\widehat{\theta}(1-\widehat{\theta})]^{1/2}}$$

as an approximate pivot, which leads to an approximate 95% confidence interval whose endpoints are $\widehat{\theta} \pm 1.96[\widehat{\theta}(1-\widehat{\theta})/n]^{1/2}$. It is easy to see that the difference between these two confidence intervals will be small when n is large. (We can also refine the confidence intervals in this example by using the continuity correction for the Normal approximation discussed in Chapter 3; the result of this is that we replace $\widehat{\theta}$ by $\widehat{\theta} + 1/(2n)$ in the expressions for the upper limits and replace $\widehat{\theta}$ by $\widehat{\theta} - 1/(2n)$ in the expressions for the lower limits.) ◇

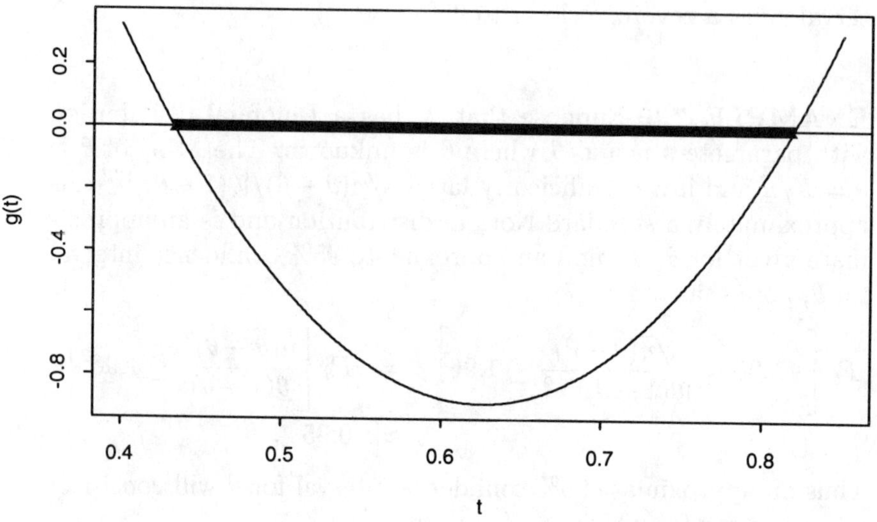

Figure 7.1 *The function $g(t)$ in Example 7.6; the approximate 95% confidence interval for θ consists of the values of t for which $g(t) \le 0$.*

It is possible to use variance stabilizing transformations (see Chapter 3) to obtain approximate pivots. For example, let $\widehat{\theta}$ be an estimator of θ and suppose that the distribution of $\sqrt{n}(\widehat{\theta} - \theta)$ is approximately Normal with mean 0 and variance $\sigma^2(\theta)$. By the Delta Method, if g is a monotone function then the distribution of $\sqrt{n}(g(\widehat{\theta}) - g(\theta))$ is approximately normal with mean 0 and variance $[g'(\theta)]^2\sigma^2(\theta)$; we choose the variance stabilizing transformation g so that $[g'(\theta)]^2\sigma^2(\theta) = 1$. Now using $\sqrt{n}(g(\widehat{\theta}) - g(\theta))$ as the approximate pivot, we obtain an approximate $100p\%$ confidence interval for $g(\theta)$ with endpoints $g(\widehat{\theta}) \pm z_p/\sqrt{n}$. Since we can take g to be strictly increasing, our approximate $100p\%$ confidence interval for θ is the interval

$$\left[g^{-1}\left(g(\widehat{\theta}) - z_p/\sqrt{n}\right), g^{-1}\left(g(\widehat{\theta}) + z_p/\sqrt{n}\right)\right].$$

It should be noted that it is not necessary to take g to be a variance stabilizing transformation. In general, it is possible to use

$$\frac{\sqrt{n}(g(\widehat{\theta}) - g(\theta))}{g'(\theta)\sigma(\theta)} \quad \text{or} \quad \frac{\sqrt{n}(g(\widehat{\theta}) - g(\theta))}{g'(\widehat{\theta})\sigma(\widehat{\theta})}$$

as approximate pivots for θ. If the parameter space Θ is not the entire real-line then it is sometimes useful to take g to be a function mapping Θ onto the real-line.

EXAMPLE 7.7: Suppose that $X_1, \cdots, X_n$ are i.i.d. Exponential random variables with parameter λ. For the MLE $\widehat{\lambda} = 1/\bar{X}$, we have $\sqrt{n}(\widehat{\lambda} - \lambda)$ approximately Normal with mean 0 and variance λ^2. It is easy to verify that a variance stabilizing transformation is $g(t) = \ln(t)$; we get an approximate 95% confidence interval for $\ln(\lambda)$ with endpoints $\ln(\widehat{\lambda}) \pm 1.96/\sqrt{n}$ and so an approximate 95% confidence interval for λ is

$$\left[\widehat{\lambda}\exp(-1.96/\sqrt{n}), \widehat{\lambda}\exp(1.96/\sqrt{n})\right].$$

Note that the function $g(t) = \ln(t)$ maps the parameter space (the positive real-numbers) onto the real-line; thus both endpoints of this confidence interval always lie in the parameter space (unlike the confidence intervals given in Example 7.5). ◇

Confidence regions

Up to this point, we have considered interval estimation only for real-valued parameters. However, it is often necessary to look at two or more parameters and hence extend the notion of confidence intervals (for a single parameter) to confidence regions for multiple parameters.

DEFINITION. Let $\boldsymbol{X} = (X_1, \cdots, X_n)$ be random variables with joint distribution depending on a (possibly) vector-valued parameter $\boldsymbol{\theta} \in \Theta$ and let $R(\boldsymbol{X})$ be a subset of Θ depending on $\boldsymbol{X}$. Then $R(\boldsymbol{X})$ is called a $100p\%$ confidence region for $\boldsymbol{\theta}$ if

$$P_\theta\left[\boldsymbol{\theta} \in R(\boldsymbol{X})\right] \geq p$$

for all $\boldsymbol{\theta}$ with equality at least one value of $\boldsymbol{\theta} \in \Theta$.

Again it is important to keep in mind that it is $R(\boldsymbol{X})$ that is random and not $\boldsymbol{\theta}$ in the probability statement above.

Note that there is nothing in the definition above that dictates that $R(\boldsymbol{X})$ be a contiguous set; thus it is possible to have a confidence region that consists of two or more disjoint regions. Thus for a single parameter, the definition of a confidence region is somewhat more general than that for a confidence interval. However, in practice, confidence regions are typically (but not always) contiguous sets.

Many of the procedures for confidence intervals extend *mutatis mutandis* to confidence regions. For example, the pivotal method can be easily extended to derive confidence regions. A random variable $g(\boldsymbol{X};\boldsymbol{\theta})$ is called a pivot (as before) if its distribution is independent of $\boldsymbol{\theta}$; if so, we have

$$\begin{aligned} p &= P_\theta\left[a \le g(\boldsymbol{X};\boldsymbol{\theta}) \le b\right] \\ &= P_\theta\left[\boldsymbol{\theta} \in R(\boldsymbol{X})\right] \end{aligned}$$

where $R(\boldsymbol{x}) = \{\boldsymbol{\theta} : a \le g(\boldsymbol{x};\boldsymbol{\theta}) \le b\}$.

EXAMPLE 7.8: Suppose that $\boldsymbol{X}_1, \cdots, \boldsymbol{X}_n$ are i.i.d. k-variate Normal random vectors with mean $\boldsymbol{\mu}$ and variance-covariance matrix C (where we assume that C^{-1} exists). Unbiased estimators of $\boldsymbol{\mu}$ and C are given by

$$\begin{aligned} \widehat{\boldsymbol{\mu}} &= \frac{1}{n}\sum_{i=1}^{n} \boldsymbol{X}_i \\ \widehat{C} &= \frac{1}{n-1}\sum_{i=1}^{n}(\boldsymbol{X}_i - \widehat{\boldsymbol{\mu}})(\boldsymbol{X}_i - \widehat{\boldsymbol{\mu}})^T, \end{aligned}$$

which are simply the natural analogs of the unbiased estimators in the univariate case. (We have assumed here that the $\boldsymbol{X}_i$'s and $\boldsymbol{\mu}$ are column vectors.) To obtain a confidence region for $\boldsymbol{\mu}$, we will use the pivot

$$g(\widehat{\boldsymbol{\mu}}, \widehat{C}; \boldsymbol{\mu}) = \frac{n(n-k)}{k(n-1)}(\widehat{\boldsymbol{\mu}} - \boldsymbol{\mu})^T \widehat{C}^{-1}(\widehat{\boldsymbol{\mu}} - \boldsymbol{\mu}),$$

which (Johnson and Wichern, 1992) has an F distribution with k and $n-k$ degrees of freedom. Let f_p be the p quantile of the $\mathcal{F}(k, n-k)$ distribution. Then a $100p\%$ confidence region for $\boldsymbol{\mu}$ is given by the set

$$R(\widehat{\boldsymbol{\mu}}, \widehat{C}) = \left\{\boldsymbol{\mu} : \frac{n(n-k)}{k(n-1)}(\widehat{\boldsymbol{\mu}} - \boldsymbol{\mu})^T \widehat{C}^{-1}(\widehat{\boldsymbol{\mu}} - \boldsymbol{\mu}) \le f_p\right\}.$$

This confidence region contains all the points lying within a certain ellipsoid in k-dimensional Euclidean space. If n is large then we can obtain an approximate $100p\%$ confidence region for $\boldsymbol{\mu}$ that will be valid for i.i.d. (not necessarily k-variate Normal) $\boldsymbol{X}_i$'s having (finite) variance-covariance matrix C. In this case,

$$g^*(\widehat{\boldsymbol{\mu}}, \widehat{C}; \boldsymbol{\mu}) = n(\widehat{\boldsymbol{\mu}} - \boldsymbol{\mu})^T \widehat{C}^{-1}(\widehat{\boldsymbol{\mu}} - \boldsymbol{\mu})$$

is approximately χ^2 distributed with k degrees of freedom and so is an approximate pivot. If c_p is the p quantile of the $\chi^2(k)$ distribution then an approximate $100p\%$ confidence region for $\boldsymbol{\mu}$ is given by the set

$$R^*(\widehat{\boldsymbol{\mu}}, \widehat{C}) = \left\{\boldsymbol{\mu} : n(\widehat{\boldsymbol{\mu}} - \boldsymbol{\mu})^T \widehat{C}^{-1}(\widehat{\boldsymbol{\mu}} - \boldsymbol{\mu}) \leq c_p\right\}.$$

The validity of this confidence region depends on the fact that the distribution of $\sqrt{n}(\widehat{\boldsymbol{\mu}} - \boldsymbol{\mu})$ is approximately $N_k(\boldsymbol{0}, C)$ and that $\widehat{C}$ is a consistent estimator of C. ◇

While it is as conceptually simple to construct confidence regions as it is to construct confidence intervals, confidence regions lose the ease of interpretation that confidence intervals have as the dimension of the parameter increases; it is straightforward to graphically represent a confidence region in two dimensions and feasible in three dimensions but for four or more dimensions, it is practical impossible to give a useful graphical representation of a confidence region. One exception to this is when a confidence region is a rectangle in the k-dimensional space; unfortunately, such regions do not seem to arise naturally! However, it is possible to construct such regions by combining confidence intervals for each of the parameters.

Suppose that $\boldsymbol{\theta} = (\theta_1, \cdots, \theta_k)$ and suppose that $[L_i(\boldsymbol{X}), U_i(\boldsymbol{X})]$ is a $100p_i\%$ confidence interval for θ_i. Now define

$$\begin{aligned} R(\boldsymbol{X}) &= [L_1(\boldsymbol{X}), U_1(\boldsymbol{X})] \times \cdots \times [L_k(\boldsymbol{X}), U_k(\boldsymbol{X})] \\ &= \{\boldsymbol{\theta} : L_i(\boldsymbol{X}) \leq \theta_i \leq U_i(\boldsymbol{X}) \quad \text{for } i = 1, \cdots, k\}. \end{aligned}$$

$R(\boldsymbol{X})$ is a confidence region for $\boldsymbol{\theta}$ but the coverage of this region is unclear. However, it is possible to give a lower bound for the coverage of $R(\boldsymbol{X})$. Using Bonferroni's inequality (see Example 1.2), it follows that

$$P_\theta[\boldsymbol{\theta} \in R(\boldsymbol{X})] \geq 1 - \sum_{i=1}^{k} P[\theta_i \notin [L_i(\boldsymbol{X}), U_i(\boldsymbol{X})]]$$

$$= 1 - \sum_{i=1}^{k}(1 - p_i).$$

This suggests a simple procedure for obtaining conservative confidence regions. Given a desired coverage probability p, we construct confidence intervals for each of the k parameters with coverage $p_i = 1-(1-p)/k$. Then the resulting region $R(\boldsymbol{X})$ (which is called a Bonferroni confidence region) will have coverage of at least p.

7.2 Highest posterior density regions

Confidence intervals represent the classic frequentist approach to interval estimation. Within the Bayesian framework, posterior distributions provide a natural analog to frequentist confidence intervals.

Suppose that $\boldsymbol{X} = (X_1, \cdots, X_n)$ are random variables with joint density or frequency function $f(\boldsymbol{x};\theta)$ where θ is a real-valued parameter. We will take $\pi(\theta)$ to be a prior density for the parameter θ; recall that this density reflects the statistician's beliefs about the parameter prior to observing the data. The statistician's beliefs after observing $\boldsymbol{X} = \boldsymbol{x}$ are reflected by the posterior density

$$\pi(\theta|\boldsymbol{x}) = \frac{\pi(\theta)f(\boldsymbol{x};\theta)}{\int_{-\infty}^{\infty} \pi(t)f(\boldsymbol{x};t)\,dt};$$

the posterior density of θ is proportional to the product of the prior density and the likelihood function.

DEFINITION. Let $\pi(\theta|\boldsymbol{x})$ be a posterior density for θ on $\Theta \subset R$. A region $C = C(\boldsymbol{x})$ is called a highest posterior density (HPD) region of content p if
(a) $\int_C \pi(\theta|\boldsymbol{x})\,d\theta = p$;
(b) for any $\theta \in C$ and $\theta^* \notin C$, we have

$$\pi(\theta|\boldsymbol{x}) \geq \pi(\theta^*|\boldsymbol{x}).$$

HPD regions are not necessarily contiguous intervals; however, if the posterior density is unimodal (as is typically the case) then the HPD region will be an interval.

It is important to note that HPD regions and confidence intervals (or confidence regions) are very different notions derived from different philosophies of statistical inference. More precisely, the confidence interval is constructed using the distribution of $\boldsymbol{X}$: if $C(\boldsymbol{X}) = [L(\boldsymbol{X}), U(\boldsymbol{X})]$ is the confidence interval, we have

$$P_\theta(\theta \in C(\boldsymbol{X})) = p \quad \text{for all } \theta$$

where P_θ is the joint probability distribution of $\boldsymbol{X}$ for a given parameter value θ. On the other hand, a HPD region is constructed using the posterior distribution of θ given $\boldsymbol{X} = \boldsymbol{x}$: if $C = C(\boldsymbol{x})$ is a HPD region, we have

$$\int_C \pi(\theta|\boldsymbol{x})\, d\theta = P(\theta \in C|\boldsymbol{X} = \boldsymbol{x}) = p.$$

Interestingly enough, in many cases, confidence intervals and HPD regions show a remarkable agreement even for modest sample sizes.

EXAMPLE 7.9: Suppose that $X_1, \cdots, X_n$ are i.i.d. Poisson random variables with mean θ and assume a Gamma prior density for θ:

$$\pi(\theta) = \frac{\lambda^\alpha \theta^{\alpha-1} \exp(-\lambda\theta)}{\Gamma(\alpha)} \quad \text{for } \theta > 0$$

(λ and α are hyperparameters). The posterior density of θ is also Gamma:

$$\pi(\theta|\boldsymbol{x}) = \frac{(n+\lambda)^{t+\alpha}\theta^{t+\alpha+1} \exp(-(n+\lambda)\theta)}{\Gamma(t+\alpha)} \quad \text{for } \theta > 0$$

where $t = x_1 + \cdots + x_n$. It is easy to verify that the posterior mode is $\widehat{\theta} = (t + \alpha - 1)/(n + \lambda)$.

What happens to the posterior density when n is large? First of all, note that the variance of the posterior distribution is $(t + \alpha)/(n + \lambda)^2$, which tends to 0 as n tends to ∞; thus the posterior density becomes more and more concentrated around the posterior mode as n increases. We will define $\psi(\theta) = \ln \pi(\theta|\boldsymbol{x})$; making a Taylor series expansion of $\psi(\theta)$ around the posterior mode $\widehat{\theta}$, we have (for large n)

$$\begin{aligned}\psi(\theta) &\approx \psi(\widehat{\theta}) + (\theta - \widehat{\theta})\psi'(\widehat{\theta}) + \frac{1}{2}(\theta - \widehat{\theta})^2\psi''(\widehat{\theta}) \\ &= \psi(\widehat{\theta}) - \frac{1}{2}(\theta - \widehat{\theta})^2 \left(\frac{n+\lambda}{\widehat{\theta}}\right)\end{aligned}$$

since $\psi'(\widehat{\theta}) = 0$ and $\psi''(\widehat{\theta}) = -(n + \lambda)/\widehat{\theta}$. We also have

$$\psi(\widehat{\theta}) = (t + \alpha)\ln(n + \lambda) + (t + \alpha - 1)\ln(\widehat{\theta}) - (n - \lambda)\widehat{\theta} - \ln\Gamma(t + \alpha)$$

and (using Stirling's approximation)

$$\begin{aligned}
\ln\Gamma(t+\alpha) &= \ln(t+\alpha-1) + \ln\Gamma(t+\alpha-1) \\
&= \ln(t+\alpha-1) + \ln\Gamma((n-\lambda)\widehat{\theta}) \\
&\approx \ln(t+\alpha-1) + \frac{1}{2}\ln(2\pi) \\
&\quad + \left[(n-\lambda)\widehat{\theta} - \frac{1}{2}\right]\ln((n-\lambda)\widehat{\theta}) - (n-\lambda)\widehat{\theta}.
\end{aligned}$$

Putting the pieces together, we get

$$\psi(\theta) \approx -\frac{1}{2}\ln(2\pi) + \frac{1}{2}\ln\left(\frac{n+\lambda}{\widehat{\theta}}\right) - \frac{1}{2}(\theta-\widehat{\theta})^2\left(\frac{n+\lambda}{\widehat{\theta}}\right)$$

and since $\pi(\theta|\boldsymbol{x}) = \exp(\psi(\theta))$,

$$\pi(\theta|\boldsymbol{x}) \approx \left(\frac{n+\lambda}{2\pi\widehat{\theta}}\right)^{1/2} \exp\left[-\frac{n+\lambda}{2\widehat{\theta}}(\theta-\widehat{\theta})^2\right].$$

Thus, for large n, the posterior distribution is approximately Normal with mean $\widehat{\theta}$ and variance $\widehat{\theta}/(n+\lambda)$. This suggests, for example, that an approximate 95% HPD interval for θ is

$$\left[\widehat{\theta} - 1.96\left(\frac{\widehat{\theta}}{n+\lambda}\right)^{1/2}, \widehat{\theta} + 1.96\left(\frac{\widehat{\theta}}{n+\lambda}\right)^{1/2}\right];$$

note that this interval is virtually identical to the approximate 95% confidence interval for θ whose endpoints are $\bar{x} \pm 1.96\sqrt{\bar{x}/n}$ since $\widehat{\theta} \approx \bar{x}$ and $n/(n+\lambda) \approx 1$ for large n. ◇

Example 7.9 illustrates that there may exist a connection (at least for large sample sizes) between confidence intervals and HPD intervals. We will now try to formalize this connection. Suppose that $X_1, \cdots, X_n$ are i.i.d. random variables with density or frequency function $f(x;\theta)$ where $f(x;\theta)$ satisfies the regularity conditions for asymptotic normality of MLEs given in Chapter 5. If θ has prior density $\pi(\theta)$ then the posterior density of θ is given by

$$\pi(\theta|\boldsymbol{x}) = \frac{\pi(\theta)\mathcal{L}_n(\theta)}{\int_{-\infty}^{\infty}\pi(t)\mathcal{L}_n(t)\,dt}$$

where $\mathcal{L}_n(\theta) = \prod_{i=1}^n f(x_i;\theta)$. We want to try to show that the posterior density can be approximated by a Normal density when n is large; the key result needed to do this is the following.

PROPOSITION 7.1 (Laplace's approximation) *Define*

$$\mathcal{I}_n = \int_{-\infty}^{\infty} g(\theta)\exp(nh(\theta))\,d\theta$$

where g and h are "smooth" functions and $h(\theta)$ is maximized at $\widehat{\theta}$. Then $\mathcal{I}_n = \widehat{\mathcal{I}}_n(1 + r_n)$ where

$$\widehat{\mathcal{I}}_n = g(\widehat{\theta})\left(-\frac{2\pi}{nh''(\widehat{\theta})}\right)^{1/2}\exp(nh(\widehat{\theta}))$$

and $|nr_n| \le M < \infty$.

We have not precisely specified the conditions on the functions g and h necessary to give a rigorous proof of Laplace's approximation. However, it is quite easy to give a heuristic proof by expanding $h(\theta)$ in Taylor series around $\widehat{\theta}$. Then

$$h(\theta) \approx h(\widehat{\theta}) + \frac{1}{2}(\theta - \widehat{\theta})^2 h''(\widehat{\theta})$$

where $h''(\widehat{\theta}) < 0$ since $\widehat{\theta}$ maximizes h. Then

$$\mathcal{I}_n \approx \exp(nh(\widehat{\theta}))\int_{-\infty}^{\infty} g(\theta)\exp\left(\frac{n}{2}h''(\widehat{\theta})(\theta - \widehat{\theta})^2\right)\,d\theta.$$

Laplace's approximation follows if we make the change of variable $s = [-nh''(\widehat{\theta})]^{1/2}(\theta - \widehat{\theta})$ and assume sufficient smoothness for g.

Laplace's approximation is quite crude but will be sufficient for our purposes; we will use it to approximate the integral

$$\int_{-\infty}^{\infty} \pi(t)\mathcal{L}_n(t)\,dt.$$

Note that we can write

$$\pi(\theta)\mathcal{L}_n(\theta) = \pi(\theta)\exp\left[n\left(\frac{1}{n}\sum_{i=1}^{n}\ln f(x_i;\theta)\right)\right]$$

and so if $\widehat{\theta}_n$ maximizes $\mathcal{L}_n(\theta)$, we have by Laplace's approximation

$$\int_{-\infty}^{\infty} \pi(t)\mathcal{L}_n(t)\,dt \approx \pi(\widehat{\theta}_n)\left(\frac{2\pi}{H_n(\widehat{\theta}_n)}\right)^{1/2}\exp(\mathcal{L}_n(\widehat{\theta}_n))$$

where

$$H_n(\theta) = -\frac{d^2}{d\theta^2}\ln\mathcal{L}_n(\theta).$$

Thus

$$\pi(\theta|\boldsymbol{x}) \approx \left(\frac{\pi(\theta)}{\pi(\widehat{\theta}_n)}\right)\left(\frac{H_n(\widehat{\theta}_n)}{2\pi}\right)^{1/2} \exp\left[\ln \mathcal{L}_n(\theta) - \ln \mathcal{L}_n(\widehat{\theta}_n)\right].$$

Since the posterior is concentrated around $\widehat{\theta}_n$ for large n, we need only worry about values of θ close to $\widehat{\theta}_n$; if the prior density is continuous everywhere then $\pi(\theta)/\pi(\widehat{\theta}_n) \approx 1$ for θ close to $\widehat{\theta}_n$. We also have (under the regularity conditions of Chapter 5) that

$$\exp\left[\ln \mathcal{L}_n(\theta) - \ln \mathcal{L}_n(\widehat{\theta}_n)\right] \approx -\frac{1}{2}(\theta - \widehat{\theta}_n)H_n(\widehat{\theta}_n),$$

which leads to the final approximation

$$\pi(\theta|\boldsymbol{x}) \approx \left(\frac{H_n(\widehat{\theta}_n)}{2\pi}\right)^{1/2} \exp\left[-\frac{H_n(\widehat{\theta}_n)}{2}(\theta - \widehat{\theta}_n)^2\right].$$

This heuristic development suggests that the posterior density can be approximated by a Normal density with mean $\widehat{\theta}_n$ and variance $1/H_n(\widehat{\theta}_n)$ when n is sufficiently large; note that $H_n(\widehat{\theta}_n)$ is simply the observed information for θ (defined in Chapter 5) and that $1/H_n(\widehat{\theta}_n)$ is an estimate of the variance of the maximum likelihood estimate $\widehat{\theta}_n$. Thus the interval who endpoints are $\widehat{\theta}_n \pm z_p H_n(\widehat{\theta}_n)^{-1/2}$ is an approximate $100p\%$ HPD interval for θ (where z_p satisfies $\Phi(z_p) - \Phi(-z_p) = p$); this is exactly the same as the approximate $100p\%$ confidence interval for θ based on the MLE.

7.3 Hypothesis testing

Suppose that $\boldsymbol{X} = (X_1, \cdots, X_n)$ are random variables with joint distribution density or frequency function $f(\boldsymbol{x};\theta)$ for some $\theta \in \Theta$. Let $\Theta = \Theta_0 \cup \Theta_1$ for two disjoint sets Θ_0 and Θ_1; given the outcome of $\boldsymbol{X}$, we would like to decide if θ lies in Θ_0 or Θ_1. In practice, Θ_0 is typically taken to be a lower dimensional sub-space of the parameter space Θ. Thus $\theta \in \Theta_0$ represents a simplification of the model in the sense that the model contains fewer parameters.

EXAMPLE 7.10: Suppose that $X_1, \cdots, X_m$ and $Y_1, \cdots, Y_n$ are independent random variables where $X_i \sim N(\mu_1, \sigma^2)$ and $Y_i \sim N(\mu_2, \sigma^2)$. The parameter space is then

$$\Theta = \{(\mu_1, \mu_2, \sigma) : -\infty < \mu_1, \mu_2 < \infty, \sigma > 0\}.$$

In many applications, it is of interest to determine whether the X_i's and Y_i's have the same distribution (that is, $\mu_1 = \mu_2$), for example, when the X_i's and Y_i's represent measurements from two different groups. In this case, we can represent Θ_0 as

$$\Theta_0 = \{(\mu_1, \mu_2, \sigma) : -\infty < \mu_1 = \mu_2 < \infty, \sigma > 0\}.$$

Note that the parameter space Θ is three-dimensional while Θ_0 is only two-dimensional. ◇

Given $\boldsymbol{X} = (X_1, \cdots, X_n)$, we need to find a rule (based on $\boldsymbol{X}$) that determines if we decide that θ lies in Θ_0 or in Θ_1. This rule essentially is a two-valued function ϕ and (without loss of generality) we will assume that ϕ can take the values 0 and 1; if $\phi(\boldsymbol{X}) = 0$ then we will decide that $\theta \in \Theta_0$ while if $\phi(\boldsymbol{X}) = 1$, we will decide that $\theta \in \Theta_1$. The function ϕ will be called a test function. In many cases, ϕ will depend on $\boldsymbol{X}$ only through some real-valued statistic $T = T(\boldsymbol{X})$, which we will call the test statistic for the test.

It is unlikely that any given test function will be perfect. Thus for a given test function ϕ, we must examine the probability of making an erroneous decision as θ varies over Θ. If $\theta \in \Theta_0$ then an error will occur if $\phi(\boldsymbol{X}) = 1$ and the probability of this error (called a type I error) is

$$P_\theta[\phi(\boldsymbol{X}) = 1] = E_\theta[\phi(\boldsymbol{X})] \quad (\theta \in \Theta_0).$$

Likewise if $\theta \in \Theta_1$ then an error will occur if $\phi(\boldsymbol{X}) = 0$ and the probability of this error (called a type II error) is

$$P_\theta[\phi(\boldsymbol{X}) = 0] = 1 - E_\theta[\phi(\boldsymbol{X})] \quad (\theta \in \Theta_1).$$

It is tempting to try to find a test function ϕ whose error probabilities are uniformly small over the parameter space. While in certain problems this is possible to do, it should be realized that there is necessarily a trade-off between the probabilities of error for $\theta \in \Theta_0$ and $\theta \in \Theta_1$. For example, let ϕ_1 and ϕ_2 be two test functions and define

$$R_1 = \{\boldsymbol{x} : \phi_1(\boldsymbol{x}) = 1\}$$

and

$$R_2 = \{(\boldsymbol{x}) : \phi_2(\boldsymbol{x}) = 1\}$$

where $R_1 \subset R_2$; note that this implies that

$$E_\theta[\phi_1(\boldsymbol{X})] \le E_\theta[\phi_2(\boldsymbol{X})]$$

for all $\theta \in \Theta$ and so

$$1 - E_\theta\left[\phi_1(\boldsymbol{X})\right] \geq 1 - E_\theta\left[\phi_2(\boldsymbol{X})\right].$$

Hence by attempting to decrease the probability of error when $\theta \in \Theta_0$, we run the risk of increasing the probability of error for $\theta \in \Theta_1$.

The classic approach to testing is to specify a test function $\phi(\boldsymbol{X})$ such that for some specified $\alpha > 0$,

$$E_\theta[\phi(\boldsymbol{X})] \leq \alpha \quad \text{for all } \theta \in \Theta_0.$$

The hypothesis that θ belongs to Θ_0 is called the null hypothesis and will be denoted by H_0; likewise, the hypothesis that θ lies in Θ_1 is called the alternative hypothesis and will be denoted by H_1. The constant α given above is called the level (or size) of the test; if $\phi(\boldsymbol{X}) = 1$, we say that H_0 is rejected at level α. The level of the test is thus the maximum probability of "rejection" of the null hypothesis H_0 when H_0 is true. This particular formulation gives us a reasonably well-defined mathematical problem for finding a test function ϕ: for a given level α, we would like to find a test function ϕ so that $E_\theta[\phi(\boldsymbol{X})]$ is maximized for $\theta \in \Theta_1$.

For a given test function ϕ, we define

$$\pi(\theta) = E_\theta[\phi(\boldsymbol{X})]$$

to be the power of the test at θ; for a specified level α, we require $\pi(\theta) \leq \alpha$ for all $\theta \in \Theta_0$ and so we are most interested in $\pi(\theta)$ for $\theta \in \Theta_1$.

The rationale for the general procedure given above is as follows. Suppose that we test H_0 versus H_1 at level α where α is small and suppose that, given data $\boldsymbol{X} = \boldsymbol{x}$, $\phi(\boldsymbol{x}) = 1$. If H_0 is true then this event is quite rare (it occurs with probability at most α) and so this gives us some evidence to believe that H_0 is false (and hence that H_1 is true). Of course, this "logic" assumes that the test is chosen so that $P_\theta(\phi(\boldsymbol{X}) = 1)$ is larger when H_1 is true. Conversely, if $\phi(\boldsymbol{x}) = 0$ then the test is very much inconclusive; this may tell us that H_0 is true or, alternatively, that H_1 is true but that the test used does not have sufficient power to detect this. Since the dimension of Θ_0 is typically lower than that of Θ (and so the model under H_0 is simpler), this approach to testing protects us against choosing unnecessarily complicated models (since the probability of doing so is at most α when the simpler model holds) but, depending on the power of the test, may prevent us from identifying more complicated models when such models are appropriate.

To get some idea of how to find "good" test functions, we will consider the simple case where the joint density or frequency function of $X_1, \cdots, X_n$ is either f_0 or f_1 where both f_0 and f_1 are known and depend on no unknown parameters. We will then test

$$H_0 : f = f_0 \quad \text{versus} \quad H_1 : f = f_1$$

at some specified level α; the null and alternative hypothesis are called simple in this case as they both consist of a single density or frequency function. The problem now becomes a straightforward optimization problem: find a test function ϕ with level α to maximize the power under H_1. The following result, the Neyman-Pearson Lemma, is important because it suggests an important principle for finding "good" test functions.

THEOREM 7.2 (Neyman-Pearson Lemma) *Suppose that* $\boldsymbol{X} = (X_1, \cdots, X_n)$ *have joint density or frequency function* $f(\boldsymbol{x})$ *where* f *is one of* f_0 *or* f_1 *and suppose we test*

$$H_0 : f = f_0 \quad \textit{versus} \quad H_1 : f = f_1.$$

Then the test whose test function is

$$\phi(\boldsymbol{X}) = \begin{cases} 1 & \textit{if } f_1(\boldsymbol{X}) \geq k f_0(\boldsymbol{X}) \\ 0 & \textit{otherwise} \end{cases}$$

(for some $0 < k < \infty$*) is a most powerful (MP) test of* H_0 *versus* H_1 *at level*

$$\alpha = E_0[\phi(\boldsymbol{X})].$$

Proof. In this proof, we will let P_0, P_1 and E_0, E_1 denote probability and expectation under H_0 and H_1. It suffices to show that if ψ is any function with $0 \leq \psi(\boldsymbol{x}) \leq 1$ and

$$E_0[\psi(\boldsymbol{X})] \leq E_0[\phi(\boldsymbol{X})]$$

then

$$E_1[\psi(\boldsymbol{X})] \leq E_1[\phi(\boldsymbol{X})].$$

We will assume that f_0 and f_1 are density functions; the same proof carries over to frequency functions with obvious modifications. First of all, note that

$$f_1(\boldsymbol{x}) - k f_0(\boldsymbol{x}) \begin{cases} \geq 0 & \text{if } \phi(\boldsymbol{x}) = 1 \\ < 0 & \text{if } \phi(\boldsymbol{x}) = 0. \end{cases}$$

Thus

$$\psi(\boldsymbol{x})(f_1(\boldsymbol{x}) - kf_0(\boldsymbol{x})) \leq \phi(\boldsymbol{x})(f_1(\boldsymbol{x}) - kf_0(\boldsymbol{x}))$$

and so

$$\int \cdots \int \psi(\boldsymbol{x})(f_1(\boldsymbol{x}) - kf_0(\boldsymbol{x}))\, d\boldsymbol{x} \leq \int \cdots \int \phi(\boldsymbol{x})(f_1(\boldsymbol{x}) - kf_0(\boldsymbol{x}))\, d\boldsymbol{x}$$

or rearranging terms,

$$\int \cdots \int (\psi(\boldsymbol{x}) - \phi(\boldsymbol{x}))f_1(\boldsymbol{x})\, d\boldsymbol{x} \leq k \int \cdots \int (\psi(\boldsymbol{x}) - \phi(\boldsymbol{x}))f_0(\boldsymbol{x})\, d\boldsymbol{x}.$$

The left-hand side above is simply $E_1[\psi(\boldsymbol{X})] - E_1[\phi(\boldsymbol{X})]$ and the right-hand side is $k(E_0[\psi(\boldsymbol{X})] - E_0[\phi(\boldsymbol{X})])$. Since

$$E_0[\psi(\boldsymbol{X})] - E_0[\phi(\boldsymbol{X})] \leq 0,$$

it follows that $E_1[\psi(\boldsymbol{X})] - E_1[\phi(\boldsymbol{X})] \leq 0$ and so $\phi(\boldsymbol{X})$ is the test function of an MP test of H_0 versus H_1. □

The Neyman-Pearson Lemma essentially states that an optimal test statistic for testing $H_0 : f = f_0$ versus $H_1 : f = f_1$ is

$$T(\boldsymbol{X}) = \frac{f_1(\boldsymbol{X})}{f_0(\boldsymbol{X})}$$

and that for a given level α, we should reject the null hypothesis H_0 is $T(\boldsymbol{X}) \geq k$ where k is chosen so that the test has level α. However, note that the Neyman-Pearson Lemma as stated here does not guarantee the existence of an MP α level test but merely states that the test that rejects H_0 for $T(\boldsymbol{X}) \geq k$ will be an MP test for some level α. Moreover, the Neyman-Pearson Lemma does guarantee uniqueness of an MP test when one exists; indeed, there may be infinitely many test functions having the same power as the MP test function prescribed by the Neyman-Pearson Lemma.

A more general form of the Neyman-Pearson Lemma gives a solution to the following optimization problem: Suppose we want to maximize $E_1[\phi(\boldsymbol{X})]$ subject to the constraints

$$E_0[\phi(\boldsymbol{X})] = \alpha \quad \text{and} \quad 0 \leq \phi(\boldsymbol{X}) \leq 1.$$

The optimal ϕ is given by

$$\phi(\boldsymbol{X}) = \begin{cases} 1 & \text{if } f_1(\boldsymbol{X}) > kf_0(\boldsymbol{X}) \\ c & \text{if } f_1(\boldsymbol{X}) = kf_0(\boldsymbol{X}) \\ 0 & \text{if } f_1(\boldsymbol{X}) < kf_0(\boldsymbol{X}) \end{cases}$$

where k and $0 \leq c \leq 1$ are chosen so that the constraints are satisfied. The function ϕ described above need not be a test function since it can possibly take values other than 0 and 1; however, in the case where the statistic

$$T(\boldsymbol{X}) = \frac{f_1(\boldsymbol{X})}{f_0(\boldsymbol{X})}$$

is a continuous random variable (which implies that $X_1, \cdots, X_n$ are continuous random variables), we can take the optimal ϕ to be either 0 or 1 for all possible values of $X_1, \cdots, X_n$ and so an MP test of $H_0 : f = f_0$ versus $H_1 : f = f_1$ exists for all levels $\alpha > 0$. Moreover, even if $\phi(\boldsymbol{X})$ given above is not a test function, it may be possible to find a valid test function $\phi^*(\boldsymbol{X})$ such that

$$E_k[\phi^*(\boldsymbol{X})] = E_k[\phi(\boldsymbol{X})] \quad \text{for } k = 0, 1$$

(see Example 7.12 below). However, for a given level α, an MP test of H_0 versus H_1 need not exist if $T(\boldsymbol{X})$ is discrete unless we are willing to consider so-called randomized tests. (Randomized tests are tests where, for some values of a test statistic, rejection or acceptance of H_0 is decided by some external random mechanism that is independent of the data.)

EXAMPLE 7.11: Suppose that $X_1, \cdots, X_n$ are i.i.d. Exponential random variables with parameter λ that is either λ_0 or λ_1 (where $\lambda_1 > \lambda_0$). We want to test

$$H_0 : \lambda = \lambda_0 \quad \text{versus} \quad H_1 : \lambda = \lambda_1$$

at level α. For a given λ, the joint density of $\boldsymbol{X} = (X_1, \cdots, X_n)$ is

$$f(\boldsymbol{x}; \lambda) = \lambda^n \exp\left(-\lambda \sum_{i=1}^{n} x_i\right)$$

and we will base our test on the statistic

$$T = \frac{f(\boldsymbol{X}; \lambda_1)}{f(\boldsymbol{X}; \lambda_0)} = (\lambda_1/\lambda_0)^n \exp\left[(\lambda_0 - \lambda_1) \sum_{i=1}^{n} X_i\right],$$

rejecting H_0 if $T \geq k$ where k is chosen so that the test has level α. Note, however, that T is a decreasing function of $S = \sum_{i=1}^{n} X_i$ (since $\lambda_0 - \lambda_1 < 0$) and so $T \geq k$ if, and only if, $S \leq k'$ for some constant k' (which will depend on k). We can choose k' so that

$$P_{\lambda_0}\left[\sum_{i=1}^{n} X_i \leq k'\right] = \alpha.$$

For given values of λ_0 and α, it is quite feasible to find k' in this problem. Under the null hypothesis, the test statistic $S = \sum_{i=1}^{n} X_i$ has a Gamma distribution with parameters n and λ_0 and $2\lambda_0 S$ has a χ^2 distribution with $2n$ degrees of freedom. For example, if $n = 5$, $\alpha = 0.01$ and $\lambda_0 = 5$ then the 0.01 quantile of a χ^2 distribution with 10 degrees of freedom is 2.56 and so

$$k' = \frac{2.56}{2 \times 5} = 0.256.$$

Thus we would reject H_0 at the 1% level if $S \leq 0.256$. ◇

EXAMPLE 7.12: Suppose that $X_1, \cdots, X_n$ are i.i.d. Uniform random variables on the interval $[0, \theta]$ where θ is either θ_0 or θ_1 (where $\theta_0 > \theta_1$). We want to test

$$H_0 : \theta = \theta_0 \quad \text{versus} \quad H_1 : \theta = \theta_1$$

at level α. The joint density of $\boldsymbol{X} = (X_1, \cdots, X_n)$ is

$$f(\boldsymbol{x}; \theta) = \frac{1}{\theta^n} I(\max(x_1, \cdots, x_n) \leq \theta)$$

and an MP test of H_0 versus H_1 will be based on the test statistic

$$T = \frac{f(\boldsymbol{X}; \theta_1)}{f(\boldsymbol{X}; \theta_0)} = (\theta_0/\theta_1)^n \, I(X_{(n)} \leq \theta_1).$$

Note that T can take only two possible values, 0 or $(\theta_0/\theta_1)^n$ depending on whether $X_{(n)}$ is greater than θ_1 or not. It follows then that the test that rejects H_0 when $X_{(n)} \leq \theta_1$ will be an MP test of H_0 versus H_1 with level

$$\alpha = P_{\theta_0}\left[X_{(n)} \leq \theta_1\right] = \left(\frac{\theta_1}{\theta_0}\right)^n$$

and the power of this test under H_1 is

$$P_{\theta_1}\left[X_{(n)} \leq \theta_1\right] = 1;$$

note that this test will also be the MP test of for any level $\alpha > (\theta_1/\theta_0)^n$ since its power is 1. If we want to find the MP test for $\alpha < (\theta_1/\theta_0)^n$, the situation is more complicated since the Neyman-Pearson Lemma does not tell us what to do. Nonetheless, intuition suggests that the appropriate test statistic is $X_{(n)}$ (which is sufficient for θ in this model) and that we should reject H_0 for $X_{(n)} \le k$ where

$$P_{\theta_0}\left[X_{(n)} \le k\right] = \left(\frac{k}{\theta_0}\right)^n = \alpha.$$

Solving the equation above, we get $k = \theta_0 \alpha^{1/n}$ and so the power of this test is

$$P_{\theta_1}\left[X_{(n)} \le \theta_0 \alpha^{1/n}\right] = \left(\frac{\theta_0 \alpha^{1/n}}{\theta_1}\right)^n = \alpha \left(\frac{\theta_0}{\theta_1}\right)^n.$$

To show that this is the MP test for $\alpha < (\theta_1/\theta_0)^n$, we use the more general form of the Neyman-Pearson Lemma; a function ϕ that maximizes $E_{\theta_1}[\phi(\boldsymbol{X})]$ subject to the constraints

$$E_{\theta_0}[\phi(\boldsymbol{X})] = \alpha < \left(\frac{\theta_1}{\theta_0}\right)^n \quad \text{and} \quad 0 \le \phi(\boldsymbol{X}) \le 1$$

is

$$\phi(\boldsymbol{X}) = \begin{cases} \alpha(\theta_0/\theta_1)^n & \text{if } X_{(n)} \le \theta_1 \\ 0 & \text{otherwise.} \end{cases}$$

It is easy to verify that

$$E_{\theta_1}[\phi(\boldsymbol{X})] = \alpha \left(\frac{\theta_0}{\theta_1}\right)^n = P_{\theta_1}\left[X_{(n)} \le \theta_0 \alpha^{1/n}\right]$$

and so the test that rejects H_0 if $X_{(n)} \le \theta_0 \alpha^{1/n}$ is an MP test for level $\alpha < (\theta_1/\theta_0)^n$. ◇

Uniformly most powerful tests

The Neyman-Pearson Lemma gives us a simple criterion for determining the test function of an MP test of a simple null hypothesis when the alternative hypothesis is also simple. However, as we noted before, this particular testing scenario is rarely applicable in practice as the alternative hypothesis (and often the null hypothesis) are usually composite.

Let $\boldsymbol{X} = (X_1, \cdots, X_n)$ be random variables with joint density or frequency function $f(\boldsymbol{x}; \theta)$ and suppose we want to test

$$H_0 : \theta \in \Theta_0 \quad \text{versus} \quad H_1 : \theta \in \Theta_1$$

at level α. Suppose that ϕ is a test function such that

$$E_\theta[\phi(\boldsymbol{X})] \le \alpha \quad \text{for all } \theta \in \Theta_0$$

and that ϕ is the MP α level test of

$$H_0' : \theta = \theta_0 \quad \text{versus} \quad H_1' : \theta = \theta_1$$

for some $\theta_0 \in \Theta_0$ and all $\theta_1 \in \Theta_1$; then the test function ϕ describes the uniformly most powerful (UMP) α level test of H_0 versus H_1.

Unfortunately, UMP tests only exist for certain testing problems. For example, if we want to test

$$H_0 = \theta = \theta_0 \quad \text{versus} \quad H_1 : \theta \neq \theta_0$$

then UMP tests typically do not exist. The reason for this is quite simple: in order for a given test to be UMP, it must be an MP test of $H_0' : \theta = \theta_0$ versus $H_1' : \theta = \theta_1$ for any $\theta_1 \neq \theta_0$. However, the form of the MP test (given by the Neyman-Pearson Lemma) typically differs for $\theta_1 > \theta_0$ and $\theta_1 < \theta_0$ as the following examples indicate.

EXAMPLE 7.13: Let X be a Binomial random variable with parameters n and θ and suppose we want to test

$$H_0 : \theta = \theta_0 \quad \text{versus} \quad H_1 : \theta \neq \theta_0$$

at some level α. First consider testing

$$H_0' : \theta = \theta_0 \quad \text{versus} \quad H_1' : \theta = \theta_1$$

where $\theta_1 \neq \theta_0$. The Neyman-Pearson Lemma suggests that the MP test of H_0' versus H_1' will be based on the statistic

$$T = \frac{f(X;\theta_1)}{f(X;\theta_0)} = \left(\frac{1-\theta_0}{1-\theta_1}\right)^n \left(\frac{\theta_1(1-\theta_0)}{\theta_0(1-\theta_1)}\right)^X.$$

If $\theta_1 > \theta_0$, it is easy to verify that T is an increasing function of X; hence, an MP test of H_0' versus H_1' will reject H_0' for large values of X. On the other hand, if $\theta_1 < \theta_0$ then T is a decreasing function of X and so an MP test will reject H_0' for small values of X. From this, we can see that no UMP test of H_0 versus H_1 will exist. ◇

EXAMPLE 7.14: Let $X_1, \cdots, X_n$ be i.i.d. Exponential random variables with parameter λ and suppose that we want to test

$$H_0 : \lambda \leq \lambda_0 \quad \text{versus} \quad H_1 : \lambda > \lambda_0$$

at level α. In Example 7.11, we saw that the MP α level test of

$$H_0' : \lambda = \lambda_0 \quad \text{versus} \quad H_1' : \lambda = \lambda_1$$

rejects H_0' for $\sum_{i=1}^n X_i \leq k$ when $\lambda_1 > \lambda_0$ where k is determined by

$$P_{\lambda_0}\left[\sum_{i=1}^n X_i \leq k\right] = \alpha.$$

It is also easy to verify that if $\lambda < \lambda_0$ then

$$P_{\lambda}\left[\sum_{i=1}^n X_i \leq k\right] < \alpha.$$

Thus the test that rejects H_0 when $\sum_{i=1}^n X_i \leq k$ is a level α test and since it is an MP α level test of H_0' versus H_1' for every $\lambda_1 > \lambda_0$, it is a UMP test of H_0 versus H_1. ◇

Examples 7.13 and 7.14 give us some insight as to when UMP tests can exist; typically, we need the following conditions:

- θ is a real-valued parameter (with no other unknown parameters), and
- the testing problem is "one-sided"; that is, we are testing

$$H_0 : \theta \leq \theta_0 \quad (\theta \geq \theta_0) \quad \text{versus} \quad H_1 : \theta > \theta_0 \quad (\theta < \theta_0)$$

  for some specified value θ_0. (In fact, UMP tests may also exist if $H_1 : \theta_L < \theta < \theta_U$ for some specified θ_L and θ_U.)

Suppose that $\boldsymbol{X} = (X_1, \cdots, X_n)$ are random variables with joint density or frequency function depending on some real-valued θ and suppose that we want to test

$$H_0 : \theta \leq \theta_0 \quad \text{versus} \quad H_1 : \theta > \theta_0$$

at level α. We noted above that this testing setup is essentially necessary for the existence of a UMP test; however, a UMP test need not exist for a particular model. We would thus like to find a sufficient condition for the existence of a UMP test of H_0 versus H_1; this is guaranteed if the family $\{f(\boldsymbol{x};\theta) : \theta \in \Theta\}$ has a property known as monotone likelihood ratio.

DEFINITION. A family of joint density (frequency) functions $\{f(\boldsymbol{x};\theta)\}$ (where $\theta \in \Theta \subset R$) is said to have monotone likelihood ratio if there exists a real-valued function $T(\boldsymbol{x})$ such that for any $\theta_1 < \theta_2$,

$$\frac{f(\boldsymbol{x};\theta_2)}{f(\boldsymbol{x};\theta_1)}$$

is a non-decreasing function of $T(\boldsymbol{x})$. (By the Factorization Criterion, the statistic $T(\boldsymbol{X})$ is necessarily sufficient for θ.)

If the family $\{f(\boldsymbol{x};\theta)\}$ has monotone likelihood ratio then the test statistic for the UMP test is $T(\boldsymbol{X})$; for example, if we test

$$H_0 : \theta \le \theta_0 \quad \text{versus} \quad H_1 : \theta > \theta_0$$

then the test that rejects H_0 (that is, $\phi(\boldsymbol{X}) = 1$) if $T(\boldsymbol{X}) \ge k$ will be a UMP α level test where

$$\alpha = P_{\theta_0}\left[T(\boldsymbol{X}) \ge k\right].$$

Similarly, for $H_0 : \theta \ge \theta_0$, the test rejecting H_0 if $T(\boldsymbol{X}) \le k$ is a UMP test.

EXAMPLE 7.15: Suppose the joint density or frequency function of $\boldsymbol{X} = (X_1, \cdots, X_n)$ is a one-parameter exponential family

$$f(\boldsymbol{x};\theta) = \exp\left[c(\theta)T(\boldsymbol{x}) - b(\theta) + S(\boldsymbol{x})\right] \quad \text{for } \boldsymbol{x} \in A$$

and assume (with loss of generality) that $c(\theta)$ is strictly increasing in θ. Then for $\theta_1 < \theta_2$, we have

$$\frac{f(\boldsymbol{x};\theta_2)}{f(\boldsymbol{x};\theta_1)} = \exp\left[(c(\theta_2) - c(\theta_1))T(\boldsymbol{x}) + b(\theta_1) - b(\theta_2)\right],$$

which is an increasing function of $T(\boldsymbol{x})$ since $c(\theta_2) - c(\theta_1) > 0$. Thus this one-parameter exponential family has monotone likelihood ratio and so if we test $H_0 : \theta \le \theta_0$ versus $H_1 : \theta > \theta_0$, we would reject H_0 for $T(\boldsymbol{X}) \ge k$ where k is chosen so that $P_{\theta_0}[T(\boldsymbol{X}) \ge k] = \alpha$. ◇

Other most powerful tests

How do we find "good" tests if a UMP test does not exist? One approach is to find the most powerful test among some restricted class of tests. One possible restriction that can be applied is unbiasedness: a test with test function ϕ is said to be an unbiased level α test if

$$E_\theta\left[\phi(\boldsymbol{X})\right] \le \alpha \quad \text{for all } \theta \in \Theta_0$$

and

$$E_\theta\left[\phi(\boldsymbol{X})\right] \ge \alpha \quad \text{for all } \theta \in \Theta_1.$$

Essentially, unbiasedness of a test requires that the power of the test for $\theta \in \Theta_1$ is greater than the level of the test. This would seem to be a very reasonable criterion to expect of a test; however, in many situations, unbiased tests do not exist. When they do, it is sometimes possible to find uniformly most powerful unbiased (UMPU) tests.

Suppose we want to test $H_0 : \theta = \theta_0$ versus $H_1 : \theta \neq \theta_0$ at level α, it may be possible to construct a UMPU test by combining UMP tests of $H_0' : \theta \leq \theta_0$ and $H_0'' : \theta \geq \theta_0$. More precisely, suppose that $\phi_1(\boldsymbol{X})$ is a UMP level α_1 test function of H_0' and $\phi_2(\boldsymbol{X})$ is a UMP level α_2 test function of H_0'' such that $\alpha_1 + \alpha_2 = \alpha$. Then $\phi = \phi_1 + \phi_2$ will be a level α test function provided that

$$\phi_1(\boldsymbol{X}) + \phi_2(\boldsymbol{X}) \leq 1$$

and by judiciously choosing α_1 and α_2, it may be possible to make ϕ an unbiased α level test function; the resulting test will typically be a UMPU test. The natural choice for α_1 and α_2 is $\alpha_1 = \alpha_2 = \alpha/2$; in general, however, this will not lead to an unbiased test as the following example indicates.

EXAMPLE 7.16: Let X be a continuous random variable with density function

$$f(x;\theta) = \theta x^{\theta-1} \quad \text{for } 0 \leq x \leq 1$$

and suppose we want to test

$$H_0 : \theta = 1 \quad \text{and} \quad H_1 : \theta \neq 1$$

at the 5% level. We will reject H_0 if either $X \leq 0.025$ or $X \geq 0.975$; clearly this is a 5% level test since

$$P_1(X \leq 0.025) = P_1(X \geq 0.975) = 0.025.$$

The power function is then

$$\begin{aligned}\pi(\theta) &= \int_0^{0.025} \theta x^{\theta-1}\,dx + \int_{0.975}^{1} \theta x^{\theta-1}\,dx \\ &= 1 + 0.025^{\theta} - 0.975^{\theta}.\end{aligned}$$

Evaluating $\pi(\theta)$ for θ close to 1 reveals that this test is not unbiased; in fact, $\pi(\theta) < 0.05$ for $1 < \theta < 2$. However, it is possible to find an unbiased 5% level test of H_0 versus H_1. This test rejects H_0 if either $X \leq 0.008521$ or if $X \geq 0.958521$. The power functions for both tests are shown in Figure 7.2. ◇

In Example 7.16, the unbiased test has higher power for $\theta > 1$

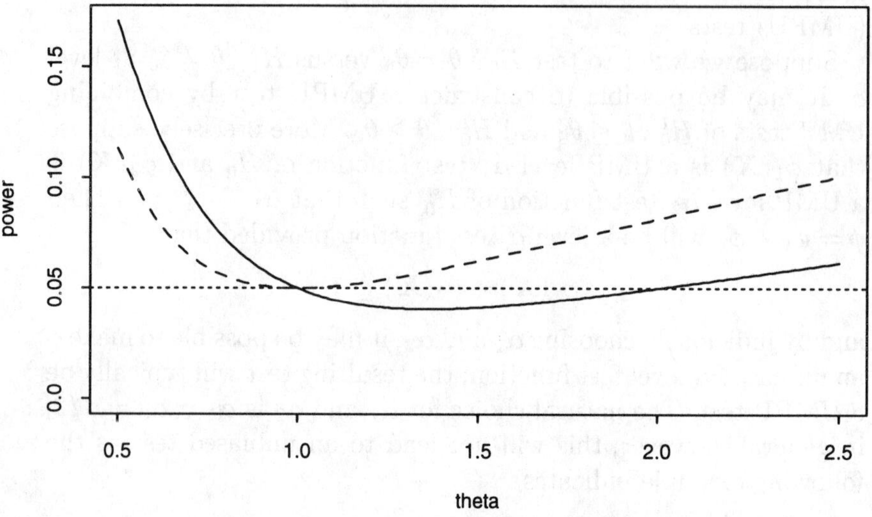

Figure 7.2 *The power functions of the test in Example 7.16; the power function for the unbiased test is indicated by the dashed line.*

but lower power for $\theta < 1$. This illustrates an important point, namely that in choosing an unbiased test, we are typically sacrificing power in some region of the parameter space relative to other (biased) tests.

The following example illustrates how to determine a UMPU test.

EXAMPLE 7.17: Let $X_1, \cdots, X_n$ be i.i.d. Exponential random variables with parameter λ and suppose we want to test

$$H_0 : \lambda = \lambda_0 \quad \text{versus} \quad H_1 : \lambda \neq \lambda_0$$

at some level α. We know, of course, that we can find UMP tests for the null hypotheses

$$H_0' : \lambda \leq \lambda_0 \quad \text{and} \quad H_0'' : \lambda \geq \lambda_0$$

using the test statistic $T = \sum_{i=1}^n X_i$; we will reject H_0' for $T \leq k_1$

and H_0'' for $T \geq k_2$. If $k_1 < k_2$, we can define the test function

$$\phi(T) = \begin{cases} 1 & \text{if } T \leq k_1 \text{ or } T \geq k_2 \\ 0 & \text{otherwise} \end{cases},$$

which will give a level α test if

$$P_{\lambda_0}[k_1 < T < k_2] = 1 - \alpha.$$

We would now like to choose k_1 and k_2 to make this test unbiased. The power of the test is given by

$$\begin{aligned} \pi(\lambda) &= 1 - P_\lambda[k_1 < T < k_2] \\ &= 1 - \int_{k_1}^{k_2} \frac{\lambda^n x^{n-1} \exp(-\lambda x)}{(n-1)!}\, dx \end{aligned}$$

since T has a Gamma distribution with parameters n and λ. For given k_1 and k_2, the power function $\pi(\lambda)$ is differentiable with derivative

$$\pi'(\lambda) = \int_{k_1}^{k_2} \frac{(\lambda^n x - n\lambda^{n-1}) x^{n-1} \exp(-\lambda x)}{(n-1)!}\, dx$$

and it is easy to see that the requirement of unbiasedness is equivalent to requiring that $\pi'(\lambda_0) = 0$. Thus k_1 and k_2 must satisfy the equations

$$\begin{aligned} 1 - \alpha &= \frac{1}{(n-1)!} \int_{k_1}^{k_2} \lambda_0^n x^{n-1} \exp(-\lambda_0 x)\, dx \\ \text{and } \quad 0 &= \frac{1}{(n-1)!} \int_{k_1}^{k_2} (\lambda_0^n x - n\lambda_0^{n-1}) x^{n-1} \exp(-\lambda_0 x)\, dx. \end{aligned}$$

(The equations given here for k_1 and k_2 are similar to the equations needed to obtain the shortest confidence interval in Example 7.3.) For example, if $n = 5$, $\lambda_0 = 3$, and $\alpha = 0.05$, we obtain $k_1 = 0.405$ and $k_2 = 3.146$. This test turns out to be the UMPU 5% level test of H_0 versus H_1. ◇

UMPU tests also exist for certain tests in k-parameter exponential families. Suppose $\boldsymbol{X} = (X_1, \cdots, X_n)$ are random variables with joint density or frequency function

$$f(\boldsymbol{x}; \theta, \boldsymbol{\eta}) = \exp\left[\theta T_0(\boldsymbol{x}) + \sum_{i=1}^{k-1} \eta_i T_i(\boldsymbol{x}) - d(\theta, \boldsymbol{\eta}) + S(\boldsymbol{x})\right]$$

where θ is real-valued and $\boldsymbol{\eta}$ is a vector of “nuisance” parameters;

we want to test

$$H_0 : \theta \le \theta_0 \quad \text{versus} \quad H_1 : \theta > \theta_0$$

for some specified value θ_0 where the value of η is arbitrary. It turns out that a UMPU α level test of H_0 versus H_1 will reject H_0 if $T_1(\boldsymbol{X}) \ge c$ where

$$\alpha = P_{\theta_0}\left[T_0(\boldsymbol{X}) \ge c | T_1 = t_1, \cdots, T_{k-1} = t_{k-1}\right]$$

where $t_1, \cdots, t_{k-1}$ are the observed values of $T_1, \cdots, T_{k-1}$; that is, $t_i = T_i(\boldsymbol{x})$ where $\boldsymbol{x}$ is the observed value of $\boldsymbol{X}$. Thus the test function is based not on the marginal distribution of the statistic T_0 but rather the conditional distribution of T_0 given the observed values of the other sufficient statistics $T_1, \cdots, T_{k-1}$. See Lehmann (1991) for more details.

In the case where θ is real-valued, it is often possible to find locally most powerful (LMP) tests. For example, suppose that we want to test

$$H_0 : \theta \le \theta_0 \quad \text{versus} \quad H_1 : \theta > \theta_0.$$

Our intuition tells us that if the true value of θ is sufficiently far from θ_0, any reasonable test of H_0 versus H_1 will have power close to 1. Thus we should be concerned mainly about finding a test whose power is as large as possible for values of θ close to θ_0. Let $\phi(\boldsymbol{X})$ be a test function and define the power function

$$\pi(\theta) = E_\theta\left[\phi(\boldsymbol{X})\right];$$

we will require that $\pi(\theta_0) = \alpha$. We would like to find ϕ so that $\pi(\theta)$ is maximized for θ close to θ_0. As a mathematical problem, this is somewhat ambiguous; however, if θ is very close to θ_0 then we have

$$\begin{aligned}\pi(\theta) &\approx \pi(\theta_0) + \pi'(\theta_0)(\theta - \theta_0) \\ &= \alpha + \pi'(\theta_0)(\theta - \theta_0)\end{aligned}$$

(provided, of course, that $\pi(\theta)$ is differentiable). This suggests that we try to find a test function ϕ to maximize $\pi'(\theta_0)$ subject to the constraint $\pi(\theta_0) = \alpha$. (In the case where $H_0 : \theta \ge \theta_0$ and $H_1 : \theta < \theta_0$, we would want to minimize $\pi'(\theta_0)$.)

Suppose that $f(\boldsymbol{x}; \theta)$ is the joint density function of continuous random variables $\boldsymbol{X} = (X_1, \cdots, X_n)$. Then for a given test function

ϕ, we have

$$\pi(\theta) = \int \cdots \int \phi(\boldsymbol{x}) f(\boldsymbol{x};\theta)\, d\boldsymbol{x}$$

and if we can differentiate under the integral sign, we have

$$\begin{aligned} \pi(\theta) &= \int \cdots \int \phi(\boldsymbol{x}) \frac{\partial}{\partial\theta} f(\boldsymbol{x};\theta)\, d\boldsymbol{x} \\ &= \int \cdots \int \phi(\boldsymbol{x}) \left(\frac{\partial}{\partial\theta} \ln f(\boldsymbol{x};\theta)\right) f(\boldsymbol{x};\theta)\, d\boldsymbol{x} \\ &= E_\theta \left[\phi(\boldsymbol{X}) \left(\frac{\partial}{\partial\theta} \ln f(\boldsymbol{X};\theta)\right)\right]. \end{aligned}$$

(The same result holds if $X_1, \cdots, X_n$ are discrete provided we can differentiate inside the summation sign.)

The development given above suggests that to find the form of the LMP test we must solve the following optimization problem: maximize

$$E_{\theta_0}\left[\phi(\boldsymbol{X}) S(\boldsymbol{X};\theta_0)\right]$$

subject to the constraint

$$E_{\theta_0}\left[\phi(\boldsymbol{X})\right] = \alpha$$

where

$$S(\boldsymbol{x};\theta) = \frac{\partial}{\partial\theta} \ln f(\boldsymbol{x};\theta).$$

The following result provides a solution to this problem.

PROPOSITION 7.3 *Suppose that $\boldsymbol{X} = (X_1, \cdots, X_n)$ has joint density or frequency function $f(\boldsymbol{x};\theta)$ and define the test function*

$$\phi(\boldsymbol{X}) = \begin{cases} 1 & \text{if } S(\boldsymbol{X};\theta_0) \geq k \\ 0 & \text{otherwise} \end{cases}$$

where k is such that $E_{\theta_0}[\phi(\boldsymbol{X})] = \alpha$. Then $\phi(\boldsymbol{X})$ maximizes

$$E_{\theta_0}\left[\psi(\boldsymbol{X}) S(\boldsymbol{X};\theta_0)\right]$$

over all test functions ψ with $E_{\theta_0}[\psi(\boldsymbol{X})] = \alpha$.

Proof. This proof parallels the proof of the Neyman-Pearson Lemma. Suppose $\psi(\boldsymbol{x})$ is a function with $0 \leq \psi(\boldsymbol{x}) \leq 1$ for all $\boldsymbol{x}$ such that $E_{\theta_0}[\psi(\boldsymbol{X})] = \alpha$. Note that

$$\phi(\boldsymbol{X}) - \psi(\boldsymbol{X}) \begin{cases} \geq 0 & \text{if } S(\boldsymbol{X};\theta_0) \geq k \\ \leq 0 & \text{if } S(\boldsymbol{X};\theta_0) \leq k \end{cases}$$

and so

$$E_{\theta_0}\left[(\phi(\boldsymbol{X}) - \psi(\boldsymbol{X}))(S(\boldsymbol{X};\theta_0) - k)\right] \geq 0.$$

Since $E_{\theta_0}\left[\phi(\boldsymbol{X}) - \psi(\boldsymbol{X})\right] = 0$, it follows that

$$E_{\theta_0}\left[\phi(\boldsymbol{X})S(\boldsymbol{X};\theta_0)\right] \geq E_{\theta_0}\left[\psi(\boldsymbol{X})S(\boldsymbol{X};\theta_0)\right],$$

which completes the proof. □

The test described in Proposition 7.3 is often called the score test since it is based on the score function (see Chapter 5); we will discuss a more general form of the score test in section 7.5. In deriving this LMP test, we have been concerned about the behaviour of the power function only for values of θ near θ_0; it is, in fact, conceivable that this LMP test function does not result in a level α test since $E_\theta[\phi(\boldsymbol{X})]$ may be greater than α for some $\theta < \theta_0$.

In the case where $X_1, \cdots, X_n$ are i.i.d. with common density or frequency function $f(x;\theta)$ then

$$S(\boldsymbol{X};\theta) = \sum_{i=1}^{n} \ell'(X_i;\theta)$$

where $\ell'(x;\theta)$ is the partial derivative with respect to θ of $\ln f(x;\theta)$. Subject to the regularity conditions given in Chapter 5, we have

$$E_{\theta_0}\left[\ell'(X_i;\theta_0)\right] = 0 \quad \text{and} \quad \text{Var}_{\theta_0}\left[\ell'(X_i;\theta_0)\right] = I(\theta)$$

and so, if n is sufficiently large, the distribution (under $\theta = \theta_0$) of the test statistic $S(\boldsymbol{X};\theta_0)$ will be approximately Normal with mean 0 and variance $nI(\theta)$. Thus the critical values for the LMP test of $H_0 : \theta \leq \theta_0$ versus $H_1 : \theta > \theta_0$ can be determined approximately from this result.

EXAMPLE 7.18: Let $X_1, \cdots, X_n$ be i.i.d. random variables with a Cauchy distribution with density

$$f(x;\theta) = \frac{1}{\pi(1 + (x-\theta)^2}$$

and suppose that we want to test

$$H_0 : \theta \geq 0 \quad \text{versus} \quad H_1 : \theta < 0$$

at the 5% level. We then have

$$S(\boldsymbol{X};0) = \sum_{i=1}^{n} \frac{2X_i}{1 + X_i^2}$$

and the LMP test of H_0 versus H_1 rejects H_0 if $S(\boldsymbol{X};0) \leq k$ where k is chosen so that

$$P_0\left[S(\boldsymbol{X};0) \leq k\right] = 0.05.$$

The exact distribution is difficult to obtain; however, since each of the summands of $S(\boldsymbol{X};0)$ has mean 0 and variance $1/2$, the distribution of $S(\boldsymbol{X};0)$ (for $\theta = 0$) will be approximately Normal with mean 0 and variance $n/2$ if n is sufficiently large. Thus we will have $k\sqrt{2/n} \approx -1.645$ and so we can take $k = -1.645\sqrt{n/2}$ to obtain a test whose level is approximately 5%. ◇

7.4 Likelihood ratio tests

Our discussion of UMP and LMP tests has involved only one parameter models since these type of optimal tests do not generally exist for models with more than one parameter. It is therefore desirable to develop a general purpose method for developing reasonable test procedures for more general situations.

We return to our general hypothesis testing problem where $\boldsymbol{X} = (X_1, \cdots, X_n)$ has a joint density or frequency function $f(\boldsymbol{x};\theta)$ where $\theta \in \Theta$ with $\Theta = \Theta_0 \cup \Theta_1$; we wish to test

$$H_0 : \theta \in \Theta_0 \quad \text{versus} \quad H_1 : \theta \in \Theta_1.$$

Earlier in this chapter, we saw that the Neyman-Pearson Lemma was useful in one-parameter problems for finding various optimal tests based on the ratio of joint density (frequency) functions. In the more general testing problem, we will use the Neyman-Pearson paradigm along with maximum likelihood estimation to give us a general purpose testing procedure.

DEFINITION. The likelihood ratio (LR) statistic Λ is defined to be

$$\Lambda = \frac{\sup_{\theta \in \Theta} f(\boldsymbol{X};\theta)}{\sup_{\theta \in \Theta_0} f(\boldsymbol{X};\theta)} = \frac{\sup_{\theta \in \Theta} \mathcal{L}(\theta)}{\sup_{\theta \in \Theta_0} \mathcal{L}(\theta)}$$

where $\mathcal{L}(\theta)$ is the likelihood function. A likelihood ratio test of $H_0 : \theta \in \Theta_0$ versus $H_1 : \theta \in \Theta_1$ will reject H_0 for large values of Λ.

In our definition of Λ, we take the supremum of $\mathcal{L}(\theta) = f(\boldsymbol{X};\theta)$ over $\theta \in \Theta$ rather than $\theta \in \Theta_1$ in the numerator of Λ. We do this mainly for convenience; as mentioned earlier, Θ_0 is often a lower dimensional subspace of Θ, which makes the calculation of the MLE under H_0 typically no more difficult than the calculation of the unrestricted MLE.

In order to use LR tests, we must know (either exactly or approximately) the distribution of the statistic Λ when H_0 is true. In some cases, Λ is a function of some other statistic T whose distribution is known; in such cases, we can use T (rather than Λ) as our test statistic. In other cases, we can approximate the distribution of Λ (or some function of Λ) by a standard distribution such as the χ^2 distribution.

EXAMPLE 7.19: Let $X_1, \cdots, X_n$ be i.i.d. Normal random variables with mean μ and variance σ^2 (both unknown) and suppose that we want to test

$$H_0 : \mu = \mu_0 \quad \text{versus} \quad H_1 : \mu \neq \mu_0.$$

In general, the MLEs of μ and σ^2 are

$$\widehat{\mu} = \bar{X} \quad \text{and} \quad \widehat{\sigma}^2 = \frac{1}{n}\sum_{i=1}^{n}(X_i - \bar{X})^2$$

while under H_0 the MLE of σ^2 is

$$\widehat{\sigma}_0^2 = \frac{1}{n}\sum_{i=1}^{n}(X_i - \mu_0)^2.$$

Substituting the respective MLEs yields

$$\Lambda = \left(\frac{\widehat{\sigma}_0^2}{\widehat{\sigma}^2}\right)^{n/2}$$

and so the LR test will reject H_0 when $\Lambda \geq k$ where k is chosen so the level of the test is some specified α. The distribution of Λ is not obvious; however, note that Λ is a monotone function of $\widehat{\sigma}_0^2/\widehat{\sigma}^2$ and

$$\begin{aligned}
\frac{\widehat{\sigma}_0^2}{\widehat{\sigma}^2} &= \frac{\sum_{i=1}^{n}(X_i - \mu_0)^2}{\sum_{i=1}^{n}(X_i - \bar{X})^2} \\
&= 1 + \frac{n(\bar{X} - \mu_0)^2}{\sum_{i=1}^{n}(X_i - \bar{X})^2} \\
&= 1 + \frac{1}{n-1}\left(\frac{n(\bar{X} - \mu_0)^2}{S^2}\right)
\end{aligned}$$

$$= 1 + \frac{T^2}{n-1}$$

where

$$S^2 = \frac{1}{n-1}\sum_{i=1}^{n}(X_i - \bar{X})^2$$

and

$$T = \sqrt{n}(\bar{X} - \mu_0)/S.$$

From this, we can conclude that Λ is a monotone function of T^2 and we know (from Chapter 2) that, when H_0 is true, T has Student's t distribution with $(n-1)$ degrees of freedom and so T^2 has an F distribution with 1 and $(n-1)$ degrees of freedom. ◇

EXAMPLE 7.20: Let $X_1, \cdots, X_m$ be i.i.d. Exponential random variables with parameter λ and $Y_1, \cdots, Y_n$ be i.i.d. random variables with parameter θ; we also assume that X_i's are independent of the Y_i's. Suppose we want to test

$$H_0 : \lambda = \theta \quad \text{versus} \quad H_1 : \lambda \neq \theta$$

The (unrestricted) MLEs of λ and θ are

$$\widehat{\lambda} = 1/\bar{X} \quad \text{and} \quad \widehat{\theta} = 1/\bar{Y}$$

while the MLEs under H_0 are

$$\widehat{\lambda}_0 = \widehat{\theta}_0 = \left[\frac{m\bar{X} + n\bar{Y}}{n+m}\right]^{-1}.$$

Substituting these MLEs, we obtain the LR statistic

$$\Lambda = \left(\frac{m}{m+n} + \frac{n}{m+n}\frac{\bar{Y}}{\bar{X}}\right)^m \left(\frac{n}{m+n} + \frac{m}{m+n}\frac{\bar{X}}{\bar{Y}}\right)^n.$$

Clearly, Λ depends only on the statistic $T = \bar{X}/\bar{Y}$ and we can make Λ large by making T large or T small. Moreover, it is quite easy to see that, when H_0 is true, T has an F distribution with $2m$ and $2n$ degrees of freedom and so, for a given value of α, it is quite simple to base a test of H_0 versus H_1 on the statistic T. ◇

Asymptotic distribution of the LR statistic

As we mentioned previously, we can often approximate the distribution of the LR test statistic. In this section, we will assume that Θ is an open subset of R^p and the H_0 parameter space Θ_0 is either a single point or an open subset of R^s where $s < p$. We will concentrate on the case where $X_1, \cdots, X_n$ are i.i.d. random variables although many of the results will hold under more general conditions.

First of all, suppose that $X_1, \cdots, X_n$ are i.i.d. random variables with density or frequency function $f(x;\theta)$ where the parameter space Θ is an open subset of the real line. We will consider the testing

$$H_0 : \theta = \theta_0 \quad \text{and} \quad H_1 : \theta \neq \theta_0.$$

In this case, the LR statistic is simply

$$\Lambda_n = \prod_{i=1}^{n} \frac{f(X_i;\widehat{\theta}_n)}{f(X_i;\theta_0)}$$

where $\widehat{\theta}_n$ is the MLE of θ. Assuming the regularity conditions in Chapter 5, we obtain the following result.

THEOREM 7.4 *Suppose that $X_1, \cdots, X_n$ are i.i.d. random variables with a density or frequency function satisfying conditions A1 to A6 in Chapter 5 with $I(\theta) = J(\theta)$. If the MLE $\widehat{\theta}_n$ satisfies $\sqrt{n}(\widehat{\theta}_n - \theta) \to_d N(0, 1/I(\theta)$ then the LR statistic Λ_n for testing $H_0 : \theta = \theta_0$ satisfies*

$$2\ln(\Lambda_n) \to_d V \sim \chi^2(1)$$

when H_0 is true.

Proof. Let $\ell(x;\theta) = \ln f(x;\theta)$ and $\ell'(x;\theta)$, $\ell''(x;\theta)$ be its derivatives with respect to θ. Under the conditions of the theorem,

$$\sqrt{n}(\widehat{\theta}_n - \theta_0) \to_d N(0, 1/I(\theta_0))$$

when H_0 is true. Taking logarithms and doing a Taylor series expansion, we get

$$\begin{aligned}
\ln(\Lambda_n) &= \sum_{i=1}^{n}[\ell(X_i;\widehat{\theta}_n) - \ell(X_i;\theta_0)] \\
&= (\theta_0 - \widehat{\theta}_n)\sum_{i=1}^{n}\ell'(X_i;\widehat{\theta}_n) - \frac{1}{2}(\widehat{\theta}_n - \theta_0)^2\sum_{i=1}^{n}\ell''(X_i;\theta_n^*) \\
&= -\frac{1}{2}n(\widehat{\theta}_n - \theta_0)^2\frac{1}{n}\sum_{i=1}^{n}\ell''(X_i;\theta_n^*)
\end{aligned}$$

where θ_n^* lies between θ_0 and $\widehat{\theta}_n$. Now under assumptions A5 and

A6 of Chapter 5, it follows that when H_0 is true

$$\frac{1}{n}\sum_{i=1}^{n} \ell''(X_i;\theta_n^*) \to_p -E_{\theta_0}[\ell''(X_i;\theta_0)] = I(\theta_0).$$

Since

$$n(\widehat{\theta}_n - \theta_0)^2 \to_d \frac{V}{I(\theta_0)}$$

the conclusion follows by applying Slutsky's Theorem. □

Henceforth, we will refer to both Λ_n and $2\ln(\Lambda_n)$ as LR statistics depending on the situation; there is no real ambiguity in doing so since they are equivalent from a hypothesis testing viewpoint.

Theorem 7.4 can be extended fairly easily to the multiparameter case. Let $\boldsymbol{\theta} = (\theta_1, \cdots, \theta_p)$ and consider testing

$$H_0 : \theta_1 = \theta_{10}, \cdots, \theta_r = \theta_{r0}$$

where $r \le p$ for some specified $\theta_{10}, \cdots, \theta_{r0}$. The LR statistic is defined by

$$\Lambda_n = \prod_{i=1}^{n} \frac{f(X_i;\widehat{\boldsymbol{\theta}}_n)}{f(X_i;\widehat{\boldsymbol{\theta}}_{n0})}$$

where $\widehat{\boldsymbol{\theta}}_{n0}$ is the MLE of $\boldsymbol{\theta}$ under H_0 (and thus whose first r components are $\theta_{10}, \cdots, \theta_{r0}$). Again we will assume the regularity conditions B1 to B6 in Chapter 5 to obtain the following theorem.

THEOREM 7.5 *Suppose that $X_1, \cdots, X_n$ are i.i.d. random variables with a density or frequency function satisfying conditions B1 to B6 in Chapter 5 with $I(\boldsymbol{\theta}) = J(\boldsymbol{\theta})$ where $\boldsymbol{\theta} = (\theta_1, \cdots, \theta_p)$. If the MLE $\widehat{\boldsymbol{\theta}}_n$ satisfies $\sqrt{n}(\widehat{\boldsymbol{\theta}}_n - \boldsymbol{\theta}) \to_d N(0, I^{-1}(\boldsymbol{\theta}))$ then the LR statistic Λ_n for testing $H_0 : \theta_1 = \theta_{10}, \cdots, \theta_r = \theta_{r0}$ satisfies*

$$2\ln(\Lambda_n) \to_d V \sim \chi^2(r)$$

when H_0 is true.

The proof of Theorem 7.5 will be left as an exercise. However, note that when $r = p$, the proof of Theorem 7.4 can be easily adapted to give a proof of Theorem 7.5. In the general case, the result can be deduced from the fact that the log-likelihood function can be approximated by a quadratic function in a neighbourhood of the true parameter value; this quadratic approximation is discussed briefly below.

Theorem 7.5 can be applied to testing null hypotheses of the form

$$H_0 : g_1(\boldsymbol{\theta}) = a_1, \cdots, g_r(\boldsymbol{\theta}) = a_r$$

for some real-valued functions $g_1, \cdots, g_r$. To see this, define parameters $\phi_1, \cdots, \phi_p$ such that $\phi_k = g_k(\boldsymbol{\theta})$ where $g_1, \cdots, g_r$ are as given in H_0 and $g_{r+1}, \cdots, g_p$ are defined so that the vector-valued function $\boldsymbol{g}(\boldsymbol{\theta}) = (g_1(\boldsymbol{\theta}), \cdots, g_p(\boldsymbol{\theta}))$ is a one-to-one function. Then provided that this function is differentiable, Theorem 7.5 can be applied using the parameters $\phi_1, \cdots, \phi_p$.

EXAMPLE 7.21: Suppose that $(X_1, Y_1), \cdots, (X_n, Y_n)$ are i.i.d. pairs of continuous random variables with the joint density function of (X_i, Y_i) given by

$$f(x, y; \theta, \lambda, \alpha) = \frac{2\theta\lambda\alpha}{(\theta x + \lambda y + \alpha)^3} \quad \text{for } x, y > 0$$

where $\theta, \lambda, \alpha > 0$. The marginal densities of X_i and Y_i are

$$\begin{aligned} f_X(x; \theta, \alpha) &= \frac{\theta\alpha}{(\theta x + \alpha)^2} \quad \text{for } x > 0 \\ \text{and} \quad f_Y(y; \lambda, \alpha) &= \frac{\lambda\alpha}{(\lambda y + \alpha)^2} \quad \text{for } y > 0, \end{aligned}$$

which are equal if $\theta = \lambda$. Thus, we may be interested in testing

$$H_0 : \theta = \lambda.$$

We can reparametrize in a number of ways. For example, we could define $\eta_1 = \theta - \lambda$, $\eta_2 = \theta$, and $\eta_3 = \alpha$, or alternatively, $\eta_1 = \theta/\lambda$ with η_2, η_3 defined as before. With either reparametrization, we could express H_0 in terms of η_1 and so we would expect our likelihood test to have an asymptotic χ^2 distribution with one degree of freedom. ◇

Other likelihood based tests

While LR tests are motivated by the Neyman-Pearson paradigm, there are in fact other commonly used tests based on the likelihood function.

Let $X_1, \cdots, X_n$ be i.i.d. random variables with density or frequency function $f(x; \boldsymbol{\theta})$ where $\boldsymbol{\theta} = (\theta_1, \cdots, \theta_p)$ and suppose that we want to test the null hypothesis

$$H_0 : \theta_1 = \theta_{10}, \cdots, \theta_r = \theta_{r0}.$$

To make the notation more compact, we will set $\boldsymbol{\phi} = (\theta_1, \cdots, \theta_r)$ and $\boldsymbol{\tau} = (\theta_{r+1}, \cdots, \theta_p)$ so that $\boldsymbol{\theta} = (\boldsymbol{\phi}, \boldsymbol{\tau})$; thus, H_0 becomes

$$H_0 : \boldsymbol{\phi} = \boldsymbol{\phi}_0$$

where $\boldsymbol{\phi}_0 = (\theta_{10}, \cdots, \theta_{r0})$.

The Wald test of H_0 compares the (unrestricted) MLE of $\boldsymbol{\phi}$ to its value under the null hypothesis; if the distance between the two is large, this might indicate that H_0 is false and so our test should reflect this. If $\widehat{\boldsymbol{\phi}}_n$ is the MLE (based on $X_1, \cdots, X_n$) then if H_0 is true, we have

$$\sqrt{n}(\widehat{\boldsymbol{\phi}}_n - \boldsymbol{\phi}_0) \to_d N_r(\mathbf{0}, C(\boldsymbol{\phi}_0, \boldsymbol{\tau}))$$

where the variance-covariance matrix $C(\boldsymbol{\phi}_0, \boldsymbol{\tau})$ can be obtained from the Fisher information matrix

$$I(\boldsymbol{\phi}_0, \boldsymbol{\tau}) = \begin{pmatrix} I_{11}(\boldsymbol{\phi}_0, \boldsymbol{\tau}) & I_{12}(\boldsymbol{\phi}_0, \boldsymbol{\tau}) \\ I_{21}(\boldsymbol{\phi}_0, \boldsymbol{\tau}) & I_{22}(\boldsymbol{\phi}_0, \boldsymbol{\tau}) \end{pmatrix}$$

by

$$C(\boldsymbol{\phi}_0, \boldsymbol{\tau}) = \left[I_{11}(\boldsymbol{\phi}_0, \boldsymbol{\tau}) - I_{21}(\boldsymbol{\phi}_0, \boldsymbol{\tau}) I_{22}^{-1}(\boldsymbol{\phi}_0, \boldsymbol{\tau}) I_{12}(\boldsymbol{\phi}_0, \boldsymbol{\tau})\right]^{-1}.$$

The Wald test statistic is

$$W_n = n(\widehat{\boldsymbol{\phi}}_n - \boldsymbol{\phi}_0)^T \widehat{C}_n (\widehat{\boldsymbol{\phi}}_n - \boldsymbol{\phi}_0)$$

where $\widehat{C}_n$ is some estimator of $C(\boldsymbol{\phi}_0, \boldsymbol{\tau})$ that is consistent under H_0. There are several possibilities for $\widehat{C}_n$; for example, we could set $\widehat{C}_n = C(\widehat{\boldsymbol{\phi}}_n, \widehat{\boldsymbol{\tau}}_n)$ (or $C(\boldsymbol{\phi}_0, \widehat{\boldsymbol{\tau}}_n)$) or we could set $\widehat{C}_n$ equal to the observed information matrix. Under H_0, $W_n \to_d \chi^2(r)$ and we reject H_0 for large values of this statistic.

The score test (or Lagrange multiplier test as it is called by econometricians) uses the fact that if the null hypothesis is false then the gradient of the log-likelihood function should not be close to the $\mathbf{0}$ vector. To be more precise, let $\boldsymbol{S}_i(\boldsymbol{\phi}, \boldsymbol{\tau})$ be the gradient of $\ln f(X_i; \boldsymbol{\phi}, \boldsymbol{\tau})$ with respect to $\boldsymbol{\phi}$. Then under H_0, we have (subject to the regularity conditions of Chapter 5),

$$\frac{1}{\sqrt{n}} \sum_{i=1}^{n} \boldsymbol{S}_i(\boldsymbol{\phi}_0, \widehat{\boldsymbol{\tau}}_{n0}) \to_d N_r(\mathbf{0}, I_{11}(\boldsymbol{\phi}_0, \boldsymbol{\tau}))$$

where $\widehat{\boldsymbol{\tau}}_{n0}$ is the MLE of $\boldsymbol{\tau}$ under H_0. The score statistic is

$$S_n = \frac{1}{n} \left(\sum_{i=1}^{n} \boldsymbol{S}_i(\boldsymbol{\phi}_0, \widehat{\boldsymbol{\tau}}_{n0})\right)^T I_{11}^{-1}(\boldsymbol{\phi}_0, \widehat{\boldsymbol{\tau}}_{n0}) \left(\sum_{i=1}^{n} \boldsymbol{S}_i(\boldsymbol{\phi}_0, \widehat{\boldsymbol{\tau}}_{n0})\right)$$

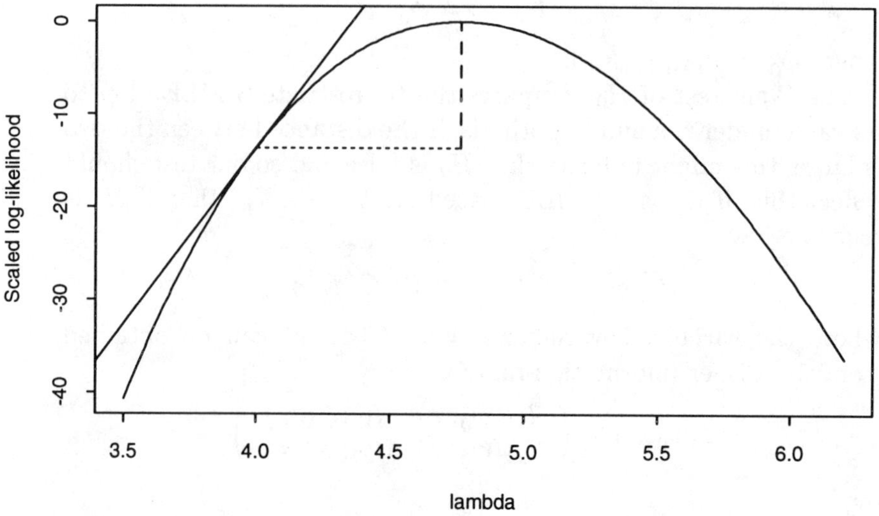

Figure 7.3 *The log-likelihood (multiplied by two and rescaled to have a maximum of 0) in Example 7.22; for testing $H_0 : \lambda = 4$, the LR statistic is the length of the vertical line, the Wald statistic is proportional to the square of the length of the horizontal line while the score statistic is proportional to the square of the slope of the tangent line at $\lambda = 4$.*

As for the Wald statistic, we reject H_0 for large values of S_n and, under H_0, we have $S_n \to_d \chi^2(r)$.

EXAMPLE 7.22: We will give an illustration of the three tests in a very simple setting. Suppose that $X_1, \cdots, X_n$ are i.i.d. Poisson random variables with mean λ and we want to test the null hypothesis $H_0 : \lambda = \lambda_0$ versus the alternative hypothesis $H_1 : \lambda \neq \lambda_0$. For the Poisson distribution, we have $I(\lambda) = 1/\lambda$ and the MLE is $\widehat{\lambda}_n = \bar{X}_n$. Therefore, the LR, Wald, and score statistics have the following formulas:

$$
\begin{aligned}
2\ln(\Lambda_n) &= 2n\left(\bar{X}_n \ln(\bar{X}_n/\lambda_0) - (\bar{X}_n - \lambda_0)\right) \\
W_n &= \frac{n(\bar{X}_n - \lambda_0)^2}{\lambda_0} \\
S_n &= \frac{n(\bar{X}_n - \lambda_0)^2}{\lambda_0}.
\end{aligned}
$$

Note that S_n and W_n are the same in this case. Figure 7.3 shows a graph of the (scaled) likelihood function for a sample of 100 observations. For the purpose of illustration, we take $\lambda_0 = 4$ in which case the value of the LR statistic (for these data) is 13.60 while the value for the score and Wald statistic is 14.44; using the asymptotic $\chi^2(1)$ null distribution, we would reject H_0 at the 0.05 level for values greater than 3.84 and at the 0.01 level for values greater than 6.63. ◇

Although both the Wald and the score statistics have the same limiting distribution (under H_0) as the LR statistic, there are some practical differences in the use of these statistics. First, suppose that we reparametrize the model by setting $\boldsymbol{\theta}^* = \boldsymbol{g}(\boldsymbol{\theta})$ where $\boldsymbol{g}$ is a one-to-one transformation on Θ. In this case, the LR statistic remains unchanged since the maxima (both restricted and unrestricted) of the likelihood function remain unchanged. On the other hand, the Wald and score statistics will not remain invariant to reparametrization although the dependence on the parametrization of the model becomes smaller as the sample size increases. Second, both the LR and Wald statistics require the computation of restricted and unrestricted MLEs; on the other hand, the score statistic requires only computation of the restricted MLE (that is, the MLE under H_0). Thus the score statistic is potentially simpler from a computational point of view and for this reason, the score test is often used in deciding whether or not parameters should be added to a model.

As mentioned above, the score test is often called the Lagrange multiplier test, particularly in econometrics. The reason for this is the following. Suppose that we want to maximize a log-likelihood function $\ln L(\boldsymbol{\theta})$ subject to the constraint $\boldsymbol{g}(\boldsymbol{\theta}) = \mathbf{0}$ for some function $\boldsymbol{g}$. To solve this problem, we can introduce a vector of Lagrange multipliers $\boldsymbol{\lambda}$ and maximize the function

$$h(\boldsymbol{\theta}, \boldsymbol{\lambda}) = \ln \mathcal{L}(\boldsymbol{\theta}) + \boldsymbol{\lambda}^T \boldsymbol{g}(\boldsymbol{\theta}).$$

If $(\widehat{\boldsymbol{\theta}}, \widehat{\boldsymbol{\lambda}})$ maximizes $h(\boldsymbol{\theta}, \boldsymbol{\lambda})$ then $\widehat{\boldsymbol{\lambda}} \approx \mathbf{0}$ implies that the constrained maximum and unconstrained maximum of $\ln \mathcal{L}(\boldsymbol{\theta})$ are close; on the other hand, if $\widehat{\boldsymbol{\lambda}}$ is not close to the $\mathbf{0}$ vector then the two maxima can be very different. This suggests that a test statistic can be based on $\widehat{\boldsymbol{\lambda}}$; in the problem considered above, this statistic is simply the score statistic.

EXAMPLE 7.23: Suppose that $X_1, \cdots, X_n$ are i.i.d. Gamma random variables with shape parameter α and scale parameter λ. We want to test

$$H_0 : \alpha = 1 \quad \text{versus} \quad H_1 : \alpha \neq 1.$$

Under H_0, note that the X_i's have an Exponential distribution. For this model, estimating both α and λ via maximum likelihood is non-trivial (although not difficult) as there is no closed-form expression for the MLEs (see Example 5.15); however, for fixed α, there is a simple closed-form expression for the MLE of λ. Thus the score test seems an attractive approach to testing H_0. For $\alpha = 1$, the MLE of λ is $\widehat{\lambda}_n = 1/\bar{X}_n$ and so the score statistic is

$$S_n = \frac{1}{n\psi'(1)} \left(\sum_{i=1}^{n} \left(\ln(X_i/\bar{X}_n) - \psi(1)\right) \right)^2$$

where $\psi(\alpha)$ and $\psi'(\alpha)$ are the first and second derivatives of $\ln\Gamma(\alpha)$ with $\psi(1) = -0.57722$ and $\psi'(1) = 1.64493$; see Example 5.15 for more details. The limiting distribution of S_n is χ^2 with 1 degree of freedom. ◇

In addition to having the same limiting distribution, the LR statistic as well as the Wald and score statistics are asymptotically equivalent (under the null hypothesis) in the sense that the difference between any two of them tends in probability to 0 as $n \to \infty$. This asymptotic equivalence is a consequence of the fact that the log-likelihood function is a quadratic function in a neighbourhood of the true parameter value (assuming the regularity conditions of Theorem 7.5). More precisely, if $\ln \mathcal{L}_n(\boldsymbol{\theta})$ is the log-likelihood function and $\boldsymbol{\theta}_0$ is the true value of the parameter then we have

$$\begin{aligned} Z_n(\boldsymbol{u}) &= \ln(\mathcal{L}_n(\boldsymbol{\theta}_0 + \boldsymbol{u}/\sqrt{n})/\mathcal{L}_n(\boldsymbol{\theta}_0)) \\ &= \boldsymbol{u}^T \boldsymbol{V}_n - \frac{1}{2}\boldsymbol{u}^T I(\boldsymbol{\theta}_0)\boldsymbol{u} + R_n(\boldsymbol{u}) \end{aligned}$$

where $R_n(\boldsymbol{u}) \to_p 0$ for each $\boldsymbol{u}$ and $\boldsymbol{V}_n \to_d N_p(\boldsymbol{0}, I(\boldsymbol{\theta}_0))$; $Z_n(\boldsymbol{u})$ is maximized at $\boldsymbol{u} = \sqrt{n}(\widehat{\boldsymbol{\theta}}_n - \boldsymbol{\theta}_0)$ and the quadratic approximation to Z_n is maximized at $\boldsymbol{u} = I^{-1}(\boldsymbol{\theta}_0)\boldsymbol{V}_n$. Thus if we are interested in testing $H_0 : \boldsymbol{\theta} = \boldsymbol{\theta}_0$, the quadratic approximation to Z_n suggests the following approximations to our test statistics:

$$\begin{aligned} 2\ln(\Lambda_n) &= 2Z_n(\sqrt{n}(\widehat{\boldsymbol{\theta}}_n - \boldsymbol{\theta}_0)) \\ &\approx \boldsymbol{V}_n^T I^{-1}(\boldsymbol{\theta}_0)\boldsymbol{V}_n \end{aligned}$$

$$\begin{aligned}
W_n &= n(\widehat{\boldsymbol{\theta}}_n - \boldsymbol{\theta}_0)^T I(\boldsymbol{\theta}_0)(\widehat{\boldsymbol{\theta}}_n - \boldsymbol{\theta}_0) \\
&= \boldsymbol{V}_n^T I^{-1}(\boldsymbol{\theta}_0)\boldsymbol{V}_n \\
S_n &= [\nabla Z_n(\mathbf{0})]^T I^{-1}(\boldsymbol{\theta}_0)[\nabla Z_n(\mathbf{0})] \\
&\approx \boldsymbol{V}_n I^{-1}(\boldsymbol{\theta}_0)\boldsymbol{V}_n
\end{aligned}$$

where $\nabla Z_n(\mathbf{0})$ is the gradient of $Z_n(\boldsymbol{u})$ at $\boldsymbol{u} = \mathbf{0}$, or equivalently, the score vector divided by $\sqrt{n}$. Thus $\boldsymbol{V}_n I^{-1}(\boldsymbol{\theta}_0)\boldsymbol{V}_n$ serves as an approximation for each of $2\ln(\Lambda_n)$ (the LR statistic), W_n (the Wald statistic), and S_n (the score statistic); note that $\boldsymbol{V}_n I^{-1}(\boldsymbol{\theta}_0)\boldsymbol{V}_n \to_d \chi^2(p)$. Similarly approximations hold for these test statistics applied to other null hypotheses considered in this section.

7.5 Other issues

P-values

Up to this point, we have assume a fixed level α when discussing hypothesis tests. That is, given α, we define a test function ϕ; then given data $\boldsymbol{X} = \boldsymbol{x}$, we reject the null hypothesis at level α if $\phi(\boldsymbol{x})$.

An alternative approach (which is more in line with current practice) is to consider a family of test functions ϕ_α for $0 < \alpha < 1$ where the test function ϕ_α has level α. We will assume the test functions $\{\phi_\alpha\}$ satisfy the condition

$$\phi_{\alpha_1}(\boldsymbol{x}) = 1 \quad \text{implies} \quad \phi_{\alpha_2}(\boldsymbol{x}) = 1$$

for any $\alpha_1 < \alpha_2$. We then define the p-value (or observed significance level) to be

$$p(\boldsymbol{x}) = \inf\{\alpha : \phi_\alpha(\boldsymbol{x}) = 1\}.$$

The p-value $p(\boldsymbol{x})$ is the smallest value of α for which the null hypothesis would be rejected at level α given $\boldsymbol{X} = \boldsymbol{x}$.

In the case where the hypothesis test is framed in terms of a single test statistic $T = T(\boldsymbol{X})$ such that $\phi_\alpha(\boldsymbol{X}) = 1$ for $T > k_\alpha$ then it is straightforward to evaluate p-values. If $G(x)$ is the null distribution function of T, then given $T(\boldsymbol{x}) = t$, the p-value is $p(\boldsymbol{x}) = 1 - G(t)$.

EXAMPLE 7.24: Suppose that $X_1, \cdots, X_m$ and $Y_1, \cdots, Y_n$ are two samples of i.i.d. random variables with $X_i \sim N(\mu_1, \sigma^2)$ and $Y_i \sim N(\mu_2, \sigma^2)$. The LR test of the null hypothesis

$$H_0 : \mu_1 = \mu_2$$

(versus the alternative $H_1 = \mu_1 \neq \mu_2$) rejects H_0 for large values of the test statistic $|T|$ where

$$T = \frac{\bar{X} - \bar{Y}}{S\sqrt{m^{-1} + n^{-1}}}$$

and S^2 is the so-called "pooled" estimator of variance:

$$S^2 = \frac{1}{m+n-2}\left[\sum_{i=1}^{m}(X_i - \bar{X})^2 + \sum_{i=1}^{n}(Y_i - \bar{Y})^2\right].$$

It is easy to verify that $T \sim \mathcal{T}(m+n-2)$ under the null hypothesis. Thus given $T = t$, the p-value is $p(t) = 1 + G(-|t|) - G(|t|)$ where $G(x)$ is the distribution function of the t distribution. However, if the alternative hypothesis is $H_1 : \mu_1 > \mu_2$, we would typically reject H_0 for large values of T. In this case, given $T = t$, the p-value is $p(t) = 1 - G(t)$. ◇

The p-value is often used as a measure of evidence against the null hypothesis: the smaller $p(\boldsymbol{x})$, the more evidence against the null hypothesis. While this use of p-values is quite common in statistical practice, its use as a measure of evidence is quite controversial. In particular, it is difficult to calibrate p-values as measures of evidence (Goodman, 1999a).

P-values are often (erroneously) interpreted as the probability (given the observed data) that the null hypothesis is true. However, if we put a prior distribution on the parameter space then it may be possible to compute such a probability (from a Bayesian perspective) using the posterior distribution. Some care is required in interpreting these probabilities though; for example, the posterior probability that the null hypothesis is true necessarily depends on the prior probability of the null hypothesis. As a Bayesian alternative to p-values, some authors have proposed using Bayes factors, which essentially measure the change in the odds of the null hypothesis from the prior to the posterior; see, for example, Kass and Raftery (1995), and DiCiccio *et al* (1997). It is often argued that Bayes factors are more easily intepretable than p-values (Goodman, 1999b) although this view is not universally shared.

We can also view p-values, in their own right, as test statistics. Suppose that we want to test

$$H_0 : \theta \in \Theta_0 \quad \text{versus} \quad H_1 : \theta \in \Theta_1$$

using a family of tests whose test functions are $\{\phi_\alpha\}$. Suppose that

$\boldsymbol{X} = (X_1, \cdots, X_n)$ are continuous random variables and that, for some $\theta_0 \in \Theta_0$,

$$E_{\theta_0}[\phi_\alpha(\boldsymbol{X})] = \alpha$$

for all $0 < \alpha < 1$. Then if the p-value $p(\boldsymbol{X})$ has a continuous distribution, it follows that

$$P_{\theta_0}\left[p(\boldsymbol{X}) \leq x\right] = x$$

for $0 < x < 1$; that is, $p(\boldsymbol{X})$ has a Uniform distribution on $[0, 1]$ when θ_0 is the true value of the parameter. (Note that, for given $\theta \in \Theta_0$, we have

$$P_\theta\left[p(\boldsymbol{X}) \leq x\right] \geq x$$

for $0 < x < 1$.)

The fact that $p(\boldsymbol{X})$ is uniformly distributed under H_0 can be useful in practice. For example, suppose that we have p-values $P_1, \cdots, P_k$ from k independent tests of the same null hypothesis. Assuming that the P_i's are uniformly distributed when the null hypothesis is true, we can combine the p-values using the test statistic

$$T = -2\sum_{i=1}^{k} \ln(P_i);$$

under the null hypothesis, T has a χ^2 with $2k$ degrees of freedom. This simple approach to meta-analysis (that is, combining results of different studies) is due to R.A. Fisher.

Obtaining confidence regions from hypothesis tests

Our discussion of confidence intervals and regions gave essentially no theoretical guidance on how to choose a "good" confidence procedure; in contrast, for hypothesis testing, the Neyman-Pearson Lemma provides a useful paradigm for deriving "good" hypothesis tests in various situations. In fact, there turns out to be a very close relationship between confidence intervals (or regions) and hypothesis tests; we can exploit this relationship to turn "good" hypothesis tests into "good" confidence procedures.

Suppose that $R(\boldsymbol{X})$ is an exact $100p\%$ confidence region for a parameter $\boldsymbol{\theta}$ and we want to test the null hypothesis

$$H_0 : \boldsymbol{\theta} = \boldsymbol{\theta}_0 \quad \text{versus} \quad H_1 : \boldsymbol{\theta} \neq \boldsymbol{\theta}_0.$$

Define the test function

$$\phi(\boldsymbol{X}) = \begin{cases} 1 & \text{if } \boldsymbol{\theta}_0 \notin R(\boldsymbol{X}) \\ 0 & \text{if } \boldsymbol{\theta}_0 \in R(\boldsymbol{X}). \end{cases}$$

It is easy to verify that the size of this test is

$$\begin{aligned} E_{\boldsymbol{\theta}_0}[\phi(\boldsymbol{X})] &= 1 - P_{\boldsymbol{\theta}_0}[\boldsymbol{\theta}_0 \in R(\boldsymbol{X})] \\ &\leq 1 - p. \end{aligned}$$

Thus we can use a $100p\%$ confidence region to construct a test of H_0 whose level is at most $1-p$. On the other hand, suppose that we have α-level tests of H_0 for each $\boldsymbol{\theta}_0 \in \Theta$; define $\phi(\boldsymbol{X}; \boldsymbol{\theta}_0)$ to be the test function for a given θ_0. Now define

$$R^*(\boldsymbol{X}) = \{\boldsymbol{\theta}_0 : \phi(\boldsymbol{X}; \boldsymbol{\theta}_0) = 0\};$$

the coverage of $R^*(\boldsymbol{X})$ is

$$\begin{aligned} P_{\boldsymbol{\theta}}[\boldsymbol{\theta} \in R^*(\boldsymbol{X})] &= P_{\boldsymbol{\theta}}[\phi(\boldsymbol{X}; \boldsymbol{\theta}) = 0] \\ &\geq 1 - \alpha. \end{aligned}$$

Thus we can construct a (possibly conservative) $100p\%$ confidence region for $\boldsymbol{\theta}$ by considering a family of $\alpha = 1-p$ level tests and defining the confidence region to be the set of $\boldsymbol{\theta}$'s for which we cannot reject the null hypothesis at level α.

This "duality" between hypothesis tests and confidence intervals or regions can be very useful in practice. For example, suppose that $X_1, \cdots, X_n$ are i.i.d. random variables with density or frequency function $f(x; \boldsymbol{\theta})$ and we want to find a confidence interval for a single parameter (call it ϕ) in $\boldsymbol{\theta}$. Writing $\boldsymbol{\theta} = (\phi, \boldsymbol{\tau})$, the LR statistic for $H_0 : \phi = \phi_0$ is

$$2\ln(\Lambda_n) = 2\sum_{i=1}^{n} \ln[f(X_i; \widehat{\phi}_n, \widehat{\boldsymbol{\tau}}_n)/f(X_i; \phi_0, \widehat{\boldsymbol{\tau}}_n(\phi_0))]$$

where $\widehat{\boldsymbol{\tau}}_n(\phi_0)$ is the MLE of $\boldsymbol{\tau}$ under H_0 (that is, assuming that $\phi = \phi_0$ is known). According to Theorem 7.5, under H_0, $2\ln(\Lambda_n) \to_d \chi^2(1)$ and H_0 is rejected for large values of the statistic $2\ln(\Lambda_n)$. Thus if k_p is the p quantile of a $\chi^2(1)$ distribution, an approximate $100p\%$ confidence interval for ϕ is

$$R(\boldsymbol{X}) = \{\phi : g(\boldsymbol{X}; \phi) \leq k_p\}$$

where $g(\boldsymbol{X};\phi)$ is the "likelihood ratio" pivot:

$$g(\boldsymbol{X};\phi) = 2\sum_{i=1}^{n} \ln[f(X_i;\widehat{\phi}_n,\widehat{\boldsymbol{\tau}}_n)/f(X_i;\phi,\widehat{\boldsymbol{\tau}}_n(\phi))].$$

Note that $g(\boldsymbol{X};\phi) \geq 0$ with $g(\boldsymbol{X};\widehat{\phi}_n) = 0$.

EXAMPLE 7.25: Suppose that $X_1, \cdots, X_n$ are i.i.d. Gamma random variables with shape parameter α and scale parameter λ. We will derive an approximate 95% confidence interval for α based on the LR test procedure. First of all, we start with the log-likelihood function

$$\ln\mathcal{L}(\alpha,\lambda) = n\alpha\ln(\lambda) + (\alpha-1)\sum_{i=1}^{n}\ln(X_i) - \lambda\sum_{i=1}^{n}X_i - n\ln\Gamma(\alpha).$$

The LR test of $H_0 : \alpha = \alpha_0$ compares the maximized likelihood with $\alpha = \alpha_0$ to the unrestricted maximized likelihood; if α is assumed known then the MLE of λ is $\widehat{\lambda}(\alpha) = \alpha/\bar{X}$. Substituting $\widehat{\lambda}(\alpha)$ for λ in the log-likelihood, we obtain the profile log-likelihood

$$\begin{aligned}\ln\mathcal{L}_p(\alpha) &= \ln\mathcal{L}(\alpha,\widehat{\lambda}(\alpha)) \\ &= n\alpha\left[\ln(\alpha/\bar{X}) - 1\right] + (\alpha-1)\sum_{i=1}^{n}\ln(X_i) - n\ln\Gamma(\alpha).\end{aligned}$$

The profile log-likelihood is maximized at $\widehat{\alpha}$, which is the MLE of α; to obtain a confidence interval for α we look at the approximate pivot

$$g(\boldsymbol{X};\alpha) = 2\ln[\mathcal{L}_p(\widehat{\alpha})/\mathcal{L}_p(\alpha)].$$

We know that, for a given α, $g(\boldsymbol{X};\alpha)$ is approximately χ^2 distributed with 1 degree of freedom; thus an approximate 95% confidence interval for α is

$$R(\boldsymbol{X}) = \{\alpha : g(\boldsymbol{X};\alpha) \leq 3.841\}.$$

For a sample of 50 i.i.d. Gamma random variables, a graph of $g(\boldsymbol{X};\alpha)$ and the approximate 95% confidence interval for α are shown in Figure 7.4. ◇

A similar approach can be followed to obtain confidence intervals (or, indeed, confidence regions) from the Wald and score tests.

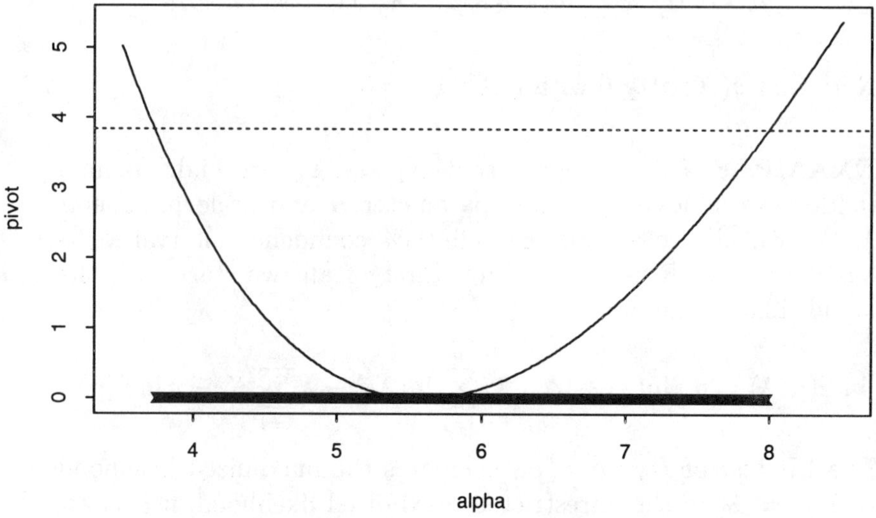

Figure 7.4 *The profile likelihood pivot in Example 7.25 and an approximate 95% confidence for α.*

Confidence intervals and tests based on non-parametric likelihood

Suppose that $X_1, \cdots, X_n$ are i.i.d. random variables with unknown distribution function F. In section 5.6, we defined a notion of non-parametric maximum likelihood estimation and noted that the empirical distribution function was the non-parametric MLE of F under this formulation. We will indicate here how the non-parametric likelihood function can be used to obtain a confidence interval for a functional parameter $\theta(F)$ or to test a null hypothesis of the form $H_0 : \theta(F) = \theta_0$. The idea is to extend the idea of LR tests to the non-parametric setting.

We will consider the hypothesis testing problem first. According to our discussion in section 5.6, we consider only distributions putting all their probability mass at the points $X_1, \cdots, X_n$. For a given vector of probabilities $\boldsymbol{p} = (p_1, \cdots, p_n)$, we define $F_{\boldsymbol{p}}$ to be the (discrete) distribution with probability p_i at X_i ($i = 1, \cdots, n$).

The non-parametric (or empirical) log-likelihood function is

$$\ln \mathcal{L}(\boldsymbol{p}) = \sum_{i=1}^{n} \ln(p_i)$$

and so to implement the LR test procedure, we must consider the unrestricted MLEs of $p_1, \cdots, p_n$ (which are $\widehat{p}_i = 1/n$ for $i = 1, \cdots, n$) as well as the restricted MLEs assuming that the null hypothesis is true; the restricted MLEs maximize $\ln \mathcal{L}(\boldsymbol{p})$ subject to the constraints $p_1 + \cdots + p_n = 1$ and $\theta(F_p) = \theta_0$. If $\widehat{\boldsymbol{p}}_0$ is this restricted MLE of $\boldsymbol{p}$ then the LR test statistic is given by

$$\ln(\Lambda_n) = -\sum_{i=1}^{n} \ln(n\widehat{p}_{i0});$$

as before, we would reject H_0 for large values of $\ln(\Lambda_n)$ (or, equivalently, Λ_n).

EXAMPLE 7.26: Suppose that $\theta(F) = \int x\, dF(x) = E(X_i)$. Under $H_0 : \theta(F) = \theta_0$, the non-parametric MLEs of $p_1, \cdots, p_n$ can be obtained by maximizing $\ln \mathcal{L}(\boldsymbol{p})$ subject to the constraints

$$\begin{aligned} p_1 + \cdots + p_n &= 1 \\ p_1 X_1 + \cdots + p_n X_n &= \theta_0. \end{aligned}$$

The MLEs (under H_0) can be determined by introducing two Lagrange multipliers, λ_1 and λ_2, and maximizing

$$\sum_{i=1}^{n} \ln(p_i) + \lambda_1(p_1 + \cdots + p_n - 1) + \lambda_2(p_1 X_1 + \cdots + p_n X_n - \theta_0)$$

with respect to $p_1, \cdots, p_n$, λ_1 and λ_2. After differentiating and setting the partial derivatives to 0, we obtain

$$\widehat{p}_i = \frac{1}{n - \lambda_2(X_i - \theta_0)}$$

where λ_2 is defined so that the constraints on the p_i's are satisfied. ◇

The limiting null distribution of the LR test statistic is not clear; we do not have the standard conditions for this statistic to have a limiting χ^2 distribution. Nonetheless, if $\theta(F)$ is a sufficiently "smooth" functional parameter then the null distribution $2\ln(\Lambda_n)$ will be approximately χ^2 with 1 degree of freedom for large n. For example, the following result can be found in Owen (1988) for the case where $\theta(F) = \int g(x)\,dF(x)$.

THEOREM 7.6 *Suppose that $\theta(F) = \int g(x)\,dF(x)$ and we are testing $H_0 : \theta(F) = \theta_0$. If $\int |g(x)|^3\,dF(x) < \infty$ then*

$$2\ln(\Lambda_n) \to_d \chi^2(1)$$

if H_0 is true.

Similar results can be given for other functional parameters $\theta(F)$; see Owen (1988) for details.

We can use the result of Theorem 7.6 to derive approximate confidence intervals for functional parameters $\theta(F)$ satisfying its conditions. For example, suppose that $\theta(F) = \int x\,dF(x)$ and define

$$g(\boldsymbol{X};\theta) = -2\sum_{i=1}^{n} \ln(n\widehat{p}_i(\theta))$$

where $\widehat{\boldsymbol{p}}(\theta)$ maximizes the non-parametric likelihood subject to the constraint

$$\sum_{i=1}^{n} p_i X_i = \theta.$$

Then $g(\boldsymbol{X};\theta(F))$ is an approximate pivot for $\theta(F)$ and hence we can define an approximate $100p\%$ confidence interval for $\theta(F)$ to be

$$R(\boldsymbol{X}) = \{\theta : g(\boldsymbol{X};\theta) \le k_p\}$$

where k_p is the p quantile of a χ^2 distribution with 1 degree of freedom.

7.6 Problems and complements

7.1: Suppose that $X_1, \cdots, X_n$ are i.i.d. Normal random variables with unknown mean μ and variance σ^2.

(a) Using the pivot $(n-1)S^2/\sigma^2$ where

$$S^2 = \frac{1}{n-1}\sum_{i=1}^{n}(X_i - \bar{X})^2,$$

we can obtain a 95% confidence interval $[k_1 S^2, k_2 S^2]$ for some constants k_1 and k_2. Find expressions for k_1 and k_2 if this confidence interval has minimum length. Evaluate k_1 and k_2 when $n = 10$.

(b) When n is sufficiently large, we can approximate the distribution of the pivot by a Normal distribution (why?). Find approximations for k_1 and k_2 that are valid for large n.

7.2: Suppose that $X_1, \cdots, X_n$ are i.i.d. Exponential random variables with parameter λ.

(a) Show that $2\lambda X_i$ has a χ^2 distribution with 2 degrees of freedom and hence that $2\lambda \sum_{i=1}^n X_i$ has a χ^2 distribution with $2n$ degrees of freedom.

(b) Suppose that $n = 5$. Give a 90% confidence interval for θ using the result of part (a).

(c) Let $\widehat{\theta}$ be the MLE of θ. Find a function g such that the distribution of $\sqrt{n}(g(\widehat{\theta}) - g(\theta))$ is approximately standard Normal and use this to give an approximate 90% confidence interval for θ valid for large n.

7.3: Suppose that $X_1, \cdots, X_n$ are i.i.d. continuous random variables with median θ.

(a) What is the distribution of $\sum_{i=1}^n I(X_i \le \theta)$?

(b) Let $X_{(1)} < \cdots < X_{(n)}$ be the order statistics of $X_1, \cdots, X_n$. Show that the interval $[X_{(\ell)}, X_{(u)}]$ is a $100p\%$ confidence interval for θ and find an expression for p in terms of ℓ and u. (Hint: use the random variable in part (a) as a pivot for θ.)

(c) Suppose that for large n, we set

$$\ell = \left\lfloor \frac{n}{2} - \frac{0.98}{\sqrt{n}} \right\rfloor \quad \text{and} \quad u = \left\lceil \frac{n}{2} + \frac{0.98}{\sqrt{n}} \right\rceil .$$

Show that the confidence interval $[X_{(\ell)}, X_{(u)}]$ has coverage approximately 95%. ($\lfloor x \rfloor$ is the largest integer less than or equal to x while $\lceil x \rceil$ is the smallest integer greater than or equal to x.)

7.4: Suppose that $X_1, \cdots, X_m, Y_1, \cdots, Y_n$ are independent Exponential random variables with $X_i \sim \text{Exp}(\lambda)$ and $Y_i \sim \text{Exp}(\theta\lambda)$.

(a) Find the MLEs of λ and θ.

(b) Let $\widehat{\theta}_{m,n}$ be the MLE of θ. Find an expression for an approximate standard error of $\widehat{\theta}_{m,n}$. (Hint: assume m and n are "large" and use asymptotic theory.)

(c) Show that $\theta \sum_{i=1}^n Y_i / \sum_{i=1}^m X_i$ is a pivot for θ. What is the distribution of this pivot? (Hint: Find the distributions of $2\lambda\theta \sum Y_i$ and $2\lambda \sum X_i$.)

(d) Using part (c), show how to construct a 95% confidence interval for θ. Give the upper and lower confidence limits for $m = 5$ and $n = 10$.

7.5: Suppose that $X_1, \cdots, X_n$ are i.i.d. Uniform random variables on $[0, \theta]$ and let $X_{(1)}, \cdots, X_{(n)}$ be the order statistics.

(a) Show that for any r, $X_{(r)}/\theta$ is a pivot for θ.

(b) Use part (a) to derive a 95% confidence interval for θ based on $X_{(r)}$. Give the exact upper and lower confidence limits when $n = 10$ and $r = 5$.

7.6: Suppose that $X_1, \cdots, X_n$ are i.i.d. nonnegative random variables whose hazard function is

$$\lambda(x) = \begin{cases} \lambda_1 & \text{for } x \le x_0 \\ \lambda_2 & \text{for } x > x_0 \end{cases}$$

where λ_1, λ_2 are unknown parameters and x_0 is a known constant.

(a) Consider testing $H_0 : \lambda_1/\lambda_2 = r$ versus $H_0 : \lambda_1/\lambda_2 \neq r$ for some specified $r > 0$. Find the form of the LR test of H_0 versus H_1.

(b) Use the result of (a) to find an approximate 95% confidence interval for λ_1/λ_2.

7.7: Suppose that $X_1, X_2, \cdots$ are i.i.d. Normal random variables with mean μ and variance σ^2, both unknown. With a fixed sample size, it is not possible to find a fixed length $100p\%$ confidence interval for μ. However, it is possible to construct a fixed length confidence interval by allowing a random sample size. Suppose that $2d$ is the desired length of the confidence interval. Let n_0 be a fixed integer with $n_0 \ge 2$ and define

$$\bar{X}_0 = \frac{1}{n_0}\sum_{i=1}^{n_0} X_i \quad \text{and} \quad S_0^2 = \frac{1}{n_0 - 1}\sum_{i=1}^{n_0}(X_i - \bar{X}_0)^2.$$

Now given S_0^2, define a random integer N to be the smallest integer greater than n_0 and greater than or equal to $[S_0 t_\alpha/d]^2$ where $\alpha = (1-p)/2$ and t_α is the $1-\alpha$ quantile of a t-distribution with $n_0 - 1$ degrees of freedom). Sample $N - n_0$ additional random variables and let $\bar{X} = N^{-1}\sum_{i=1}^{N} X_i$.

(a) Show that $\sqrt{N}(\bar{X} - \mu)/S_0$ has a t-distribution with $n_0 - 1$ degrees of freedom.

(b) Use the result of part (a) to construct a $100p\%$ confidence interval for μ and show that this interval has length at most $2d$.

7.8: Suppose that $X_1, \cdots, X_n$ are i.i.d. random variables with Cauchy density

$$f(x;\theta) = \frac{1}{\pi} \frac{1}{1+(x-\theta)^2}.$$

(a) Let $\widetilde{\theta}_n$ be the median of $X_1, \cdots, X_n$. It can be shown that $\sqrt{n}(\widetilde{\theta}_n - \theta) \to_d N(0, \sigma^2)$. Find the value of σ^2 and use this result to construct an approximate 95% confidence interval for θ.

(b) Find an approximate 95% confidence interval for θ using the maximum likelihood estimator. (You may assume that the MLE is consistent.)

(c) Which of these two (approximate) confidence intervals is narrower?

7.9: Suppose that $X_1, \cdots, X_n$ are i.i.d. random variables with density function

$$f(x;\mu) = \exp[-\lambda(x-\mu)] \quad \text{for } x \geq \mu.$$

Let $X_{(1)} = \min(X_1, \cdots, X_n)$.

(a) Show that

$$S(\lambda) = 2\lambda \sum_{i=1}^{n} (X_i - X_{(1)}) \sim \chi^2(2(n-1))$$

and hence is a pivot for λ. (Hint: Note that the distribution of $X_i - X_{(1)}$ does not depend on μ and see Problem 2.26.)

(b) Describe how to use the pivot in (a) to give an exact 95% confidence interval for λ.

(c) Give an approximate 95% confidence interval for λ based on $S(\lambda)$ for large n.

7.10: Smith (1988) discusses Bayesian methods for estimating the population size in multiple mark/recapture experiments (see Example 2.13). In such experiments, we assume a fixed population size N. Initially (stage 0), n_0 items are sampled (captured) without replacement and marked. Then at stage i ($i = 1, \cdots, k$), n_i items are sampled and the number of marked items m_i is observed; any unmarked items are marked before being returned to the population. Thus the total number of marked items in the population at stage i is

$$M_i = \sum_{j=0}^{i-1}(n_j - m_j) \quad \text{where } m_0 = 0.$$

The conditional distribution of the number of marked items at stage i (given M_i, which depends on $m_1, \cdots, m_{i-1}$) is Hypergeometric with frequency function

$$f(m_i|M_i, N) = \binom{M_i}{m_i}\binom{N - M_i}{n_i - m_i} \Big/ \binom{N}{n_i}.$$

(a) To make the problem somewhat more analytically tractable, we can approximate the Hypergeometric (conditional) frequency function $f(m_i|M_i, N)$ by the Poisson approximation

$$f(m_i|M_i, N) \approx \frac{\exp(-M_i/N)(M_i/N)^{m_i}}{m_i!}.$$

Justify this approximation; specifically, what assumptions are being made about M_i and N in making this approximation.

(b) Using the Poisson approximation in (a), we can now pretend that N is a "continuous" rather than discrete parameter. Set $\omega = 1/N$ and assume the following prior density for ω:

$$\pi(\omega) = 2(1 - \omega) \quad \text{for } 0 \leq \omega \leq 1$$

For the data given in Table 7.1, find the posterior density of N (not ω) and find the 95% HPD interval for N.

(d) In this paper, we treat ω (or equivalently N) as a continuous parameter even though it is discrete. Given a prior frequency function $\pi(N)$ for N, give an expression for the posterior frequency function.

7.11: Consider a random sample of n individuals who are classified into one of three groups with probabilities θ^2, $2\theta(1-\theta)$, and $(1-\theta)^2$. If Y_1, Y_2, Y_3 are the numbers in each group then $\boldsymbol{Y} = (Y_1, Y_2, Y_3)$ has a Multinomial distribution:

$$f(\boldsymbol{y}; \theta) = \frac{n!}{y_1!y_2!y_3!}\theta^{2y_1}\left[2\theta(1-\theta)\right]^{y_2}(1-\theta)^{2y_3}$$

for $y_1, y_2, y_3 \geq 0$; $y_1 + y_2 + y_3 = n$ where $0 < \theta < 1$. (This model is the Hardy-Weinberg equilibrium model from genetics.)

Table 7.1 *Data for Problem 7.10; n_i is the number of fish caught in sample i and m_i is the number of marked fish caught.*

i	n_i	m_i
0	10	0
1	27	0
2	17	0
3	7	0
4	1	0
5	5	0
6	6	2
7	15	1
8	9	5
9	18	5
10	16	4
11	5	2
12	7	2
12	19	3

(a) Find the maximum likelihood estimator of θ and give the asymptotic distribution of $\sqrt{n}(\widehat{\theta}_n - \theta)$ as $n \to \infty$.

(b) Consider testing $H_0 : \theta = \theta_0$ versus $H_1 : \theta > \theta_0$. Suppose that for some k

$$P_{\theta_0}\left[2Y_1 + Y_2 \geq k\right] = \alpha$$

Then the test that rejects H_0 when $2Y_1 + Y_2 \geq k$ is a UMP level α test of H_0 versus H_1.

(c) Suppose n is large and $\alpha = 0.05$. Find an approximate value for k in the UMP test in part (b). (Hint: Approximate distribution of $2Y_1 + Y_2$ by a Normal distribution; the approximation will depend on n and θ_0.)

(d) Suppose that $\theta_0 = 1/2$ in part (b). How large must n so that a 0.05 level test has power at least 0.80 when $\theta = 0.6$? (Hint: Use the approximation in (c) to evaluate k and then approximate the distribution of $2Y_1 + Y_2$ by a Normal distribution when $\theta = 0.6$.)

7.12: Suppose that $X_1, \cdots, X_n$ are i.i.d. random variables with density function

$$f(x;\theta) = \theta x^{\theta-1} \quad \text{for } 0 \le x \le 1$$

where $\theta > 0$.

(a) Show that the UMP level α test of $H_0 : \theta \le 1$ versus $H_1 : \theta > 1$ rejects H_0 for

$$T = \sum_{i=1}^{n} \ln(X_i) \ge k.$$

What is the distribution of T under H_0?

(b) Suppose $\alpha = 0.05$ and $n = 50$. Find either an exact or an approximate value for k.

(c) Suppose we use the following test statistic for testing H_0 versus H_1:

$$S = \sum_{i=1}^{n} \sin(\pi X_i/2).$$

We reject H_0 when $S \ge k$. If $\alpha = 0.05$, find an approximate value for k (assuming n is reasonably large). (Hint: Approximate the distribution of S when $\theta = 1$ using the CLT.)

(d) Suppose that $\theta = 2$. Determine how large (approximately) n should be so that the test in part (c) has power 0.90. How large (approximately) must n be for the UMP test of part (a)? (Hint: Find the mean and variance of S and T when $\theta = 2$ and apply the CLT.)

7.13: Suppose that $X \sim \text{Bin}(m,\theta)$ and $Y \sim \text{Bin}(n,\phi)$ are independent random variables and consider testing

$$H_0 : \theta \ge \phi \quad \text{versus} \quad H_1 : \theta < \phi.$$

(a) Show that the joint frequency function of X and Y can be written in the form

$$f(x,y;\theta,\phi) = \left(\frac{\theta(1-\phi)}{\phi(1-\theta)}\right)^x \left(\frac{\phi}{1-\phi}\right)^{x+y} \exp\left[d(\theta,\phi) + S(x,y)\right]$$

and that H_0 is equivalent to

$$H_0 : \ln\left(\frac{\theta(1-\phi)}{\phi(1-\theta)}\right) \ge 0.$$

(b) The UMPU test of H_0 versus H_1 rejects H_1 at level α if $X \geq k$ where k is determined from the conditional distribution of X given $X + Y = z$ (assuming that $\theta = \phi$). Show that this conditional distribution is Hypergeometric. (This conditional test is called Fisher's exact test.)

(c) Show that the conditional frequency function of X given $X + Y = z$ is given by

$$P(X = x|X + Y = z) = \binom{m}{x}\binom{n}{z-x}\psi^x \Big/ \sum_s \binom{m}{s}\binom{n}{z-s}\psi^s$$

where the summation extends over s from $\max(0, z - n)$ to $\min(m, z)$ and

$$\psi = \frac{\theta(1-\phi)}{\phi(1-\theta)}.$$

(This is called a non-central Hypergeometric distribution.)

7.14: Let $X_1, \cdots, X_m$ be i.i.d. Poisson random variables with parameter λ and $Y_1, \cdots, Y_n$ be i.i.d. Poisson random variables (independent of the X_i's) with parameter μ. Suppose we want to test

$$H_0 : \mu = \lambda \quad \text{versus} \quad H_1 : \mu \neq \lambda$$

at level α.

(a) Find the LR statistic $\Lambda = \Lambda_{m,n}$ for testing H_0. Assuming that $m, n \to \infty$, show that the null distribution of $2\ln(\Lambda_{m,n})$ tends to a χ^2 distribution with 1 degree of freedom.

(b) An alternative approach to testing in this problem is a conditional test. Define $S = X_1 + \cdots + X_m$ and $T = Y_1 + \cdots + Y_n$. Show that the conditional distribution of S given $S + T = y$ is Binomial and give the values of the parameters.

(c) Let $\phi = \mu/\lambda$. Using the conditional distribution in (b), show that the MP conditional test of

$$H_0' : \phi = \phi_0 \quad \text{versus} \quad H_1' : \phi = \phi_1$$

rejects H_0' for large (small) values of S if $\phi_1 > \phi_0$ ($\phi_1 < \phi_0$).

(d) Use the result of (c) to give a reasonable conditional test of H_0 versus H_1 in (a).

7.15: Suppose that $X_1, \cdots, X_{10}$ are i.i.d. Uniform random variables on $[0, \theta]$ and consider testing

$$H_0 : \theta = 1 \quad \text{versus} \quad H_1 : \theta \neq 1$$

at the 5% level. Consider a test that rejects H_0 if $X_{(10)} < a$ or $X_{(10)} > b$ where $a < b \leq 1$.

(a) Show that a and b must satisfy the equation

$$b^{10} - a^{10} = 0.95.$$

(b) Does an unbiased test of H_0 versus H_1 of this form exist? If so, find a and b to make the test unbiased. (Hint: evaluate the power function and note that it must be minimized at $\theta = 1$ in order for the test to be unbiased.)

7.16: Let $\boldsymbol{X}_n = (X_1, \cdots, X_n)$ and suppose that we are testing $H_0 : \theta \in \Theta_0$ versus $H_1 : \theta \in \Theta_1$. A sequence of α level test functions $\{\phi_n\}$ is consistent for testing H_0 versus H_1,

$$E_\theta[\phi_n(\boldsymbol{X}_n)] \to 1$$

for all $\theta \in \Theta_1$.

(a) Let $X_1, \cdots, X_n$ be i.i.d. random variables with density or frequency function $f(x;\theta)$ and suppose that we test $H_0 : \theta = \theta_0$ versus $H_1 : \theta = \theta_1$. Show that the sequence of MP α level tests is consistent.

(b) Let $X_1, \cdots, X_n$ be i.i.d. random variables with density or frequency function $f(x;\theta)$ and suppose we want to test $H_0 : \theta = \theta_0$ versus $H_1 : \theta \neq \theta_0$ using the LR test statistic

$$T_n = \sum_{i=1}^{n} \ln[f(X_i; \widehat{\theta}_n)/f(X_i;\theta_0)]$$

where $\widehat{\theta}_n$ is the MLE of θ and H_0 is rejected for large values of T_n. If $T_n/n \to_p 0$ under H_0, show that the sequence of tests is consistent.

7.17: Suppose that $\boldsymbol{X} = (X_1, \cdots, X_n)$ are continuous random variables with joint density $f(\boldsymbol{x})$ where $f = f_0$ or $f = f_1$ are the two possibilities for f. Based on $\boldsymbol{X}$, we want to decide between f_0 and f_1 using a non-Neyman-Pearson approach. Let $\phi(\boldsymbol{X})$ be an arbitrary test function where f_0 is chosen if $\phi = 0$ and f_1 is chosen if $\phi = 1$. Let $E_0(T)$ and $E_1(T)$ be expectations of a statistic $T = T(\boldsymbol{X})$ assuming the true joint densities are f_0 and f_1 respectively.

(a) Show that the test function ϕ that minimizes

$$\alpha E_0[\phi(\boldsymbol{X})] + (1-\alpha)E_1[1-\phi(\boldsymbol{X})]$$

(where $0 < \alpha < 1$ is a known constant) has the form

$$\phi(\boldsymbol{X}) = 1 \quad \text{if} \quad \frac{f_1(\boldsymbol{X})}{f_0(\boldsymbol{X})} \geq k$$

and 0 otherwise. Specify the value of k.

(b) Suppose that $X_1, \cdots, X_n$ are i.i.d. continuous random variables with common density f where $f = f_0$ or $f = f_1$ ($f_0 \neq f_1$). Let $\phi_n(\boldsymbol{X})$ be the optimal test function (for some α) based on $X_1, \cdots, X_n$ as described in part (a). Show that

$$\lim_{n\to\infty} (\alpha E_0[\phi_n(\boldsymbol{X})] + (1-\alpha)E_1[1-\phi_n(\boldsymbol{X})]) = 0.$$

(Hint: Use the facts that

$$\begin{aligned} E_0[\ln(f_1(X_i)/f_0(X_i))] &< 0 \\ \text{and} \quad E_1[\ln(f_1(X_1)/f_0(X_1))] &> 0 \end{aligned}$$

and apply the WLLN.)

7.18: Suppose that $\boldsymbol{X} = (X_1, \cdots, X_n)$ are continuous random variables with joint density $f(\boldsymbol{x};\theta)$ where θ is a real-valued parameter. We want to test

$$H_0 : \theta = \theta_0 \quad \text{versus} \quad H_1 : \theta \neq \theta_0$$

at level α. For any test function $\phi(\boldsymbol{X})$, define the power function

$$\pi(\theta) = E_\theta[\phi(\boldsymbol{X})]$$

and assume that $\pi(\theta)$ may be differentiated twice under the integral sign so that, for example,

$$\pi'(\theta) = \int \cdots \int \phi(\boldsymbol{x}) \frac{\partial}{\partial\theta} f(\boldsymbol{x};\theta)\, d\boldsymbol{x}.$$

(a) Show that the test function maximizing $\pi''(\theta_0)$ subject to the constraints $\pi'(\theta_0) = 0$ and $\pi(\theta_0) = \alpha$ satisfies

$$\phi(\boldsymbol{X}) = 1 \quad \text{if} \quad \ell''(\boldsymbol{X};\theta_0) + [\ell'(\boldsymbol{X};\theta_0)]^2 + k_1\ell'(\boldsymbol{X};\theta_0) \geq k_2$$

and $\phi(\boldsymbol{X}) = 0$ otherwise where k_1, k_2 are constants so that the constraints are satisfied and ℓ', ℓ'' are the first two partial derivatives of $\ln f$ with respect to θ.

(b) Suppose that $X_1, \cdots, X_n$ are i.i.d. random variables. For large n, argue that the "locally most powerful unbiased" test described in part (a) can be approximated by the test function satisfying

$$\phi^*(\boldsymbol{X}) = 1 \quad \text{if} \quad \left(\frac{1}{\sqrt{n}}\sum_{i=1}^{n}\frac{\partial}{\partial\theta}\ln f(X_i;\theta_0)\right)^2 \geq k$$

and 0 otherwise. (Hint: Divide the test statistic in part (a) by n and consider its behaviour as $n \to \infty$.)

(c) Suppose that $X_1, \cdots, X_{100}$ are i.i.d. Exponential random variables with parameter λ and we test

$$H_0 : \lambda = 2 \quad \text{versus} \quad H_1 : \lambda \neq 2$$

at the 5% level. Use the result of (b) to approximate the locally most powerful test, explicitly evaluating all constants.

7.19: Consider a simple classification problem. An individual belongs to exactly one of k populations. Each population has a known density $f_i(x)$ $(i = 1, \cdots, k)$ and it is known that a proportion p_i belong to population i $(p_1 + \cdots + p_k = 1)$. Given disjoint sets $R_1, \cdots, R_k$, a general classification rule is

$$\text{classify as population } i \text{ if } x \in R_i \ (i = 1, \cdots, k).$$

The total probability of correct classification is

$$C(R_1, \cdots, R_k) = \sum_{i=1}^{k} p_i \int_{R_i} f_i(x)\, dx.$$

We would like to find the classification rule (that is, the sets $R_1, \cdots, R_k$) that maximizes the total probability of correct classification.

(a) Suppose that $k = 2$. Show that the optimal classification rule has

$$\begin{aligned} R_1 &= \left\{x : \frac{f_1(x)}{f_2(x)} \geq \frac{p_1}{p_2}\right\} \\ R_2 &= \left\{x : \frac{f_1(x)}{f_2(x)} < \frac{p_1}{p_2}\right\}. \end{aligned}$$

(b) Suppose that f_1 and f_2 are Normal densities with different means but equal variances. Find the optimal classification rule using the result of part (a) (that is, find the regions R_1 and R_2).

(c) Find the form of the optimal classification rule for general k.

7.20: Let $X_1, \cdots, X_n$ and $Y_1, \cdots, Y_n$ be independent random variables where X_i is Exponential with parameter $\lambda_i \theta$ and Y_i is Exponential with parameter λ_i. Suppose that we wish to test $H_0 : \theta = 1$ versus $H_1 : \theta \neq 1$ at the 5% level.

(a) Show that the LR test of H_0 versus H_1 rejects H_0 when

$$T_n = \sum_{i=1}^{n} \left(\ln(\widehat{\theta}_n) - 2 \ln \left(\frac{\widehat{\theta}_n R_i + 1}{R_i + 1} \right) \right) \geq k$$

where $R_i = X_i / Y_i$ and $\widehat{\theta}_n$ satisfies

$$\frac{n}{\widehat{\theta}_n} - 2 \sum_{i=1}^{n} \frac{R_i}{\widehat{\theta}_n R_i + 1} = 0.$$

(b) Find the limiting distribution of the statistic T_n as $n \to \infty$ when $\theta = 1$. (Note that the standard result cannot be applied here since the dimension of the parameter space is not fixed but growing with n.)

7.21: A heuristic (but almost rigorous) proof of Theorem 7.5 can be given by using the fact the the log-likelihood function is approximately quadratic in a neighbourhood of the true parameter value. Suppose that we have i.i.d. random variables $X_1, \cdots, X_n$ with density or frequency function $f(x; \boldsymbol{\theta})$ where $\boldsymbol{\theta} = (\theta_1, \cdots, \theta_p)$, define

$$\begin{aligned} Z_n(\boldsymbol{u}) &= \ln(\mathcal{L}_n(\boldsymbol{\theta} + \boldsymbol{u}/\sqrt{n})/\mathcal{L}_n(\boldsymbol{\theta})) \\ &= \boldsymbol{u}^T \boldsymbol{V}_n - \frac{1}{2} \boldsymbol{u}^T I(\boldsymbol{\theta}) \boldsymbol{u} + R_n(\boldsymbol{u}) \end{aligned}$$

where $R_n(\boldsymbol{u}) \to_p 0$ for each $\boldsymbol{u}$ and $\boldsymbol{V}_n \to_d N_p(\boldsymbol{0}, I(\boldsymbol{\theta}))$.

(a) Suppose we want to test the null hypothesis

$$H_0 : \theta_1 = \theta_{10}, \cdots, \theta_0 = \theta_{r0}.$$

Show that, if H_0 is true, the LR statistic is

$$2 \ln(\Lambda_n) = 2 \left[Z_n(\widehat{\boldsymbol{U}}_n) - Z_n(\widehat{\boldsymbol{U}}_{n0}) \right]$$

where $\widehat{\boldsymbol{U}}_n$ maximizes $Z_n(\boldsymbol{u})$ and $\widehat{\boldsymbol{U}}_{n0}$ maximizes $Z_n(\boldsymbol{u})$ subject to the constraint that $u_1 = \cdots = u_r = 0$.

(b) Suppose that $Z_n(\boldsymbol{u})$ is exactly quadratic (that is, $R_n(\boldsymbol{u}) = 0$). Show that

$$\begin{aligned}\widehat{\boldsymbol{U}}_n &= I^{-1}(\boldsymbol{\theta})\boldsymbol{V}_n \\ \widehat{\boldsymbol{U}}_{n0} &= \begin{pmatrix} \boldsymbol{0} \\ I_{22}^{-1}(\boldsymbol{\theta})\boldsymbol{V}_{n2} \end{pmatrix}\end{aligned}$$

where $\boldsymbol{V}_n$ and $I(\boldsymbol{\theta})$ are expressed as

$$\begin{aligned}\boldsymbol{V}_n &= \begin{pmatrix} \boldsymbol{V}_{n1} \\ \boldsymbol{V}_{n2} \end{pmatrix} \\ I(\boldsymbol{\theta}) &= \begin{pmatrix} I_{11}(\boldsymbol{\theta}) & I_{12}(\boldsymbol{\theta}) \\ I_{21}(\boldsymbol{\theta}) & I_{22}(\boldsymbol{\theta}) \end{pmatrix}.\end{aligned}$$

(c) Assuming that nothing is lost asymptotically in using the quadratic approximation, deduce Theorem 7.5 from parts (a) and (b).

7.22: Suppose that $X_1, \cdots, X_n$ are independent Exponential random variables with parameters $\lambda_1, \cdots, \lambda_n$, respectively. We want to test the null hypothesis

$$H_0 : \lambda_1 = \cdots = \lambda_n$$

versus the alternative hypothesis that at least two of the λ_i's are different.

(a) Derive the LR test of H_0. If Λ_n is the LR test statistic, show that $2\ln(\Lambda_n) \to_p \infty$ under H_0.

(b) Find b_n such that $[\ln(\Lambda_n) - b_n]/\sqrt{n}$ converges in distribution to a Normal distribution.

7.23: Suppose that $X_1, \cdots, X_n$ are independent Exponential random variables with $E(X_i) = \beta t_i$ where $t_1, \cdots, t_n$ are known positive constants and β is an unknown parameter.

(a) Show that the MLE of β is

$$\widehat{\beta}_n = \frac{1}{n}\sum_{i=1}^{n} X_i/t_i.$$

(b) Show that

$$\sqrt{n}(\widehat{\beta} - \beta) \to_d N(0, \beta^2).$$

(Hint: note that $X_1/t_1, \cdots, X_n/t_n$ are i.i.d. random variables.)

(c) Suppose we want to test

$$H_0 : \beta = 1 \quad \text{versus} \quad H_1 : \beta \neq 1.$$

Show that the LR test of H_0 versus H_1 rejects H_0 for large values of

$$T_n = n(\widehat{\beta}_n - \ln(\widehat{\beta}_n) - 1)$$

where $\widehat{\beta}_n$ is defined as in part (a).

(d) Show that when H_0 is true, $2T_n \to_d \chi^2(1)$.

CHAPTER 8

Linear and Generalized Linear Models

8.1 Linear models

Linear models include an extremely wide class of models and are possibly the most widely used models in applied statistics. The reasons for this popularity are obvious - linear models are simple in form, easy to interpret, and (under appropriate assumptions) statistical inference for linear models is remarkably elegant.

In this chapter, we will mainly apply some of the concepts developed in earlier chapters to the linear model; we will not go into any particular depth on the theory of linear (and generalized linear) models as there are numerous texts that do this in some depth; see, for example, Seber (1977) as well as Sen and Srivastava (1990) for more detailed treatment of linear model theory and practice.

The general form of the linear model is

$$\begin{aligned} Y_i &= \beta_0 + \beta_1 x_{i1} + \cdots + \beta_p x_{ip} + \varepsilon_i \quad (i = 1, \cdots, n) \\ &= \boldsymbol{x}_i^T \boldsymbol{\beta} + \varepsilon_i \end{aligned}$$

where $\boldsymbol{x}_i = (1, x_{i1}, \cdots, x_{ip})^T$ is a vector of known constants (called covariates or predictors), $\boldsymbol{\beta} = (\beta_0, \beta_1, \cdots, \beta_p)^T$ is a vector of unknown parameters, and $\varepsilon_1, \cdots, \varepsilon_n$ are i.i.d. Normal random variables with mean 0 and unknown variance σ^2. Alternatively, we can say that $Y_1, \cdots, Y_n$ are independent Normal random variables with $E(Y_i) = \boldsymbol{x}_i^T \boldsymbol{\beta}$ and $\text{Var}(Y_i) = \sigma^2$. (It is possible to write the linear model without the intercept β_0 and none of the theory developed in this chapter is contingent on the presence of β_0 in the model. However, the intercept is almost always included in practice; unless there is a substantive reason to delete it from the model, it may be dangerous to do so.)

Linear models include simple and multiple regression models (where the $\boldsymbol{x}_i$'s are typically vectors of covariates) as well as fixed effects analysis of variance (ANOVA) models.

EXAMPLE 8.1: Consider a single factor ANOVA model where we have k treatments and n_i observations for treatment i:

$$Y_{ij} = \mu + \alpha_i + \varepsilon_{ij} \quad (i = 1, \cdots, k;\ j = 1, \cdots, n_i).$$

This can be written in the form $Y_{ij} = \boldsymbol{x}_{ij}^T \boldsymbol{\beta} + \varepsilon_{ij}$ with $\boldsymbol{\beta} = (\mu, \alpha_1, \cdots, \alpha_k)^T$ and

$$\begin{aligned} \boldsymbol{x}_{1j} &= (1, 1, 0, \cdots, 0)^T \\ \boldsymbol{x}_{2j} &= (1, 0, 1, 0, \cdots, 0)^T \\ \vdots\ &\ \vdots\quad \vdots \\ \boldsymbol{x}_{kj} &= (1, 0, \cdots, 0, 1)^T. \end{aligned}$$

Note that the parametrization as given above is not identifiable; typically, we put some constraint on the α_i's (for example, $\alpha_1 = 0$ or $\alpha_1 + \cdots + \alpha_k = 0$) to yield an identifiable parametrization. ◇

8.2 Estimation in linear models

Under the assumptions given in the previous section, namely that the Y_i's are independent random variables with Normal distributions, we can easily derive the MLEs of the unknown parameters $\beta_0, \beta_1, \cdots, \beta_p$ and σ^2. Given $Y_1 = y_1, \cdots, Y_n = y_n$, the log-likelihood function is

$$\ln \mathcal{L}(\boldsymbol{\beta}, \sigma) = -n \ln(\sigma) - \frac{1}{2\sigma^2} \sum_{i=1}^{n} (y_i - \boldsymbol{x}_i^T \boldsymbol{\beta})^2 - \frac{n}{2} \ln(2\pi).$$

Differentiating with respect to the unknown parameters, we obtain

$$\begin{aligned} \frac{\partial}{\partial \boldsymbol{\beta}} \ln \mathcal{L}(\boldsymbol{\beta}, \sigma) &= \frac{1}{\sigma^2} \sum_{i=1}^{n} (y_i - \boldsymbol{x}_i^T \boldsymbol{\beta}) \boldsymbol{x}_i \\ \frac{\partial}{\partial \sigma} \ln \mathcal{L}(\boldsymbol{\beta}, \sigma) &= -\frac{n}{\sigma} + \frac{1}{\sigma^3} \sum_{i=1}^{n} (y_i - \boldsymbol{x}_i^T \boldsymbol{\beta})^2. \end{aligned}$$

Setting these derivatives to 0, it follows that the MLE of $\boldsymbol{\beta}$ satisfies the so-called normal equations

$$\sum_{i=1}^{n} (Y_i - \boldsymbol{x}_i^T \widehat{\boldsymbol{\beta}}) \boldsymbol{x}_i = \boldsymbol{0}$$

while the MLE of σ^2 is

$$\widehat{\sigma}^2 = \frac{1}{n}\sum_{i=1}^{n}(Y_i - \boldsymbol{x}_i^T\widehat{\boldsymbol{\beta}})^2.$$

The MLE of $\boldsymbol{\beta}$ is, in fact, a least squares estimator. That is, $\widehat{\boldsymbol{\beta}}$ minimizes

$$\sum_{i=1}^{n}(Y_i - \boldsymbol{x}_i^T\boldsymbol{\beta})^2.$$

This fact, along with the fact that the Y_i's are Normal random variables, allows us to exploit the geometrical properties of the multivariate Normal distribution to derive the properties of the estimators $\widehat{\boldsymbol{\beta}}$ and $\widehat{\sigma}^2$. To do this, it is convenient to write the linear model in matrix form.

Define random vectors $\boldsymbol{Y} = (Y_1, \cdots, Y_n)^T$ and $\boldsymbol{\varepsilon} = (\varepsilon_1, \cdots, \varepsilon_n)^T$ as well as the matrix

$$X = \begin{pmatrix} 1 & x_{11} & \cdots & x_{1p} \\ 1 & x_{21} & \cdots & x_{2p} \\ \vdots & \vdots & \ddots & \vdots \\ 1 & x_{n1} & \cdots & x_{np} \end{pmatrix} = \begin{pmatrix} \boldsymbol{x}_1^T \\ \boldsymbol{x}_2^T \\ \vdots \\ \boldsymbol{x}_n^T \end{pmatrix};$$

X is called the design matrix. We can then rewrite the linear model as

$$\boldsymbol{Y} = X\boldsymbol{\beta} + \boldsymbol{\varepsilon}$$

so that $\boldsymbol{Y}$ has a multivariate Normal distribution with mean vector $X\boldsymbol{\beta}$ and variance-covariance matrix $\sigma^2 I$.

Using the matrix formulation of the linear model, we can rewrite the normal equations (which determine $\widehat{\boldsymbol{\beta}}$) as

$$\sum_{i=1}^{n} Y_i\boldsymbol{x}_i = \sum_{i=1}^{n}\boldsymbol{x}_i\boldsymbol{x}_i^T\widehat{\boldsymbol{\beta}}$$

or

$$X^T\boldsymbol{Y} = (X^TX)\widehat{\boldsymbol{\beta}}.$$

Hence if $(X^TX)^{-1}$ exists (as is the case if the parametrization is identifiable) then

$$\widehat{\boldsymbol{\beta}} = (X^TX)^{-1}X^T\boldsymbol{Y}.$$

Likewise, the MLE of σ^2 is given by the formula

$$\widehat{\sigma}^2 = \frac{1}{n}\sum_{i=1}^{n}(Y_i - \boldsymbol{x}_i^T\widehat{\boldsymbol{\beta}})^2$$

$$
\begin{aligned}
&= \frac{1}{n}\left\|\boldsymbol{Y} - X\widehat{\boldsymbol{\beta}}\right\|^2 \\
&= \frac{1}{n}\left\|\boldsymbol{Y} - X(X^TX)^{-1}X^T\boldsymbol{Y}\right\|^2 \\
&= \frac{1}{n}\|(I-H)\boldsymbol{Y}\|
\end{aligned}
$$

where $H = X(X^TX)^{-1}X^T$ is a projection matrix onto the space spanned by the columns of X. The matrix H is often called the "hat" matrix since the "fitted values" of $\boldsymbol{Y}$, $\widehat{\boldsymbol{Y}} = X\widehat{\boldsymbol{\beta}}$, are obtained via the equation $\widehat{\boldsymbol{Y}} = H\boldsymbol{Y}$.

The following properties of $\widehat{\boldsymbol{\beta}}$ and $\widehat{\sigma}^2$ now follow easily from multivariate Normal theory.

PROPOSITION 8.1 *Assume* $(X^TX)^{-1}$ *exists. Then*
(a) $\widehat{\boldsymbol{\beta}} \sim N_{p+1}(\boldsymbol{\beta}, \sigma^2(X^TX)^{-1})$;
(b) $n\widehat{\sigma}^2/\sigma^2 \sim \chi^2(n-p-1)$;
(c) $\widehat{\boldsymbol{\beta}}$ *and* $\widehat{\sigma}^2$ *are independent.*

Proof. (a) Recall that $\boldsymbol{Y} \sim N_n(X\boldsymbol{\beta}, \sigma^2 I)$ and

$$\widehat{\boldsymbol{\beta}} = (X^TX)^{-1}X^T\boldsymbol{Y} = A\boldsymbol{Y}.$$

Hence $\widehat{\boldsymbol{\beta}} \sim N_{p+1}(AX\boldsymbol{\beta}, \sigma^2 AA^T)$ with

$$
\begin{aligned}
AX\boldsymbol{\beta} &= (X^TX)^{-1}X^TX\boldsymbol{\beta} = \boldsymbol{\beta} \\
AA^T &= (X^TX)^{-1}X^TX(X^TX)^{-1} = (X^TX)^{-1}.
\end{aligned}
$$

(b) Let $H = X(X^TX)^{-1}X^T$ and note that

$$
\begin{aligned}
H\widehat{\boldsymbol{\beta}} &= H\boldsymbol{Y} \\
&= H(X\boldsymbol{\beta} + \boldsymbol{\varepsilon}) \\
&= X\boldsymbol{\beta} + H\boldsymbol{\varepsilon}
\end{aligned}
$$

since $X\boldsymbol{\beta}$ lies in the column space of X onto which H projects. Thus

$$
\begin{aligned}
n\frac{\widehat{\sigma}^2}{\sigma^2} &= \frac{1}{\sigma^2}\|(I-H)\boldsymbol{Y}\|^2 \\
&= \frac{1}{\sigma^2}\|(I-H)\boldsymbol{\varepsilon}\|^2 \\
&= \frac{1}{\sigma^2}\boldsymbol{\varepsilon}^T(I-H)\boldsymbol{\varepsilon}.
\end{aligned}
$$

The rank of H is $(p+1)$ so that the rank of $I - H$ is $(n-p-1)$;

thus

$$\frac{1}{\sigma^2}\varepsilon^T(I-H)\varepsilon \sim \chi^2(n-p-1).$$

(c) To show that $\widehat{\boldsymbol{\beta}}$ and $\widehat{\sigma}^2$ are independent it suffices to show that $\widehat{\boldsymbol{\beta}}$ and $\boldsymbol{Y} - X\widehat{\boldsymbol{\beta}}$ are independent. Note that

$$\widehat{\boldsymbol{\beta}} = (X^TX)^{-1}X^T\boldsymbol{Y} = A\boldsymbol{Y}$$

and

$$\boldsymbol{Y} - X\widehat{\boldsymbol{\beta}} = (I-H)\boldsymbol{Y} = B\boldsymbol{Y}.$$

It suffices then to show that AB equals a matrix of 0's:

$$\begin{aligned} AB &= (X^TX)^{-1}X^T(I-H) \\ &= (X^TX)^{-1}X^T - (X^TX)^{-1}X^TH \\ &= (X^TX)^{-1}X^T - (X^TX)^{-1}X^T \\ &= 0 \end{aligned}$$

since $X^TH = (HX)^T = X^T$. □

Of course, the result of Proposition 8.1 assumes i.i.d. normally distributed errors. However, if we remove the assumption of normality of the errors, we still have

$$E(\widehat{\boldsymbol{\beta}}) = \boldsymbol{\beta} \quad \text{and} \quad \text{Cov}(\widehat{\boldsymbol{\beta}}) = \sigma^2(X^TX)^{-1}.$$

In fact, $E(\widehat{\boldsymbol{\beta}}) = \boldsymbol{\beta}$ if $\text{Cov}(\varepsilon) = \sigma^2 C$ for any C.

EXAMPLE 8.2: Consider a simple linear regression model

$$Y_i = \beta_0 + \beta_1 x_i + \varepsilon_i \quad (i = 1, \cdots, n)$$

where $\varepsilon_i \sim N(0, \sigma^2)$. The design matrix in this case is

$$X = \begin{pmatrix} 1 & x_1 \\ 1 & x_2 \\ \vdots & \vdots \\ 1 & x_n \end{pmatrix}$$

and

$$X^TX = \begin{pmatrix} n & \sum_{i=1}^n x_i \\ \sum_{i=1}^n x_i & \sum_{i=1}^n x_i^2 \end{pmatrix}.$$

The least squares estimators of β_0 and β_1 are

$$\begin{aligned} \widehat{\beta}_1 &= \frac{\sum_{i=1}^n (x_i - \bar{x})^2 Y_i}{\sum_{i=1}^n (x_i - \bar{x})^2} \\ \widehat{\beta}_0 &= \bar{Y} - \widehat{\beta}_1 \bar{x}. \end{aligned}$$

The distributions of $\widehat{\beta}_0$ and $\widehat{\beta}_1$ can be obtained from Proposition 8.1; in particular, we have

$$\widehat{\beta}_1 \sim N\left(\beta_1, \frac{\sigma^2}{\sum_{i=1}^n (x_i - \bar{x})^2}\right).$$

Note that the variance of $\widehat{\beta}_1$ effectively decreases as the x_i's become more dispersed. ◇

8.3 Hypothesis testing in linear models

Again consider the linear model

$$\begin{aligned} Y_i &= \beta_0 + \beta_1 x_{i1} + \cdots + \beta_p x_{ip} + \varepsilon_i \quad (i = 1, \cdots, n) \\ &= \boldsymbol{x}_i^T \boldsymbol{\beta} + \varepsilon_i. \end{aligned}$$

Suppose that we want to test the null hypothesis

$$H_0 : \beta_{r+1} = \beta_{r+2} = \cdots = \beta_p = 0$$

against the alternative hypothesis that all parameters are unrestricted. We will consider the likelihood ratio (LR) test procedure.

To implement the LR test, we need to find the MLEs under H_0 as well as the unrestricted MLEs. We first define the "reduced" design matrix

$$X_r = \begin{pmatrix} 1 & x_{11} & \cdots & x_{1r} \\ 1 & x_{21} & \cdots & x_{2r} \\ \vdots & \vdots & \ddots & \vdots \\ 1 & x_{n1} & \cdots & x_{nr} \end{pmatrix}.$$

Then the MLEs of $\boldsymbol{\beta}$ are

$$\begin{aligned} \widehat{\boldsymbol{\beta}}_r &= (X_r^T X_r)^{-1} X_r^T \boldsymbol{Y} \quad \text{(under } H_0\text{)} \\ \widehat{\boldsymbol{\beta}} &= (X^T X)^{-1} X^T \boldsymbol{Y} \quad \text{(unrestricted)} \end{aligned}$$

while the MLEs of σ^2 are

$$\begin{aligned} \widehat{\sigma}_r^2 &= \frac{1}{n} \left\| \boldsymbol{Y} - X_r \widehat{\boldsymbol{\beta}}_r \right\|^2 \\ &= \frac{1}{n} \boldsymbol{Y}^T (I - H_r) \boldsymbol{Y} \quad \text{(under } H_0\text{)} \\ \widehat{\sigma}^2 &= \frac{1}{n} \left\| \boldsymbol{Y} - X \widehat{\boldsymbol{\beta}} \right\|^2 \\ &= \frac{1}{n} \boldsymbol{Y}^T (I - H) \boldsymbol{Y} \quad \text{(unrestricted)} \end{aligned}$$

where H and H_r are projection matrices:

$$H = X(X^T X)^{-1} X^T \quad \text{and} \quad H_r = X_r(X_r^T X_r)^{-1} X_r^T.$$

Now substituting into the log-likelihood function, we get

$$\begin{aligned} \ln \mathcal{L}(\widehat{\boldsymbol{\beta}}_r, \widehat{\sigma}_r^2) &= -\frac{n}{2} \ln \left(\boldsymbol{Y}^T (I - H_r) \boldsymbol{Y} \right) + \text{constant} \\ \ln \mathcal{L}(\widehat{\boldsymbol{\beta}}, \widehat{\sigma}^2) &= -\frac{n}{2} \ln \left(\boldsymbol{Y}^T (I - H) \boldsymbol{Y} \right) + \text{constant} \end{aligned}$$

where the constant term is the same for both likelihoods.

The LR statistic for testing H_0 is now

$$\begin{aligned} \Lambda &= \frac{\mathcal{L}(\widehat{\boldsymbol{\beta}}, \widehat{\sigma}^2)}{\mathcal{L}(\widehat{\boldsymbol{\beta}}_r, \widehat{\sigma}_r^2)} \\ &= \left(\frac{\boldsymbol{Y}^T (I - H_r) \boldsymbol{Y}}{\boldsymbol{Y}^T (I - H) \boldsymbol{Y}} \right)^{n/2} \\ &= \left(\frac{\text{RSS}_r}{\text{RSS}} \right)^{n/2} \end{aligned}$$

where $\text{RSS}_r = \boldsymbol{Y}^T(I - H_r)\boldsymbol{Y}$ and $\text{RSS} = \boldsymbol{Y}^T(I - H)\boldsymbol{Y}$ are the residual sums of squares for the reduced and full models respectively.

The LR criterion suggests that we should reject the null hypothesis for large values of the LR statistic Λ; since Λ is an increasing function of RSS_r/RSS, this is equivalent to rejecting for large values of RSS_r/RSS or (equivalently) $(\text{RSS}_r - \text{RSS})/\text{RSS}$. In fact, the test statistic we will use to test H_0 is

$$F = \frac{(\text{RSS}_r - \text{RSS})/(p - r)}{\text{RSS}/(n - p - 1)}.$$

PROPOSITION 8.2 *Under the null hypothesis*

$$H_0 : \beta_{r+1} = \cdots = \beta_p = 0$$

the test statistic F has an F distribution with $(p - r), (n - p - 1)$ degrees of freedom.

Proof. We need to show that $(\text{RSS}_r - \text{RSS})/\sigma^2 \sim \chi^2(p - r)$ and $\text{RSS}/\sigma^2 \sim \chi^2(n - p - 1)$ as well as the independence of the two random variables. Note that we proved that $\text{RSS}/\sigma^2 \sim \chi^2(n-p-1)$ in Proposition 8.1.

To show that $(\text{RSS}_r - \text{RSS})/\sigma^2 \sim \chi^2(p-r)$, we note that

$$\begin{aligned}\text{RSS}_r - \text{RSS} &= \boldsymbol{Y}^T(H - H_r)\boldsymbol{Y} \\ &= \boldsymbol{\varepsilon}^T(H - H_r)\boldsymbol{\varepsilon}\end{aligned}$$

since $\boldsymbol{Y} = X_r\boldsymbol{\beta}_r + \boldsymbol{\varepsilon}$ and $HX_r = H_rX_r = X_r$. Next note that $H - H_r$ is a projection matrix; clearly $H - H_r$ is symmetric and $(H - H_r)^2 = H - H_r$ since $HH_r = H_rH = H_r$. Thus

$$\frac{1}{\sigma^2}(\text{RSS}_r - \text{RSS}) \sim \chi^2(q)$$

where $q = \text{trace}(H - H_r) = p - r$. Finally, to show independence, it suffices to show that $(H - H_r)\boldsymbol{Y}$ is independent of $(I - H)\boldsymbol{Y}$. This holds since $(H - H_r)(I - H) = 0$. □

EXAMPLE 8.3: Consider a single factor ANOVA model

$$Y_{ij} = \mu + \alpha_i + \varepsilon_i \quad (i = 1, \cdots, k; j = 1, \cdots, n_i)$$

where k is the number of treatment groups. To make the parametrization identifiable, we will set $\alpha_1 = 0$ so that $\boldsymbol{\beta} = (\mu, \alpha_2, \cdots, \alpha_k)$. Suppose we want to test the null hypothesis of no treatment effect:

$$H_0 : \alpha_1 = \alpha_2 = \cdots = \alpha_k = 0$$

Under H_0, the MLE of μ is $\widehat{\mu}_r = \bar{Y}$ while the unrestricted MLEs are

$$\begin{aligned}\widehat{\mu} &= \bar{Y}_1 \\ &= \frac{1}{n_1}\sum_{j=1}^{n_1} Y_{ij} \\ \text{and} \quad \widehat{\alpha}_i &= \bar{Y}_i - \bar{Y}_1 \quad \text{for } i \geq 2.\end{aligned}$$

The residual sums of squares for the restricted and unrestricted models are

$$\begin{aligned}\text{RSS}_r &= \sum_{i=1}^{k}\sum_{j=1}^{n_i}(Y_{ij} - \bar{Y})^2 \\ \text{RSS} &= \sum_{i=1}^{k}\sum_{j=1}^{n_i}(Y_{ij} - \bar{Y}_i)^2\end{aligned}$$

and so it follows (after some algebra) that

$$\text{RSS}_r - \text{RSS} = \sum_{i=1}^{k} n_i(\bar{Y}_i - \bar{Y})^2.$$

Thus the F statistic for testing H_0 is

$$F = \frac{\sum_{i=1}^{k} n_i(\bar{Y}_i - \bar{Y})^2/(k-1)}{\sum_{i=1}^{k}\sum_{j=1}^{n_i}(Y_{ij} - \bar{Y}_i)^2/(n-k)}$$

where $n = n_1 + \cdots + n_k$; under H_0, $F \sim \mathcal{F}(k-1, n-k)$. ◇

Proposition 8.2 can be extended to F tests of the null hypothesis

$$H_0 : A\boldsymbol{\beta} = \boldsymbol{c}$$

where A is an $s \times (p+1)$ matrix with rank s and $\boldsymbol{c}$ is a vector of length s. By introducing a vector of Lagrange multipliers $\boldsymbol{\lambda}$, the least squares estimator of $\boldsymbol{\beta}$ under H_0 can be determined to be

$$\widehat{\boldsymbol{\beta}}_r = \widehat{\boldsymbol{\beta}} + (X^T X)^{-1} A^T \left[A(X^T X)^{-1} A^T\right]^{-1} (\boldsymbol{c} - A\widehat{\boldsymbol{\beta}})$$

where $\widehat{\boldsymbol{\beta}}$ is the least squares estimator under the full model. Setting $\text{RSS}_r = \|\boldsymbol{Y} - X\widehat{\boldsymbol{\beta}}_r\|^2$, it can be shown that if H_0 is true then

$$F = \frac{(\text{RSS}_r - \text{RSS})/s}{\text{RSS}/(n-p-1)} \sim \mathcal{F}(s, n-p-1)$$

See Problem 8.8 for details.

Power of the F test

We showed above that the LR test of the null hypothesis $H_0 : \beta_{r+1} = \cdots = \beta_p = 0$ reduces to a test whose test statistic has an F distribution under the null hypothesis. As with any hypothesis testing procedure, the power of this test will depend on the distribution of the test statistic when H_0 does not hold. For most test statistics, the distribution of the test statistic is often quite difficult to determine (even approximately) when H_0 is false. However, it turns out to be straightforward to determine the distribution of the F statistic in general. To do this, we need to define the non-central χ^2 and non-central F distributions.

DEFINITION. Let $X_1, \cdots, X_n$ be independent Normal random variables with $E(X_i) = \mu_i$ and $\text{Var}(X_i) = 1$ and define

$$V = \sum_{i=1}^{n} X_i^2.$$

Then V has a non-central χ^2 distribution with n degrees of freedom and non-centrality parameter

$$\theta^2 = \sum_{i=1}^{n} \mu_i^2$$

$(V \sim \chi^2(n; \theta^2))$. The density function of V is

$$f_V(x) = \sum_{k=0}^{\infty} g_{2k+n}(x) \frac{\exp(-\theta^2/2)(\theta^2/2)^k}{k!}$$

where g_{2k+n} is the density function of a (central) χ^2 distribution with $2k + n$ degrees of freedom.

Figure 8.1 shows the densities of central and non-central χ^2 distributions with 10 degrees of freedom.

DEFINITION. Let V and W be independent random variables with $V \sim \chi^2(n; \theta^2)$ and $W \sim \chi^2(m)$, and define

$$U = \frac{V/n}{W/m}.$$

Then U has a non-central F distribution with n, m degrees of freedom and non-centrality parameter θ^2 $(U \sim \mathcal{F}(n, m; \theta^2))$. The density function of U is

$$f_U(x) = \sum_{k=0}^{\infty} h_{2k+n,m}(x) \frac{\exp(-\theta^2/2)(\theta^2/2)^k}{k!}$$

where $h_{2k+n,m}$ is the density function of a (central) F distribution with $2k + n, m$ degrees of freedom.

THEOREM 8.3 *Let* $\boldsymbol{X} \sim N_n(\boldsymbol{\mu}, I)$. *If* H *is a projection matrix with* $tr(H) = p$ *then*

$$\boldsymbol{X}^T H \boldsymbol{X} \sim \chi^2(p; \boldsymbol{\mu}^T H \boldsymbol{\mu}).$$

Proof. First of all, note that $\boldsymbol{X}^T H \boldsymbol{X} = \|H\boldsymbol{X}\|^2$ and that $H\boldsymbol{X} \sim N_n(H\boldsymbol{\mu}, H)$. Now take an orthogonal matrix O such that OH is a diagonal matrix with p 1's and $n - p$ 0's on the diagonal. The conclusion now follows since $\|H\boldsymbol{X}\|^2 = \|OH\boldsymbol{X}\|^2$. $\square$

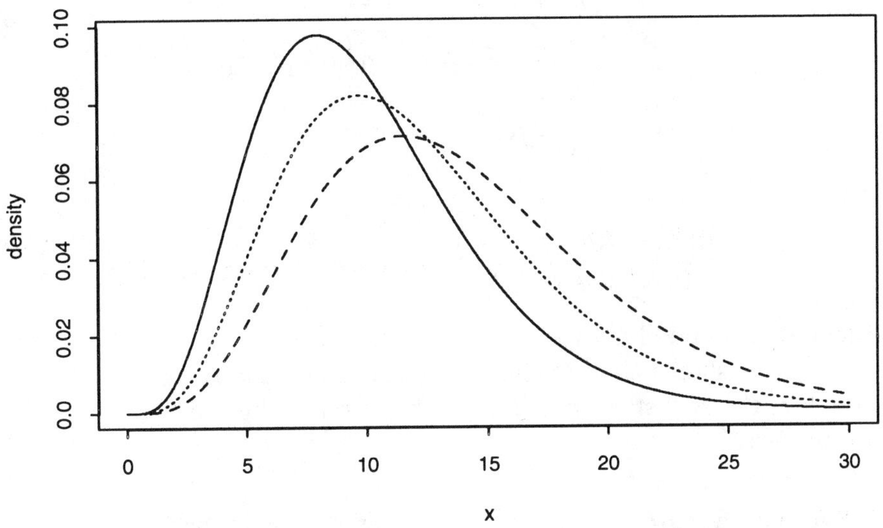

Figure 8.1 *Densities of the central and non-central χ^2 distributions with 10 degrees of freedom; the solid line is the $\chi^2(10)$ density, the dotted line is the $\chi^2(10;2)$ density and the dashed line is the $\chi^2(10;4)$ density.*

How do non-central distributions arise in the context of the linear model? The easiest way to see this is to look at the numerator of the F statistic for testing the adequacy of a reduced model. From before we have

$$\begin{aligned} \text{RSS} &= \boldsymbol{Y}^T(I - H)\boldsymbol{Y} \\ \text{RSS}_r &= \boldsymbol{Y}^T(I - H_r)\boldsymbol{Y} \end{aligned}$$

and so

$$\text{RSS}_r - \text{RSS} = \boldsymbol{Y}^T(H - H_r)\boldsymbol{Y}.$$

Note that

$$\frac{1}{\sigma^2}\boldsymbol{Y}^T(I - H)\boldsymbol{Y} \sim \chi^2(n - p - 1)$$

even under the alternative hypothesis. On the other hand,

$$\frac{1}{\sigma^2}\boldsymbol{Y}^T(H - H_r)\boldsymbol{Y} \sim \chi^2(p - r; \theta^2)$$

where

$$\begin{aligned}\theta^2 &= \frac{1}{\sigma^2}\boldsymbol{\beta}^T X^T (H - H_r) X\boldsymbol{\beta} \\ &= \frac{1}{\sigma^2}\left[\boldsymbol{\beta}^T X^T H X\boldsymbol{\beta} - \boldsymbol{\beta}^T X^T H_r X\boldsymbol{\beta}\right] \\ &= \frac{1}{\sigma^2}\left[\|X\boldsymbol{\beta}\|^2 - \|H_r X\boldsymbol{\beta}\|^2\right].\end{aligned}$$

Thus

$$F = \frac{(\text{RSS}_r - \text{RSS})/(p-r)}{\text{RSS}/(n-p-1)} \sim \mathcal{F}(p-r, n-p-1; \theta^2)$$

where θ^2 is defined above.

It can be shown that, for fixed values of $p - r$ and $n - p - 1$, the power of the F test is an increasing function of the non-centrality parameter θ^2.

EXAMPLE 8.4: Consider the simple linear regression model

$$Y_i = \beta_0 + \beta_1 x_i + \varepsilon_i \quad (i = 1, \cdots, n)$$

where we will assume (for simplicity) that $\sum_{i=1}^n x_i = 0$. We want to test the null hypothesis

$$H_0 : \beta_1 = 0 \quad \text{versus} \quad H_1 : \beta_1 \neq 0.$$

In this case, the full and reduced design matrices are

$$X = \begin{pmatrix} 1 & x_1 \\ 1 & x_2 \\ \vdots & \vdots \\ 1 & x_n \end{pmatrix} \quad \text{and} \quad X_r = \begin{pmatrix} 1 \\ 1 \\ \vdots \\ 1 \end{pmatrix}.$$

The projection matrix H_r is simply

$$H_r = \begin{pmatrix} 1/n & \cdots & 1/n \\ \vdots & \ddots & \vdots \\ 1/n & \cdots & 1/n \end{pmatrix}$$

and so

$$H_r X\boldsymbol{\beta} = \begin{pmatrix} 1/n & \cdots & 1/n \\ \vdots & \ddots & \vdots \\ 1/n & \cdots & 1/n \end{pmatrix} \begin{pmatrix} \beta_0 + \beta_1 x_1 \\ \vdots \\ \beta_0 + \beta_1 x_n \end{pmatrix}$$

$$= \begin{pmatrix} \beta_0 \\ \vdots \\ \beta_0 \end{pmatrix}$$

since $\sum_{i=1}^{n} x_i = 0$; thus $\|H_r X\boldsymbol{\beta}\|^2 = n\beta_0^2$. Likewise,

$$\begin{aligned} \|X\boldsymbol{\beta}\|^2 &= \sum_{i=1}^{n}(\beta_0 + \beta_1 x_i)^2 \\ &= n\beta_0^2 + \beta_1^2 \sum_{i=1}^{n} x_i^2. \end{aligned}$$

Thus the non-centrality parameter for the distribution of the F statistics for testing H_0 is

$$\theta^2 = \frac{\beta_1^2}{\sigma^2} \sum_{i=1}^{n} x_i^2;$$

more generally, if $\sum_{i=1}^{n} x_i \neq 0$ then

$$\theta^2 = \frac{\beta_1^2}{\sigma^2} \sum_{i=1}^{n} (x_i - \bar{x})^2.$$

Given that the power of the F test increases as the non-centrality parameter increases, the form of θ^2 makes sense from an intuitive point of view; for fixed σ^2, the power increases as $|\beta_1|$ increases while for fixed β_1, the power decreases as σ^2 increases. Also note that the power increases as the x_i's become more spread out; this is a potentially important point from a design perspective. ◇

EXAMPLE 8.5: Consider a single factor ANOVA model

$$Y_{ij} = \mu + \alpha_i + \varepsilon_i \quad (i = 1, \cdots, k; j = 1, \cdots, n_i)$$

where to make the parametrization identifiable, we assume (as in Example 8.1) that $\alpha_1 = 0$. The form of the F test of $H_0 : \alpha_1 = \cdots = \alpha_k = 0$ was given in Example 8.3. To evaluate the non-centrality parameter of the F statistic, we first need to evaluate $\|H_r X\boldsymbol{\beta}\|^2$ for this model; as in Example 8.4, H_r is an $n \times n$ matrix whose entries are all $1/n$ (where $n = n_1 + \cdots n_k$) and so

$$\|H_r X\boldsymbol{\beta}\|^2 = n \left(\frac{1}{n} \sum_{i=1}^{k} \sum_{j=1}^{n_i} (\mu + \alpha_i) \right)^2$$

$$= n\left(\mu + \frac{1}{n}\sum_{i=1}^{k} n_i\alpha_i\right)^2.$$

Thus the non-centrality parameter for the F statistic is

$$\begin{aligned}
\theta^2 &= \frac{1}{\sigma^2}\left[\sum_{i=1}^{k}\sum_{j=1}^{n_i}(\mu+\alpha_i)^2 - n\left(\mu + \frac{1}{n}\sum_{i=1}^{k} n_i\alpha_i\right)^2\right] \\
&= \frac{1}{\sigma^2}\left[\sum_{i=1}^{k} n_i\alpha_i^2 - \frac{1}{n}\left(\sum_{i=1}^{k} n_i\alpha_i\right)^2\right].
\end{aligned}$$

Note that this non-centrality parameter is proportional to the variance of a probability distribution putting probability mass of n_i/n at the point α_i (for $i = 1, \cdots, k$); hence, the more dispersed the α_i's the greater the non-centrality parameter and hence the power of the F test. ◇

8.4 Non-normal errors

To this point, we have assumed normally distributed errors in the linear model. It is interesting to consider how much of the preceding theory remains valid under i.i.d. finite variance (but non-normal errors). Clearly, all the preceding results about distribution theory for estimators and test statistics will not hold in the more general setting; these results are very much dependent on the errors (and hence the responses Y_i) being normally distributed.

However, many of the results from the previous section do carry over to the case of finite variance i.i.d. errors. For example, for normally distributed errors, we have

$$\widehat{\boldsymbol{\beta}} \sim N_{p+1}\left(\boldsymbol{\beta}, \sigma^2(X^TX)^{-1}\right)$$

while in general, we can say

$$E(\widehat{\boldsymbol{\beta}}) = \boldsymbol{\beta} \quad \text{and} \quad \text{Cov}(\widehat{\boldsymbol{\beta}}) = \sigma^2(X^TX)^{-1}.$$

Likewise, the MLE $\widehat{\sigma}^2$ will remain a biased estimator of σ^2 while $\text{RSS}/(n-p-1)$ will be an unbiased estimator of σ^2.

There is also an optimality result for the least squares estimator of $\boldsymbol{\beta}$ that holds for finite variance i.i.d. errors. Consider the linear model

$$Y_i = \boldsymbol{x}_i^T\boldsymbol{\beta} + \varepsilon_i \quad (i = 1, \cdots, n)$$

and define a parameter

$$\theta = \sum_{j=0}^{p} a_j \beta_j = \boldsymbol{a}^T \boldsymbol{\beta}$$

where $a_0, a_1, \cdots, a_p$ are some (known) constants. We want to consider unbiased estimators of θ of the form

$$\widehat{\theta} = \sum_{i=1}^{n} c_i Y_i = \boldsymbol{c}^T \boldsymbol{Y};$$

such estimators are called linear estimators of θ (since they are linear in the Y_i's).

THEOREM 8.4 (Gauss-Markov Theorem) *Assume that the design matrix has full rank and let $\widehat{\boldsymbol{\beta}} = (X^T X)^{-1} X^T \boldsymbol{Y}$ be the least squares estimator of $\boldsymbol{\beta}$. Then*

$$\widehat{\theta} = \boldsymbol{a}^T \widehat{\boldsymbol{\beta}}$$

has the minimum variance of all linear, unbiased estimators of $\theta = \boldsymbol{a}^T \boldsymbol{\beta}$. (The estimator $\widehat{\theta}$ is often called the best linear unbiased estimator (BLUE) of θ.)

Proof. For any $\boldsymbol{c}$, $E(\boldsymbol{c}^T \boldsymbol{Y}) = \boldsymbol{c}^T X \boldsymbol{\beta}$. Thus if $E(\boldsymbol{c}^T \boldsymbol{Y}) = \boldsymbol{a}^T \boldsymbol{\beta}$ for all $\boldsymbol{\beta}$, it follows that $\boldsymbol{a}^T = \boldsymbol{c}^T X$. It suffices then to show that if $\boldsymbol{a}^T = \boldsymbol{c}^T X$ then

$$\text{Var}(\boldsymbol{c}^T \boldsymbol{Y}) \geq \text{Var}(\boldsymbol{a}^T \widehat{\boldsymbol{\beta}}).$$

Note that $\text{Var}(\boldsymbol{c}^T \boldsymbol{Y}) = \sigma^2 \boldsymbol{c}^T \boldsymbol{c}$ while $\text{Var}(\boldsymbol{a}^T \widehat{\boldsymbol{\beta}}) = \sigma^2 \boldsymbol{c}^T H \boldsymbol{c}$ where $H = X(X^T X)^{-1} X^T$. Thus

$$\begin{aligned} \text{Var}(\boldsymbol{c}^T \boldsymbol{Y}) - \text{Var}(\boldsymbol{a}^T \widehat{\boldsymbol{\beta}}) &= \sigma^2 \left(\boldsymbol{c}^T \boldsymbol{c} - \boldsymbol{c}^T H \boldsymbol{c}\right) \\ &= \sigma^2 \left(\boldsymbol{c}^T (I - H) \boldsymbol{c}\right) \\ &\geq 0 \end{aligned}$$

since $I - H$ is a projection matrix and hence positive definite. □

At first glance, the conclusion of the Gauss-Markov Theorem seems to be very strong. However, notice that the class of estimators considered in the Gauss-Markov (namely linear, unbiased estimators) is very small. If one considers biased estimators (for example, ridge estimators; see Hoerl and Kennard (1970) and Problem 8.9) then it may be possible to achieve a smaller mean square error. Moreover, for non-normal error distributions, it is often possible to find better estimators of $\theta = \boldsymbol{a}^T \boldsymbol{\beta}$ using non-linear estimators of $\boldsymbol{\beta}$.

Some large sample theory

In Proposition 8.1, we showed that the least squares estimator of $\boldsymbol{\beta}$ is exactly normally distributed when the errors are normally distributed. In this section, we will show that the least squares estimator has an asymptotic Normal distribution under appropriate conditions on the design.

Consider the linear model

$$Y_i = \boldsymbol{x}_i^T \boldsymbol{\beta} + \varepsilon_i \quad (i = 1, \cdots, n)$$

where $\varepsilon_1, \cdots, \varepsilon_n$ are i.i.d. random variables with mean 0 and variance σ^2. Define the least squares estimator

$$\widehat{\boldsymbol{\beta}}_n = (X_n^T X_n)^{-1} X_n^T \boldsymbol{Y}_n$$

where the subscript n has been added to make explicit the dependence on the sample size. Since $\boldsymbol{Y}_n = X_n \boldsymbol{\beta} + \boldsymbol{\varepsilon}_n$, it follows that

$$\widehat{\boldsymbol{\beta}}_n - \boldsymbol{\beta} = (X_n^T X_n)^{-1} X_n^T \boldsymbol{\varepsilon}_n.$$

To obtain a limiting distribution, we need to normalize $\widehat{\boldsymbol{\beta}}_n - \boldsymbol{\beta}$ by multiplying it by a sequence of constants or matrices. Define a symmetric matrix A_n to be a "square root" of $X_n^T X_n$; that is, $A_n^2 = X_n^T X_n$. We will consider the limiting distribution of

$$A_n(\widehat{\boldsymbol{\beta}}_n - \boldsymbol{\beta}) = A_n^{-1} X_n^T \boldsymbol{\varepsilon}_n.$$

Note that $\text{Cov}(A_n(\widehat{\boldsymbol{\beta}}_n - \boldsymbol{\beta})) = \sigma^2 I$.

THEOREM 8.5 *Suppose that*

$$\max_{1 \le i \le n} \boldsymbol{x}_i^T (X_n^T X_n)^{-1} \boldsymbol{x}_i \to 0$$

as $n \to \infty$. *Then*

$$A_n(\widehat{\boldsymbol{\beta}}_n - \boldsymbol{\beta}) \to_d N_{p+1}(\boldsymbol{0}, \sigma^2 I).$$

Proof. The idea here is to use the Cramér-Wold device together with the CLT for weighted sums of i.i.d. random variables. By the Cramér-Wold device, it suffices to show that

$$\boldsymbol{a}^T A_n^{-1} X_n^T \boldsymbol{\varepsilon}_n \to_d N(0, \sigma^2 \boldsymbol{a}^T \boldsymbol{a})$$

for all vectors $\boldsymbol{a}$. Note that

$$\boldsymbol{a}^T A_n^{-1} X_n^T \boldsymbol{\varepsilon}_n = \sum_{i=1}^n c_{ni} \varepsilon_i = \boldsymbol{c}_n^T \boldsymbol{\varepsilon}_n$$

where
$$c_{ni} = \boldsymbol{a}^T A_n^{-1} \boldsymbol{x}_i.$$
By the CLT for weighted sums, it suffices to show that
$$\max_{1 \le i \le n} \frac{c_{ni}^2}{\sum_{k=1}^n c_{nk}^2} \to 0.$$
First of all, we have
$$\begin{aligned} c_{ni}^2 &\le (\boldsymbol{a}^T \boldsymbol{a})(\boldsymbol{x}_i^T A_n^{-2} \boldsymbol{x}_i) \\ &= (\boldsymbol{a}^T \boldsymbol{a})(\boldsymbol{x}_i^T (X_n^T X_n)^{-1} \boldsymbol{x}_i) \end{aligned}$$
and also
$$\begin{aligned} \sum_{k=1}^n c_{nk}^2 = \boldsymbol{c}_n^T \boldsymbol{c}_n &= \boldsymbol{a}^T A_n^{-1} (X_n^T X_n) A_n^{-1} \boldsymbol{a} \\ &= \boldsymbol{a}^T \boldsymbol{a}. \end{aligned}$$
Thus
$$\frac{\max_{1 \le i \le n} c_{ni}^2}{\sum_{k=1}^n c_{nk}^2} \le \max_{1 \le i \le n} \boldsymbol{x}_i^T (X_n^T X_n)^{-1} \boldsymbol{x}_i \to 0$$
by hypothesis. □

What is the practical interpretation of the condition
$$\max_{1 \le i \le n} \boldsymbol{x}_i^T (X_n^T X_n)^{-1} \boldsymbol{x}_i \to 0?$$

If we define $H_n = X_n (X_n^T X_n)^{-1} X_n^T$ to be the "hat" matrix then $h_{ni} = \boldsymbol{x}_i^T (X_n^T X_n)^{-1} \boldsymbol{x}_i$ is the i-th diagonal element of H_n; since H_n is a projection matrix, the sum of the h_{ni}'s is $(p+1)$. The condition above implies that the h_{ni}'s tend uniformly to 0 as $n \to \infty$; since there are n diagonal elements and their sum must be $(p+1)$, this does not seem to be a terribly stringent condition to fulfill. In regression analysis, the diagonal elements h_{ni} can be interpreted as describing the potential influence of the point $\boldsymbol{x}_i$ on the estimate of $\boldsymbol{\beta}$; the larger h_{ni}, the greater the potential influence (or leverage) of $\boldsymbol{x}_i$. Thus in practical terms, we can interpret the condition above to mean that the leverages of all the $\boldsymbol{x}_i$'s are small. If this is the case and n is sufficient large then $\widehat{\boldsymbol{\beta}}_n$ is approximately Normal with mean $\boldsymbol{\beta}$ and variance-covariance matrix $\sigma^2 (X_n^T X_n)^{-1}$.

EXAMPLE 8.6: Consider a simple linear regression model
$$Y_i = \beta_0 + \beta_1 x_i + \varepsilon_i \quad (i = 1, \cdots, n).$$

For this model, it can be shown that the diagonals of the "hat" matrix are

$$h_{ni} = \frac{1}{n} + \frac{(x_i - \bar{x})^2}{\sum_{k=1}^n (x_k - \bar{x})^2}.$$

We will consider two simple design scenarios. First of all, suppose that $x_i = i$ for $i = 1, \cdots, n$. Then

$$h_{ni} = \frac{1}{n} + \frac{(i - (n+1)/2)^2}{(n^3 - n)/12}$$

and so

$$\max_{1 \le i \le n} h_{ni} = h_{nn} = \frac{1}{n} + \frac{6(n-1)}{n(n+1)},$$

which tends to 0 as $n \to \infty$; thus asymptotic normality holds. Next suppose that $x_i = 2^i$ for $i = 1, \cdots, n$. In this case, $\bar{x} = 2(2^n - 1)/n$ and

$$\sum_{i=1}^n (x_i - \bar{x})^2 = \frac{4^{n+1} - 4}{3} - \frac{4^{n+1} + 4 - 2^{n+3}}{n}.$$

Thus we have

$$\max_{1 \le i \le n} h_{ni} = h_{nn}$$

with $h_{nn} \to 3/4$ as $n \to \infty$. For this design (or, more correctly, sequence of designs), we will not have asymptotic normality. ◇

Other estimation methods

Perhaps not surprisingly (given the popularity of the linear model in practice), a vast number of alternatives to least squares estimation have been proposed. Most of these alternative estimation methods are motivated by the fact that least squares estimation is not particularly robust to deviations from its nominal assumptions; for example, a single observation $(\boldsymbol{x}_i, Y_i)$ can have effectively an unbounded influence on the value of the least squares estimator. We will briefly outline some of the alternatives to least squares estimation here. Figure 8.2 shows how the least squares line can be affected by a small number of points.

The simplest alternatives to least squares estimation replace the "sum of squares" objective function by an objective function that penalizes large deviations less severely. For example, we might define $\widehat{\boldsymbol{\beta}}$ to minimize

$$\sum_{i=1}^n \rho(Y_i - \boldsymbol{x}_i^T \boldsymbol{\beta})$$

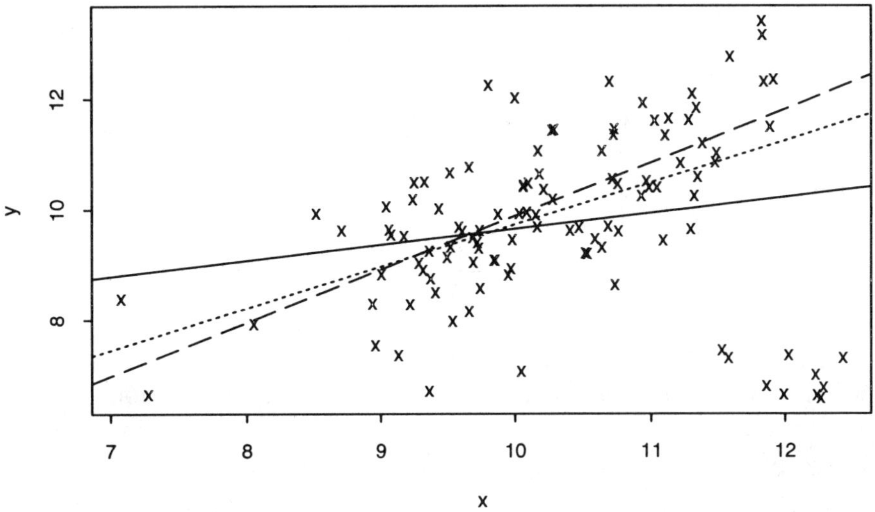

Figure 8.2 *Estimated regression lines; the solid line is the least squares line, the dotted line is the L_1 line and the dashed line is the LMS line. Notice how the least squares line is pulled more towards the 10 "outlying" observations than are the other two lines.*

where $\rho(x)$ is a function with $\rho(x)/x^2 \to 0$ as $x \to \pm\infty$; typically, $\rho(x) \to \infty$ as $x \to \pm\infty$ although $\rho(x)$ could be a bounded function. Such estimators are referred to as M-estimators where the "M" is an allusion to the fact that these estimators could be viewed as MLEs were the errors to come from the appropriate distribution. In many cases, the function ρ depends on a scale parameter σ (so that $\rho = \rho_\sigma$) that must also be estimated.

In the case where ρ is a convex function, we can prove an analogous result to Theorem 8.5.

THEOREM 8.6 *Suppose that ρ is a convex function with*

$$\rho(x) = \int_0^x \psi(t)\, dt$$

where ψ is a non-decreasing function with $E[\psi(\varepsilon_i)] = 0$, $E[\psi^2(\varepsilon_i)]$ finite and $\lambda(t) = E[\psi(\varepsilon_1 + t) - \psi(\varepsilon_i)]$ is differentiable at $t = 0$ with *derivative* $\lambda'(0) > 0$. *Suppose also that*

$$\max_{1 \le i \le n} \boldsymbol{x}_i^T (X_n^T X_n)^{-1} \boldsymbol{x}_i \to 0$$

as $n \to \infty$. *Then*

$$A_n(\widehat{\boldsymbol{\beta}}_n - \boldsymbol{\beta}) \to_d N_{p+1}(\mathbf{0}, \gamma^2 I)$$

where

$$\gamma^2 = \frac{E[\psi^2(\varepsilon_i)]}{[\lambda'(0)]^2}.$$

A proof of this result is sketched in Problem 8.10. Note that when there is an intercept in the model then the assumption that $E[\psi(\varepsilon_i)] = 0$ is no restriction since we can always redefine the intercept so this condition holds. When ρ is twice differentiable then typically we have $\lambda'(0) = E[\psi'(\varepsilon_i)]$ where ψ' is the derivative of ψ, or equivalently, the second derivative of ρ; since ρ is convex, $\psi'(x) \ge 0$. If $\rho(x) = |x|$ (in which case, we have L_1-estimators) then $\psi(x) = I(x \ge 0) - I(x \le 0)$ and so $\lambda(t) = 1 - 2F(-t)$; if $F'(0) = f(0) > 0$ then $\lambda'(0) = 2f(0)$.

M-estimators are generally robust against non-normality of the errors, particularly heavy-tailed error distributions where M-estimators can be more efficient than least squares estimators. However, M-estimators are less robust against more general outliers, for example, situations similar to that illustrated in Figure 8.2. Numerous (more robust) alternatives have been proposed, such as GM-estimators (Krasker and Welsch, 1982), which bound the influence that any observation can have, and the least median of squares (LMS) estimator of Rousseeuw (1984), for which $\widehat{\boldsymbol{\beta}}$ minimizes

$$\text{median}\left\{|Y_i - \boldsymbol{x}_i^T \boldsymbol{\beta}| : 1 \le i \le n\right\}$$

over $\boldsymbol{\beta}$. The LMS estimator has a breakdown point (roughly defined to be the fraction of "bad" observations needed to drive an estimator to the boundary of the parameter space; see Donoho and Huber (1983)) of 50%; the breakdown point of M-estimators is effectively 0. However, while the LMS estimator is quite robust, it is extremely inefficient when the classical model (with Normal errors) is true; see Kim and Pollard (1990) for details.

8.5 Generalized linear models

Generalized linear models represent a generalization of classical linear models (where the response is nominally assumed to be normally distributed) to situations where the response has a non-normal distribution, for example, a Binomial, Poisson or Gamma distribution. The standard reference for generalized linear models is the book by McCullagh and Nelder (1989).

In classical linear models theory (as described in sections 8.2 and 8.3), we are given responses $Y_1, \cdots, Y_n$ that we assume to be normally distributed with means $\mu_1, \cdots, \mu_n$ and constant variance σ^2 where

$$\begin{aligned} \mu_i &= \beta_0 + \beta_1 x_{i1} + \cdots + \beta_p x_{ip} \\ &= \boldsymbol{x}_i^T \boldsymbol{\beta}; \end{aligned}$$

$\boldsymbol{\beta} = (\beta_0, \beta_1, \cdots, \beta_p)^T$ are unknown parameters and $x_{i1}, \cdots, x_{ip}$ are known constants (for $i = 1, \cdots, n$). Thus $\mu_i = E(Y_i)$ is a linear function of the parameter vector $\boldsymbol{\beta}$ and $\text{Var}(Y_i)$ is a constant (typically unknown). From this specification, it is possible to find the likelihood function for $\boldsymbol{\beta}$ and σ^2; the MLE of $\boldsymbol{\beta}$ turns out to be the least squares estimator.

It is possible to generalize the notion of linear models to non-normal response variables. Let $Y_1, \cdots, Y_n$ be independent random variables from some family of distributions with means $\mu_1, \cdots, \mu_n$ and variances $\sigma_1^2, \cdots, \sigma_n^2$ where $\sigma_i^2 \propto V(\mu_i)$; it will be shown that this property holds for random variables belonging to a certain class of distributions that includes one-parameter exponential families. Given covariates $\boldsymbol{x}_i$ $(i = 1, \cdots, n)$, we will assume that some function of μ_i is a linear function of $\boldsymbol{x}_i$; that is,

$$g(\mu_i) = \boldsymbol{x}_i^T \boldsymbol{\beta}$$

for some strictly increasing function g, which is called the link function. The classical (considered in sections 8.1 to 8.3) linear model has $g(\mu) = \mu$ and $V(\mu) = \sigma^2$, a constant. The following two examples illustrate possible models for discrete responses.

EXAMPLE 8.7: (Poisson regression) Assume that $Y_1, \cdots, Y_n$ are independent Poisson random variables. For the Poisson distribution, it is well-known that the variance is equal to the mean so $V(\mu) = \mu$. The standard Poisson regression model uses a logarithmic link function; that is,

$$\ln(\mu_i) = \boldsymbol{x}_i^T \boldsymbol{\beta}$$

where $\mu_i = E(Y_i)$. (This is often called a log-linear Poisson model.) One advantage of the logarithmic link is the fact that the function $\ln(x)$ maps the interval $(0, \infty)$ onto the entire real line. ◇

EXAMPLE 8.8: (Binary regression) Here we assume that $Y_1, \cdots, Y_n$ are independent random variables taking the values 0 and 1 with

$$P(Y_i = 1) = \theta_i \quad \text{and} \quad P(Y_i = 0) = 1 - \theta_i.$$

In this case, $\mu_i = E(Y_i) = \theta_i$ and $\text{Var}(Y_i) = \theta_i(1 - \theta_i)$ so that $V(\mu) = \mu(1 - \mu)$. The most commonly used link function is the logistic or logit link

$$g(\mu) = \ln\left(\frac{\mu}{1 - \mu}\right);$$

the model $g(\mu) = \boldsymbol{x}_i^T \boldsymbol{\beta}$ is called a logistic regression model. Other commonly used link functions include the so-called probit link

$$g(\mu) = \Phi^{-1}(\mu)$$

(where Φ^{-1} is the inverse of the standard Normal distribution function) and the complementary log-log link

$$g(\mu) = \ln(-\ln(1 - \mu)).$$

Note that these three link functions map the interval $(0, 1)$ onto the entire real line; in each case, g is the inverse of the distribution function of a continuous random variable. ◇

Likelihood functions and estimation

We will now assume that $Y_1, \cdots, Y_n$ have density or frequency functions that belong to a one-parameter exponential family, possibly enriched by an additional scale parameter. In particular, we will assume that the density (frequency) function of Y_i is

$$f(y; \theta_i, \phi) = \exp\left[\frac{\theta_i y - b(\theta_i)}{\phi} + c(y, \phi)\right] \quad \text{for } y \in A$$

where θ_i and ϕ are parameters, and the set A does not depend on θ_i or ϕ. When ϕ is known, this family of density (frequency) functions is an exponential family; when ϕ is unknown, this family may or may not be an exponential family. In any event, if Y_i has the distribution given above, it can be shown that

$$E(Y_i) = b'(\theta_i)$$

and

$$\text{Var}(Y_i) = \phi\, b''(\theta_i)$$

where b' and b'' are the first two derivatives of b. If b' is a one-to-one function (either strictly increasing or strictly decreasing) then θ_i can be uniquely determined from $\mu_i = E(Y_i)$ and so it follows that

$$\text{Var}(Y_i) = \phi V(\mu_i).$$

We will refer to the function V as the variance function as it gives the variance of any Y_i up to a constant multiple that depends on the parameter ϕ.

We now assume that $\boldsymbol{Y} = (Y_1, \cdots, Y_n)$ are independent random variables with the density (frequency) function of Y_i given above where

$$g(\mu_i) = g(b'(\theta_i)) = \boldsymbol{x}_i^T \boldsymbol{\beta}.$$

Given $\boldsymbol{Y} = \boldsymbol{y}$, the log-likelihood function of $\boldsymbol{\beta}$ is simply

$$\ln \mathcal{L}(\boldsymbol{\beta}, \phi) = \sum_{i=1}^{n} \left[\frac{\theta_i(\boldsymbol{\beta}) y_i - b(\theta_i(\boldsymbol{\beta}))}{\phi} + c(y_i, \phi) \right].$$

The MLE of $\boldsymbol{\beta}$ can be determined by maximizing the log-likelihood function given above; differentiating the log-likelihood with respect to the elements of $\boldsymbol{\beta}$, we obtain the following equations for the MLE $\widehat{\boldsymbol{\beta}}$:

$$\sum_{i=1}^{n} \frac{Y_i - \mu_i(\widehat{\boldsymbol{\beta}})}{g'(\mu_i(\widehat{\boldsymbol{\beta}}) V(\mu_i(\widehat{\boldsymbol{\beta}}))} \boldsymbol{x}_i = \boldsymbol{0}$$

where $\mu_i(\boldsymbol{\beta}) = g^{-1}(\boldsymbol{x}_i^T \boldsymbol{\beta})$.

There are two points to be made regarding the maximum likelihood estimator of $\boldsymbol{\beta}$. The first point is the fact that this estimator remains the same regardless of whether the scale parameter ϕ is known or unknown. The second, and more interesting, point is that the estimating equations depend only on the distribution of the Y_i's via the link function g and the "variance function" V, which expresses the relationship (up to a constant multiple) between the variance and mean of the response. This fact suggests the possibility of formulating generalized linear models by specifying only the relationship between the variance and mean of the response and not the distribution of the response itself. This will be pursued below.

The link function plays a very important role in the formulation of a generalized linear model. Frequently, the link function g is chosen so that $g(\mu_i) = \theta_i$ in which case the log-likelihood function becomes

$$\ln \mathcal{L}(\boldsymbol{\beta}, \phi) = \sum_{i=1}^{n} \left[\frac{y_i \boldsymbol{x}_i^T \boldsymbol{\beta} - b(\boldsymbol{x}_i^T \boldsymbol{\beta})}{\phi} + c(y_i, \phi) \right].$$

In this case, the family of distributions of $\boldsymbol{Y}$ are a $(p+1)$-parameter exponential family if ϕ is known; the link g is called the natural or canonical link function.

EXAMPLE 8.9: Suppose that $Y_1, \cdots, Y_n$ are independent Exponential random variables with means $\mu_i > 0$. In this case, the density of Y_i is

$$f(y; \mu_i) = \exp[-y/\mu_i - \ln(\mu_i)] \quad \text{for } y \geq 0$$

and so the natural link function is $g(\mu) = -1/\mu$. Note that the range of this link function is not the entire real line (unlike the link functions in Examples 8.7 and 8.8); this fact makes the use of the natural link function somewhat undesirable in practice and the link function $g(\mu) = \ln(\mu)$ is usually preferred in practice. ◇

The natural link functions in the Poisson and binary regression models are the log and logistic links respectively (see Examples 8.7 and 8.8). Despite its name, there is no really compelling practical reason to prefer the natural link to any other link function. (Nonetheless, there are some theoretical advantages to using the natural link which relate to the fact that joint distribution of $\boldsymbol{Y}$ is an exponential family.)

Inference for generalized linear models

In the classical linear model, the assumption of normality makes it possible to give exact sampling distributions of parameter estimators and test statistics. Except in special cases, this is not true for most generalized linear models.

There are several options available for approximating sampling distributions. For example, computer intensive approaches such as

Monte Carlo simulation and resampling (for example, the bootstrap; see Efron and Tibshirani (1993)) may be used in general. If the sample size is sufficiently large, it is often possible to approximate the joint distribution of the MLEs by a multivariate Normal distribution. In particular, under regularity conditions on the design (analogous to those in Theorems 8.5 and 8.6), it can be shown that the MLE $\widehat{\boldsymbol{\beta}}$ is approximately multivariate Normal with mean vector $\boldsymbol{\beta}$ and variance-covariance matrix $\phi(X^T W(\boldsymbol{\mu})X)^{-1}$ where X is an $n \times (p+1)$ matrix whose i-th row is $\boldsymbol{x}_i^T$ and $W(\boldsymbol{\mu})$ is a diagonal matrix whose i-th diagonal element is $[V(\mu_i)]^{-1}[g'(\mu_i)]^{-2}$. An estimator of the variance-covariance matrix of $\widehat{\boldsymbol{\beta}}$ can be obtained by substituting the estimators of $\boldsymbol{\beta}$ and ϕ:

$$\widehat{\mathrm{Cov}}(\widehat{\boldsymbol{\beta}}) = \widehat{\phi}\left(X^T W(\widehat{\boldsymbol{\mu}})X\right)^{-1}$$

where $\widehat{\boldsymbol{\mu}}$ depends on $\widehat{\boldsymbol{\beta}}$. Given $\widehat{\mathrm{Cov}}(\widehat{\boldsymbol{\beta}})$, estimated standard errors of the parameter estimates will be the square roots of the diagonal elements of $\widehat{\mathrm{Cov}}(\widehat{\boldsymbol{\beta}})$.

Likewise, hypothesis tests and confidence intervals are typically based on the asymptotic normality of the MLEs; for example, we can often adapt the likelihood testing theory outlined in section 7.4 to obtain χ^2 approximations for the null distributions of likelihood based test statistics, such as the LR and score tests. Some care should be exercised in using these approximations though.

Numerical computation of parameter estimates

The MLE of $\boldsymbol{\beta}$ are the solutions of the estimating equations given above. Unfortunately, no explicit representation of these estimators exists and hence maximum likelihood estimates must generally be obtained using some iterative numerical method such as the Newton-Raphson or Fisher scoring algorithm. It can be shown that the Fisher scoring algorithm is equivalent to solving a sequence of weighted least squares problems. In the case where the natural link is used, the Fisher scoring algorithm coincides exactly with the Newton-Raphson algorithm since the model is a $(p+1)$-parameter exponential family.

We assume a generalized linear model with link function $g(\mu)$ and variance function $V(\mu)$ with $g(\mu_i) = \boldsymbol{x}_i^T\boldsymbol{\beta}$ for $i = 1, \cdots, n$. Recall that the Fisher scoring algorithm iteratively updates the estimate of $\boldsymbol{\beta}$ using the score function (that is, the gradient of the log-likelihood function) and the expected Fisher information matrix evaluated at the previous estimate of $\boldsymbol{\beta}$. Let $\widehat{\boldsymbol{\beta}}^{(k)}$ be the estimate of $\boldsymbol{\beta}$ after k iterations of the algorithm and

$$\widehat{\mu}_i^{(k)} = g^{-1}\left(\boldsymbol{x}_i^T \widehat{\boldsymbol{\beta}}^{(k)}\right).$$

Then the $(k+1)$ iterate in the Fisher scoring algorithm is defined to be

$$\widehat{\boldsymbol{\beta}}^{(k+1)} = \widehat{\boldsymbol{\beta}}^{(k)} + H^{-1}\left(\widehat{\boldsymbol{\beta}}^{(k)}\right) S\left(\widehat{\boldsymbol{\beta}}^{(k)}\right)$$

where

$$H(\boldsymbol{\beta}) = \sum_{i=1}^{n} \frac{\boldsymbol{x}_i \boldsymbol{x}_i^T}{V(\mu_i(\boldsymbol{\beta}))[g'(\mu_i(\boldsymbol{\beta}))]^2}$$

is the expected Fisher information matrix (evaluated at $\boldsymbol{\beta}$) and

$$S(\boldsymbol{\beta}) = \sum_{i=1}^{n} \frac{y_i - \mu_i(\boldsymbol{\beta})}{V(\mu_i(\boldsymbol{\beta}))g'(\mu_i(\boldsymbol{\beta}))} \boldsymbol{x}_i$$

is the score function. Rearranging the expression for $\widehat{\boldsymbol{\beta}}^{(k+1)}$, we get

$$\widehat{\boldsymbol{\beta}}^{(k+1)} = (X^T W^{(k)} X)^{-1} X^T W^{(k)} \boldsymbol{z}^{(k)}$$

where X is the matrix whose i-th row is $\boldsymbol{x}_i^T$, $W^{(k)}$ is a diagonal matrix whose i-th diagonal element is

$$w_i^{(k)} = \frac{1}{V\left(\widehat{\mu}_i^{(k)}\right)\left[g'\left(\widehat{\mu}_i^{(k)}\right)\right]^2}$$

and $\boldsymbol{z}^{(k)}$ is a vector whose i-th element is

$$\begin{aligned} z_i^{(k)} &= g\left(\widehat{\mu}_i^{(k)}\right) + g'\left(\widehat{\mu}_i^{(k)}\right)\left(y_i - \widehat{\mu}_i^{(k)}\right) \\ &= \boldsymbol{x}_i^T \widehat{\boldsymbol{\beta}}^{(k)} + g'\left(\widehat{\mu}_i^{(k)}\right)\left(y_i - \widehat{\mu}_i^{(k)}\right). \end{aligned}$$

This formulation of the Fisher scoring algorithm suggests that the sequence of estimates $\{\widehat{\boldsymbol{\beta}}^{(k)}\}$ is simply a sequence of weighted least squares estimates; that is, $\widehat{\boldsymbol{\beta}}^{(k+1)}$ minimizes the weighted least squares objective function

$$\sum_{i=1}^{n} w_i^{(k)} (z_i^{(k)} - \boldsymbol{x}_i^T \boldsymbol{\beta})^2$$

over all $\boldsymbol{\beta}$; for this reason, the Fisher scoring algorithm described above is often called iteratively reweighted least squares. One attractive feature of this algorithm is that it can be implemented quite easily using a weighted least squares algorithm.

8.6 Quasi-Likelihood models

We noted earlier that the estimating equations defining MLE of $\boldsymbol{\beta}$ depended on the distribution of the Y_i's only on the link function g and the variance function V (where $\text{Var}(Y_i) = \phi V(\mu_i)$). This fact suggests that it may be possible to estimate $\boldsymbol{\beta}$ in the model

$$g(\mu_i) = \boldsymbol{x}_i^T \boldsymbol{\beta} \quad (i = 1, \cdots, n)$$

merely by specifying the relationship between the variance and mean of the Y_i's.

Suppose that $Y_1, \cdots, Y_n$ are independent random variables with $\mu_i = E(Y_i)$ where $g(\mu_i) = \boldsymbol{x}_i^T \boldsymbol{\beta}$ and $\text{Var}(Y_i) = \phi V(\mu_i)$ for $i = 1, \cdots, n$; the variance function $V(\mu)$ is a known function and ϕ is a "dispersion" parameter whose value may be unknown. Note that we are not specifying the distribution of the Y_i's, only the relationship between the mean and variance.

To estimate $\boldsymbol{\beta}$, we will introduce the quasi-likelihood function (Wedderburn, 1974). Define a function $\psi(\mu; y)$ so that

$$\frac{\partial}{\partial \mu} \psi(\mu; y) = \frac{(y - \mu)}{V(\mu)}.$$

Then given $\boldsymbol{Y} = \boldsymbol{y}$, we define the quasi-likelihood function (or perhaps more correctly, the quasi-log-likelihood function) by

$$\mathcal{Q}(\boldsymbol{\beta}) = \sum_{i=1}^{n} \psi(\mu_i; y_i)$$

where, of course, μ_i depends on $\boldsymbol{\beta}$ via the relationship $g(\mu_i) = \boldsymbol{x}_i^T \boldsymbol{\beta}$.

Parameter estimates may now be obtained by maximizing the quasi-likelihood function. Taking partial derivatives of $\mathcal{Q}(\boldsymbol{\beta})$ with respect to $\boldsymbol{\beta}$, we get the same estimating equations for $\widehat{\boldsymbol{\beta}}$ as before:

$$\sum_{i=1}^{n} \frac{Y_i - \widehat{\mu}_i}{g'(\widehat{\mu}_i) V(\widehat{\mu}_i)} \boldsymbol{x}_i = \boldsymbol{0}.$$

The estimate $\widehat{\boldsymbol{\beta}}$ can be computed numerically as before by using the reweighted least squares algorithm given above.

EXAMPLE 8.10: Suppose that $V(\mu) = \mu$. Then the function $\psi(\mu; y)$ satisfies

$$\frac{\partial}{\partial \mu}\psi(\mu; y) = \frac{(y-\mu)}{\mu},$$

which gives

$$\psi(\mu; y) = y\ln(\mu) - \mu + c(y)$$

(where $c(y)$ is an arbitrary function of y). Note that the logarithm of the Poisson frequency function with mean μ is $y\ln(\mu)-\mu-\ln(y!)$, which has the form given above; this makes sense since the mean and the variance are equal for the Poisson distribution. ◇

EXAMPLE 8.11: Suppose that $V(\mu) = \mu^2(1-\mu)^2$ for $0 < \mu < 1$. Then

$$\frac{\partial}{\partial \mu}\psi(\mu; y) = \frac{(y-\mu)}{\mu^2(1-\mu)^2},$$

which gives

$$\psi(\mu; y) = (2y-1)\ln\left(\frac{\mu}{1-\mu}\right) - \frac{y}{\mu} - \frac{1-y}{1-\mu} + c(y).$$

This particular variance function can be useful when the data are continuous proportions; however, the function $\psi(\mu; y)$ is not equal to the logarithm of any known density or frequency function. See Wedderburn (1974) for an application that uses this variance function. ◇

EXAMPLE 8.12: Suppose that $V(\mu) = \mu + \alpha\mu^2$ where $\alpha > 0$ is a known constant. In this case,

$$\psi(\mu; y) = y\ln(\mu) - (y + 1/\alpha)\ln(1 + \alpha\mu) + c(y).$$

The logarithm of the Negative Binomial frequency function

$$f(y; \mu) = \frac{\Gamma(y + 1/\alpha)}{y!\Gamma(1/\alpha)} \frac{(\alpha\mu)^y}{(1+\alpha\mu)^{y+1/\alpha}} \quad \text{for } y = 0, 1, 2, \cdots$$

has the same form as $\psi(\mu; y)$. However, typically α is an unknown parameter and so quasi-likelihood estimation is not equivalent to maximum likelihood estimation in this case. More details of regression models for Negative Binomial data can be found in Dean and Lawless (1989); these models are often used as an alternative to Poisson regression models. ◇

It should be noted that quasi-likelihood will not generally provide the most efficient estimation of $\boldsymbol{\beta}$ unless the quasi-likelihood function is the true log-likelihood function of the data. However, more or less correct inference for $\boldsymbol{\beta}$ can still be carried out using the quasi-likelihood estimator of $\boldsymbol{\beta}$. In particular, $\widehat{\boldsymbol{\beta}}$ will be approximately multivariate Normal with mean vector $\boldsymbol{\beta}$ and variance-covariance matrix $\phi(X^T W(\boldsymbol{\mu})X)^{-1}$ where X and $W(\boldsymbol{\mu})$ are as previously defined.

The following example illustrates how we can obtain more efficient estimates if the error distribution is known.

EXAMPLE 8.13: Consider an ordinary linear regression model where $\mu_i = \boldsymbol{x}_i^T \boldsymbol{\beta}$ and $V(\mu_i)$ is constant. It is easy to see that maximum quasi-likelihood estimator of $\boldsymbol{\beta}$ is simply the least squares estimator. However, suppose that the density of Y_i were given by

$$f(y; \mu_i, \sigma) = \frac{\exp[(y-\mu_i)/\sigma]}{\sigma\left(1+\exp[(y-\mu_i)/\sigma]\right)^2}.$$

Then the MLEs of $\boldsymbol{\beta}$ and σ satisfy the equations

$$\sum_{i=1}^{n} \left(1 - \frac{2\exp[(Y_i - \boldsymbol{x}_i^T\widehat{\boldsymbol{\beta}})/\widehat{\sigma})]}{1+\exp[(Y_i - \boldsymbol{x}_i^T\widehat{\boldsymbol{\beta}})/\widehat{\sigma})]}\right) \boldsymbol{x}_i = \boldsymbol{0}$$

and

$$\frac{n}{\widehat{\sigma}} + \sum_{i=1}^{n} \left(1 - \frac{2\exp[(Y_i - \boldsymbol{x}_i^T\widehat{\boldsymbol{\beta}})/\widehat{\sigma})]}{1+\exp[(Y_i - \boldsymbol{x}_i^T\widehat{\boldsymbol{\beta}})/\widehat{\sigma})]}\right) (Y_i - \boldsymbol{x}_i^T\widehat{\boldsymbol{\beta}}) = 0.$$

In this case, the MLE $\widehat{\boldsymbol{\beta}}$ will be approximately multivariate Normal with mean vector $\boldsymbol{\beta}$ and variance-covariance matrix $3\sigma^2(X^TX)^{-1}$ while the least squares estimator will be approximately multivariate Normal with mean vector $\boldsymbol{\beta}$ and variance-covariance matrix $3.2899\sigma^2(X^TX)^{-1}$ since $\text{Var}(Y_i) = \frac{1}{3}\pi^2\sigma^2 = 3.2899\sigma^2$. ◇

Estimation of the dispersion parameter

So far, we have not discussed the estimation of the dispersion parameter ϕ if it is unknown. When the distribution of the Y_i's is specified then ϕ can be estimated using maximum likelihood estimation. In the quasi-likelihood setting, if $\text{Var}(Y_i) = \phi V(\mu_i)$, there are several approaches to estimating ϕ; perhaps the simplest approach is to use the fact that

$$\phi = \frac{E[(Y_i - \mu_i)^2]}{V(\mu_i)}$$

for $i = 1, \cdots, n$. This suggests that ϕ can be estimated by the method of moments estimator

$$\widehat{\phi} = \frac{1}{n^*} \sum_{i=1}^{n} \frac{(Y_i - \widehat{\mu}_i)^2}{V(\widehat{\mu}_i)}$$

where $\widehat{\mu}_i$ depends on the maximum likelihood or maximum quasi-likelihood estimator $\widehat{\boldsymbol{\beta}}$ and n^* is either the sample size n or the "residual degrees of freedom", $n - p - 1$. An alternative estimator of ϕ is

$$\widetilde{\phi} = \frac{1}{n^*} \sum_{i=1}^{n} \left(\psi(Y_i; Y_i) - \psi(\widehat{\mu}_i; Y_i) \right).$$

While the origin of $\widetilde{\phi}$ as an estimator of ϕ is far from clear, the connection between ϕ and $\widetilde{\phi}$ can be seen by making a Taylor series expansion of $\psi(\mu; Y_i)$ around $\mu = \widehat{\mu}_i$:

$$\begin{aligned} \psi(Y_i; Y_i) &= \psi(\widehat{\mu}_i; Y_i) + \frac{\partial}{\partial \mu} \psi(\widehat{\mu}_i; Y_i)(Y_i - \widehat{\mu}_i) + \cdots \\ &= \psi(\widehat{\mu}_i; Y_i) + \frac{(Y_i - \widehat{\mu}_i)^2}{V(\widehat{\mu}_i)} + \cdots. \end{aligned}$$

Thus

$$\begin{aligned} \widetilde{\phi} &= \frac{1}{n^*} \sum_{i=1}^{n} \left(\psi(Y_i; Y_i) - \psi(\widehat{\mu}_i; Y_i) \right) \\ &= \frac{1}{n^*} \sum_{i=1}^{n} \frac{(Y_i - \widehat{\mu}_i)^2}{V(\widehat{\mu}_i)} + \cdots \\ &= \widehat{\phi} + \cdots, \end{aligned}$$

which suggests that $\widehat{\phi} \approx \widetilde{\phi}$ provided that the remainder terms in the Taylor series expansion are negligible.

8.7 Problems and complements

8.1: Suppose that $\boldsymbol{Y} = X\boldsymbol{\beta} + \boldsymbol{\varepsilon}$ where $\boldsymbol{\varepsilon} \sim N_n(\boldsymbol{0}, \sigma^2 I)$ and X is $n \times (p+1)$. Let $\widehat{\boldsymbol{\beta}}$ be the least squares estimator of $\boldsymbol{\beta}$.

(a) Show that

$$S^2 = \frac{\|\boldsymbol{Y} - X\widehat{\boldsymbol{\beta}}\|^2}{n-p-1}$$

is an unbiased estimator of σ^2.

(b) Suppose that the random variables in $\boldsymbol{\varepsilon}$ are uncorrelated with common variance σ^2. Show that S^2 is an unbiased estimator of σ^2.

(c) Suppose that $(X^TX)^{-1}$ can be written as

$$(X^TX)^{-1} = \begin{pmatrix} c_{00} & c_{01} & c_{02} & \cdots & c_{0p} \\ c_{10} & c_{11} & c_{12} & \cdots & c_{1p} \\ c_{20} & c_{21} & c_{22} & \cdots & c_{2p} \\ \vdots & \vdots & \ddots & \ddots & \vdots \\ c_{p0} & c_{p1} & c_{p2} & \cdots & c_{pp} \end{pmatrix}.$$

Show that

$$\frac{\widehat{\beta}_j - \beta_j}{S\sqrt{c_{jj}}} \sim \mathcal{T}(n-p-1)$$

for $j = 0, 1, \cdots, p$.

8.2: Suppose that $\boldsymbol{Y} = X\boldsymbol{\beta} + \boldsymbol{\varepsilon}$ where $\boldsymbol{\varepsilon} \sim N_n(\boldsymbol{0}, \sigma^2 C)$ for some known non-singular matrix C.

(a) Show that the MLE of $\boldsymbol{\beta}$ is

$$\widehat{\boldsymbol{\beta}} = (X^TC^{-1}X)^{-1}X^TC^{-1}\boldsymbol{Y}.$$

(This estimator is called the generalized least squares estimator of $\boldsymbol{\beta}$.)

(b) Find the (exact) distribution of $\widehat{\boldsymbol{\beta}}$.

(c) Find the distribution of the (ordinary) least squares estimator of $\boldsymbol{\beta}$ for this model. Compare its distribution to that of the generalized least squares estimator.

8.3: Consider the linear model

$$Y_i = \beta_0 + \beta_1 x_{i1} + \cdots + \beta_p x_{ip} + \varepsilon_i \quad (i = 1, \cdots, n)$$

where for $j = 1, \cdots, p$, we have

$$\sum_{i=1}^{n} x_{ji} = 0.$$

(a) Show that the least squares estimator of β_0 is $\widehat{\beta}_0 = \bar{Y}$.

(b) Suppose that, in addition, we have

$$\sum_{i=1}^{n} x_{ji} x_{ki} = 0$$

for $1 \leq j \neq k \leq p$. Show that the least squares estimator of β_j is

$$\widehat{\beta}_j = \frac{\sum_{i=1}^{n} x_{ji} Y_i}{\sum_{i=1}^{n} x_{ji}^2}.$$

8.4: Consider the linear model

$$Y_{1i} = \alpha_1 + \beta_1 x_{1i} + \varepsilon_{1i} \quad \text{for } i = 1, \cdots, n$$

and

$$Y_{2i} = \alpha_2 + \beta_2 x_{2i} + \varepsilon_{2i} \quad \text{for } i = 1, \cdots, n$$

where the ϵ_{ji}'s are independent $N(0, \sigma^2)$ random variables. Define the response vector $\boldsymbol{Y} = (Y_{11}, \cdots, Y_{1n}, Y_{21}, \cdots, Y_{2n})^T$ and the parameter vector $\gamma = (\alpha_1, \alpha_2, \beta_1, \beta_2)^T$.

(a) The linear model above can be written as $\boldsymbol{Y} = X\gamma + \varepsilon$. Give the design matrix X for this model.

(b) Suppose we wish to test the null hypothesis $H_0 : \beta_1 = \beta_2$ versus the alternatve hypothesis $H_1 : \beta_1 \neq \beta_2$. Let RSS_0 and RSS_1 be the residual sums of squares under H_0 and H_1 respectively and give the F statistic for testing H_0 in terms of RSS_0 and RSS_1. What is the distribution of this statistic under H_0? What is the distribution of the test statistic under H_1?

8.5: Suppose that $\boldsymbol{Y} = \boldsymbol{\theta} + \varepsilon$ where $\boldsymbol{\theta}$ satisfies $A\boldsymbol{\theta} = \boldsymbol{0}$ for some known $q \times n$ matrix A having rank q. Define $\widehat{\boldsymbol{\theta}}$ to minimize $\|\boldsymbol{Y} - \boldsymbol{\theta}\|^2$ subject to $A\boldsymbol{\theta} = \boldsymbol{0}$. Show that

$$\widehat{\boldsymbol{\theta}} = (I - A^T(AA^T)^{-1}A)\boldsymbol{Y}.$$

8.6: Consider the linear model

$$\boldsymbol{Y} = X_1\boldsymbol{\beta} + X_2\gamma + \varepsilon$$

where $\varepsilon \sim N_n(\boldsymbol{0}, \sigma^2 I)$; we will also assume that the columns of X_2 are linearly independent of the columns of X_1 to guarantee identifiability of the model. Suppose that $\boldsymbol{\beta}$ is estimated using only X_1:

$$\widehat{\boldsymbol{\beta}} = X_1(X_1^T X_1)^{-1} X_1^T \boldsymbol{Y}.$$

(a) Show that

$$E(\widehat{\boldsymbol{\beta}}) = \boldsymbol{\beta} + (X_1^T X_1)^{-1} X_1^T X_2 \boldsymbol{\gamma}.$$

When would $\widehat{\boldsymbol{\beta}}$ be unbiased?

(b) Assume X_1 has q columns and define

$$S^2 = \frac{\boldsymbol{Y}^T (I - H_1) \boldsymbol{Y}}{n - q}.$$

Show that $E(S^2) \geq \sigma^2$ with equality if, and only if, $\boldsymbol{\gamma} = \mathbf{0}$.

8.7: (a) Suppose that $U \sim \chi^2(1; \theta_1^2)$ and $V \sim \chi^2(1; \theta_2^2)$ where $\theta_1^2 > \theta_2^2$. Show that U is stochastically greater than V. (Hint: Let $X \sim N(0,1)$ and show that $P(|X + \theta| > x)$ is an increasing function of θ for each x.)

(b) Suppose that $U_n \sim \chi^2(n; \theta_1^2)$ and $V_n \sim \chi^2(1; \theta_2^2)$ where $\theta_1^2 > \theta_2^2$. Show that U_n is stochastically greater than V_n.

8.8: Consider the linear model $\boldsymbol{Y} = X\boldsymbol{\beta} + \varepsilon$ and suppose we want to test the null hypothesis

$$H_0 : A\boldsymbol{\beta} = \boldsymbol{c}$$

where A is an $s \times (p+1)$ matrix with rank s and $\boldsymbol{c}$ is a vector of length s.

(a) Show that the least squares estimator of $\boldsymbol{\beta}$ under H_0 is given by

$$\widehat{\boldsymbol{\beta}}_r = \widehat{\boldsymbol{\beta}} + (X^T X)^{-1} A^T \left[A (X^T X)^{-1} A^T \right]^{-1} (\boldsymbol{c} - A\widehat{\boldsymbol{\beta}})$$

where $\widehat{\boldsymbol{\beta}}$ is the least squares estimator under the full model. (Hint: Let $\boldsymbol{\lambda}$ be a vector of Lagrange multipliers and minimize the objective function

$$g(\boldsymbol{\beta}, \boldsymbol{\lambda}) = \|\boldsymbol{Y} - X\boldsymbol{\beta}\|^2 + \boldsymbol{\lambda}^T (A\boldsymbol{\beta} - \boldsymbol{c})$$

over both $\boldsymbol{\beta}$ and $\boldsymbol{\lambda}$.

(b) Let RSS_r be the residual sum of squares under H_0. Show that

$$\mathrm{RSS}_r - \mathrm{RSS} = (A\widehat{\boldsymbol{\beta}} - \boldsymbol{c})^T \left[A (X^T X)^{-1} A^T \right]^{-1} (A\widehat{\boldsymbol{\beta}} - \boldsymbol{c}).$$

(c) Show that

$$\frac{1}{\sigma^2} (\mathrm{RSS}_r - \mathrm{RSS}) \sim \chi^2(s)$$

when H_0 is true. (Hint: Note that when H_0 is true $A\widehat{\boldsymbol{\beta}} \sim N_s(\boldsymbol{c}, \sigma^2 A(X^T X)^{-1} A^T)$.)

(d) Show that

$$F = \frac{(\text{RSS}_r - \text{RSS})/s}{\text{RSS}/(n-p-1)} \sim \mathcal{F}(s, n-p-1)$$

when H_0 is true.

(e) Suppose that H_0 is false so that $A\boldsymbol{\beta} = \boldsymbol{a} \neq \boldsymbol{c}$. Find the distribution of the F statistic in part (d). (Hint: Note that $A\widehat{\boldsymbol{\beta}} - \boldsymbol{c} \sim N_s(\boldsymbol{a} - \boldsymbol{c}, \sigma^2 A(X^T X)^{-1} A^T)$.)

8.9: Suppose that $\boldsymbol{Y} = X\boldsymbol{\beta} + \boldsymbol{\varepsilon}$ where $\boldsymbol{\varepsilon} \sim N_n(\boldsymbol{0}, \sigma^2 I)$ and define the ridge estimator (Hoerl and Kennard, 1970) $\widehat{\boldsymbol{\beta}}_\lambda$ to minimize

$$\|\boldsymbol{Y} - X\boldsymbol{\beta}\|^2 + \lambda\|\boldsymbol{\beta}\|^2$$

for some $\lambda > 0$. (Typically in practice, the columns of X are centred and scaled, and $\boldsymbol{Y}$ is centred.)

(a) Show that

$$\begin{aligned} \widehat{\boldsymbol{\beta}}_\lambda &= (X^T X + \lambda I)^{-1} X^T \boldsymbol{Y} \\ &= (I + \lambda(X^T X)^{-1})^{-1} \widehat{\boldsymbol{\beta}} \end{aligned}$$

where $\widehat{\boldsymbol{\beta}}$ is the least squares estimator of $\boldsymbol{\beta}$. Conclude that $\widehat{\boldsymbol{\beta}}_\lambda$ is a biased estimator of $\boldsymbol{\beta}$.

(b) Consider estimating $\theta = \boldsymbol{a}^T\boldsymbol{\beta}$ for some known $\boldsymbol{a} \neq \boldsymbol{0}$. Show that

$$\text{MSE}_\theta(\boldsymbol{a}^T\widehat{\boldsymbol{\beta}}_\lambda) \le \text{MSE}_\theta(\boldsymbol{a}^T\widehat{\boldsymbol{\beta}})$$

for some $\lambda > 0$.

8.10: In this problem, we will sketch a proof of Theorem 8.6. We start by defining the function

$$Z_n(\boldsymbol{u}) = \sum_{i=1}^n \left[\rho(\varepsilon_i - \boldsymbol{x}_i^T A_n^{-1}\boldsymbol{u}) - \rho(\varepsilon_i)\right],$$

which is a convex function since $\rho(x)$ is convex. Note that $Z_n(\boldsymbol{u})$ is minimized at $u = \sqrt{n}(\widehat{\boldsymbol{\beta}}_n - \boldsymbol{\beta})$; thus if

$$(Z_n(\boldsymbol{u}_1), \cdots, Z_n(\boldsymbol{u}_k)) \to_d (Z(\boldsymbol{u}_1), \cdots, Z(\boldsymbol{u}_k))$$

where $Z(\boldsymbol{u})$ is uniquely minimized then $\sqrt{n}(\widehat{\boldsymbol{\beta}}_n - \boldsymbol{\beta})$ converges in distribution to the minimizer of Z (Davis *et al*, 1992).

(a) Let $v_{ni} = \boldsymbol{x}_i^T A_n^{-1}\boldsymbol{u}$ for a given $\boldsymbol{u}$. Show that

$$\begin{aligned} Z_n(\boldsymbol{u}) &= -\sum_{i=1}^n \boldsymbol{x}_i^T A_n^{-1}\boldsymbol{u}\,\psi(\varepsilon_i) \\ &\quad + \sum_{i=1}^n \int_0^{v_{ni}} [\psi(\varepsilon_i) - \psi(\varepsilon_i - t)]\,dt \\ &= Z_{n1}(\boldsymbol{u}) + Z_{n2}(\boldsymbol{u}). \end{aligned}$$

(b) Show that

$$(Z_{n1}(\boldsymbol{u}_1), \cdots, Z_{n1}(\boldsymbol{u}_k)) \to_d \left(\boldsymbol{u}_1^T \boldsymbol{W}, \cdots, \boldsymbol{u}_k^T \boldsymbol{W}\right)$$

where

$$\boldsymbol{W} \sim N_{p+1}\left(\boldsymbol{0}, E[\psi^2(\varepsilon_1)]I\right).$$

(c) Show that as $n \to \infty$,

$$\begin{aligned} E[Z_{n2}(\boldsymbol{u})] &\to \frac{\lambda(0)}{2}\boldsymbol{u}^T\boldsymbol{u} \\ \text{and}\quad \mathrm{Var}[Z_{n2}(\boldsymbol{u})] &\to 0. \end{aligned}$$

(d) Deduce from parts (b) and (c) that

$$(Z_n(\boldsymbol{u}_1), \cdots, Z_n(\boldsymbol{u}_k)) \to_d (Z(\boldsymbol{u}_1), \cdots, Z(\boldsymbol{u}_k))$$

where $Z(\boldsymbol{u}) = \boldsymbol{u}^T\boldsymbol{W} + \lambda(0)\boldsymbol{u}^T\boldsymbol{u}/2$.

(e) Show that $Z(\boldsymbol{u})$ in (d) is minimized at $\boldsymbol{u} = -\boldsymbol{W}/\lambda(0)$.

8.11: Suppose that $Y_i = \boldsymbol{x}_i^T\boldsymbol{\beta} + \varepsilon_i$ $(i = 1, \cdots, n)$ where the ε_i's are i.i.d. with mean 0 and finite variance. Consider the F statistic (call it F_n) for testing

$$H_0 : \beta_{r+1} = \cdots = \beta_p = 0$$

where $\boldsymbol{\beta} = (\beta_0, \cdots, \beta_p)^T$.

(a) Under H_0 and assuming the conditions of Theorem 8.5 on the $\boldsymbol{x}_i$'s, show that

$$(p - r)F_n \to_d \chi^2(p - r).$$

(b) If H_0 is not true, what happens to $(p - r)F_n$ as $n \to \infty$? (Hint: Look at the "non-centrality" of F_n when H_0 is not true.)

8.12: Consider the linear regression model

$$Y_i = \boldsymbol{x}_i^T\boldsymbol{\beta} + \varepsilon_i \quad (i = 1, \cdots, n)$$

where $\varepsilon_1, \cdots, \varepsilon_n$ are i.i.d. random variables with density function

$$f(x) = \frac{\lambda}{2} \exp(-\lambda|x|)$$

and $\lambda > 0$ is an unknown parameter. (This distribution is sometimes called the Laplace distribution.)

(a) Show that the MLE of $\boldsymbol{\beta}$ minimizes the function

$$g(\boldsymbol{\beta}) = \sum_{i=1}^{n} |Y_i - \boldsymbol{x}_i^T \boldsymbol{\beta}|.$$

This estimator is called the L_1 estimator of $\boldsymbol{\beta}$.

(b) Because the absolute value function is not differentiable at 0, numerical methods based on derivatives will not work for this problem; however, linear programming algorithms (such as the simplex algorithm or interior point algorithms) can be used to find estimate of $\boldsymbol{\beta}$. Consider the following linear programming problem:

$$\text{minimize} \quad \sum_{i=1}^{n} (e_i^+ + e_i^-)$$

subject to the constraints

$$\begin{aligned} \boldsymbol{x}_i^T \boldsymbol{\beta} + e_i^+ - e_i^- &= Y_i \quad \text{for } i = 1, \cdots, n \\ e_i^+ &\geq 0 \quad \text{for } i = 1, \cdots, n \\ e_i^- &\geq 0 \quad \text{for } i = 1, \cdots, n \end{aligned}$$

(The unknowns in this problem are $\boldsymbol{\beta}$ as well as the e_i^+'s and the e_i^-'s.) Show that if $\widehat{\boldsymbol{\beta}}$ is a solution to this linear programming problem then it also minimizes $g(\boldsymbol{\beta})$ in part (a). (See Portnoy and Koenker (1997) for more discussion of computational methods for L_1 estimation.)

8.13: Consider the linear regression model

$$Y_i = \boldsymbol{x}_i^T \boldsymbol{\beta} + \varepsilon_i \quad (i = 1, \cdots, n)$$

where $\varepsilon_1, \cdots, \varepsilon_n$ are i.i.d. Exponential random variables with unknown parameter λ.

(a) Show that the density function of Y_i is

$$f_i(y) = \lambda \exp[-\lambda(y - \boldsymbol{x}_i^T \boldsymbol{\beta})] \quad \text{for } y \geq \boldsymbol{x}_i^T \boldsymbol{\beta}.$$

(b) Show that the MLE of $\boldsymbol{\beta}$ for this model maximizes the function

$$g(\boldsymbol{u}) = \sum_{i=1}^{n} \boldsymbol{x}_i^T \boldsymbol{u}$$

subject to the constraints

$$Y_i \geq \boldsymbol{x}_i^T \boldsymbol{u} \quad \text{for } i = 1, \cdots, n.$$

(c) Suppose that $Y_i = \beta x_i + \varepsilon_i$ $(i = 1, \cdots, n)$ where $\varepsilon_1, \cdots, \varepsilon_n$ i.i.d. Exponential random variables with parameter λ and $x_i > 0$ for all i. If $\widehat{\beta}_n$ is the MLE of β, show that $\widehat{\beta}_n - \beta$ has an Exponential distribution with parameter $\lambda \sum_{i=1}^n x_i$. (Hint: Show that $\widehat{\beta}_n = \min_{1 \leq i \leq n} Y_i / x_i$.)

8.14: Suppose that Y has a density or frequency function of the form

$$f(y; \theta, \phi) = \exp\left[\frac{\theta y - b(\theta)}{\phi} + c(y, \phi)\right]$$

for $y \in A$, which is independent of the parameters θ and ϕ.

(a) Show that $E_\theta(Y) = b'(\theta)$.

(b) Show that $\text{Var}_\theta(Y) = \phi b''(\theta)$.

8.15: Suppose that Y has a density or frequency function of the form

$$f(y; \theta, \phi) = \exp\left[\theta y - \frac{b(\theta)}{\phi} + c(y, \phi)\right]$$

for $y \in A$, which is independent of the parameters θ and ϕ. This is an alternative to the general family of distributions considered in Problem 8.14. and is particularly appropriate for discrete distributions.

(a) Show that the Negative Binomial distribution of Example 8.12 has this form.

(b) Show that $E_\theta(Y) = \phi^{-1} b'(\theta)$ and $\text{Var}_\theta(Y) = \phi^{-1} b''(\theta)$.

8.16: The Inverse Gaussian distribution is a continuous distribution whose density function is

$$f(y; \delta, \sigma) = \frac{1}{\sigma\sqrt{2\pi y^3}} \exp\left[\frac{1}{2\sigma^2 y}(1 - \delta y)^2\right]$$

for $y > 0$.

(a) If Y has the density function above, show that $E(Y) = 1/\delta$ and $\text{Var}(Y) = \sigma^2/\delta^3$. (Thus $\text{Var}(Y) = \sigma^2[E(Y)]^3$.)

(b) Suppose that $Y_1, \cdots, Y_n$ are independent Inverse Gaussian random variables with $E(Y_i) = \mu_i$ and $\ln(\mu_i) = \boldsymbol{x}_i^T\boldsymbol{\beta}$ where $\boldsymbol{\beta}$ is unknown. Outline the iteratively reweighted least squares algorithm for computing maximum likelihood estimate of $\boldsymbol{\beta}$.

(c) Find an expression for the MLE of σ^2.

8.17: Lambert (1992) describes an approach to regression modelling of count data using a zero-inflated Poisson distribution. That is, the response variables $\{Y_i\}$ are nonnegative integer-valued random variables with the frequency function of Y_i given by

$$P(Y_i = y) = \begin{cases} \theta_i + (1-\theta_i)\exp(-\lambda_i) & \text{for } y = 0 \\ (1-\theta_i)\exp(-\lambda_i)\lambda_i^y/y! & \text{for } y = 1, 2, \cdots \end{cases}$$

where θ_i and λ_i depend on some covariates; in particular, it is assumed that

$$\begin{aligned} \ln\left(\frac{\theta_i}{1-\theta_i}\right) &= \boldsymbol{x}_i^T\boldsymbol{\beta} \\ \ln(\lambda_i) &= \boldsymbol{x}_i^T\boldsymbol{\phi}. \end{aligned}$$

where $\boldsymbol{x}_i$ $(i = 1, \cdots, n)$ are covariates and $\boldsymbol{\beta}$, $\boldsymbol{\phi}$ are vectors of unknown parameters.

(a) The zero-inflated Poisson model can viewed as a mixture of a Poisson distribution and a distribution concentrated at 0. That is, let Z_i be a Bernoulli random variable with $P(Z_i = 1) = \theta_i$ such that $P(Y_i = 0|Z_i = 0) = 1$ and given $Z_i = 1$, Y_i is Poisson distributed with mean λ_i. Show that

$$P(Z_i = 0|Y_i = y) = \begin{cases} \theta_i/[\theta_i + (1-\theta_i)\exp(-\lambda_i)] & \text{for } y = 0 \\ 0 & \text{for } y \geq 1 \end{cases}$$

(b) Suppose that we could observe $(Y_1, Z_1), \cdots, (Y_n, Z_n)$ where the Z_i's are defined in part (a). Show that the MLE of $\boldsymbol{\beta}$ depends only on the Z_i's

(c) Use the "complete data" likelihood in part (b) to describe an EM algorithm for computing maximum likelihood estimates of $\boldsymbol{\beta}$ and $\boldsymbol{\phi}$.

(d) In the spirit of the zero-inflated Poisson model, consider the following simple zero-inflated Binomial model: for $i = 1, \cdots, n$, $Y_1, \cdots, Y_n$ are independent random variables with

$$P(Y_i = 0) \quad = \quad \phi + (1-\phi)(1-\theta_i)^m$$

Table 8.1 *Data for Problem 8.17; for each observation* $m = 6$.

x_i	y_i	x_i	y_i	x_i	y_i	x_i	y_i
0.3	0	0.6	0	1.0	0	1.1	0
2.2	1	2.2	0	2.4	0	2.5	0
3.0	4	3.2	0	3.4	4	5.8	5
6.2	0	6.5	5	7.1	4	7.6	6
7.7	4	8.2	4	8.6	4	9.8	0

$$P(Y_i = y) \quad = \quad (1-\theta)\binom{m}{y}\theta_i^y(1-\theta_i)^{m-y} \quad \text{for } y = 1, \cdots, m$$

where $0 < \phi < 1$ and

$$\ln\left(\frac{\theta_i}{1-\theta_i}\right) = \beta_0 + \beta_1 x_i$$

for some covariates $x_1, \cdots, x_n$.

Derive an EM algorithm for estimating ϕ and $\boldsymbol{\beta}$ and use it to estimate the parameters for the data in Table 8.1; for each observation, $m = 6$. with $m = 6$:

(e) Carry out a likelihood ratio test for $H_0 : \beta_1 = 0$ versus $H_1 : \beta_1 \neq 0$. (Assume that the standard χ^2 approximation can be applied.)

8.18: Suppose that $Y_1, \cdots, Y_n$ are independent Bernoulli random variables with parameters $\theta_i = P(Y_i = 1)$ $(i = 1, \cdots, n)$ and assume the logistic regression model (see Example 8.8)

$$\ln\left(\frac{\theta_i}{1-\theta_i}\right) = \beta_0 + \beta_1 x_{i1} + \cdots + \beta_p x_{ip}.$$

In many epidemiological studies, the data are sampled retrospectively; that is, there is an additional (unobservable) variable that determines whether or not an individual is sampled and the sampling probability may depend on the value of Y_i. Let Z_i be a Bernoulli random variable with $Z_i = 1$ if individual i is sampled and 0 otherwise and suppose that

$$P(Z_i = 1 | Y_i = 0) = p_0 \quad \text{and} \quad P(Z_i = 1 | Y_i = 1) = p_1$$

where p_0 and p_1 are constant for $i = 1, \cdots, n$.

(a) Assuming the logistic regression model, show that

$$P(Y_i = 1|Z_i = 1) = \frac{p_1\theta_i}{p_1\theta_i + p_0(1-\theta_i)}.$$

(b) Let $\phi_i = P(Y_i = 1|Z_i = 1)$. Show that

$$\ln\left(\frac{\phi_i}{1-\phi_i}\right) = \beta_0^* + \beta_1 x_{i1} + \cdots + \beta_p x_{ip}$$

where $\beta_0^* = \beta_0 + \ln(p_1/p_0)$. This indicates that $\beta_1, \cdots, \beta_p$ may still be estimated under retrospective sampling.

8.19: Consider finding a quasi-likelihood function based on the variance function $V(\mu) = \mu^r$ for some specified $r > 0$.

(a) Find the function $\psi(\mu; y)$ for $V(\mu)$.

(b) Show that

$$\psi(\mu; y) = \frac{\partial}{\partial\mu} \ln f(y; \mu)$$

for some density or frequency function $f(y; \mu)$ when $r = 1, 2, 3$.

8.20: The Multinomial logit model is sometimes useful when a response is discrete-valued and takes more than two possible values. Given a covariate vector $\boldsymbol{x}_i$, we assume that $\boldsymbol{Y}_i$ has the following Multinomial distribution (see Problem 2.28):

$$P_\theta(\boldsymbol{Y}_i = \boldsymbol{y}) = \theta_{1i}^{y_1} \times \cdots \times \theta_{ri}^{y_r}$$

where $y_1 + \cdots + y_r = 1$ and

$$\ln\left(\frac{\theta_{ki}}{\theta_{1i}}\right) = \boldsymbol{x}_i^T \boldsymbol{\beta}_k \quad \text{for } k = 2, \cdots, r$$

(a) Show that

$$\theta_{ki} = \frac{\exp(\boldsymbol{x}_i^T\boldsymbol{\beta}_k)}{1 + \exp(\boldsymbol{x}_i^T\boldsymbol{\beta}_2) + \cdots + \exp(\boldsymbol{x}_i^T\boldsymbol{\beta}_r)}$$

for $k = 2, \cdots, r$ with

$$\theta_{1i} = \frac{1}{1 + \exp(\boldsymbol{x}_i^T\boldsymbol{\beta}_2) + \cdots + \exp(\boldsymbol{x}_i^T\boldsymbol{\beta}_r)}.$$

(b) Suppose that we model the components of $\boldsymbol{Y}_i$ as independent Poisson random variables with means $\lambda_{1i}, \cdots, \lambda_{r1}$ with

$$\ln(\lambda_{ki}) = \psi_i + \boldsymbol{x}_i^T\boldsymbol{\beta}_k \quad (k = 1, \cdots, r).$$

Show that the MLEs of $\boldsymbol{\beta}_1, \cdots, \boldsymbol{\beta}_r$ for the Poisson model are the same as those for the Multinomial logit model.

CHAPTER 9

Goodness-of-Fit

9.1 Introduction

Up to this point, we have assumed a particular statistical model for given observations. For example, given i.i.d. random variables $X_1, \cdots, X_n$, we have assumed that they have a common density (frequency) function $f(x;\theta)$ where only the parameter θ is unknown; that is, we have assumed the form of the distribution up to the value of an unknown parameter, which allows us to focus our energies on inference for this parameter.

However, there are many situation where we want to test whether a particular distribution "fits" our observed data. In some cases, these tests are informal; for example, in linear regression modeling, a statistician usually examines diagnostic plots (or other procedures) that allow him to determine whether the particular model assumptions (for example, normality and/or independence of the errors) are satisfied. However, in other cases where the form of the model has more significance, statisticians tend to rely more on formal hypothesis testing methods. We will concentrate on these methods in this chapter.

Roughly speaking, we can put goodness-of-fit tests into two classes. The first class of tests divides the range of the data into disjoint "bins"; the number of observations falling in each bin is compared to the expected number under the hypothesized distribution. These tests can be used for both discrete and continuous distribution although they are most natural for discrete distributions as the definition of the bins tends to be less arbitrary for discrete distributions than it is for continuous distributions.

The second class of tests are used almost exclusively for testing continuous distributions. For these tests, we compare an empirical distribution function of the data to the hypothesized distribution function; the test statistic for these tests is based either on some measure of distance being the two distributions or on a measure of "correlation" between the distributions.

9.2 Tests based on the Multinomial distribution

Suppose that we observe i.i.d. random variables $X_1, \cdots, X_n$ from an unknown distribution (which may be continuous or discrete). Our immediate goal is to find a test of the null hypothesis

$$H_0 : \text{the } X_i\text{'s have density (frequency) function } f(x; \theta)$$

where the alternative hypothesis is

$$H_1 : \text{the } X_i\text{'s have an arbitrary distribution.}$$

The parameter θ above may be known or unknown; if it is unknown, we will assume that it is finite dimensional.

Let S be a set such that $P(X_i \in S) = 1$. Then define disjoint sets $A_1, \cdots, A_k$ such that

$$S = \bigcup_{j=1}^{k} A_j.$$

The idea now is to count the number of X_i's that fall into each of the sets $A_1, \cdots, A_k$; we can define random variables

$$Y_j = \sum_{i=1}^{n} I(X_i \in A_j) \quad (j = 1, \cdots, k).$$

In practice, the number of sets k is chosen to be much sample than the sample size n. Because the X_i's are i.i.d., the random vector $\boldsymbol{Y} = (Y_1, \cdots, Y_k)$ has a Multinomial distribution (see Problem 2.28):

$$P(\boldsymbol{Y} = y) = \frac{n!}{y_1! \times \cdots \times y_k!} \prod_{j=1}^{k} \phi_j^{y_j}$$

where $\phi_j = P(X_i \in A_j)$ for $j = 1, \cdots, k$.

Now we assume that the X_i's have a density or frequency function $f(x; \boldsymbol{\theta})$ where $\boldsymbol{\theta} = (\theta_1, \cdots, \theta_p)$ is unknown. In this case, we can express the ϕ_j's more precisely in terms of the parameter $\boldsymbol{\theta}$:

$$\phi_j = P_\theta(X_i \in A_j) = p_j(\boldsymbol{\theta}) \quad (j = 1, \cdots, k).$$

Thus if we are interested in testing the null hypothesis that the X_i's have density (frequency) function $f(x; \boldsymbol{\theta})$ for some $\boldsymbol{\theta} \in \Theta$, we can express this null hypothesis in terms of the parameters of the Multinomial random vector $\boldsymbol{Y}$:

$$H_0 : \phi_j = p_j(\boldsymbol{\theta}) \quad (j = 1, \cdots, n) \quad \text{for some } \theta \in \Theta.$$

EXAMPLE 9.1: Let $X_1, \cdots, X_n$ be i.i.d. random variables with density function

$$f(x;\theta) = \theta x^{\theta-1} \quad \text{for } 0 \le x \le 1$$

where $\theta > 0$ is unknown. We can take $S = [0,1]$ and define (arbitrarily) $A_1 = [0,1/4]$, $A_2 = (1/4,1/2]$, $A_3 = (1/2,3/4]$, $A_4 = (3/4,1]$. Integrating the density, we get

$$p_j(\theta) = \int_{(j-1)/4}^{j/4} \theta x^{\theta-1}\,dx = \left(\frac{j}{4}\right)^\theta - \left(\frac{j-1}{4}\right)^\theta$$

for $j = 1, \cdots, 4$. Defining $Y_1, \cdots, Y_4$ to be the counts as above, we have

$$\begin{aligned} &P_\theta(Y_1 = y_1, \cdots, Y_4 = y_4) \\ &= \frac{n!}{y_1!y_2!y_3!y_4!}\prod_{j=1}^{4}\left[\left(\frac{j}{4}\right)^\theta - \left(\frac{j-1}{4}\right)^\theta\right]^{y_j}. \end{aligned}$$

Viewing this joint frequency function as a function of θ (given $bY = \boldsymbol{y}$), we have a likelihood function for θ based on the counts $Y_1, \cdots, Y_4$. ◇

At this point, we should note that in going from the original data $\boldsymbol{X} = (X_1, \cdots, X_n)$ to the counts $\boldsymbol{Y} = (Y_1, \cdots, Y_k)$, we do typically lose information in that $\boldsymbol{Y} = \boldsymbol{Y}(\boldsymbol{X})$ is typically neither sufficient for θ (when H_0 is true) nor in any nonparametric sense (when H_0 is false). This is an inevitable consequence of goodness-of-fit testing, particularly for continuous distributions. However, if $\boldsymbol{X}$ is discrete with $S = \{x_1, \cdots, x_k\}$ (a finite set) then we could define $A_j = \{x_j\}$ and $\boldsymbol{Y}$ is still sufficient for θ (although not necessarily minimal sufficient).

We noted in Example 9.1 that the joint frequency function of $\boldsymbol{Y}$ under the null hypothesis can be interpreted as a likelihood function for the unknown parameter θ. In fact, this is generally true if one regards the probabilities $\boldsymbol{\phi} = (\phi_1, \cdots, \phi_k)$ as unknown parameters. This suggests that we can carry out a likelihood ratio (LR) test of the null hypothesis by comparing the maximized H_0 likelihood (that is, maximizing over θ) to the "unrestricted" maximized likelihood (that is, maximizing over $\boldsymbol{\phi}$ subject to $\phi_j \ge 0$ (for $j = 1, \cdots, k$) and $\phi_1 + \cdots + \phi_k = 1$).

In principle, the LR test is quite simple. Given $\boldsymbol{Y} = \boldsymbol{y}$, define the log-likelihoods

$$\ln \mathcal{L}_1(\boldsymbol{\phi}) = \sum_{j=1}^{k} y_j \ln(\phi_j) + \ln\left(\frac{n!}{y_1! \times \cdots \times y_k!}\right)$$

and

$$\ln \mathcal{L}_0(\boldsymbol{\theta}) = \sum_{j=1}^{k} y_j \ln(p_j(\boldsymbol{\theta})) + \ln\left(\frac{n!}{y_1! \times \cdots \times y_k!}\right).$$

The maximum likelihood estimator of $\boldsymbol{\phi}$ is $\widehat{\boldsymbol{\phi}} = \boldsymbol{Y}/n$ while the maximum likelihood estimator of $\boldsymbol{\theta}$ satisfies the equation

$$\sum_{j=1}^{k} \frac{Y_j}{p_j(\widehat{\boldsymbol{\theta}})} p_j'(\widehat{\boldsymbol{\theta}}) = 0$$

where $p_j'(\boldsymbol{\theta})$ is the derivative or gradient of p_j with respect to $\boldsymbol{\theta}$. If Λ is the LR statistic, we have

$$\begin{aligned}\ln(\Lambda) &= \ln \mathcal{L}_1(\widehat{\boldsymbol{\phi}}) - \ln \mathcal{L}_0(\widehat{\boldsymbol{\theta}}) \\ &= \sum_{j=1}^{k} Y_j \ln\left(\frac{Y_j}{np_j(\widehat{\boldsymbol{\theta}})}\right);\end{aligned}$$

the null hypothesis will be rejected for large values of Λ (or equivalently $\ln(\Lambda)$). Of course, to implement the LR test, we need to know the distribution of Λ under the null hypothesis; given this null distribution, we can determine the "rejection region" for Λ to make it (at least approximately) an α level test or, alternatively, compute a p-value for the test.

In Chapter 7, we showed that, under i.i.d. sampling, we could approximate the null distribution of $2\ln(\Lambda)$ by a χ^2 distribution. It is quite easy to see that this theory also applies to this situation since $\boldsymbol{Y}$ can be viewed as a sum of n independent Multinomial random vectors. Given this fact, we can apply the standard asymptotic theory for LR tests; in doing so, we are assuming that n (the sample size) tends to infinity while k (the number of bins) as well as p (the dimension of $\boldsymbol{\theta}$) remain fixed.

The asymptotic theory for the MLE of $\boldsymbol{\theta}$ under the Multinomial model is quite simple if notationally somewhat cumbersome. If $\boldsymbol{Y}_n \sim \text{Mult}(n, \boldsymbol{p}(\boldsymbol{\theta}))$ then

$$\text{Cov}(\boldsymbol{Y}_n) = nC(\boldsymbol{\theta})$$

where $C(\boldsymbol{\theta})$ is a $k \times k$ matrix whose diagonal elements are $C_{ii}(\boldsymbol{\theta}) = p_i(\boldsymbol{\theta})(1 - p_i(\boldsymbol{\theta}))$ and whose off-diagonal elements are $C_{ij}(\boldsymbol{\theta}) = -p_i(\boldsymbol{\theta})p_j(\boldsymbol{\theta})$; $C(\boldsymbol{\theta})$ will have rank $k-1$ provided that $0 < p_j(\boldsymbol{\theta}) < 1$ for $j = 1, \cdots, k$. If $P(\boldsymbol{\theta})$ is a $k \times p$ matrix whose j-th row is the gradient of $\ln(p_j(\boldsymbol{\theta})$ then under mild regularity conditions, the MLE of $\boldsymbol{\theta}$, $\widehat{\boldsymbol{\theta}}_n$ satisfies

$$\sqrt{n}(\widehat{\boldsymbol{\theta}}_n - \boldsymbol{\theta}) \to_d N_p(\mathbf{0}, I(\boldsymbol{\theta})^{-1})$$

where

$$I(\boldsymbol{\theta}) = P(\boldsymbol{\theta})^T C(\boldsymbol{\theta}) P(\boldsymbol{\theta}).$$

The inverse of $I(\boldsymbol{\theta})$ exists provided that $p \leq k - 1$ and $P(\boldsymbol{\theta})$ has rank p.

THEOREM 9.1 *Suppose that* $\boldsymbol{Y}_n \sim$ *Mult*$(n, p_1(\boldsymbol{\theta}), \cdots, p_k(\boldsymbol{\theta}))$ *where* $p_1, \cdots, p_k$ *are twice continuously differentiable functions on* Θ, *an open subset of* R^p *where* $p \leq k - 2$. *Define*

$$\ln(\Lambda_n) = \sum_{j=1}^{k} Y_{nj} \ln\left(\frac{Y_{nj}}{np_j(\widehat{\boldsymbol{\theta}}_n)}\right).$$

If $\sqrt{n}(\widehat{\boldsymbol{\theta}}_n - \boldsymbol{\theta}) \to_d N_p(\mathbf{0}, I(\boldsymbol{\theta})^{-1})$ *then*

$$2\ln(\Lambda_n) \to_d \chi^2(k - 1 - p).$$

Theorem 9.1 can be viewed as a special case of Theorem 7.5 applied to Multinomial distributions. It is important to note that $\widehat{\boldsymbol{\theta}}_n$ is the MLE of $\boldsymbol{\theta}$ under the Multinomial model and *not* the MLE under the original model; however, the difference between these two MLEs is typically not large.

An alternative statistic to the LR statistic is Pearson's χ^2 statistic:

$$K_n^2 = \sum_{j=1}^{k} \frac{(Y_{nj} - np_j(\widehat{\boldsymbol{\theta}}_n))^2}{np_j(\widehat{\boldsymbol{\theta}}_n)}.$$

This statistic turns out to be closely related to the LR statistic. Note that K_n^2 compares the "observed" number of observations falling in bin j (Y_{nj}) to the (estimated) "expected" number of observations under the null hypothesis ($np_j(\widehat{\boldsymbol{\theta}}_n)$). As with the LR statistic, we will reject the null hypothesis for large values of K_n^2.

THEOREM 9.2 *Assume the same conditions on* $\boldsymbol{Y}_n$ *and* $\widehat{\boldsymbol{\theta}}_n$ *as in Theorem 9.1. Then*

$$K_n^2 - 2\ln(\Lambda_n) \to_p 0$$

and so

$$K_n^2 \to_d \chi^2(k-1-p).$$

Proof. It follows from the CLT and the asymptotic normality of the MLE $\widehat{\boldsymbol{\theta}}_n$ that for each j,

$$\sqrt{n}\left(\frac{Y_{nj}}{n} - p_j(\boldsymbol{\theta})\right) \to_d N(0, p_j(\boldsymbol{\theta})(1-p_j(\boldsymbol{\theta})))$$

$$\sqrt{n}(p_j(\widehat{\boldsymbol{\theta}}_n) - p_j(\boldsymbol{\theta})) \to_d N(0, p_j'(\boldsymbol{\theta})^T I(\boldsymbol{\theta})^{-1} p_j'(\boldsymbol{\theta}))$$

where $p_j'(\boldsymbol{\theta})$ is the gradient of $p_j(\boldsymbol{\theta})$ (which we assume here to be a column vector). Thus

$$\frac{Y_{nj}}{n} - p_j(\widehat{\boldsymbol{\theta}}_n) \to_p 0$$

$$n\left|\frac{Y_{nj}}{n} - p_j(\widehat{\boldsymbol{\theta}}_n)\right|^r \to_p 0$$

for any $r > 2$. A Taylor series expansion gives

$$\begin{aligned}\ln(Y_{nj}/n) - \ln(p_j(\widehat{\boldsymbol{\theta}}_n)) &= \frac{1}{p_j(\widehat{\boldsymbol{\theta}}_n)}\left(\frac{Y_{nj}}{n} - p_j(\widehat{\boldsymbol{\theta}}_n)\right) \\ &\quad -\frac{1}{2p_j^2(\widehat{\boldsymbol{\theta}}_n)}\left(\frac{Y_{nj}}{n} - p_j(\widehat{\boldsymbol{\theta}}_n)\right)^2 \\ &\quad +\frac{1}{3p_j^3(\boldsymbol{\theta}_n^*)}\left(\frac{Y_{nj}}{n} - p_j(\widehat{\boldsymbol{\theta}}_n)\right)^3\end{aligned}$$

where $\boldsymbol{\theta}_n^*$ lies on the line segment joining Y_{nj} and $\widehat{\boldsymbol{\theta}}_n$. We also have

$$\begin{aligned}2\ln(\Lambda_n) &= 2n\sum_{j=1}^{k}\left(\frac{Y_{nj}}{n} - p_j(\widehat{\boldsymbol{\theta}}_n)\right)\ln\left(\frac{Y_{nj}}{np_j(\widehat{\boldsymbol{\theta}}_n)}\right) \\ &\quad +2n\sum_{j=1}^{k} p_j(\widehat{\boldsymbol{\theta}}_n)\ln\left(\frac{Y_{nj}}{np_j(\widehat{\boldsymbol{\theta}}_n)}\right).\end{aligned}$$

Now substituting the Taylor series expansion above for $\ln(Y_{nj}/n) - \ln(p_j(\widehat{\boldsymbol{\theta}}_n))$, it follows that $2\ln(\Lambda_n) - K_n^2 \to_p 0$ and so $K_n \to_d \chi^2(k-1-p)$. □

Theorem 9.2 suggests that Pearson's χ^2 statistic should be nearly identical to the LR statistic when the null hypothesis is true (at least if n is sufficiently large). In fact, even when the null hypothesis does not hold, the two statistics can still be very close; the following example illustrates this.

Table 9.1 *Frequency of goals in First Division matches and "expected" frequency under Poisson model in Example 9.2*

Goals	0	1	2	3	4	≥ 5
Frequency	252	344	180	104	28	16
Expected	248.9	326.5	214.1	93.6	30.7	10.2

EXAMPLE 9.2: Consider the data on goals scored in soccer (football) games given in Example 5.25; the complete data are summarized in Table 5.6. We want to test the hypothesis that the number of goals scored in a game follows a Poisson distribution. Because there are relatively few games in which more than 6 goals were scored, we will consider the games with 5 or more goals scored as a single bin.

If we set Y_j to be the number of games in which j goals were scored for $j = 0, \cdots, 4$ and Y_5 to be the number of games in which 5 or more goals were scored then $(Y_0, \cdots, Y_5)$ has a Multinomial distribution where the probabilities $p_0(\lambda), \cdots, p_5(\lambda)$ under the Poisson model are

$$p_j(\lambda) = \frac{\exp(-\lambda)\lambda^j}{j!} \quad (j = 0, \cdots, 4)$$

$$p_5(\lambda) = 1 - \sum_{j=0}^{4} p_j(\lambda).$$

Given the data in Table 9.1, we can determine the maximum likelihood estimate of λ numerically (a closed-form solution does not exist); for these data, $\widehat{\lambda} = 1.3118$. The expected frequencies in Table 9.1 are obtained simply $924\, p_j(\widehat{\lambda})$ for $j = 0, \cdots, 5$. We then obtain

$$2\ln(\Lambda) = 2\sum_{j=0}^{5} y_j \ln\left(\frac{y_j}{924\, p_j(\widehat{\lambda})}\right) = 10.87$$

$$K^2 = \sum_{j=0}^{5} \frac{(y_j - 924\, p_j(\widehat{\lambda}))^2}{924\, p_j(\widehat{\lambda})} = 11.09.$$

Under the null hypothesis (Poisson model), both statistics should be approximately $\chi^2(4)$ distributed as there are six bins and the Poisson model has one unknown parameter. Using the $\chi^2(4)$ distribution, we obtain (approximate) p-values of 0.028 (LR test) and 0.026 (Pearson's χ^2 test). The small p-values suggest that the Poisson model may not be appropriate; however, a comparison of the observed and "expected" frequencies suggests that the Poisson model may not be that bad. ◇

Log-linear models

Log-linear models describe the structure of dependence or association in data that has been cross-classified according to several discrete or categorical variables. More precisely, suppose we take a random sample from a population where each member of the population is described by some attributes, each of which take a finite number of values or levels. Then for each combination of levels, we count the number of individuals in the sample having this combination of levels. One simple goal is to determine if the variables are independent or, failing that, to determine the structure of the dependence (or association) between the variables.

To simplify the discussion, we will restrict ourselves to the situation where three categorical variables (which we will call U, V, and W) are measured for each individual, and we will assume that these have, respectively, u, v, and w possible values or levels. (All of the discussion below can be extended to four or more variables without difficulty.) Define

$$\theta_{ijk} = P(U = i, V = j, W = k)$$

for $i = 1, \cdots, u$, $j = 1, \cdots, v$, and $k = 1, \cdots, w$. Define Y_{ijk} to be the number of individuals in the sample with $U = i$, $V = j$, and $W = k$; then the random vector $\boldsymbol{Y} = (Y_{111}, \cdots, Y_{uvw})$ has a Multinomial distribution with probabilities θ_{ijk} ($i = 1, \cdots, u$; $j = 1, \cdots, v$; $k = 1, \cdots, w$). It follows that the MLE of θ_{ijk} is

$$\widehat{\theta}_{ijk} = \frac{Y_{ijk}}{n}.$$

This general model (where the θ_{ijk}'s are unconstrained) is often called the saturated model since there are as many parameters as there are observations.

The goal in log-linear modeling is to find a simpler model for the probabilities θ_{ijk}'s; note that under the saturated model described above, the dimension of the parameter space is effectively $uvw - 1$. Simplifications of the general model result from expressing the θ_{ijk}'s in terms of marginal probabilities. For example, we can define

$$\begin{aligned}
\theta_{ij+} &= P(U = i, V = j) \\
&= \sum_{k=1}^{w} \theta_{ijk} \\
\theta_{+j+} &= P(V = j) \\
&= \sum_{i=1}^{u} \sum_{k=1}^{w} \theta_{ijk}
\end{aligned}$$

where a + in the subscript indicates that we have summed over that index. Given these probabilities, we can define a number of possible models that describe the dependence structure between the variables U, V and W; some of these are:

$$\begin{aligned}
\theta_{ijk} &= \theta_{i++}\theta_{+j+}\theta_{++k} \\
&\quad \text{(mutual independence of } U, V \text{ and } W) \\
\theta_{ijk} &= \theta_{i++}\theta_{+jk} \\
&\quad ((V, W) \text{ is independent of } U) \\
\theta_{ijk} &= \frac{\theta_{i+k}\theta_{+jk}}{\theta_{++k}} \\
&\quad (U \text{ and } V \text{ are conditionally independent given } W).
\end{aligned}$$

These models for θ_{ijk} are called log-linear models since $\ln(\theta_{ijk})$ can expressed as a sum of terms; for example, in the model where U and V are conditionally independent given W, we can write

$$\begin{aligned}
\ln(\theta_{ijk}) &= \ln(\theta_{i+k}) + \ln(\theta_{+jk}) - \ln(\theta_{++k}) \\
&= \alpha_{ik} + \beta_{jk} + \gamma_k.
\end{aligned}$$

Generally, we can write a log-linear model as

$$\ln(\theta_{ijk}) = \boldsymbol{x}_{ijk}^T \boldsymbol{\beta}$$

where $\boldsymbol{x}_{ijk}$ is typically a vector of 0's and 1's (depending on the structure of the log-linear model) and $\boldsymbol{\beta}$ is a vector of unknown parameters. To insure identifiability, we must assume that the design matrix

$$X = \begin{pmatrix} \boldsymbol{x}_{111}^T \\ \vdots \\ \boldsymbol{x}_{uvw} \end{pmatrix}$$

has rank equal to the length of the parameter vector $\boldsymbol{\beta}$. Note, however, that since the θ_{ijk}'s are probabilities, we must also impose the constraint

$$\sum_{i=1}^{u}\sum_{j=1}^{v}\sum_{k=1}^{w}\theta_{ijk} = \sum_{i=1}^{u}\sum_{j=1}^{v}\sum_{k=1}^{w}\exp(\boldsymbol{x}_{ijk}^T\boldsymbol{\beta}) = 1.$$

Because of this constraint, the effective number of parameters in the log-linear model is one less than the length of $\boldsymbol{\beta}$. Given $\boldsymbol{Y} = \boldsymbol{y}$, the log-likelihood function for $\boldsymbol{\beta}$ is

$$\ln \mathcal{L}(\boldsymbol{\beta}) = \sum_{i=1}^{u}\sum_{j=1}^{v}\sum_{k=1}^{w} y_{ijk}(\boldsymbol{x}_{ijk}^T\boldsymbol{\beta})$$

subject to the constraint

$$\sum_{i=1}^{u}\sum_{j=1}^{v}\sum_{k=1}^{w}\exp(\boldsymbol{x}_{ijk}^T\boldsymbol{\beta}) = 1.$$

Maximum likelihood estimates can be computed using either the iterative proportional fitting algorithm or the Newton-Raphson algorithm; see Agresti (1990) and Fienberg (1980) for details.

Given the MLE $\widehat{\boldsymbol{\beta}}$, we can define the expected count $\widehat{Y}_{ijk} = n\exp(\boldsymbol{x}_{ijk}^T\widehat{\boldsymbol{\beta}})$. To test the null hypothesis

$$H_0 : \theta_{ijk} = \exp(\boldsymbol{x}_{ijk}^T\boldsymbol{\beta}) \quad \text{versus} \quad H_1 : \theta_{ijk}\text{'s unspecified}$$

we can use either the LR test or Pearson's χ^2 test whose test statistics are;

$$\begin{aligned} 2\ln(\Lambda_n) &= 2\sum_{i=1}^{u}\sum_{j=1}^{v}\sum_{k=1}^{w} Y_{ijk}\ln(Y_{ijk}/\widehat{Y}_{ijk}) \\ K_n^2 &= \sum_{i=1}^{u}\sum_{j=1}^{v}\sum_{k=1}^{w}\frac{(Y_{ijk}-\widehat{Y}_{ijk})^2}{\widehat{Y}_{ijk}}. \end{aligned}$$

If n is sufficiently large then the null distributions of both statistics is χ^2 with $(uvw - p)$ where p is the length of $\boldsymbol{\beta}$. (As before, an informal rule-of-thumb says that n is large enough to apply the χ^2 approximation if $\min_{i,j,k}\widehat{Y}_{ijk} \geq 5$.)

EXAMPLE 9.3: Suppose that we want to test the null hypothesis that U, V, and W are independent. This hypothesis can be expressed via the log-linear model

$$\ln(\theta_{ijk}) = \mu + \alpha_i + \beta_j + \gamma_k$$

(for $i = 1, \cdots, u$, $j = 1, \cdots, v$, and $k = 1, \cdots, w$) where, to make the model identifiable, we assume that $\alpha_1 = \beta_1 = \gamma_1 = 0$; the number of parameters in this model is $p = 1 + (u-1) + (v-1) + (w-1) = u + v + w - 2$ although due to the constraint on the θ_{ijk}'s the effective number of parameters is $p - 1$. The expected counts $\widehat{Y}_{ijk}$ have a very simple form in this model:

$$\widehat{Y}_{ijk} = n\left(\frac{Y_{i++}}{n}\right)\left(\frac{Y_{+j+}}{n}\right)\left(\frac{Y_{++k}}{n}\right)$$

where as before the + in the subscript indicates that summation over the corresponding index. Thus under the null hypothesis, both the LR statistic and Pearson's χ^2 statistic have χ^2 distributions with $(uvw - u - v - w + 2)$ degrees of freedom. ◇

We can also use LR and Pearson's χ^2 tests to compare nested log-linear models; see Agresti (1990) for more details.

9.3 Smooth goodness-of-fit tests

Suppose that $X_1, \cdots, X_n$ are i.i.d. continuous random variables with unknown distribution function F. We want to test the null hypothesis

$$H_0 : F = F_\theta$$

where F_θ may depend on some unknown parameters.

To start, we will consider a somewhat simpler testing problem in which $F_\theta = F_0$ is a fixed continuous distribution function and does not depend on any unknown parameters. Admittedly, this is not a very useful problem in practice since we would like to test whether the X_i's come from a particular family of distributions; however, this simplication is very useful in illustrating the general principles used in constructing "smooth" goodness-of-fit tests. D'Agostino and Stephens (1986) gives a comprehensive survey of smooth goodness-of-fit tests.

Given the fixed continuous distribution F_0, define $Y_i = F_0(X_i)$ $(i = 1, \cdots, n)$. Since F_0 is continuous, the Y_i's have a Uniform distribution on $[0, 1]$ if the distribution function of the X_i's is F_0

(that is, H_0 is true; on the other hand, if H_0 is false (that is, F_0 is not the distribution function of the X_i's) then the Y_i's will not have a Uniform distribution although their distribution will be concentrated on the interval $[0, 1]$. Therefore (at least in this simple problem), we can assume that the distribution function F of the X_i's is concentrated on $[0, 1]$ and the null hypothesis becomes

$$H_0 : F(x) = x \quad \text{for } 0 \le x \le 1.$$

Our test statistics for testing H_0 will be based on the empirical distribution function

$$\widehat{F}_n(x) = \frac{1}{n} \sum_{i=1}^{n} I(X_i \le x),$$

which was earlier used to define substitution principle estimators. Recall that

$$\sup_{0 \le x \le 1} |\widehat{F}_n(x) - F(x)| \to_p 0$$

as $n \to \infty$ and so this suggests that a reason test statistic should compare the empirical distribution function $\widehat{F}_n$ to the Uniform distribution function in some way, for example,

$$\sup_{0 \le x \le 1} |\widehat{F}_n(x) - x| \quad \text{or} \quad \int_0^1 (\widehat{F}_n(x) - x)^2 \, dx.$$

As one might imagine, the null distributions of such test statistics are quite complicated to determine. However, it is fairly straightforward to determine the asymptotic null distributions (as $n \to \infty$). To do this, we need to take a closer look at the limiting distribution of $B_n(x) = \sqrt{n}(\widehat{F}_n(x) - x)$ as a random function on the interval $[0, 1]$. While the mathematics of this is somewhat beyond the scope of this book, it is fairly easy to give a heuristic development of the limiting distribution of B_n.

We will limit our discussion to the limiting distribution of B_n at a finite number of points. By the Multivariate CLT, we have for $0 \le x_1 < x_2 < \cdots < x_k \le 1$,

$$\begin{pmatrix} B_n(x_1) \\ B_n(x_2) \\ \vdots \\ B_n(x_k) \end{pmatrix} \to_d N_k(\mathbf{0}, C)$$

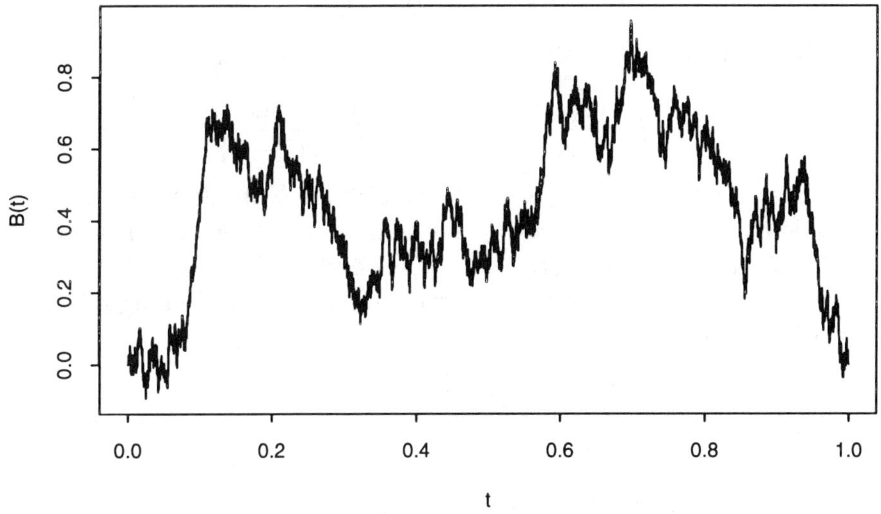

Figure 9.1 *A simulated realization of a Brownian bridge process.*

where the variance-covariance matrix C has (i, j) element

$$\begin{aligned} C(i,j) &= \text{Cov}(I(X_1 \leq x_i), I(X_1 \leq x_j)) \\ &= E[I(X_1 \leq x_i)I(X_1 \leq x_j)] \\ &\quad -E[I(X_1 \leq x_i)]EI(X_1 \leq x_j)] \\ &= \min(x_i, x_j) - x_i x_j. \end{aligned}$$

The trick now is to find a random function (or stochastic process) B on $[0, 1]$ such that the random vector $(B(x_1), \cdots, B(x_k))$ has the limiting distribution given above (for any finite number of points $x_1, \cdots, x_k$). Such a random function exists and is called a Brownian bridge. A simulated realization of a Brownian bridge is given in Figure 9.1.

Now define the test statistics

$$K_n = \sqrt{n} \sup_{0 \leq x \leq 1} |\widehat{F}_n(x) - x| = \sup_{0 \leq x \leq 1} |B_n(x)|,$$

(which is called the Kolmogorov-Smirnov statistic),

$$W_n^2 = n \int_0^1 (\widehat{F}_n(x) - x)^2\, dx = \int_0^1 B_n^2(x)\, dx,$$

(which is called the Cramér-von Mises statistic), and

$$A_n^2 = n \int_0^1 \frac{(\widehat{F}_n(x) - x)^2}{x(1-x)}\, dx = \int_0^1 \frac{B_n^2(x)}{x(1-x)}\, dx$$

(which is called the Anderson-Darling statistic). Note that each test statistic can be written as $\phi(B_n)$ for some ϕ; the "convergence" of B_n to B suggests that $\phi(B_n) \to_d \phi(B)$ and, in fact, this can be proved rigorously. Thus we have, for example, for the Kolmogorov-Smirnov statistic

$$K_n \to_d \sup_{0 \le x \le 1} |B(x)| = K$$

where

$$P(K > x) = 2 \sum_{j=1}^{\infty} (-1)^{j+1} \exp(-2j^2x^2).$$

Likewise, representations of the limiting distributions of the Cramér-von Mises and Anderson-Darling statistics can be obtained; both limiting random variables can be represented as an infinite weighted sum of independent χ^2 random variables with 1 degree of freedom. If $Z_1, Z_2, \cdots$ are independent $N(0,1)$ random variables then

$$\begin{aligned} W_n^2 &\to_d \sum_{j=1}^{\infty} \frac{Z_j^2}{j^2\pi^2} \\ A_n^2 &\to_d \sum_{j=1}^{\infty} \frac{Z_j^2}{j(j+1)}. \end{aligned}$$

The limiting distribution functions of W_n^2 and A_n^2 can be obtained by first obtaining the characteristic function of the limiting distribution and then inverting the characteristic function.

Both the Cramér-von Mises and the Anderson-Darling tests are quite powerful for alternatives that are close to the Uniform distribution and tend to be better for these alternatives (in terms of power) than the Kolmogorov-Smirnov test. However, in practice, we are typically interested in detecting larger departures from the Uniform distribution and, in such cases, all three tests perform well.

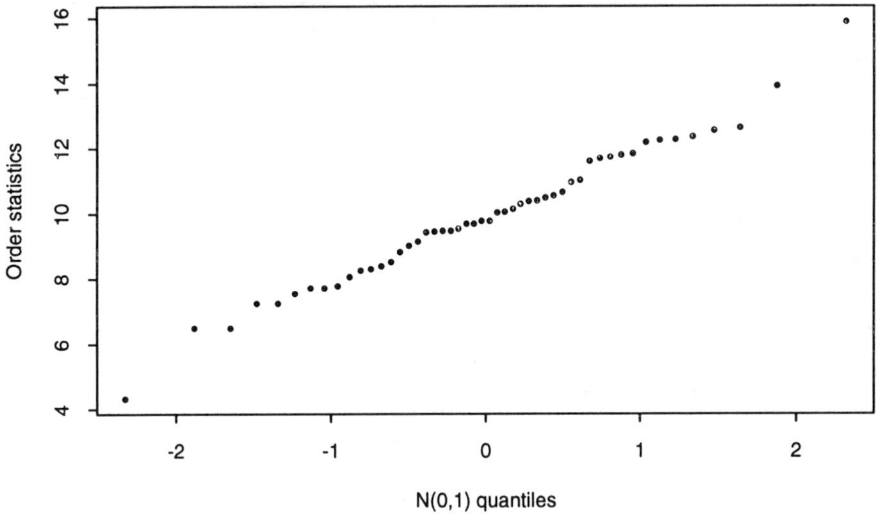

Figure 9.2 *Normal probability plot with normally distributed data.*

Tests based on probability plots

Suppose that $X_1, \cdots, X_n$ are i.i.d. continuous random variables with distribution function F and we want to test the null hypothesis

$$H_0 : F(x) = F_0\left(\frac{x-\mu}{\sigma}\right)$$

where μ and $\sigma > 0$ are unknown parameters.

There is a simple *ad hoc* approach to checking the validity of H_0 which involves plotting the order statistics $X_{(1)}, \cdots, X_{(n)}$ against values of the inverse of F_0; these plots are known as probability plots (or quantile-quantile plots). The idea behind probability plots is quite simple. If H_0 is true and n is reasonably large then we have

$$X_{(i)} \approx \mu + \sigma E_0(X_{(i)}) \quad (i = 1, \cdots, n)$$

where $E_0(X_{(i)})$ is the expected value of the order statistic $X_{(i)}$ under sampling from F_0. Thus if H_0 is true and we plot $X_{(i)}$ versus $E_0(X_{(i)})$ for $i = 1, \cdots, n$, we should expect to see these points falling close to a line whose slope is σ and whose y-intercept is μ. However, if the X_i's do not come from the family of distributions

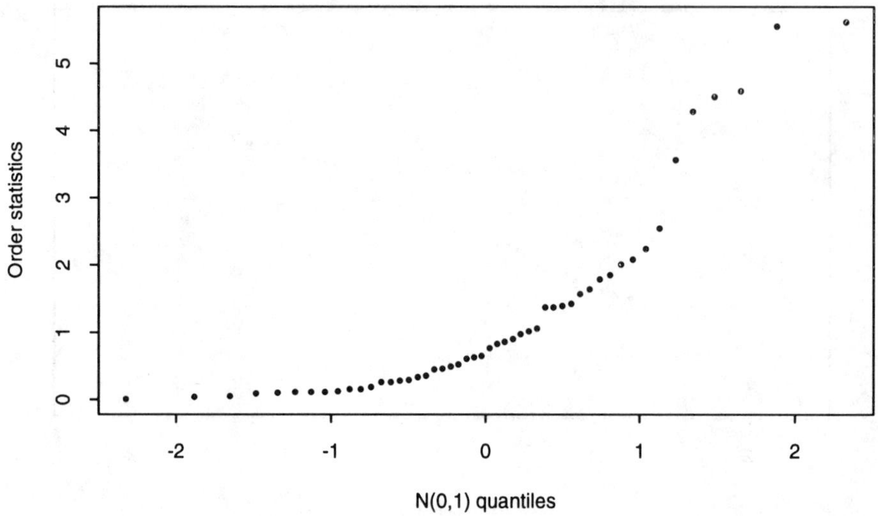

Figure 9.3 *Normal probability plot with non-normally distributed data.*

in H_0 then we would expect to see some curvature (or other non-linearity) in the points. In practice, $E_0(X_{(i)})$ is typically not easy to evaluate but may be approximated; two common approximations for $E_0(X_{(i)})$ are $F_0^{-1}((i - 1/2)/n)$ and $F_0^{-1}(i/(n + 1))$.

Two Normal probability plots are given in Figures 9.2 and 9.3 for sample sizes of $n = 50$; Figure 9.2 shows a Normal probability plot with normally distributed data while Figure 9.3 shows a Normal probability plot for non-normal (in fact, Exponential) data.

While probability plots are typically used merely as informal "eyeball" tests of distributional assumptions, we can use them to carry out formal goodness-of-fit tests. The standard approach to goodness-of-fit testing based on probability plots is to use as a test statistic an estimator of the correlation between the order statistics and their expected values (or suitable approximations thereof) under i.i.d. sampling from the distribution function F_0; for example, see Lockhart and Stephens (1998). Given vectors x

and $\boldsymbol{y}$ of length n, define

$$r(\boldsymbol{x}, \boldsymbol{y}) = \frac{\sum_{i=1}^{n}(x_i - \bar{x})(y_i - \bar{y})}{\left(\sum_{i=1}^{n}(x_i - \bar{x})^2\right)^{1/2}\left(\sum_{i=1}^{n}(y_i - \bar{y})^2\right)^{1/2}}$$

to be the correlation between $\boldsymbol{x}$ and $\boldsymbol{y}$. Now given

$$\boldsymbol{X}_n = \left(X_{(1)}, \cdots, X_{(n)}\right)$$

and

$$E(\boldsymbol{X}_n) = (E_0(X_{(1)}), \cdots, E_0(X_{(n)}))$$

(where, as before, E_0 denotes expected value under sampling from F_0), we define the test statistic

$$R_n = r(\boldsymbol{X}_n, E_0(\boldsymbol{X}_n)).$$

(Alternatively, we can replace $E_0(X_{(i)})$ by one of the approximations given above.) We can then use R_n as a test statistic to test the null hypothesis

$$H_0 : F(x) = F_0\left(\frac{x - \mu}{\sigma}\right) \quad \text{for some } \mu \text{ and } \sigma.$$

It is easy to see that the null distribution of R_n is independent of the unknown parameters μ and σ. Moreover, since both $\boldsymbol{X}_n$ and $E_0(\boldsymbol{X}_n)$ have non-decreasing elements, the correlation R_n is necessarily nonnegative; when H_0 is true, this correlation is typically close to 1. The null distribution of R_n can (for any n) be approximated quite well by Monte Carlo sampling; it suffices to generate independent random variables from F_0. It is also possible to derive the asymptotic distribution of R_n, which will depend on F_0; the derivation of these limiting distributions turns out to be quite difficult from a technical point of view.

In the next two examples, we will give the limiting distribution for R_n in the case of the extreme value distributions and Normal distributions.

EXAMPLE 9.4: Suppose that

$$F_0(x) = \exp[-\exp(-x)];$$

F_0 is called a type I extreme value distribution. If $X_1, \cdots, X_n$ are i.i.d. with distribution function $F(x) = F((x - \mu)/\sigma)$, we have

$$\frac{n(1 - R_n^2) - \ln(n)}{2\sqrt{\ln(n)}} \to_d N(0, 1).$$

Thus for n sufficiently large, the null distribution of R_n^2 is approximately Normal with mean $1 - \ln(n)/n$ and variance $4\ln(n)/n^2$. ◇

EXAMPLE 9.5: Suppose that F_0 is the standard Normal distribution. A correlation test related to the statistic R_n was proposed by Shapiro and Francia (1972). However, another correlation-type test was proposed earlier by Shapiro and Wilk (1965); the so-called Shapiro-Wilk test takes into account the correlation between the order statistics in defining the correlation. Interestingly, the limiting behaviour of both test statistics is more or less identical, if somewhat bizarre. Defining

$$a_n = \frac{1}{(n+1)^3} \sum_{i=1}^{n} \frac{i(n+1-i)}{\phi^2(\Phi^{-1}(i/(n+1)))} - \frac{3}{2}$$

(where ϕ is the density function and Φ^{-1} is the quantile function of a standard Normal distribution), we have

$$n(1 - R_n^2) - a_n \to_d \sum_{k=1}^{\infty} \frac{1}{k+2}(Z_k^2 - 1)$$

where $Z_1, Z_2, \cdots$ are i.i.d. standard Normal random variables. This limiting distribution is sufficiently complicated that it is probably easier to simply approximate the distribution of R_n using simulation! ◇

The power of correlation tests based on probability plots seems to vary according to the family of distributions being tested under the null hypothesis. The general rule-of-thumb seems to be (Lockhart and Stephens, 1998) that correlation tests work quite well for "short-tailed" distributions (such as the Normal) but less well for longer-tailed distributions such as the type I extreme value distribution of Example 9.4. As before, these power considerations are for alternatives that are close to the null; for gross departures from the null hypothesis, correlation tests are adequate although the same information is usually available from the corresponding probability plot.

Table 9.2 *Data for Problem 9.1.*

3.5	7.9	8.5	9.2	11.4	17.4	20.8	21.2
21.4	22.5	25.3	25.7	25.9	26.2	26.6	27.8
28.7	30.1	30.2	30.9	35.0	36.0	39.0	39.0
39.6	43.2	44.8	47.7	57.5	62.5	72.8	83.1
96.6	106.6	115.3	118.1	152.5	169.2	202.2	831.0

9.4 Problems and complements

9.1: The distribution of personal incomes is sometimes modelled by a distribution whose density function is

$$f(x;\alpha,\theta) = \frac{\alpha}{\theta}\left(1+\frac{x}{\theta}\right)^{-(\alpha+1)} \quad \text{for } x \geq 0$$

for some unknown parameters $\alpha > 0$ and $\theta > 0$. The data given in Table 9.2 are a random sample of incomes (in 1000s of dollars) as declared on income tax forms. Thinking of these data as outcomes of i.i.d. random variables $X_1, \cdots, X_{40}$, define

$$\begin{aligned} Y_1 &= \sum_{i=1}^{40} I(X_i \leq 25), \quad Y_2 = \sum_{i=1}^{40} I(25 < X_i \leq 40), \\ Y_3 &= \sum_{i=1}^{40} I(40 < X_i \leq 90) \quad \text{and} \quad Y_4 = \sum_{i=1}^{40} I(X_i > 90). \end{aligned}$$

(a) What is the likelihood function for the parameters α and θ based on $(Y_1, \cdots, Y_4)$?

(b) Find the maximum likelihood estimates of α and θ based on the observed values of $(Y_1, \cdots, Y_4)$ in the sample.

(c) Test the null hypothesis that the density of the data is $f(x;\alpha,\theta)$ for some α and θ using both the LR statistic and Pearson's χ^2 statistic. Compute approximate p-values for both test statistics.

9.2: Consider testing goodness-of-fit for the Zeta distribution with frequency function is

$$f(x;\alpha) = \frac{x^{-(\alpha+1)}}{\zeta(\alpha+1)} \quad \text{for } x = 1, 2, 3, \cdots$$

Table 9.3 *Data for Problem 9.2.*

Observation	1	2	3	4	5	6	7
Frequency	128	30	12	6	3	1	1

where $\alpha > 0$ and

$$\zeta(p) = \sum_{k=1}^{\infty} k^{-p}$$

is the Zeta function.

(a) Let $X_1, \cdots, X_n$ be i.i.d. Zeta random variables. Define

$$\begin{aligned} Y_j &= \sum_{i=1}^{n} I(X_i = j) \quad \text{for } j = 1, \cdots, k \\ Y_{k+1} &= \sum_{i=1}^{n} I(X_i \geq k+1) \end{aligned}$$

Find an expression for the MLE of α based on $Y_1, \cdots, Y_{k+1}$.

(b) Let $\widehat{\alpha}_n$ be the MLE in part (a). Find the limiting distribution of $\sqrt{n}(\widehat{\alpha}_n - \alpha)$. (Note that this limiting distribution will be different than the limiting distribution of the MLE based on the X_i's.)

(c) Carry out the Pearson χ^2 and LR goodness-of-fit tests for the Zeta distribution using $Y_1, \cdots, Y_4$. What are the (approximate) p-values for the two test statistics?

9.3: Consider Theorem 9.2 where now we assume that $\widehat{\boldsymbol{\theta}}_n$ is some estimator (not necessarily the MLE from the Multinomial model) with

$$\sqrt{n}(\widehat{\boldsymbol{\theta}}_n - \boldsymbol{\theta}) \to_d N_p(\mathbf{0}, C(\boldsymbol{\theta})).$$

(a) Show that $K_n^2 - 2\ln(\Lambda_n) \to_p 0$ (under the null hypothesis).

(b) What can be said about the limiting distribution of $2\ln(\Lambda_n)$ under this more general assumption on $\widehat{\boldsymbol{\theta}}_n$?

9.4: Consider the general log-linear model for Multinomial probabilities in a three-way cross-classification

$$\ln(\theta_{ijk}) = \boldsymbol{x}_{ijk}^T \boldsymbol{\beta}.$$

for $i = 1, \cdots, u$, $j = 1, \cdots, v$, and $k = 1, \cdots, w$.

(a) Let $\boldsymbol{z}_{ijk} = \boldsymbol{x}_{ijk} - \boldsymbol{x}_{111}$. Show that the log-linear model can be rewritten as a Multinomial logit model (see Problem 8.20)

$$\ln\left(\frac{\theta_{ijk}}{\theta_{111}}\right) = \boldsymbol{z}_{ijk}^T \boldsymbol{\beta}$$

and hence

$$\theta_{ijk} = \frac{\exp(\boldsymbol{z}_{ijk}^T \boldsymbol{\beta})}{\boldsymbol{z}_{111}^T \boldsymbol{\beta} + \cdots + \boldsymbol{z}_{uvw}^T \boldsymbol{\beta}}.$$

(b) Using the result of Problem 8.20, suggest a Poisson log-linear model that can be used to estimate the parameter $\boldsymbol{\beta}$ in the log-linear model.

9.5: Suppose that $X_1, \cdots, X_n$ are i.i.d. continuous random variables whose range is the interval $(0, 1)$. To test the null hypothesis that the X_i's are uniformly distributed, we can use the statistic

$$V_n = \left(\frac{1}{\sqrt{n}} \sum_{i=1}^{n} \sin(2\pi X_i)\right)^2 + \left(\frac{1}{\sqrt{n}} \sum_{i=1}^{n} \cos(2\pi X_i)\right)^2.$$

(a) Suppose that the X_i's are Uniform random variables on $[0, 1]$. Show that as $n \to \infty$,

$$\left(\frac{1}{\sqrt{n}} \sum_{i=1}^{n} \sin(2\pi X_i), \frac{1}{\sqrt{n}} \sum_{i=1}^{n} \cos(2\pi X_i)\right) \to_d (Z_1, Z_2)$$

where Z_1 and Z_2 are independent $N(0, \sigma^2)$ random variables. Find the value of σ^2.

(b) Find the asymptotic distribution of V_n when the X_i's are uniformly distributed.

(c) Suppose that either $E[\sin(2\pi X_i)]$ or $E[\cos(2\pi X_i)]$ (or both) are non-zero. Show that $V_n \to_p \infty$ in the sense that $P(V_n \leq M) \to 0$ for any $M > 0$. (Hint: Use the WLLN.)

(d) Suppose that $\{v_{n,\alpha}\}$ is such that

$$P(V_n > v_{n,\alpha}) = \alpha$$

when the X_i's are uniformly distributed. If the X_i's satisfy the condition given in part (c), show that

$$\lim_{n \to \infty} P(V_n > v_{n,\alpha}) = 1$$

for any $\alpha > 0$.

9.6: Suppose that $(X_1, Y_1), \cdots, (X_n, Y_n)$ are i.i.d. random variables such that $X_i^2 + Y_i^2 = 1$ with probability 1; thus $(X_1, Y_1), \cdots, (X_n, Y_n)$ represent an i.i.d. sample from a distribution on the unit circle $C = \{(x, y) : x^2 + y^2 = 1\}$. In this problem, we are interested in testing if this distribution is Uniform on the unit circle. To do this, we define new random variables $\Phi_1, \cdots, \Phi_n$ on the interval $[0, 1]$ so that

$$X_i = \cos(2\pi\Phi_i) \quad \text{and} \quad Y_i = \sin(2\pi\Phi_i);$$

then the distribution of (X_i, Y_i) is Uniform on the unit circle if, and only if, the distribution of Φ_i is Uniform on $[0, 1]$.

(a) If $\widehat{F}_n(x)$ is the empirical distribution of $\Phi_1, \cdots, \Phi_n$, and $B_n(x) = \sqrt{n}(\widehat{F}_n(x) - x)$, we can define Watson's statistic to be

$$U_n^2 = \int_0^1 B_n^2(x)\, dx - \left(\int_0^1 B_n(x)\right)^2.$$

Using the heuristic approach of section 9.3, show that

$$U_n \to_d \int B^2(x)\, dx - \left(\int_0^1 B(x)\right)^2$$

where $B(x)$ is a Brownian bridge process.

(b) The definition of $\Phi_1, \cdots, \Phi_n$ above is dependent on the orientation of the coordinate system; for example, for any given θ, we could define $\Phi_1, \cdots, \Phi_n$ on $[0, 1]$ so that

$$X_i = \cos(2\pi\Phi_i + \theta) \quad \text{and} \quad Y_i = \sin(2\pi\Phi_i + \theta)$$

Now define U_n^2 as in (a) using the (new) Φ_i's. Show that U_n^2 is independent of θ.

(c) Consider using the test statistic

$$V_n = \left(\frac{1}{\sqrt{n}} \sum_{i=1}^{n} \sin(2\pi\Phi_i)\right)^2 + \left(\frac{1}{\sqrt{n}} \sum_{i=1}^{n} \cos(2\pi\Phi_i)\right)^2$$

as in Problem 9.5. Show that V_n is independent of θ.

(d) Consider using either the Kolmogorov-Smirnov, Anderson-Darling or Cramér-von Mises tests for testing uniformity on the unit circle. Show that none of these tests is independent of θ.

9.7: A Brownian Bridge process can be represented by the infinite series

$$B(x) = \frac{\sqrt{2}}{\pi} \sum_{k=1}^{\infty} \frac{\sin(\pi k x)}{k} Z_k$$

where $Z_1, Z_2, \cdots$ are i.i.d. Normal random variables with mean 0 and variance 1.

(a) Assuming that expected values can be taken inside infinite summations, show that

$$E[B(x)B(y)] = \min(x, y) - x\,y$$

for $0 \leq x, y \leq 1$.

(b) Define

$$W^2 = \int_0^1 B^2(x)\,dx$$

using the infinite series representation of $B(x)$. Show that the distribution of W^2 is simply the limiting distributions of the Cramér-von Mises statistic.

9.8: Consider the representation of the Brownian bridge $B(x)$ in Problem 9.7.

(a) Show that

$$\int_0^1 B(x)\,dx = \frac{2^{3/2}}{\pi^2} \sum_{k=1}^{\infty} \frac{Z_{2k-1}}{(2k-1)^2}$$

(b) Let $\bar{B} = \int_0^1 B(x)\,dx$. Show that

$$\int B^2(x)\,dx - \left(\int_0^1 B(x)\right)^2 = \int_0^1 [B(x) - \bar{B}]^2\,dx.$$

(c) Use parts (a) and (b) to show that the limiting distribution of Watson's statistic U_n^2 in Problem 9.6 is the distribution of the random variable

$$U^2 = \sum_{k=1}^{\infty} \frac{Z_{2k-1} + Z_{2k}}{2k^2\pi^2}.$$

9.9: Suppose that $X_1, \cdots, X_n$ are i.i.d. Exponential random variables with parameter λ. Let $X_{(1)} < \cdots < X_{(n)}$ be the order statistics and define the so-called normalized spacings (Pyke,

1965)

$$\begin{aligned} D_1 &= nX_{(1)} \\ \text{and} \quad D_k &= (n-k+1)(X_{(k)} - X_{(k-1)}) \quad (k = 2, \cdots, n). \end{aligned}$$

According to Problem 2.26, $D_1, \cdots, D_n$ are also i.i.d. Exponential random variables with parameter λ.

(a) Let $\bar{X}_n$ be the sample mean of $X_1, \cdots, X_n$ and define

$$T_n = \frac{1}{n\bar{X}_n^2} \sum_{i=1}^{n} D_i^2.$$

Show that $\sqrt{n}(T_n - 2) \to_d N(0, 20)$.

(b) Why might T_n be a useful test statistic for testing the null hypothesis that the X_i's are Exponential? (Hint: Note that $D_1 + \cdots + D_n = n\bar{X}$; show that subject to $a_1 + \cdots + a_n = k$, $a_1^2 + \cdots + a_n^2$ is minimized at $a_i = k/n$.)

References

Agresti, A. (1990) *Categorical Data Analysis.* New York: Wiley.

Bahadur, R.R. (1964) On Fisher's bound for asymptotic variances. *Annals of Mathematical Statistics.* **35**, 1545-1552.

Berger, J.O. and Wolpert, R. (1988) *The Likelihood Principle* (2nd edition). Hayward, CA: Institute of Mathematical Statistics.

Beveridge, S. and Nelson, C.R. (1981) A new approach to decomposition of economic time series into permanent and transitory components with particular attention to measurement of the "business cycle". *Journal of Monetary Economics.* **7**, 151-174.

Billingsley, P. (1995) *Probability and Measure* (3rd edition). New York: Wiley.

Bitterlich, W. (1956) Die Relaskopmessung in ihrer Bedeutung für die Forstwirtschaft. *Österreichen Vierteljahresschrift für Forstwesen.* **97**, 86-98.

Buja, A. (1990) Remarks on functional canonical variates, alternating least squares methods and ACE. *Annals of Statistics.* **18**, 1032-1069.

D'Agostino, R.B. and Stephens, M.A. (1986) *Goodness-of-Fit Techniques.* New York: Marcel Dekker.

Davis, R.A., Knight, K. and Liu, J. (1992) M-estimation for autoregressions with infinite variance. *Stochastic Processes and their Applications*, **40**, pp.145-180.

Dean, C.B. and Lawless, J.F. (1989) Tests for detecting overdispersion in Poisson regression models. *Journal of the American Statistical Association.* **84**, 467-472.

de Moivre, A. (1738) *The Doctrine of Chances.* (2nd edition).

Dempster, A.P., Laird, N. and Rubin, D.B. (1977) Maximum likelihood from incomplete data via the EM algorithm. *Journal of the Royal Statistical Society, Series B.* **39**, 1-22.

DiCiccio, T.J., Kass, R.E., Raftery, A. and Wasserman, L. (1997) Computing Bayes factors by combining simulation and asymptotic approximations. *Journal of the American Statistical Association.* **92**, 903-915.

Donoho, D. and Huber, P.J. (1983) The notion of breakdown point. In *A Festschrift for Erich L. Lehmann.* 157-184. Belmont, CA: Wadsworth.

Efron, B. (1982) *The Jackknife, the Bootstrap and Other Resampling Plans.* Philadelphia: SIAM

Efron, B. and Hinkley, D.V. (1978) Assessing the accuracy of the maximum likelihood estimator: observed versus expected Fisher information. *Biometrika.* **65**, 457-487.

Efron, B. and Tibshirani, R.J. (1993) *An Introduction to the Bootstrap.* New York: Chapman and Hall.

Ferguson, T.S. (1967) *Mathematical Statistics: A Decision Theoretic Approach.* New York: Academic Press.

Fienberg, S.E. (1980) *The Analysis of Cross-Classified Categorical Data.* (2nd edition) Cambridge, MA: MIT Press.

Fisher, R.A. (1920) A mathematical examination of determining the accuracy of an observation by the mean error, and by the mean square error. *Monthly Notices of the Royal Astronomical Society.* **80**, 758-770.

Fisher, R.A. (1922) On the mathematical foundations of theoretical statistics. *Philosophical Transactions of the Royal Society of London, Series A.* **222**, 309-368.

Geyer, C.J. (1994) On the asymptotics of constrained M-estimation. *Annals of Statistics.* **22**, 1993-2010.

Gilks, W.R., Richardson, S. and Spiegelhalter, D.J. (1996) *Markov Chain Monte Carlo in Practice.* London: Chapman and Hall.

Goodman, S.N. (1999a) Toward evidence-based medical statistics. 1: the p-value fallacy. *Annals of Internal Medicine.* **130**, 995-1004.

Goodman, S.N. (1999b) Toward evidence-based medical statistics. 2: the Bayes factor. *Annals of Internal Medicine.* **130**, 1005-1013.

Hájek, J. (1970) A characterization of limiting distributions of regular estimates. *Zeitschrift für Wahrscheinlichkeitstheorie und Verwandte Gebiete.* **14**, 323-330.

Hampel, F.R., Ronchetti, E., Rousseuw, P. and Stahel, W. (1986) *Robust Statistics: The Approach based on influence functions.* New York: Wiley.

Hoerl, A.E. and Kennard, R.W. (1970) Ridge regression. Biased estimation for non-orthogonal problems. *Technometrics.* **12**, 55-67.

Holgate, P. (1967) The angle-count method. *Biometrika.* **54**, 615-624.

Jeffreys, H. (1961) *The Theory of Probability.* Oxford: Clarendon Press.

Johnson, R.A. and Wichern, D.W. (1992) *Applied Multivariate Statistical Analysis (third edition).* Englewood Cliffs, NJ: Prentice-Hall.

Kass, R.E. and Raftery, A. (1995) Bayes factors. *Journal of the American Statistical Association.* **90**, 773-795.

Kim, J. and Pollard, D. (1990) Cube root asymptotics. *Annals of Statistics.* **18**, 191-219.

Kolmogorov, A.N. (1930) Sur la loi forte des grands nombres. *Comptes Rendus de l'Academie des Sciences, Paris* **191**, 910-912.

Kolmogorov, A.N. (1933) *Grundbegriffe der Wahrscheinlichkeitsrechnung.* Berlin: Springer.

Krasker, W.S. and Welsch, R. (1982) Efficient bounded-influence regression. *Journal of the American Statistical Assocation.* **77**, 595-604.

Lambert, D. (1992) Zero-inflated Poisson regression with an application to defects in manufacturing. *Technomatrics.* **34**, 1-14.

Laplace, P.S. (1810) Mémoire sur les intégrales définies et leur application aux probabilités. (reproduced in *Oeuvres de Laplace.* **12**, 357-412.

Lee, H.S. (1996) Analysis of overdispersed paired count data. *Canadian Journal of Statistics.* **24**, 319-326.

Lee, P.M. (1989) *Bayesian Statistics: An Introduction.* New York: Oxford.

Lehmann, E.L. (1991) *Testing Statistical Hypotheses* (2nd edition). Pacific Grove, CA: Wadsworth.

Lindeberg, J.W. (1922) Eine neue Herleitung des Exponentialgesetz in der Wahrscheinlichkeitsrechnung. *Mathematische Zeitschrift.* **15**, 211-225.

Lockhart, R.A. and Stephens, M.A. (1998) The probability plot: tests of fit based in the correlation coefficient. In *Handbook of Statistics, Volume 17.* (N. Balakrishnan and C.R. Rao, eds.) 453-473.

McCullagh, P. (1994) Does the moment-generating function characterize a distribution? *The American Statistician.* **48**, 208.

McCullagh, P. and Nelder, J.A. (1989) *Generalized Linear Models (2nd edition).* London: Chapman and Hall.

Millar, R.B. (1987) Maximum likelihood estimation of mixed stock fishery composition. *Canadian Journal of Fisheries and Aquatic Science.* **44**, 583-590.

Owen, A.B. (1988) Empirical likelihood ratio confidence intervals for a single parameter. *Biometrika.* **75**, 237-249.

Portnoy, S. and Koenker, R. (1997) The Gaussian hare and the Laplacian tortoise: computability of squared-error versus absolute-error estimators (with discussion). *Statistical Science.* **12**, 279-300.

Pyke, R. (1965) Spacings (with discussion). *Journal of the Royal Statistical Society, Series B.* **27**, 395-449.

Quenouille, M. (1949) Approximate tests of correlation in time series. *Journal of the Royal Statistical Society, Series B.* **11**, 18-44.

Rousseeuw, P.J. (1984) Least median of squares regression. *Journal of the American Statistical Association.* **79**, 871-880.

Rudin, W. (1976) *Principles of Mathematical Analysis (3rd edition).* New York: McGraw-Hill.

Savage, L.J. (1972) *The Foundations of Statistics.* New York: Dover.

Scholz, F.W. (1980) Towards a unified definition of maximum likelihood. *Canadian Journal of Statistics.* **8**, 193-203.

Seber, G.A.F. (1977) *Linear Regression Analysis.* New York: Wiley.

Sen, A. and Srivastava, M. (1990) *Regression Analysis: Theory, Methods and Applications.* New York: Springer.

Shapiro, S.S. and Francia, R.S. (1972) Approximate analysis-of-variance test for normality. *Journal of the American Statistical Association.* **67**, 215-216.

Shapiro, S.S. and Wilk, M.B. (1965) An analysis-of-variance test for normality (complete samples). *Biometrika.* **52**, 591-611.

Silverman, B.W. (1986) *Density Estimation for Statistics and Data Analysis.* London: Chapman and Hall.

Skorokhod, A.V. (1956) Limit theorems for stochastic processes. *Theory of Probability and its Applications.* **1**, 261-290.

Smith, P.J. (1988) Bayesian methods for multiple capture-recapture surveys. *Biometrics.* **44**, 1177-1189.

Thisted, R.A. (1988) *Elements of Statistical Computing.* New York: Chapman and Hall.

Tierney, L.J. (1987) An alternative regularity condition for Hájek's representation theorem. *Annals of Statistics.* **15**, 427-431.

Tukey, J.W. (1958) Bias and confidence in not quite large samples. (Abstract.) *Annals of Mathematical Statistics.* **29**, 614.

Vardi, Y. (1985) Empirical distributions in selection bias models. *Annals of Statistics.* **13**, 178-203.

von Mises, R. (1931) *Wahrscheinlichkeitsrechnung und ihre Anwendung in der Statistik und theoretischen Physik.* Leipzig.

Waller, L.A. (1995) Does the characteristic function numerically distinguish distributions? *American Statistician.* **49**, 150-152.

Wedderburn, R.W.M. (1974) Quasi-likelihood functions, generalized linear models and the Gauss-Newton method. *Biometrika.* **61**, 439-447.

World Bank (1999) *The World Development Report 1999/2000.* New York: Oxford.

Wu, C.F.J. (1983) On the convergence properties of the EM algorithm. *Annals of Statistics.* **11**, 95-103.

Index